Anaerobic Digestate Management

Anaerobic Digestate Management

Vinay Kumar Tyagi, Kaoutar Aboudi and
Cigdem Eskicioglu

IWA PUBLISHING

Published by **IWA Publishing**
Unit 104–105, Export Building
1 Clove Crescent
London E14 2BA, UK
Telephone: +44 (0)20 7654 5500
Fax: +44 (0)20 7654 5555
Email: publications@iwap.co.uk
Web: www.iwapublishing.com

First published 2022

British Library Cataloguing in Publication Data
A CIP catalogue record for this book is available from the British Library

ISBN: 9781789062748 (Paperback)
ISBN: 9781789062755 (eBook)
ISBN: 9781789062762 (ePub)

Contents

Chapter 18
Life-cycle assessment of various anaerobic digestate management technologies ***385***
Sridhar Pilli, Bella Kunnoth, Sagarika Mothe and Polisetty Venkateswara Rao

Chapter 19
Management of anaerobic digestate in a biorefinery advocating circular economy ***405***
Larissa C. Ampese, William Gustavo Sganzerla, Mauro Berni, Gilberto Martins and Tânia Forster Carneiro

Editors

(1) Dr. Vinay Kumar TYAGI, National Institute of Hydrology, Roorkee, INDIA

Vinay Kumar TYAGI is a Scientist D at the Environmental Hydrology Division, National Institute of Hydrology, Roorkee, India. His research interest includes energy-efficient wastewater treatment systems, advanced technologies for sludge pre-treatment, biomass to bioenergy recovery, and health-related water microbiology.

Dr. Tyagi has published two-digit refereed articles at SCI journals in the past several years. He has more than 100 publications to his credit (Google Scholar Citation: 3200, h-index:27) including four full-text books on sludge management" (CRC Press), anaerobic effluent treatments (IWA publishing), clean energy & resources recovery from Biomass wastes and wastewater (Elsevier), twenty three book chapters and manuals most notably are O&M of Sewerage Works, and Operation and Maintenance Guidelines for UASB-DHS System, jointly published by Government of India and Japan. Dr. Tyagi's research achievement has been receiving many academic honors, including (a) Outstanding Review Contribution, Elsevier; (b) research articles are enlisted among the Top 1% of highly cited papers of the academic field of Engineering worldwide, as well as few of them listed in the most downloaded articles; Dr. Tyagi is fellow and life member of several professional bodies and organizations. He is Management Committee Member of International Water Association (IWA) Sludge Management Specialist Group and represent South Asia Division. He is an active Reviewer of more than 80 International Peer-Reviewed Journals and editor for IWA publishing (UK), editorial board member of Environmental Science and Eco-technology journal and lead editor virtual special issue for international journal including Biomass Conversion and Biorefinery (Springer)

Dr. Tyagi has extensive international research exposure of working in Taiwan, Spain, Japan, and in Singapore in various environmental research domains of wastewater (municipal & industrial) treatment, sludge pre-treatment, and biomass to bioenergy recovery.

(2) Dr. Kaoutar Aboudi, University of Cadiz, SPAIN

Kaoutar ABOUDI is a collaborating Postdoctoral researcher at the Department of Chemical Engineering and Food Technology, Faculty of Sciences, University of Cadiz, Spain. Currently, she is Postdoctoral Fellow at IMDEA Energy Institute in Madrid, Spain, under the Unit of Excellence"María de Maeztu" Grant.

Her prime focus is towards further development of advanced technologies for the valorisation and the treatment of organic wastes through the application of different technologies such as the aerobic and anaerobic digestion processes, advanced systems for wastes and wastewater pre-treatment, biomass to bioenergy and value-added chemicals recovery, and life cycle analysis in the circular economy framework.

Dr. Aboudi is an expert in the anaerobic digestion process with several published papers in peer reviewed high-quality journals. Moreover, Dr. Aboudi co-edited a full text book entitled, "Clean energy & Resources recovery from Biomass wastes" (Elsevier), and co-authored several book chapters (Elsevier and IWA publishing editorial). She participated in more than 18 international conferences worldwide. She co-authored two patents related to the valorization of lignocellulosic biomass for bioethanol production. She is an active Reviewer of more than 15 International Peer-Reviewed Journals. Dr. Aboudi has performed several international internships at prestigious Universities such as the Delft University of Technology in the Netherlands and the New University of Lisbon, in Portugal.

(3) Dr. Cigdem Eskicioglu, University of British Columbia, Canada

Cigdem Eskicioglu is a Professor and *Natural Sciences and Engineering Research Council* of Canada (*NSERC*)/Metro Vancouver Senior Industrial Research Chair in Advanced Resource Recovery from Wastewater in the School of Engineering at University of British Columbia (UBC)'s Okanagan Campus in Canada. She is internationally recognized for her research in developing new bioreactor technologies for maximizing energy (i.e., biomethane and biohydrogen) and for resource recovery from organic waste, particularly wastewater sludge.

Over the last 14 years, Dr. Eskicioglu has attracted more than $12M in research funding and has founded/led the Bioreactor Technology Group (BTG) at (UBC)'s Okanagan Campus. The group has been operating four state-of the-art facilities (i.e., Advanced Anaerobic Digestion, Thermo-chemical Biomass Processing, Particle Sizing/Imaging and Trace Contaminant Detection). As a result of BTG's unique infrastructure/expertise, the group has developed partnerships with many wastewater utilities, industries, and government agencies for sustainable waste disposal.

Dr. Eskicioglu has established a strong publication record (over 120 peer-reviewed articles, Google Scholar Citation: 4200, h-index: 31) in the most prestigious journals, conferences and has 1 patent (pending) in the field. Dr. Eskicioglu has been recognized by the 2020 UBC Okanagan Researcher of the Year Award (Engineering and Science category), the UBC School of Engineering's Outstanding Researcher Award (2017), the Confederation of University Faculty Association's 2012 BC Early in Career Award, and by City of Kelowna's 2011 Mayor's Environmental Award. She has also served as scientific committee or panel member for many national/international grant agencies and conferences. Additionally, she is a sought-after consultant for Canadian and international firms, water utilities and non-profit organizations.

Contributors

A.E. Atabani
Alternative Fuels Research Laboratory (AFRL), Energy Division, Department of Mechanical Engineering, Faculty of Engineering, Erciyes University, 38039 Kayseri, Turkey

Ahmed Tawfik
National Research Centre, Water Pollution Research Department 12622, Dokki, Cairo, Egypt

Alberto Cerutti
DIATI, Politecnico di Torino Corso Duca degli Abruzzi 24, 10129 Torino, Italy

Angana Chaudhuri
School of Energy Science and Engineering, Indian Institute of Technology Guwahati, Guwahati, Assam, 781039, India

Barampouti Elli
Maria National Technical University of Athens, School of Chemical Engineering, Unit of Environmental Science & Technology, 9 Iroon Polytechniou Str., Zographou Campus, GR-15780 Athens, Greece

Barbara Ruffino
DIATI, Politecnico di Torino Corso Duca degli Abruzzi 24, 10129 Torino, Italy

Bella Kunnoth
Water and Environment Division, Department of Civil Engineering, National Institute of Technology - Warangal, Telangana, India

Bholu Ram Yadav
CSIR- National Environmental Engineering Research Institute (CSIR- NEERI), Nehru Marg, Nagpur – 440020, India

Cigdem Eskicioglu
UBC Bioreactor Technology Group, School of Engineering, University of British Columbia,Okanagan Campus, 3333
University Way, Kelowna, BC V1V 1V7, Canada

Cristhian Chicaiza-Ortiz
Biomass to Resources Group, Universidad Regional Amazónica IKIAM, Tena, Napo 150150, Ecuador.

China-UK Low Carbon College, Shanghai Jiao Tong University, Shanghai 201306, China.

Programa de posgrado en Cambio Climático, Sustentabilidad y Desarrollo. Área de ambiente y sustentabilidad. Universidad Andina Simón Bolívar, Quito, Ecuador

David Krisa
UBC Bioreactor Technology Group, School of Engineering, The University of British Columbia, Okanagan Campus, 3333 University Way, Kelowna, BC V1V 1V7, Canada

Dominique Patureau
INRAE, Univ. Montpellier, LBE, 102 Avenue des étangs, F-11000 Narbonne, France

Donato Scrinzi
Department of Civil, Environmental and Mechanical Engineering, University of Trento, Trento, Italy.

Elsayed Elbeshbishy
Civil Engineering, Ryerson University, 350 Victoria St. Toronto, Ontario, M5B 2K3, Canada

Farokhlaqa Kakar
Civil Engineering, Ryerson University, 350 Victoria St. Toronto, Ontario, M5B 2K3, Canada

Felipe Guilayn
SUEZ, CIRSEE, 38 rue du Président Wilson, 78230 Le Pecq, France

Frances Okoye
Civil Engineering, Ryerson University, 350 Victoria St. Toronto, Ontario, M5B 2K3, Canada

Gabriel Capson-Tojo
CRETUS, Department of Chemical Engineering, Universidade de Santiago de Compostela, 15782 Santiago de Compostela, Galicia, Spain

Gianni Andreottola
Department of Civil, Environmental and Mechanical Engineering, University of Trento, Trento, Italy

Gilberto Martins
Center for Engineering, Modeling and Applied Social Sciences, Federal University of ABC - UFABC, Av. dos Estados, 5001 Santo André, São Paulo, Brazil

Giuseppe Campo
DIATI, Politecnico di Torino Corso Duca degli Abruzzi 24, 10129 Torino, Italy

Hamdi Muratcobanoglu
Department of Environmental Engineering, Faculty of Engineering, Niğde Ömer Halisdemir University, 51240 Niğde, Turkey

Huan Liu
UBC Bioreactor Technology Group, School of Engineering, University of British Columbia,Okanagan Campus, 3333 University Way, Kelowna, BC V1V 1V7, Canada

Ibrahim Alper Basar
UBC Bioreactor Technology Group, School of Engineering, University of British Columbia,Okanagan Campus, 3333 University Way, Kelowna, BC V1V 1V7, Canada

Jakub Zdarta
Institute of Materials Science and Engineering, Faculty of Mechanical Engineering and Management, Poznan University of Technology, Poznan, Poland

Jingxin Zhang
Biomass to Resources Group, Universidad Regional Amazónica IKIAM, Tena, Napo 150150, Ecuador

Johir A. H. Mohammed
Centre for Technology in Water and Wastewater, School of Civil and Environmental Engineering, University of Technology Sydney, NSW 2007, Australia

Julie Jimenez
LBE, Univ. Montpellier, INRAE, 102 avenue des Etangs, 11100, Narbonne, France

Kaoutar Aboudi
Department of Chemical Engineering and Food Technology, University of Cadiz, 11510 Puerto Real, Cadiz, Spain

Karel Diéguez-Santana
Facultad de Ciencias de la Vida, Universidad Regional Amazónica IKIAM, Tena, Napo 150150, Ecuador.

Department of Organic and Inorganic Chemistry, University of Basque Country UPV/EHU, 48940 Leioa, Spain

Larissa C Ampese
School of Food Engineering, University of Campinas (UNICAMP), Monteiro Lobato St., n.80, Campinas, SP, Brazil

Long D. Nghiem
Centre for Technology in Water and Wastewater, School of Civil and Environmental Engineering, University of Technology Sydney, NSW 2007, Australia

Luca Fiori
Department of Civil, Environmental and Mechanical Engineering, University of Trento, Trento, Italy.

Luong N. Nguyen
Centre for Technology in Water and Wastewater, School of Civil and Environmental Engineering, University of Technology Sydney, NSW 2007, Australia

M. Mofijur
Centre for Technology in Water and Wastewater, School of Civil and Environmental Engineering, University of Technology Sydney, NSW 2007, Australia

M.R. Atelge
Department of Mechanical Engineering, Faculty of Engineering, Siirt University, 56100 Siirt, Turkey

Mai Sofia
National Technical University of Athens, School of Chemical Engineering, Unit of Environmental Science & Technology, 9 Iroon Polytechniou Str., Zographou Campus, GR-15780 Athens, Greece

Malamis Dimitris
National Technical University of Athens, School of Chemical Engineering, Unit of Environmental Science & Technology, 9 Iroon Polytechniou Str., Zographou Campus, GR-15780 Athens, Greece

Mauro Berni
Interdisciplinary Center of Energy Planning (NIPE), University of Campinas (UNICAMP), Campinas, São Paulo, Brazil

Minh T. Vu
Centre for Technology in Water and Wastewater, School of Civil and Environmental Engineering, University of Technology Sydney, NSW 2007, Australia

Moustakas Konstantinos
National Technical University of Athens, School of Chemical Engineering, Unit of Environmental Science & Technology, 9 Iroon Polytechniou Str., Zographou Campus, GR-15780 Athens, Greece

Nirenkumar Pathak
Centre for Technology in Water and Wastewater, School of Civil and Environmental Engineering, University of Technology Sydney, NSW 2007, Australia

Nitesh Machrika
CSIR- National Environmental Engineering Research Institute (CSIR- NEERI), Nehru Marg, Nagpur - 440020, India

Nongmaithem Debeni Devi
School of Energy Science and Engineering, Indian Institute of Technology Guwahati, Guwahati, Assam, 781039, India

Pallavi Gahlot
Environmental Biotechnology Group, Department of Civil Engineering, Indian Institute of Technology Roorkee, Roorkee-247667, India

Parmila Devi
UBC Bioreactor Technology Group, School of Engineering, University of British Columbia,Okanagan Campus, 3333 University Way, Kelowna, BC V1V 1V7, Canada

Polisetty Venkateswara Rao
Water and Environment Division, Department of Civil Engineering

National Institute of Technology - Warangal, Telangana, India

Roberta Ferrentino
Department of Civil, Environmental and Mechanical Engineering, University of Trento, Trento, Italy.

Rongfen Chen
Advanced Environmental Biotechnology Centre, Nanyang Environment & Water Research Institute, Nanyang Technological University, 1 Cleantech Loop, Singapore 637141, Singapore

Interdisciplinary Graduate School, Nanyang Technological University, 50 Nanyang Avenue, 639798, Singapore

Sagarika Mothe
Water and Environment Division, Department of Civil Engineering National Institute of Technology - Warangal, Telangana, India

Sridhar Pilli
Water and Environment Division, Department of Civil Engineering National Institute of Technology - Warangal, Telangana, India

Sunil Kumar
CSIR- National Environmental Engineering Research Institute (CSIR- NEERI), Nehru Marg, Nagpur - 440020, India

Sutapa Das
Department of Chemical Engineering, Indian Institute of Technology Guwahati, Guwahati, Assam, 781039, India

Tânia Forster Carneiro
School of Food Engineering, University of Campinas (UNICAMP), Monteiro Lobato St., n.80, Campinas, SP, Brazil

Umme Sharmeen Hyder
Civil Engineering, Ryerson University, 350 Victoria St. Toronto, Ontario, M5B 2K3, Canada

Vaibhav V. Goud
School of Energy Science and Engineering, Indian Institute of Technology Guwahati, Guwahati, Assam, 781039, India Department of Chemical Engineering, Indian Institute of Technology Guwahati, Guwahati, Assam, 781039, India

Vinay Kumar Tyagi
Environmental Hydrology Division, National Institute of Hydrology, Roorkee-247667, India

Vinay Pratap
Academy of Science and Innovative Research (AcSIR), Ghaziabad-201002, India, CSIR- National Environmental Engineering Research Institute (CSIR- NEERI), Nehru Marg, Nagpur - 440020, India

Washington Logroño
Department of Environmental Science and Bioeconomy, Sustainable Environment Research Division, ECARI–Ecuadorian Centre for Advanced Research and Innovation, Ecuador.

Department of Environmental Microbiology, Helmholtz Centre for Environmental Research–UFZ, 04318 Leipzig, Germany.

Grupo de Investigación y Desarrollo para el Ambiente y Cambio Climático (GIDAC-CEAA), Facultad de Ciencias, Escuela Superior Politécnica de Chimborazo, 060155 Riobamba, Ecuador

William Gustavo Sganzerla
School of Food Engineering, University of Campinas (UNICAMP), Monteiro Lobato St., n.80, Campinas, SP, Brazil

Yan Zhou
Advanced Environmental Biotechnology Centre, Nanyang Environment & Water Research Institute, Nanyang Technological University, 1 Cleantech Loop, Singapore 637141, Singapore

School of Civil and Environmental Engineering, Nanyang Technological University, 50 Nanyang Avenue, Singapore 639798, Singapore

Foreword

The minimal energy requirement and lower generation of residual solids together with the opportunity to recover energy-rich biogas make the anaerobic digestion (AD) a sustainable process for organic waste treatment. However, the anaerobic technology usually does not offer a complete solution for the treatment as it does not attain the desired discharge or reuse standards in terms of pathogens, heavy metals, trace contaminants of emerging concern, or an excess of nutrients content such as nitrogen or phosphorous.

In recent times, the circular economy model has gained increasing attention from the researchers and policy-makers. There is growing evidence on the importance of closing the loop of technological processes applied for biomass waste valorization, promoting different actions to recover the energy and value-added products. In this context, this book, dedicated to addressing updated strategies for anaerobic digestate management, is timely and will bridge the gap between the need for practical and sustainable processes for the treatment of biomass waste such as AD, and valorization of the anaerobic digestate, thus improving the cost-effectiveness and reducing the carbon footprint of the process.

The book content has been structured in four parts: A) state of the art of anaerobic digestate management and treatment; B) management and post-treatment of anaerobic digestate from a wide range of biomass wastes (OFMSW, microalgae, sludges, etc.) and examples of successful studies with insightful discussion; C) new technological approaches for energy and value-added products recovery from anaerobic digestate; and, D) anaerobic digestate management in the circular economy framework with life-cycle assessment (LCA) studies. In addition to addressing the background of digestate management, the classical treatment of anaerobic digestate leveraging its rich nutrient content for the improvement of agricultural soils has been discussed in the beginning of the book. On the other hand, several types of biomass wastes can be used in AD, and each biomass has its specific characteristics, leading to a digestate with different properties. These chapters address a wide range of organic wastes, undertaking different strategies to manage and valorize the anaerobic digestate. Currently, much attention is given to the processes aimed to recover value-added chemicals and bio-products from anaerobic digestate. In this context, five chapters are included in this book. They are about bio-products recovery from anaerobic digestate, including ammonia recovery, renewable fuels recovery (biogas, syngas, char, etc.), and polyhydroxyalkanoates production. Since anaerobic digestate may contain contaminants that hinder its use for several applications, a dedicated chapter covering the fate of emerging contaminants in anaerobic digestate has been included. Scientific efforts to optimize the sustainable technologies for the valorization of biomass and the production of bioenergy and valuable products are ceaseless. Therefore, a chapter on anaerobic digestate management using novel and optimized strategies (e.g., photobioreactor technology) is also included in this book.

All in all, this book brings the fundamentals and technological perspectives for anaerobic digestate treatment and recovery of valuable products from a wide range of organic wastes, presenting state-of-art information in a collection of meticulously selected chapters. Considering that AD is the most widely used technology worldwide for biomass wastes valorization, a book compiling up-to-date information on the strategies used in the management and valorization of anaerobic digestate was missing; thus, compliments to the editors for their effort in bringing this book in time.

Ashok Pandey
BRSI Distinguished Fellow
HTBS National Innovation Chair
Distinguished Scientist
Centre for Innovation and Translational Research
CSIR-Indian Institute for Toxicology Research
Lucknow, India

Acknowledgements

Funders
- Department of Biotechnology, DBT (Grant No. BT/RLF/Re-entry/12/2016)
- Department of Science and Technology, DST (Grant No. DST/IMRCD/India-EU/Water Call 2/ SARASWATI 2.0/2018/C)

Contributors (Corresponding Authors Alphabetically)
A.E. Atabani
Ahmed Tawfik
Barbara Ruffino
Cigdem Eskicioglu
Cristhian Chicaiza-Ortiz
Donato Scrinzi
Elsayed Elbeshbishy
Gabriel Capson-Tojo
Julie Jimenez
Luong N. Nguyen
Moustakas Konstantinos
Polisetty Venkateswara Rao
Sunil Kumar
Tânia Forster Carneiro
Vaibhav V. Goud
Vinay Kumar Tyagi
Yan Zhou

doi: 10.2166/9781789062755_0001

Chapter 1

Anaerobic digestate management: an introduction

*Felipe Guilayn[1] and Gabriel Capson-Tojo[2]**

[1]SUEZ, CIRSEE, 38 rue du Président Wilson, Le Pecq 78230, France
[2]CRETUS, Department of Chemical Engineering, Universidade de Santiago de Compostela, Santiago de Compostela 15782, Galicia, Spain
**Corresponding author e-mail: gabriel.capson-tojo@inrae.fr*

ABSTRACT

Anaerobic digestion (AD) is a key technology in the current transition from a linear to a circular economy. As such, the number of AD plants has increased considerably in the last decade, and it is expected to further increase substantially in the coming years. This, together with the implementation of policies to foster resource recovery, call for the development and implementation of digestate management approaches that allow the recovery of resources contained within digestate (e.g., water, nutrients, carbon, or energy). Traditional techniques such as thermal drying, incineration, composting, and landfilling allow a safe digestate disposal (and in some cases a certain degree of resource recovery). The development of new technologies such as enhanced precipitation, enhanced thermal conversion processes, photoautotrophic biomass production, and enhanced filtration, is opening the door to a more intensive and efficient recovery of resources. To ensure the implementation of these novel technologies, policies favouring their application must be clearly defined, and legal frameworks must be updated. This book presents a comprehensive review of the state of the art of AD digestate management. Traditional and novel resource recovery approaches are addressed, as well as the main technological challenges that these technologies face (e.g., ecotoxicity issues). To give a holistic overview, the current legal framework regarding digestate reutilisation is also assessed, as well as options for process integration and future perspectives.

Keywords: Biogas effluent, digestate, anaerobic digestion, resource recovery, circular economy

1.1 WHAT IS ANAEROBIC DIGESTION (AD) AND DIGESTATE?

1.1.1 AD as a crucial technology for waste management on a circular economy

Traditional and most current economic systems are based on a linear approach, where resources are extracted from the environment, used to produce goods,

Editor(s): Vinay Kumar Tyagi, Kaoutar Aboudi and Cigdem Eskicioglu

distributed, consumed, and eventually disposed as waste when they are no longer useful. This linear resource use is by definition unsustainable. In addition, the unprecedent economic growth that occurred since the industrial revolution has been made possible by the consumption of fossil fuels, which, although incredibly convenient, are non-renewable in a human time scale. The impact of the current linear, fossil-based economy on the global environment has been an increasing concern in the last decades. The most notorious latest example is the first instalment of the Sixth Assessment Report by the Intergovernmental Panel on Climate Change (IPCC). This document states that, if current practices in terms of fossil fuel consumption are continued, the average temperature of the planet will increase over 1.5°C due to the emission of greenhouse gases, which will have considerable impacts on the global climate, affecting both the environment and human populations (IPCC, 2021). Another example of increasing awareness on issues related to an unsustainable economic system are the sustainable development goals (SDGs) of the United Nations (UN), with 9 out of 17 being directly related to the environment or to production systems (UN Environment Management Group, 2021).

With this uncertainty in mind, a change from a linear economy towards a circular one must take place. In the latter, waste as such is minimised, and resources are recovered and recycled, entering again the production–consumption loop. To implement a more sustainable development process, it is also essential to gradually move away from fossil fuels as energy vectors, substituting them by more sustainable (and eventually renewable) energy sources. The US Green New Deal and the European Green Deal are examples of policies that are being implemented worldwide to achieve this goal (European Commission, 2019; House of Representatives, 2019). A main challenge for the implementation of renewable energy sources is the development of efficient, cost-effective energy storage alternatives. At the current technological state, energy storage is mostly carried out using batteries, which is far from being cost-effective and has serious environmental concerns due to the large quantities of non-abundant metals needed. A potential solution for this conundrum is the so-called dual gas-power network (Brémond *et al.*, 2021). In this approach, power from renewable sources would be transformed into high energy density gaseous carriers, such as hydrogen (e.g., via water electrolysis, to be used directly as fuel) or methane (e.g., via hydrogen methanation, to be injected into the gas grid). This integration of the gas and power systems is in the core of the European Green Deal, with the hydrogen and decarbonised gas market package currently being prepared, aiming at updating the energy market, and including the decarbonisation of the production and consumption of hydrogen and methane.

Other than technologies allowing the production of hydrogen, biomethane or syngas from power, processes such as AD or biomass gasification will also be crucial in the gas-based, sustainable, circular economy described above (Brémond *et al.*, 2021). Opposed to most of these technologies, AD is already at a fully commercial stage of development. AD is an anaerobic biological process offering a triple role: (1) production of biogas (a mixture mostly composed of methane and carbon dioxide), (2) waste treatment and stabilisation, and (3) generation of a nutrient-rich digestate (Appels *et al.*, 2011; Capson-Tojo *et al.*,

2016). With over 132 000 small, medium, or large-scale digesters operating worldwide (and additional 50 million micro-scale digesters serving homes/small communities), AD is already playing a main role as a technology generating a green fuel and allowing the recovery of resources other than energy (e.g., nutrients such as nitrogen and phosphorus) if the digestate is used, for example, as a fertiliser (World Biogas Association, 2021). The multiple benefits that AD has to offer become obvious when assessing how AD can help to achieve multiple UN SDGs (see Figure 1.1).

Due to its advantages and thanks to policies favouring its implementation, the number of AD plants has grown tremendously in the last decades (see Figure 1.2). In the EU, the major biogas producer with a capacity for power generation of 209 TWh from biogas in 2018, the main feedstocks are crop-derived (mostly in Germany and UK), but also manure, slurries and sewage sludge are used (e.g., in France, Denmark, and others). Agricultural energy crops are currently slowing down, but changes in policies regarding waste management/valorisation will surely boost AD (see Capson-Tojo *et al.* (2016) for an example in France, where the valorisation of commercial food waste through soil return is now mandatory). Indeed, current policies are expected to cause a 10-fold increase in biomethane demand by 2030, with up to 370 TWh coming from gas (both hydrogen and methane; see National Energy and Climate Plan by the European Commission; (Pflüger, 2020)). China is the second major biogas producer (84 TWh), having many small-scale digesters treating agricultural and food wastes in rural areas. In the coming years, China plans to expand its biogas industry by building industrial-scale digesters. Similarly, India intends to double their biogas capacity in the coming

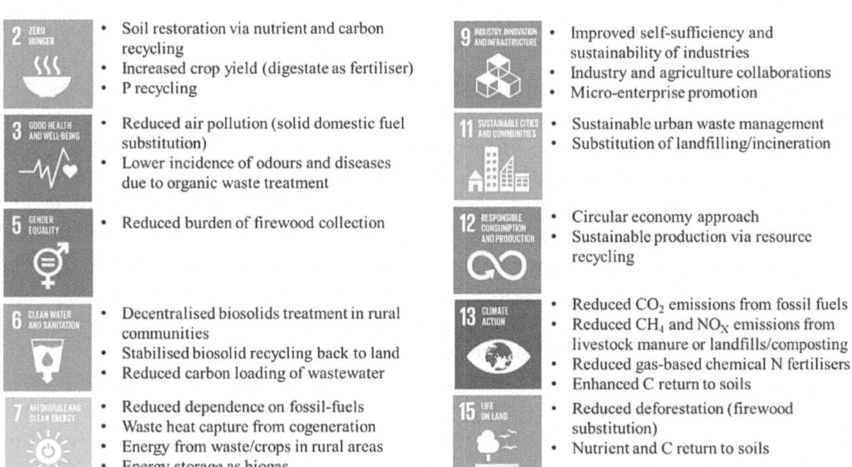

Figure 1.1 Summary of potential contributions of AD to the SDGs of the United Nations (adapted from World Biogas Association, 2018 and UN Environment Management Group, 2021). A focus on digestate management is given (vision from the authors).

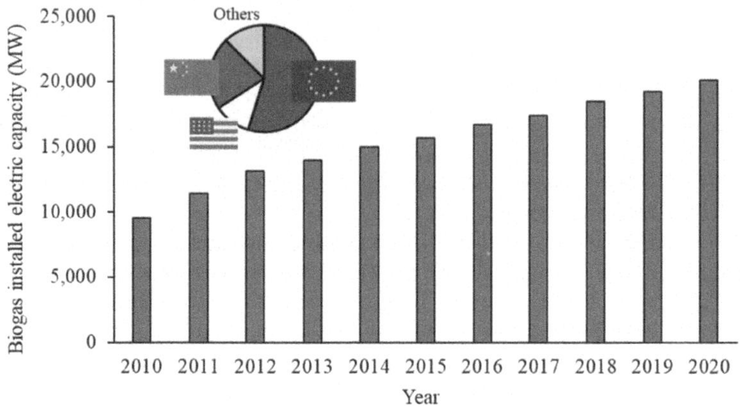

Figure 1.2 Evolution from 2010 to 2020 of the global biogas installed electric capacity. The pie chart shows the geographic distribution of the biggest biogas producers in 2018 (as percentage of total capacity).

years, and other countries are following this trend (e.g., Nepal, Malaysia, Vietnam, etc.) (Akhiar *et al.*, 2020). The third world player is the US (42 TWh), producing biogas mostly from food and municipal solid wastes. The rest of the world (47 TWh) relies mainly on small digesters fed with agricultural and food wastes. See World Biogas Association (2021) for more information about biogas production worldwide.

Although there is already a wide application of biogas technologies around the world, the industry is still in its initial stages of development, implying that there is a huge potential for future development. As of 2019, only 1.9–2.2% of the overall potential of AD was being exploited worldwide (World Biogas Association, 2019). The potential for growth of the biogas industry is thus extraordinary, and virtually involves every country. If major feedstocks (e.g. agricultural biomass, manure, food waste, or municipal solid waste) were actually recovered, we could generate 26–37% of the current natural gas consumed, or 16–22% of the electricity consumed worldwide (World Biogas Association, 2019). Africa is a clear example of untapped AD potential. Indeed, AD is still at an early stage of development in the continent. Nevertheless, several countries have implemented national biogas programmes (e.g. Kenya, Uganda, Ethiopia, Tanzania, Rwanda, Cameroon, Burkina Faso or Benin) (Roopnarain & Adeleke, 2017). Recognising the potential of AD in this region of the globe, the International Energy Agency (IEA) has proposed their sustainable development scenario (which meets the world's goals fully to tackle climate change), where biogas provides a source of clean cooking to an additional 200 million people by 2040, half of which are in Africa (International Energy Agency, 2020).

In developed countries (e.g., USA or EU), the drivers for further AD expansion will surely be based on novel policies (e.g., favouring decarbonisation or penalising traditional management processes), as well as on technological developments. An example of the latter is biogas upgrading to biomethane for direct injection in the natural gas grid (opposed to power production via co-generation). This approach is still far from widespread application, but it is a practice gaining importance, and will surely boost AD growth in the future, as upgrading technologies become more cost-effective and policies aimed at gas supply decarbonisation are implemented (International Energy Agency, 2020).

In summary, although the future of AD is not without challenges, the prospects are promising. This is particularly true when considering that, from a holistic point of view, most relevant actors (i.e., technological, economic, social, and political) are all working together towards the further implementation of AD as key technology in future, more sustainable societies.

1.1.2 AD digestate: definition and current context as fertiliser

The increase in the number of AD plants and capacity will obviously result in greater amounts of digestate to be dealt with. The European Biogas Association defines digestate as 'the solid or liquid material from controlled anaerobic fermentation processes of biodegradable material' (European Biogas Association, 2015). Approximately, for every tonne of feedstock treated in a digester, around ~50–85% by weight emerges as digestate, mostly depending on the water content of the influent (World Biogas Association (2021); the characteristics of the digestate, which mostly depend on the reactor feed, will be further discussed later). To give an idea of the global magnitudes to be dealt with, it has been estimated that the EU28 alone generates around 180 million tonnes yr^{-1} of digestate, 120 from agricultural AD, 46 from municipal solid waste AD, and 7 from source-sorted waste treatment (data from 2013 to 2018, sewage sludge apparently not included (Corden *et al.*, 2019)). Extrapolating these numbers to other countries, we could state that around 290–300 million tonnes yr^{-1} are currently produced worldwide, a value that could be increased 10-fold by 2030 if AD development predictions are fulfilled (it must be considered that worldwide comprehensive data on digestate production is still difficult to obtain and that the presented values must be taken as estimates). Therefore, efficient digestate management will be crucial in a holistic AD implementation within a circular economy strategy. This is being recognised in several countries, with policies regarding digestate management (considering it as a resource) being implemented. An example is the European Green Deal, which specifies a transition to a circular economy, with a 'zero pollution Europe', and a farm-to-fork strategy.

As most of the nutrients in the influent biomass are retained within the digestate, it is a nutrient-rich effluent. AD digestate generally contains high concentrations of nutrients (i.e., nitrogen and phosphorus), recalcitrant organic matter, and trace elements (Guilayn *et al.*, 2019b). Therefore, digestate can be considered as an organic fertiliser, recovering and recycling resources from treated feedstocks, which otherwise could end up in landfills or water bodies. Furthermore, the nutrient contents in digestate make it a potential replacement for mineral fertilisers. It has been estimated that 1 tonne of digestate used as fertiliser saves the equivalent of 1 tonne of oil, 108 tonnes of water, and 7 tonnes of CO_2 emitted (European Biogas Association, 2015). In addition, digestate has proven to be a much more efficient and safer fertiliser than raw organic materials (commonly used, such as raw livestock slurry or crop residues), offering a more extensive pathogen reduction, an improved nutrient availability for plant absorption, less odours, less invasive weeds, less gaseous emissions, and a reduced risk of water and soil pollution. Because of these advantages, the vast majority of digestate is nowadays already used directly as a fertiliser in the EU28 (Corden *et al.*, 2019). In terms of potential, the use of digestate as soil amendment could replace 5–7% of the total inorganic fertiliser currently in use (World Biogas Association, 2019).

A crucial benefit of digestate application as fertiliser that is often overlooked is the return of part of the carbon in the feedstock to the soil. Carbon recovery is critical to maintain a healthy soil, and with soil quality in many parts of the world at risk of depletion of organic carbon, replenishment of nutrients and carbon has become critical. Indeed, the worldwide intensive utilisation of synthetic fertilisers (without any carbon supply due to the lack of organic matter), together with poor land management practices, have led to soil degradation worldwide, with 30% of the world's cropland becoming unproductive in the last 40 years (FAO, 2015; World Biogas Association, 2021). Carbon recovery via digestate application as fertiliser is a sustainable solution for this problem.

Taking into account all the above, it is clear that the traditional mindset of only-energy-focused AD processes is changing, considering digestate as a secondary product, rather than as a waste stream to be disposed. Nevertheless, despite the advantages of digestate utilisation for resource recovery purposes, several challenges must be tackled to make this practice a worldwide reality. Each digestate stream must be properly managed according to its characteristics, which will ultimately depend on the AD entries/substrates (and therefore can be predicted to some extent). In addition, the AD process will also affect the digestate characteristics to some extent (e.g., lower presence of pathogens in thermophilic operation). Even the benefit of digestate application as fertiliser will also vary significantly depending on external factors other than the feedstock type, such as soil type, crop needs, climatic conditions, or relevant regulations. This might create the need of transporting the digestate over long distances, jeopardising the economic feasibility of this approach and increasing its environmental impact. Another limitation is that digestate might contain potentially harmful substances/microorganisms, such as pathogens, heavy metals, bisphenol, phthalates, pharmaceuticals, polychlorinated biphenyls (PCBs), polycyclic aromatic hydrocarbons (PAHs), polyfluoroalkyl substances (PFAs), and/or microplastics (Corden *et al.*, 2019). Furthermore, improper digestate management can also result in the spread of virulent and/or antibiotic-resistant genes and bacteria, favouring the spread of pathogenic superbugs. Therefore, risk assessment and management protocols must be developed and implemented. Technical solutions must also be developed to ensure a safe digestate utilisation, including the improvement of existing post-processing methods and the development of novel technologies. Finally, the lack of legal frameworks and the clash with existing legislation must also be addressed, aiming to establish a safe and logical system that enables an optimal use of resources (European Biogas Association, 2015).

1.1.3 What resources are contained within digestates?

As aforementioned, digestate is the stream where non-biodegradable materials and excess nutrients are concentrated. Excluding a small fraction of water that ends up as vapour collected together with biogas (whose extent will mostly depend on the digester's working temperature), the water present in the substrate will end up in the digestate. Therefore, most of the raw digestate (in weight) is composed of water (see Figure 1.3). This implies that water is the most obvious resource that can be recovered. Examples of how this can be achieved are the production of high-quality purified water (e.g., via filtration techniques) or crop irrigation with the digestate liquid fraction (with or without post-treatment). Although this practice is not currently a common objective of digestate management, recent governmental initiatives favouring water recovery and reuse will surely promote this approach in the future (see, e.g., Frijns *et al.*, 2021). It is however important to consider that long distance transportation of water is not economically and environmentally reasonable, and that the overall water flow of the digestate is modest (around 0.5%) compared to agricultural needs (40 billion m^3 in Europe in 2010 (Eurostat, 2021a)).

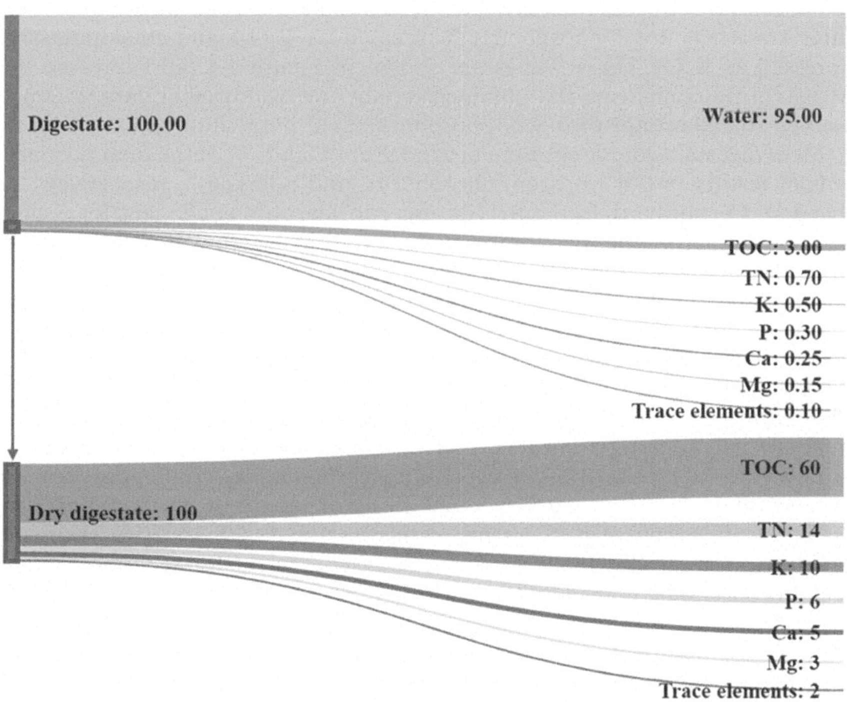

Figure 1.3 Common valuable resources contained within the digestate (data from an internal database (SUEZ) and from Monlau *et al.* (2015) and Vaneeckhaute *et al.* (2017)). The upper figure shows typical values for whole (raw) digestate and the figure below a 'zoom' after excluding water. Values represent weight percentages.

As introduced above, the digestate is a stream rich in carbon, nutrients, and macro/micronutrients (Figure 1.3). All of these are potentially valuable and their recovery is essential in a circular, sustainable economy. These compounds are commonly present in both solid and liquid forms, such as soluble and solid organics, NH_3/NH_4^+ and organic-bound N for nitrogen, or PO_4^{3-} and orthophosphate salts (e.g., struvite) for phosphorus. This implies that their maximised recovery must include a holistic, integrated process, involving different stages within a biorefinery approach. Further chapters of this book will elaborate on the recovery of each of these valuable compounds.

Although the contents of resources other than water shown in Figure 1.3 might seem low, the recovery potential can be huge considering the current digestate flows. To put these carbon and nutrient flows in perspective, a rough estimation of the resource recovery potential can be carried out by considering a figure of 200 million metric tons produced in the EU (lower end of current estimates) and an average digestate composition. To allow for the consideration of statistics, data from an internal database (SUEZ) was used, assuming an average of 5% dry matter (DM) for this digestate, as well as the interquartile ranges (IQRs; where 50% of the observations are situated) for digestate composition in terms of ammoniacal nitrogen, total nitrogen, volatile solids (as a proxy for the organic matter contents), total phosphorus (expressed as P_2O_5) and total potassium (expressed as K_2O). These values are shown in Figure 1.4 (all expressed in a DM basis). By comparing the obtained numbers with Eurostat data, it can be observed that the total nitrogen, phosphorus, and potassium contained in the 200 Mt of digestate could represent around 9%, 4% and 7% of the total European needs of fertilisers (for nitrogen, phosphorus, and potassium, respectively; see Table 1.1). Certainly, these figures do not consider any coefficient for techno-economic feasibility, neither the bioavailability of these nutrient flows for the plants. However, it must be considered that AD is a new technology, and therefore the digestate flows will surely increase (see previous section). In addition, several political stimuli will promote the reduction of nutrient losses in the food production chain, while promoting nutrient recycling, favouring resource recovery applications, and leading to a reduction of the total fertiliser needs.

Regarding organic matter contents in digestates, 1 ton of volatile solids (VS) can be considered as 0.5 tons of carbon (a usual conversion factor). Considering the flows in Table 1.1, the corresponding carbon flow would represent the CO_2-equivalent in greenhouse gas emissions of up to 1.6 million European inhabitants (average of 8.2 ton per capita in the EU28 in 2019 (Eurostat, 2021b)). As for the nutrient flow analysis, a coefficient for techno-economic feasibility should be considered, as well as a correction factor to consider only the fraction of carbon that is actually stable over long term (after soil return). In the future, the digestate-carbon flow is expected to increase, while the per capita net greenhouse gases emissions should decrease sharply, meaning that these numbers could be far more positive regarding potential digestate carbon recovery, especially if coupled to trending carbon-stabilising technologies, such as pyrolysis for biochar production.

Other than water and components, great amounts of energy are contained within the digestate. The amount of energy carried as chemical energy (COD) in

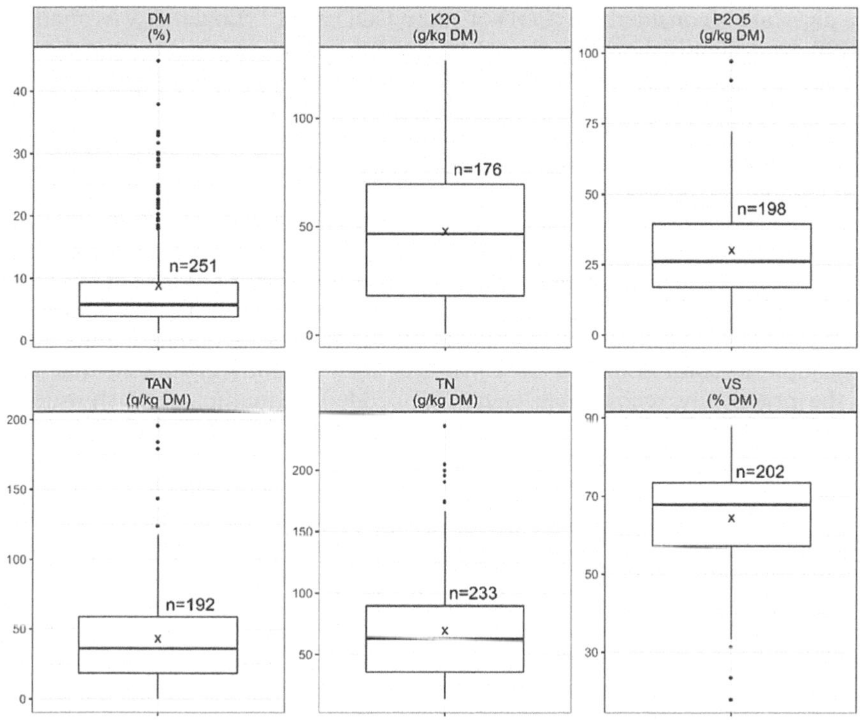

Figure 1.4 Composition of digestates based on a heterogenous large internal database (SUEZ). Tukey-style boxplots. 'x' represents the mean and 'n' the number of observations. DM stands for dry matter, TAN for total ammoniacal nitrogen, TN for total nitrogen and VS for volatile solids.

Table 1.1 Digestate potential flows in the EU27.

Parameter	Digestate range (%DM, IQR)	Flow[a] (kt yr^{-1})	EU27 reference (kt yr^{-1})
TAN	1.8–6.0 ($n=192$)	183–596	10 039[b,c]
TN	2.8–9.1 ($n=233$)	284–907	10 039[b,c]
P_2O_5	1.7–4.0 ($n=198$)	168–399	1114[b,d]
K_2O	1.8–6.9 ($n=176$)	180–693	2371[b,e]
VS	57.4–73.5 ($n=202$)	5741–7345	N/A

[a]Based on 200 Mt yr^{-1} of digestates with an average of 5% DM.
[b]Data from 2019 (Eurostat table code TAI01), EU28 not available.
[c]Nitrogen fertilisers.
[d]Phosphorus fertilisers.
[e]Potassium fertilisers.
DM, dry matter; IQR, interquartile range; TAN, total ammoniacal nitrogen; TN, total nitrogen; and VS, volatile solids.

the digestate is considerable (COD of 1.62 g COD·g VS^{-1} (Logan & Visvanathan, 2019)), and should be utilised to the greatest extent possible, either via the production of value-added reduced compounds (e.g., biofuels or fatty acids), or by the recovery of energy as heat/electricity. Several thermochemical processes exist for the latter purpose. Incineration or pelletisation of dried digestate are options allowing partial energy recovery by combustion. More efficient recovery options aim at generating a biofuel, transforming the contained energy into a useful form. Examples are gasification/pyrolysis processes for bio-oil/syngas generation or fermentation for biohydrogen or bioethanol production (Guilayn *et al.*, 2020). These technologies will be discussed in further sections of this book.

Another form of energy contained within the digestate that is commonly overlooked is thermal energy, particularly if AD is carried out under thermophilic conditions (50–55°C). It has been recently estimated that 90% of the practically recoverable energy embedded in municipal wastewater is thermal energy (being the remaining 10% present as chemical energy) (Hao *et al.*, 2019). Similar values could be expected for the digestate, as despite being a more concentrated stream than wastewater, it is also found at much higher temperatures (i.e., 35–55°C). If recovered, this thermal energy could be utilised for general heating/cooling requirements in the plant, for drying dewatered sludge, and obviously for reactor temperature control. This thermal energy could indirectly offset considerably the energy demand of the plant. As pointed out by Hao *et al.* (2019) for wastewater, the limitations in thermal energy recovery are generally not due to technical difficulties, but due to supply distances and/or governmental policies.

1.2 PROPER DIGESTATE MANAGEMENT AS A NECESSARY STEP IN A CIRCULAR ECONOMY

1.2.1 Importance of digestate management

In AD facilities, digestate management practices can vary from relatively simple processes (such as storage and direct spread) to highly complex processing lines including holistic advanced post-treatment plants. This treatment process complexity/cost compromise is closely related to the plant scale and to local factors, such as local nutrient surplus (notably nitrogen) or the land-spreading distance. As shown in previous studies (see, e.g., Fuchs & Drosg, 2013), these local factors can be key, as they might determine the economy feasibility of a defined treatment train. In any case, from the simplest to the most complex process line, any AD designer or operator must consider a complex series of environmental, economic, regulatory, and social constraints, which will drive the choice of an appropriate, feasible, digestate management strategy.

Other than post-treatment processing, proper digestate management comprises a package of good practices within the plant. For instance, any digestate storage volume should be coupled to an equivalent volume of retention, to avoid the contamination of nearby soils and water bodies in case of leaking. Indeed, the most essential digestate management practices should be framed, enforced, and controlled by regulatory bodies, such as local environmental authorities and labelling systems, being well described, and constantly evolving in the countries where AD is a well-developed technology. This is an ongoing, never-ending task that is not covered in this book in detail, as this would end up in a series of dedicated case studies almost in a country basis, which is out of the scope of this document. Entities such as the World Biogas Association or the European Biogas Association regularly publish reports on this topic. Nevertheless, the current general legal framework regarding digestate reutilisation is assessed in this book, identifying critical points, and giving recommendations for future modifications.

When digestate or any by/co-product exits the AD plant, a first glaring consideration for its management is that land spreading is associated with environmental risks (Nkoa, 2014), being usually a highly regulated activity, especially in the EU, where the digestate is usually under a waste status. Different chapters of this book will cover the current essential and emerging pollution risks. Other than obvious product-use quality criteria, regulations and environmental authorities in place must impose a series of innocuity criteria, spreading method restrictions, or land application restrictions (such as season/weather limitations and spreading limits), which commonly impose the need of an integrated environmental assessment of local and regional impacts. Under current EU regulations, land spreading of raw digestates is thus a pure cost to operators, mainly due to its waste status. Nevertheless, if managed properly, this approach can be conceived as the most economic and environmentally efficient solution for smaller plants surrounded by farmlands (common case for agriculture digesters). Digestates being a liquid-wet product (65–97% moisture contents are common), the average transportation distances as low as 10 km can already be economically prohibitive and environmentally unreasonable (Möller *et al.*, 2010). Chapter 2 covers the essential aspects of digestate 'direct' spreading in farmlands, either as a fertiliser or as a soil amendment product.

In the EU, achieving an end-of-waste status is a fundamental aspect for marketing digestate (by-)products. End-of-waste status can be achieved mainly through national standards, and more recently through EU labelling standards (CE 2019/1009), which impose a reasonable amount of innocuity/fertilising/amendment value (according to the established quality criteria) and might even impose upstream processing. Raw digestates (with no post-treatment at all) tend to be either not included in, or not conforming to such quality criteria. They are usually too diluted when compared to mineral fertiliser standards (N, P, and K contents), or too poor to fulfil organic soil amendment standards (Guilayn *et al.*, 2019a). The huge relevance of digestate management is clear in this context.

Depending on the regulatory framework, matching agricultural needs and practices can be challenging, having direct economic implications. To begin with, digestate production is relatively constant on a yearly basis, while agricultural needs are seasonal. This difference is usually buffered by large storage capacities (either in situ or ex situ), meaning that the product must be relatively stable over a long period (up to 6 months). Secondly, modern agriculture and farmers are adapted to near-perfect, stable, predictable, and consistent chemical fertilising products, and its short-term economic benefits are easily considered and difficult to beat. On the contrary, digestates and other organic fertilisers tend to present a significant quality variation over time, and the well-proved long-term benefits (also economic) can be less appealing and are indeed more difficult to quantify from an economic perspective. As of today, this short-term vision in agricultural productivities is a known worldwide problem that has already been around for decades. It is a proven fact that conventional intensive agriculture results in soil carbon depletion, erosion, desertification, and/or loss of biodiversity, which jeopardises its own economic profitability in the long run, not to mention the long list of associated environmental, social and human-health impacts (Tilman *et al.*, 2002). Resource recovery via a proper AD digestate management can be an immediate, effective palliative to this crucial challenge.

If AD is only regarded as an energy-producing technology, the fact that other renewable energies are rapidly lowering its levelised cost of energy (LCOE) might challenge its development (Brémond *et al.*, 2021), and this is causing a 'natural' impression for decreasing subsidies on AD as a renewable energy-producing technology. Research and innovation in digestate management is undoubtedly a key aspect for the future of AD as a fully sustainable and economic attractive technology. Table 1.2 illustrates this issue through a rapid

Table 1.2 Current yields and revenues needed for AD to be a competitive solution (i.e., with a combined methane and digestate revenue able to provide a net methane production cost of 0).

Parameter	Unit	AD plant in optimistic scenario today (today's selling prices)	AD plant with methane prices competitive to solar energy today	AD plant with methane prices competitive to natural gas today
Methane yield	$Nm^3\ CH_4\ t_{feedstock}^{-1}$	600	600	600
Feedstock organic matter content	$t\ t^{-1}$	20%	20%	20%
Organic matter conversion rate	$t\ t^{-1}$	90%	90%	90%
Digestate 'yield'	$t_{digestate} \cdot t_{feedstock}^{-1}$	82%	82%	82%
Imposed AD's CH_4 valorisation price	$€ \cdot MWh^{-1}$	100	30–70[c]	20–40[d]
AD's LCOE[a]	$€ \cdot MWh^{-1}$	−79	−79	−79
Minimum digestate handling revenue for feasible operation[b]	$€ \cdot t_{digestate}^{-1}$	−31	+13 to +70	+72 to +86

[a]Digestate handling excluded (Eurostat, 2021c).
[b]For achieving economic equilibrium ('CH_4 net production cost = 0'), the negative value indicates a maximum possible cost while positive values indicate minimum necessary revenues for being competitive compared to solar energy and natural gas LCOE's as reference for AD CH_4 selling price.
[c]Solar energy LCOE in different European countries (Lugo-Laguna *et al.*, 2021).
[d]Natural gas price ranges in Europe excluding taxes, first half of 2021 (Eurostat, 2021c).These values do not consider gate fees (feedstock prices) or reductions due to AD LCOE solution. Units are given per wet ton of material.

exercise. Without any compensation from AD positive externalities, and with low perspectives for significant AD cost reduction, a significant positive net result from digestate management would be necessary by considering CH_4 valorisation prices competitive with todays' LCOE for solar energy (steadily decreasing) or natural gas (could increase significantly with carbon taxation). In this context, increasing the value that can be added to digestates either by targeting value-added agricultural products or by accessing new markets is a core research topic around digestate management.

Overall, digestate management is a complex field, as several factors must be considered in a holistic approach. Nevertheless, it is essential to ensure safe waste disposal, to allow resource recovery processes, and to fulfil current and future regulations. In addition, efficient digestate valorisation might be key for the economic feasibility of future AD installations. This will only be achieved if proper digestate management practices are implemented.

1.2.2 Current state of digestate management practices

Almost every AD plant counting with digestate post-treatment will count with, at least, a phase separation step. This is the first crucial step, after which posterior post-treatment processes (if any) are applied. Considering phase separation, digestate is typically present in three forms: (a) whole digestate, being the material coming straight out of the digester, often with less than 5% DM content (no separation applied); (b) liquor, being the liquid fraction of the whole digestate, where most of the DM has been excluded; (c) fibre, which corresponds to the solid fraction after separation, containing the remaining separated DM (World Biogas Association, 2021). Digestate mechanical separation promotes a series of operational benefits that allow a more efficient digestate management and enhance resource recovery (Guilayn *et al.*, 2019b). To begin with, the liquid fraction is better pumpable, stored, transported, and spread as liquid. The solid fraction is also easier to transport, in this case as a solid and stackable material. From a resource recovery point of view, the liquor will carry soluble components, concentrating nitrogen (mostly as NH_4^+/NH_3) and potassium, both in soluble forms readily accessible to plants. The solid fraction will carry the larger particles and insoluble matter, concentrating recalcitrant organic matter, phosphorus (notably present as precipitates or adsorbed onto solid organics), and the residual slow-release organic nitrogen, also contained within solid organic compounds. Without any further post-treatment, phase separation by itself allows a better resource management by generating a product closer to an 'organic fertiliser' (liquid fraction) and a product closer to a 'soil amendment' (solid fraction), while partially separating nitrogen from phosphorus, the usual limiting parameters for land spreading.

The current full-scale post-treatment practices for the solid fraction are still quite limited, and generally involve composting (usually with a large proportion of a bulk material), landfilling (not included in Figure 1.5), and thermal drying. Full-scale liquid fraction post-treatments are even less common, as the liquid is generally simply stored in basins, land spread, and sometimes sent to wastewater treatment plants. Some post-treatment options that exist include (vacuum-) evaporation, membrane filtration, nitrogen stripping and scrubbing,

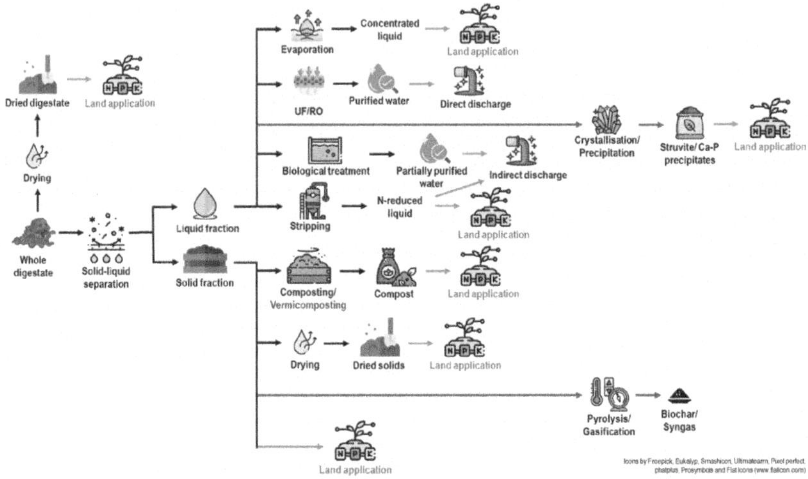

Figure 1.5 Current full-scale commercial technologies for digestate management identified in a state-of-the art review from 2013 (Fuchs & Drosg, 2013) and in a recent review focused on the same topic (Guilayn *et al.*, 2020). The new technologies identified only in the latest review are highlighted with red arrows. This shows that little has changed over 10–20 years. Adapted from Fuchs and Drosg (2013).

nitrification/denitrification, struvite recovery (quite limited to sewage sludge digestate), and some combinations of these unitary processes. Figure 1.5 presents some of the processes currently being applied for each digestate fraction.

Over the last 20 years, when it comes to digestate treatment lines/processes, academic researchers have proposed a series of innovative and promising technologies (c.f. the next section and further chapters of this book). However, many professionals in the field feel that little has changed at full-scale within the same period: the state of the art of fully developed commercial techniques (technology readiness level 9) seems quite untouched (Figure 1.5). Indeed, most of the technologies identified in a recent state-of-the art review of full-scale techniques (Guilayn *et al.*, 2020) were already a reality in the precedent decade (see Fuchs & Drosg, 2013). Moreover, the two 'new' technologies in the recent review have been around for several decades. Full-scale struvite recovery started to be reported in wastewater treatment plants in the early 2000s (Ueno & Fujii, 2001). The same can be said of industrial biomass pyrolysis/gasification (mostly wood (Meier & Faix, 1999)). It can be expected that recent political/social engagements towards a circular economy and towards fighting climate change will boost the adoption of new technologies creating more value from digestates, but this leap is still to be taken.

The AD industry is not to be blamed for this lack of mobility. How can a waste-derived product (still considered as a waste-status product) be competitive against lower-cost, better-performing, almost internationally borderless products derived from traditional petrochemical-based industries? Recent political and social plans/pressures such as the EU/USA Green Deal, net zero emission targets, consumer behaviour evolution plans, carbon taxation, and regulatory advances, must be effectively implemented to promote end-of-waste pathways, which will boost the upscaling and adoption of bio-sourced and upcycled products. In this context, regulatory/legislative advances are crucial. The implementation of more advanced technologies for digestate management, aiming towards the production of value-added products can be expected over the next decade. An overview is provided in the next section, and the most promising approaches (as well as currently applied processes) are discussed in detail in the coming chapters, covered by experts from each field.

1.2.3 Novel/promising technologies for digestate management

Despite the remarkable academic/scientific progress on digestate management that has occurred in the last decade, there is still a large room for improvement, particularly for the development of processes allowing a safe and optimal resource recovery. Research on the latter has been mainly fuelled by increasing market demands (e.g., of high-quality fertilisers) and by regulatory drivers, favouring the implementation of circular economic strategies (see Figure 1.6a for a corresponding scheme).

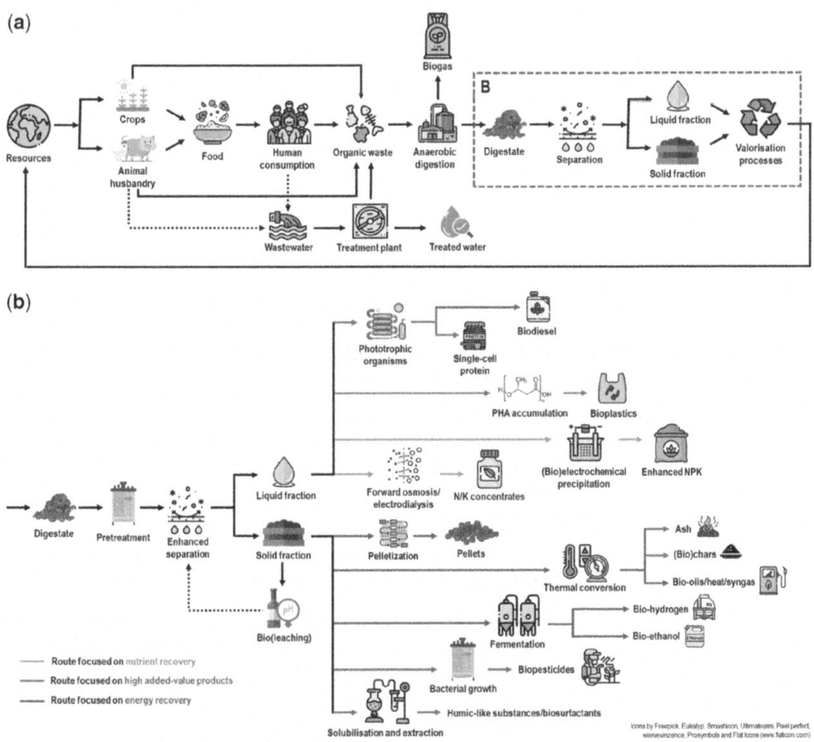

Figure 1.6 Schematic representation of (a) a general efficient digestate management focused or resource recovery, and (b) different novel options for resource recovery from digestate, including the generated products.

Regardless of the technology or approach used, the need for efficiently separating the liquid and solid digestate fractions appears essential for an effective resource recovery, as this allows at optimal resource management, with different dedicated valorisation process for each fraction (Guilayn *et al.*, 2019b). Because of this, large efforts are being invested on the development of digestate pretreatments technologies that improve digestate dewatering, favouring phase separation. The whole idea is to convert bound/interstitial water into free water, as mechanical dewatering processes can only separate the latter (Wang *et al.*, 2021). This can be achieved by, for example, destroying extracellular polymeric substances (EPS), which form gel-like substances that jeopardise dewatering. Other than dewatering, pretreatment methods also aim at improving the digestate qualities, for example, improving its stability, reducing the amount of metals in the solid fraction, or inactivating pathogens (thus producing biosolids class A, with unrestricted agricultural reuse and public contact, particularly crucial in sludge-derived digestates (Wang *et al.*, 2021)). While processes already exist to achieve these goals to some extent (see Section 1.2.2), novel technologies should aim at, not only improving dewaterability, but also provide an integrated improvement, offering multiple benefits simultaneously while also reducing the costs or current methods. Some examples of promising technologies have been reviewed in Wang *et al.* (2021), with a focus on sewage sludge (one of the worse cases/concerns in terms of presence of pathogens). Among the selected integrative technologies, iron-based advanced oxidation and acidic aerobic digestion recently appeared as promising alternatives, as both improve dewaterability, pathogen reduction, and digestate stabilisation. If needed, both processes could be coupled with novel cost-effective options for metal solubilisation, such as bioleaching (bei Li *et al.*, 2021; Yesil *et al.*, 2021). See Figure 1.6b for a scheme showing how pretreatment, dewatering and bioleaching could be coupled. It must be mentioned that, while the different technologies being developed for digestates from different origins or for particular purposes have been reviewed elsewhere (e.g., Logan & Visvanathan (2019) for municipal solid waste, Guilayn *et al.* (2020) for urban and centralised plants, Monlau *et al.* (2015) for agricultural waste, or Vaneeckhaute *et al.* (2017) for resource recovery technologies), here we present a general discussion. The authors refer the readers to the aforementioned articles for particular information on these topics.

After an efficient solid/liquid separation, most of the valorisation options for any of the fractions can be grouped according to the main resource that they aim to recover: nutrients (e.g., nitrogen, phosphorus or potassium), energy (in the form of solid fuels, diesel, oils, other reduced organics, heat, hydrogen, syngas, etc.), or high value-added products (e.g., single-cell protein, polyhydroxyalkanoates (PHAs), biopesticides, or humic-like substances).

Starting with the liquid fraction, the most researched option for its valorisation has been the recovery of nutrients in the form of either liquid concentrates or solid precipitates (e.g., struvite). Although these approaches have been considerably studied, novel processes under development have a great potential to increase recovery efficiencies (Vaneeckhaute *et al.*, 2017). Examples of promising options where further research is needed are forward osmosis (Camilleri-Rumbau *et al.*, 2021; Ferrari, 2020; Zhao *et al.*, 2012), electrodialysis (Camilleri-Rumbau *et al.*, 2021; Shi *et al.*, 2019), or enhanced precipitation via (bio)electrochemical processes (Cusick & Logan, 2012; Fraunhofer Institute for Interfacial Engineering and Biotechnology, 2012). These approaches have shown that they might enable a more efficient recovery, lowering the amounts of chemicals dosed, increasing the final nutrient concentrations, and potentially generating high-quality products due to high rejections. Research is still needed for an efficient upscaling and to investigate potential economic limitations. Another approach being researched for liquid-digestate valorisation is the generation of high value-added products, such as biomass (mostly as single-cell protein source) and/or PHA. The latter is a precursor of bioplastics, and most research so far has focused on using filtered digestate as growth media for biological PHA accumulators, such as *Cupriavidus necator* or others (Afreen *et al.*, 2021; Kovalcik *et al.*, 2017; Papa *et al.*, 2020; Passanha *et al.*, 2013). If the carbon source used is from a renewable/sustainable/profitable origin, this might be a promising option for generating a high-value product with a bright future ahead. Regarding biomass production via phototrophic organisms, most research has focused on microalgae as mediators for nutrient uptake, either for the production of biomass itself (single-cell protein), or for biodiesel generation after lipid transesterification (Guilayn *et al.*, 2020; Monlau *et al.*, 2015). Although this approach has already reached industrial pilot scale (Uggetti *et al.*, 2014), more research is needed to elucidate if process integration can help to solve a main issue of this technology, which is the instability of the cultivation system, probably due to the presence of bacteria in the influent. Coupling this approach with filtration/purification strategies might be an option to stabilise the process. Other than microalgae, photosynthetic organisms such as plants via hydroponic cultivation systems (Pelayo Lind *et al.*, 2021) or purple phototrophic bacteria (Capson-Tojo *et al.*, 2020) are also promising, although more research is needed.

Regarding solid digestate, advanced processes for energy recovery via thermal conversion (e.g., controlled pyrolysis, hydrothermal liquefaction, or hydrothermal carbonisation) are the most widely researched approach (Guilayn *et al.*, 2020; Monlau *et al.*, 2015). The traditional goal of this idea is to produce a useful form of energy (in the form of heat, oils, syngas, or other fuels) to enhance the overall energy balance of the process. Interestingly, recent research has also focused on the valorisation of the generated chars and/or ashes (as by-products), aiming at an optimised resource recovery. For example, ash valorisation as source of metals (after extraction) and phosphorus has been recently proposed ('Phos4You – We deliver Phosphorus made in Europe,' 2021). In addition, it has been shown that biochar can serve as soil improver (Al-Wabel *et al.*, 2018), as well as bio-adsorbent (Wu *et al.*, 2017). In addition, biochar can also be used to stabilise the AD process (Capson-Tojo *et al.*, 2018; Fagbohungbe *et al.*, 2017). Although biochar production from digestate is not competitive at the moment compared to wood-derived biochar, this might very well be an option in future integrated management processes allowing an optimal digestate valorisation. Another energy recovery option from solid digestate is the generation of biofuels via fermentation. Both biohydrogen and bioethanol have been produced from digestate, generating clean, renewable fuels (Monlau *et al.*, 2015). Challenges of this approach, such as the need of a pretreatment for fibre hydrolysis and/or low volumetric production rates, might limit its implementation, but recent developments are moving the field forwards (Stoumpou *et al.*, 2020). Finally, the production of high value-added compounds from solid digestate has also gained attention lately. Relevant examples are the generation of biopesticides, enzymes, or biosurfactants via solid-state fermentation (Cerda *et al.*, 2019). Although this approach is still under development, the high value of the obtained products, and increasing markets, have the potential of pushing it forwards (always coupled with other valorisation alternatives). Another value-added product that can be generated from solid digestate are humic-like substances. In this case, these compounds can be directly extracted from the digestate, and after purification they can be used as soil improvers or biostimulants (Guilayn *et al.*, 2020). To maximise the extraction of these substances from digestates, a strong alkaline treatment for their solubilisation is needed, which might limit the application of this option (Montoneri, 2017).

Obviously, the processes described above for product generation are not mutually exclusive. On the contrary, optimal digestate valorisation will surely rely on process integration, following the concept of environmental biorefinery (Capson-Tojo *et al.*, 2016; Moscoviz *et al.*, 2018; Venkata *et al.*, 2016). This facility mimics the idea of an oil refinery, where substrate valorisation is maximised via the integration of several processes. In the case of digestate management, a holistic valorisation approach will obviously include options for the conversion of both the liquid and solid digestate fractions. Moreover, sequential valorisation steps from highest to lowest valuable products are also needed to optimise recovery (e.g., extracting high value-added compounds early on the treatment train and use its residual effluent for less profitable alternatives, ideally leaving low-value energetic valorisation for the last fraction). Other than sequential integration strategies, several processes can be coupled, enhancing performances, and reducing environmental impacts. Some integration examples are: (a) the uptake of CO_2 from biogas via photoautotrophic organisms for single-cell protein production, acting as carbon source and serving for biogas upgrading; (b) recycling of the bioleaching effluent for integrated treatment with the liquid fraction; (c) the use of biochar for enhancing the digestion performance and thus the digestate quality; or (d) the use of compounds from solid digestate fermentation as carbon source for microorganisms growing using the nutrient-rich liquid fraction as growth media (e.g., purple phototrophic bacteria for single-cell protein (Capson-Tojo *et al.*, 2020)). Process integration aiming at maximising resource recovery and high value-added generation will surely be the core of future modern digestate management facilities.

The optimal technology (or integrated technologies) to be applied will depend on many factors. Other than obvious constraints such as readiness level or legal approval, local, national, and even international factors will affect the selection process (e.g., legal agreements and policies). On a local scale, the distance from potential product buyers might be a critical factor, as transportation is known to be a major cost. In addition, the nature and the stability of the plant entries will also affect considerably the post-treatment technology to be selected. A typical example is the dichotomy between small plants in rural areas and large plants for urban/centralise treatment. On the one hand, small plants are often focused on fertiliser production, as this product can be applied directly on-site (or close by), and because the plant entries are usually stable over time (e.g., agricultural waste). In addition, these regions do not have high energy demands. On the other, large plants in urban areas might move away from this option, as their scale allows for a larger capital expenditure, thus permitting more complex and ambitious processes within a full-valorisation scheme. In addition, these installations have a much larger influent variability, which makes it difficult to ensure product quality in simple digestate treatment approaches (e.g., drying and/or composting). Furthermore, high value-added products and/or energy recovery might be favoured in this situation, as potential buyers will be around, and urban regions have high energy demands. As an example, Tampio *et al.* (2016) found that, in their case, evaporation combined with reverse osmosis was the most efficient nutrient recovery technology for generating a transportable fertiliser from digestate, mainly due to the low product mass and the reduced energy consumption for transportation. In their study, the selection of the treatment technology was heavily dependent on the location of the AD plant relative to the agricultural land, and on the type of fertiliser products needed. Summarising, selecting the right post-treatment train is an arduous task, where several factors must be considered, including economical, technological, regional, and (inter)national constrains.

1.3 CONCLUSIONS

The increasing number of AD plants and the implementation of policies based on resource recovery and circular economy call for the development and implementation of digestate management approaches that allow the recovery of the resources contained within it (e.g., water, nutrients, carbon, or energy). While traditional practices such as thermal drying, incineration, composting, or landfilling allow to dispose digestate in a safe manner (and in some cases a certain degree of resource recovery), novel methods are being developed to allow a more intensive and efficient recovery of resources. Processes such as enhanced precipitation, enhanced thermal conversions, photoautotrophic biomass growth, or enhanced filtration, are promising. While there is a lot of work ahead of us, the tremendous effort currently being put on this task ensures a strong and fast development of the field. Despite research/technological efforts, a particular emphasis should be put on the translation of research development to the field, aiming at the practical implementation of novel concepts at industrial scale. As importantly, legal frameworks must be updated to avoid hindering the application of novel technologies.

This book presents a comprehensive review of the state of the art on AD digestate management. The book introduces the application of digestates as fertilisers, also addressing the challenges that the particular characteristics of digestates produced from different substrates (e.g., manure, municipal solid waste, agricultural waste, or algae biomass) pose. Afterwards, different novel processes for resource recovery from digestate are discussed, including options for nitrogen, phosphorus, or energy recovery, addressing also pre- and post-treatments allowing an enhanced recovery. Novel high value-added products that can be produced from digestates are also discussed, including hydrochar, PHAs (bioplastics), or single-cell protein. The presence of pollutants (including emerging ones) and their relevance are also discussed, together with potential ecotoxicity issues. The results from life-cycle analyses are summarised and critically discussed, and the current legal framework regarding digestate reutilisation is assessed, identifying critical points, and giving recommendations for future modifications. Finally, options for process integration within the biorefinery concept and future perspectives are also discussed. Overall, this book gives an excellent overview of the current state and the most promising advances in digestate management, focusing on emerging pollution concerns and the creation of value.

REFERENCES

Afreen R., Tyagi S., Singh G. P. and Singh M. (2021). Challenges and perspectives of polyhydroxyalkanoate production from microalgae/cyanobacteria and bacteria as microbial factories: an assessment of hybrid biological system. *Frontiers in Bioengineering and Biotechnology*, **9**, 1–14, https://doi.org/10.3389/fbioe.2021.624885

Akhiar A., Ahmad Zamri M. F. M., Torrijos M., Shamsuddin A. H., Battimelli A., Roslan E., Mohd Marzuki M. H. and Carrere H. (2020). Anaerobic digestion industries progress throughout the world. IOP Conference Series: Earth and Environmental Science, **476**, 012074. https://doi.org/10.1088/1755-1315/476/1/012074

Al-Wabel M. I., Hussain Q., Usman A. R. A., Ahmad M., Abduljabbar A., Sallam A. S. and Ok Y. S. (2018). Impact of biochar properties on soil conditions and agricultural sustainability: a review. *Land Degradation and Development*, **29**, 2124–2161, https://doi.org/10.1002/ldr.2829

Appels L., Lauwers J., Degrve J., Helsen L., Lievens B., Willems K., Van Impe J. and Dewil R. (2011). Anaerobic digestion in global bio-energy production: potential and research challenges. *Renewable and Sustainable Energy Reviews*, **15**, 4295–4301, https://doi.org/10.1016/j.rser.2011.07.121

bei Li Y., li Song J., jing Yao Q., xu Chen Z., Wei Y., long Li H., xiao Wang M., jing Wang B. and min Zhou J. (2021). Effects of dissolved oxygen on the sludge dewaterability and extracellular polymeric substances distribution by bioleaching. *Chemosphere*, **281**, 130906, https://doi.org/10.1016/j.chemosphere.2021.130906

Brémond U., Bertrandias A., Steyer J. P., Bernet N. and Carrere H. (2021). A vision of European biogas sector development towards 2030: trends and challenges. *Journal of Cleaner Production*, **287**, 125065, https://doi.org/10.1016/j.jclepro.2020.125065

Camilleri-Rumbau M. S., Briceño K., Søtoft L. F., Christensen K. V., Roda-Serrat M. C., Errico M. and Norddahl B. (2021). Treatment of manure and digestate liquid fractions using membranes: opportunities and challenges. *International Journal of Environmental Research and Public Health*, **18**, 1–30, https://doi.org/10.3390/ijerph18063107

Capson-Tojo G., Rouez M., Crest M., Steyer J.-P., Delgenès J.-P. and Escudié R. (2016). Food waste valorization via anaerobic processes: a review. *Reviews in*

Environmental Science and Bio/Technology, **15**, 499–547, https://doi.org/10.1007/s11157-016-9405-y

Capson-Tojo G., Girard C., Rouez M., Crest M., Steyer J.-P., Bernet N., Delgenès J.-P. and Escudié R. (2018). Addition of biochar and trace elements in the form of industrial FeCl3 to stabilize anaerobic digestion of food waste: dosage optimization and long-term study. *Journal of Chemical Technology & Biotechnology*, **94**(2), 505–515, https://doi.org/10.1002/jctb.5797

Capson-Tojo G., Batstone D. J., Grassino M., Vlaeminck S. E., Puyol D., Verstraete W., Kleerebezem R., Oehmen A., Ghimire A., Pikaar I., Lema J. M. and Hülsen T. (2020). Purple phototrophic bacteria for resource recovery: challenges and opportunities. *Biotechnology Advances*, **43**, 107567, https://doi.org/10.1016/j.biotechadv.2020.107567

Cerda A., Mejias L., Rodríguez P., Rodríguez A., Artola A., Font X., Gea T. and Sánchez A. (2019). Valorisation of digestate from biowaste through solid-state fermentation to obtain value added bioproducts: a first approach. *Bioresource Technology*, **271**, 409–416, https://doi.org/10.1016/j.biortech.2018.09.131

Corden C., Bougas K., Cunningham E., Tyrer D., Kreißig J., Zetti E., Gamero E., Wildey R. and Crookes M. (2019). Digestate and compost as fertilisers: risk assessment and risk management options.

Cusick R. D. and Logan B. E. (2012). Phosphate recovery as struvite within a single chamber microbial electrolysis cell. *Bioresource Technology*, **107**, 110–115, https://doi.org/10.1016/j.biortech.2011.12.038

European Biogas Association (2015). Digestate Factsheet: the value of organic fertilisers for Europe's economy, society and environment.

European Commission (2019). COM (2019) 640 final – the European green deal.

Eurostat (2021a). Irrigation: number of farms, areas and equipment by size of irrigated area and NUTS 2 regions [WWW Document]. http://appsso.eurostat.ec.europa.eu/nui/show.do?query=BOOKMARK_DS-200702_QID_-1FF19F4A_UID_-3F171EB0&layout=TIME,C,X,0;INDIC_EF,L,X,1;GEO,L,Y,0;IRRIG,L,Z,0;INDICATORS,C,Z,1;&zSelection=DS-200702INDICATORS,OBS_FLAG;DS-200702IRRIG,TOTAL;&rankName1=IRRIG_1_2 (accessed 8 November 2021).

Eurostat (2021b). Greenhouse gas emissions per capita [WWW Document]. http://appsso.eurostat.ec.europa.eu/nui/show.do?dataset=t2020_rd300&lang=en (accessed 4 November 2021)

Eurostat (2021c). Natural gas prices for non-household consumers, first half 2021 [WWW Document]. https://ec.europa.eu/eurostat/statistics-explained/index.php?title=File:Natural_gas_prices_for_non-household_consumers,_first_half_2021_v1.png (accessed 4 November 2021).

Fagbohungbe M. O., Herbert B. M. J., Hurst L., Ibeto C. N., Li H., Usmani S. Q. and Semple K. T. (2017). The challenges of anaerobic digestion and the role of biochar in optimizing anaerobic digestion. *Waste Management*, **61**, 236–249, https://doi.org/10.1016/j.wasman.2016.11.028

FAO (2015). International Year of Soil Conference Summary. International Year of Soil Conference.

Ferrari F. (2020). Combining forward osmosis and anaerobic membrane bioreactor technologies for raw municipal wastewater treatment.

Fraunhofer Institute for Interfacial Engineering and Biotechnology (2012). Recovery of plant nutrients for a sustainable agriculture.

Frijns J., Charpentier L., Malamis S., Monteleone D., Makropoulos C., Koop S., Smith H., Iurlaro C., Nordin A., Rubini A., Achilleos E. and Kuzmickaite V. (2021). Water in the circular economy policy development.

Fuchs W. and Drosg B. (2013). Assessment of the state of the art of technologies for the processing of digestate residue from anaerobic digesters. *Water Science & Technology*, **67**, 1984–1993, https://doi.org/10.2166/wst.2013.075

Guilayn F., Jimenez J., Martel J. L., Rouez M., Crest M. and Patureau D. (2019a). First fertilizing-value typology of digestates: a decision-making tool for regulation. *Waste Management*, **86**, 67–79, https://doi.org/10.1016/j.wasman.2019.01.032

Guilayn F., Jimenez J., Rouez M., Crest M. and Patureau D. (2019b). Digestate mechanical separation: efficiency profiles based on anaerobic digestion feedstock and equipment choice. *Bioresource Technology*, **274**, 180–189, https://doi.org/10.1016/j.biortech.2018.11.090

Guilayn F., Rouez M., Crest M., Patureau D. and Jimenez J. (2020). Valorization of digestates from urban or centralized biogas plants: a critical review. *Reviews in Environmental Science and Bio/Technology*, **19**(2), 419–462, https://doi.org/10.1007/s11157-020-09531-3

Hao X., Li J., van Loosdrecht M. C. M., Jiang H. and Liu R. (2019). Energy recovery from wastewater: heat over organics. *Water Research*, **161**, 74–77, https://doi.org/10.1016/j.watres.2019.05.106

House of Representatives (2019). Recognizing the duty of the Federal Government to create a Green New Deal.

International Energy Agency (2020). Outlook for biogas and biomethane. Prospects for organic growth, World Energy Outlook Special Report.

IPCC (2021). Climate Change 2021: The Physical Science Basis, IPCC Press Release.

Kovalcik A., Meixner K., Mihalic M., Zeilinger W., Fritz I., Fuchs W., Kucharczyk P., Stelzer F. and Drosg B. (2017). Characterization of polyhydroxyalkanoates produced by *Synechocystis salina* from digestate supernatant. *International Journal of Biological Macromolecules*, **102**, 497–504, https://doi.org/10.1016/j.ijbiomac.2017.04.054

Logan M. and Visvanathan C. (2019). Management strategies for anaerobic digestate of organic fraction of municipal solid waste: current status and future prospects. *Waste Management & Research*, **37**, 27–39, https://doi.org/10.1177/0734242X18816793

Lugo-Laguna D., Arcos-Vargas A. and Nuñez-hernandez F. (2021). A European assessment of the solar energy cost: key factors and optimal technology. *Sustainability*, **13**, 3238, https://doi.org/10.3390/su13063238

Meier D. and Faix O. (1999). State of the art of applied fast pyrolysis of lignocellulosic materials – a review. *Bioresource Technology*, **68**, 71–77, https://doi.org/10.1016/S0960-8524(98)00086-8

Möller K., Schulz R. and Müller T. (2010). Substrate inputs, nutrient flows and nitrogen loss of two centralized biogas plants in southern Germany. *Nutrient Cycling in Agroecosystems*, **87**, 307–325, https://doi.org/10.1007/s10705-009-9340-1

Monlau F., Sambusiti C., Ficara E., Aboulkas A., Barakat A. and Carrere H. (2015). New opportunities for agricultural digestate valorization: current situation and perspectives. *Energy & Environmental Science*, **8**, 2600–2621, https://doi.org/10.1039/C5EE01633A

Montoneri E. (2017). Municipal waste treatment, technological scale up and commercial exploitation: the case of bio-waste lignin to soluble lignin-like polymers. In: Food Waste Reduction and Valorisation, P. Morone, F. Papendiek, V. E. Tartiu (eds.), Springer, Switzerland, pp. 79–120.

Moscoviz R., Trably E., Bernet N. and Carrère H. (2018). The environmental biorefinery: state-of-the-art on the production of hydrogen and value-added biomolecules in mixed-culture fermentation. *Green Chemistry*, **20**, 3159–3179, https://doi.org/10.1039/C8GC00572A

Nkoa R. (2014). Agricultural benefits and environmental risks of soil fertilization with anaerobic digestates: a review. *Agronomy for Sustainable Development*, **34**, 473–492, https://doi.org/10.1007/s13593-013-0196-z

Papa G., Pepè Sciarria T., Carrara A., Scaglia B., D'Imporzano G. and Adani F. (2020). Implementing polyhydroxyalkanoates production to anaerobic digestion of organic fraction of municipal solid waste to diversify products and increase total energy recovery. *Bioresource Technology*, **318**, 124270, https://doi.org/10.1016/j.biortech.2020.124270

Passanha P., Esteves S. R., Kedia G., Dinsdale R. M. and Guwy A. J. (2013). Increasing polyhydroxyalkanoate (PHA) yields from *Cupriavidus necator* by using filtered digestate liquors. *Bioresource Technology*, **147**, 345–352, https://doi.org/10.1016/j.biortech.2013.08.050

Pelayo Lind O., Hultberg M., Bergstrand K. J., Larsson-Jönsson H., Caspersen S. and Asp H. (2021). Biogas digestate in vegetable hydroponic production: pH dynamics and pH management by controlled nitrification. *Waste and Biomass Valorization*, **12**, 123–133, https://doi.org/10.1007/s12649-020-00965-y

Pflüger S. (2020). Mapping the state of play of renewable gases in Europe.

Phos4You – We deliver Phosphorus made in Europe [WWW Document] 2021. https://www.nweurope.eu/projects/project-search/phos4you-phosphorus-recovery-from-waste-water-for-your-life/ (accessed 4 October 2021).

Roopnarain A. and Adeleke R. (2017). Current status, hurdles and future prospects of biogas digestion technology in Africa. *Renewable and Sustainable Energy Reviews*, **67**, 1162–1179, https://doi.org/10.1016/j.rser.2016.09.087

Shi L., Xie S., Hu Z., Wu G., Morrison L., Croot P., Hu H. and Zhan X. (2019). Nutrient recovery from pig manure digestate using electrodialysis reversal: membrane fouling and feasibility of long-term operation. *Journal of Membrane Science*, **573**, 560–569, https://doi.org/10.1016/j.memsci.2018.12.037

Stoumpou V., Novakovic J., Kontogianni N., Barampouti E. M., Mai S., Moustakas K., Malamis D. and Loizidou M. (2020). Assessing straw digestate as feedstock for bioethanol production. *Renewable Energy*, **153**, 261–269, https://doi.org/10.1016/j.renene.2020.02.021

Tampio E., Marttinen S. and Rintala J. (2016). Liquid fertilizer products from anaerobic digestion of food waste: mass, nutrient and energy balance of four digestate liquid treatment systems. *Journal of Cleaner Production*, **125**, 22–32, https://doi.org/10.1016/j.jclepro.2016.03.127

Tilman D., Cassman K. G., Matson P. A., Naylor R. and Polasky S. (2002). Agricultural sustainability and intensive production practices. *Nature*, **418**, 671–677, https://doi.org/10.1038/nature01014

Ueno Y. and Fujii M. (2001). Three years experience of operating and selling recovered struvite from full-scale plant. *Environmental Technology (United Kingdom)*, **22**, 1373–1381.

Uggetti E., Sialve B., Latrille E. and Steyer J. P. (2014). Anaerobic digestate as substrate for microalgae culture: the role of ammonium concentration on the microalgae productivity. *Bioresource Technology*, **152**, 437–443, https://doi.org/10.1016/j.biortech.2013.11.036

UN Environment Management Group (2021). The UN Sustainable Development Goals [WWW Document]. https://unemg.org/our-work/supporting-the-sdgs/the-un-sustainable-development-goals/ (accessed 17 September 2021).

Vaneeckhaute C., Lebuf V., Michels E., Belia E., Vanrolleghem P. A., Tack F. M. G. and Meers E. (2017). Nutrient recovery from digestate: systematic technology review and product classification. *Waste and Biomass Valorization*, **8**, 21–40, https://doi.org/10.1007/s12649-016-9642-x

Venkata S. M., Nikhil G. N., Chiranjeevi P., Nagendranatha Reddy C., Rohit M. V., Naresh Kumar A. and Sarkar O. (2016). Waste Biorefinery models towards sustainable bioeconomy: critical review and future perspectives.*Bioresource Technology*, **215**, 2–12, https://doi.org/10.1016/j.biortech.2016.03.130

Wang Z., Liu T., Duan H., Song Y., Lu X., Hu S., Yuan Z., Batstone D. and Zheng M. (2021). Post-treatment options for anaerobically digested sludge: current status and future prospect. *Water Research*, **205**, 117665, https://doi.org/10.1016/j.watres.2021.117665

World Biogas Association (2018). The contribution of Anaerobic Digestion and Biogas towards achieving the UN Sustainable Development Goals, World Biogas Association.

World Biogas Association (2019). Global potential of biogas, World Biogas Association.

World Biogas Association (2021). Biogas: Pathways to 2030 – Report.

Wu S., He H., Inthapanya X., Yang C., Lu L., Zeng G. and Han Z. (2017). Role of biochar on composting of organic wastes and remediation of contaminated soils – a review. *Environmental Science and Pollution Research*, **24**, 16560–16577, https://doi.org/10.1007/s11356-017-9168-1

Yesil H., Molaey R., Calli B. and Tugtas A. E. (2021). Extent of bioleaching and bioavailability reduction of potentially toxic heavy metals from sewage sludge through pH-controlled fermentation. *Water Research*, **201**, 117303, https://doi.org/10.1016/j.watres.2021.117303

Zhao S., Zou L., Tang C. Y. and Mulcahy D. (2012). Recent developments in forward osmosis: opportunities and challenges. *Journal of Membrane Science*, **396**, 1–21, https://doi.org/10.1016/j.memsci.2011.12.023

doi: 10.2166/9781789062755_0025

Chapter 2

Anaerobic digestate as fertilizer for land application

Julie Jimenez

Laboratoire de Biotechnologies de l'Environnement (LBE), University of Montpellier, INRAE, 102 avenue des Etangs, 11100, Narbonne, France
Corresponding author e-mail: julie.jimenez@inrae.fr

ABSTRACT

Biogas digestates have a high organic fertilizer potential for crop soils. However, composition variability induced by several feedstock types and post-treatments impacts the agronomical value of the digestates. After focusing on this variability definition and description, the effects on chemical (nutrients), biological and physical soil fertility are described, as well as the associated environmental and health risks. Despite several data and literature found on chemical fertility (i.e., short- and mid-term fertility for crop nutrition), missing data is still observed concerning biological and physical fertility and the wide range of experimental conditions avoid getting generic conclusions. Moreover, pedoclimatic conditions, agricultural practices, crop types together with digestate composition knowledge are crucial to anticipate digestate impacts on soil.

Keywords: Digestates, organic matter, nitrogen, phosphorus, potassium, crop fertilizer, soil fertility

2.1 INTRODUCTION

Like energy recovery from Anaerobic Digestion (AD), the recovery of digestate is a key issue for the viability of an AD project as well as for its acceptability and local integration. It is therefore necessary to have a good understanding of the agronomic potential of this co-product and to ensure its environmental safety (Reibel & Leclerc, 2018). Indeed, AD can be an interesting process by which to produce organic fertilizers considering the potential benefits: it can contribute to soil carbon storage (i.e., organic matter stabilization) and make mineral elements (N and P) available to plants. However, this agronomic interest is potentially accompanied by environmental impacts. Better knowledge about the impacts of digestate spreading allows a better and safe use of this organic fertilizer to substitute crops soil mineral fertilization.

Editor(s): Vinay Kumar Tyagi, Kaoutar Aboudi and Cigdem Eskicioglu

Soil fertilization can be defined under three categories: (i) biological, as the short-term fertilization associated with microorganisms and soil mesofauna activation by biodegradable Organic Matter (OM); (ii) chemical, as short- to mid-term fertilization associated with available nutrient uptake by crops/plants for their growth; and (iii) physical, as the long-term fertilization associated with stable and complex OM used as a bulking agent in soil structure. Following digestate composition trends, OM is composed of both stable/humidified and labile OM (non-biodegradable in anaerobic conditions, but potentially decomposable in aerobic conditions), and the mineralization of nutrients enhances their availability for plant nutrition (Fernández-Dominguez *et al.*, 2021). All the ingredients seem to be adequately present to fulfil an agrosystem's needs. However, the input AD feedstock influences the digestate composition (Fernández-Dominguez *et al.*, 2021; Guilayn *et al.*, 2019a). Therefore, several research questions arise: How can the digestate composition variability be described and understood? How does this variability impact the agronomic value? Are there toxic elements? What are the environmental impacts of digestates spreading? What are the biological and physical impacts of this practice?

This Chapter sets out to answer all these issues following state-of-the-art findings in the scientific literature and research. The agronomical use of digestates as organic fertilizers in crop soils will be described. Digestate composition and variability have first to be discussed to better understand the impact of digestate spreading. Then, the three types of soil fertility (i.e., chemical, biological and physical) will be presented before discussing associated risks and safety.

2.2 DIGESTATE CHARACTERIZATION VARIABILITY

Rationalizing the variability of digestate quality is crucial to improving the use of digestate in land applications.

2.2.1 Impact of AD on digestate characterization

Some general trends can be observed regarding the quality of digestates. A schematic view is provided in Figure 2.1.

First, mass reduction occurs during AD due to the conversion of OM into biogas and, to a lesser extent, due to water loss, during hydrolysis. Moreover, Moscoviz and Jimenez (2021) showed that this loss increases significantly with the increase of dry matter (DM) content. This increase of mass reduction can lead to a concentration of some compounds, nutrients and non-biodegradable OM in the digestate.

Concerning OM, mechanistic models such as ADM1 (Batstone *et al.*, 2002) show that at least 80–90% of OM is associated with non-biodegradable OM in AD conditions (i.e., hydraulic retention time – HRT) and with recalcitrant OM, whereas only 10–20% of OM is associated with microorganisms, because of the low range of anaerobic bacteria and archaea growth rates. Moreover, depending on the biodegradability of the substrate, 50 to more than 90% of OM from AD input is converted into biogas. Indeed, the least biodegradable

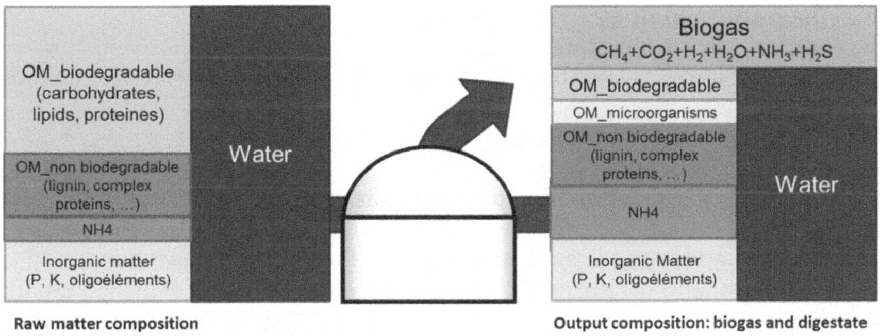

Figure 2.1 Schematic mass balance of organic and inorganic matter during AD. *OM: organic matter.

material such as fibrous feedstock (manure, crop residues, etc.) or stabilized sludge from wastewater treatment plants can have lower ranges of conversion. This conversion impacts the quality of digestate OM. Brunetti *et al.* (2012) and Fernández-Dominguez *et al.* (2021) reported that humification can occur during AD resulting in OM complexification. Fernández-Dominguez *et al.* (2021) evaluated the impact of feedstock on OM characterization and showed a general trend in which OM accessibility decreased during AD, whilst the complexity revealed by 3D fluorescence increased after AD.

The mineralization of OM leads to the mineralization of organic nitrogen and phosphorus. Organic nitrogen is converted into total ammoniacal nitrogen (TAN). This N form is readily accessible to plants meaning that digestates can be used as substitutes of mineral N sources for crop fertilization. Organic phosphorus content in digestate is very low (i.e., the inorganic fraction is about 78 to 93% (Jimenez *et al.*, 2020; Mazzini *et al.*, 2020; Regelink *et al.*, 2021)). Mineralization can lead to phosphates conversion. However, phosphates can be bounded with carbonates, calcium or metals reducing the availability of phosphorus for crop nutrition (Jimenez *et al.*, 2020).

The nutrients N, P and K are theoretically conserved in the digestate. However, during AD, P precipitation can occur (Möller & Müller, 2012), with some authors showing low P loss (10%, Schievano *et al.*, 2011) and others larger losses (36%, Marcato *et al.*, 2008).

Concerning health parameters, an overview of the state of the art of knowledge about the impact of pathogen inactivation during AD has been published (Jiang *et al.*, 2020). The main conclusions are: (i) process temperature has a significant effect (thermophilic versus mesophilic) but some pathogens (spore-forming Gram positive bacteria) can resist; (ii) the type of process (batch versus continuous stirred tank reactor – CSTR) does not have a significant impact; (iii) HRT has no impact; and (iv) substrate type has no significant impact. However, the authors highlighted that volatile fatty acids (VFA) and TAN content can inactivate pathogens in AD. This latter needs further investigation.

2.2.2 Digestate typology

The composition of digestates is a result of feedstock composition, process design, post-treatments and the overall resulting process performance (Guilayn *et al.*, 2019a).

2.2.2.1 Feedstock impact

Based on a large database of digestate composition in terms of agronomic value, Guilayn *et al.* (2019a) performed statistical analysis and highlighted two main results: (i) dry and wet AD split raw digestates into two groups depending on the DM content; and (ii) the main feedstock has an impact on digestate quality. Indeed, 6 groups of digestates were found thanks to data clustering. Each group is related to the main feedstock. Figure 2.2 extracted from this study presents the characteristics of each digestate type and the range values for each parameter are presented in Table 2.1.

Digestates related to digestate types 1 and 6 come from agricultural digesters such as those fed by fibrous feedstock (cattle manure, silage and green waste) in wet conditions (CSTR, type 1) or dry conditions (batch or plug flow reactor, type 6). Larger amounts of dry matter (>5% in Wet-AD, relating to interquartile ranges) and volatile solids (>70% DM), and a higher C/N ratio (5–20) are all related to the more fibrous characteristic of agricultural inputs. Agricultural digestates tend to be easier to commercialize from a regulatory point of view. Wastewater treatment plant (WWTP) sludge digestates are represented by type 2 in the Figure. Sludge AD in a WWTP is essentially a large mesophilic CSTR and receives a combination of primary (more biodegradable) and secondary (biological) sludge (less biodegradable). This sludge is characterized by a low DM range (typically <5%), high phosphorus (>1% DM) and low potassium

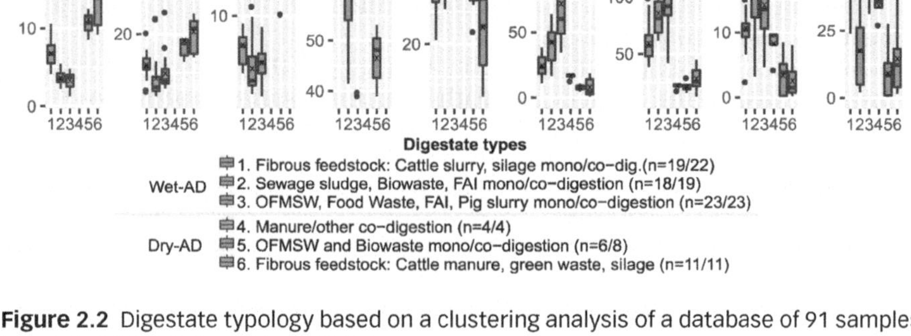

Figure 2.2 Digestate typology based on a clustering analysis of a database of 91 samples (extracted from Guilayn *et al.*, 2019a).

Table 2.1 Range of agronomic parameter values according to digestate typology, based on Guilayn *et al.* (2019a).

Digestate Type	TC/ TN	TC/ Norg	DM (% Raw matter)	VS (% DM)	TAN/TN (% NT)	TN (g/ kg MS)	P (g/kg MS)	K (g/kg MS)
1	4–11	9–16	4–8	70–81	20–60	40–100	7–15	25–75
2	3–5	5–12	2–10	41–72	30–80	60–125	4–40	2–50
3	2–5	6–13	2–8	59–74	42–81	40–175	5–25	30–90
4	14–20	55–70	12–16	64–74	70–75	15–25	8–10	28–37
5	8–12	15–20	18–28	39–54	32–45	15–30	1–8	1–26
6	9–24	15–25	12–24	69–90	5–55	15–40	1–8	2–38

Note: Concentrations of the following parameters: Dried Matter (DM in % of raw matter), Volatile Solids (VS in % DM), total carbon (TC in gC/gDM), total and organic nitrogen (TN and Norg in gN/gDM), C/TN and TC/Norg, TAN (g N/gDM), potassium (K in gK2O/gDM) and phosphorus (P in gP2O5/gDM).

(<2.5%) content. In Europe, in the case of spreading, sewage sludge digestates are usually spread under the status of waste and following national standards. Some countries rely on end-of-waste standards for composted sludge (e.g., NF U 44-095 in France); they were not included as inputs for the latest EU regulation on fertilizers (CE 2019/1009).

Centralized digesters are large scale biogas facilities (>1–2 MW and >50 ktons/year), usually mesophilic CSTRs treating a great mixture of organic waste streams from a large area, including food processing waste, municipal/ industrial sludge, and household, green and agricultural waste, amongst others. Most of these installations are wet-AD as the typical pre-treatment for depackaging and hygienization relies on heavy dilution. Given their highly variable input, they are more difficult to cluster but could be associated mainly with type 3 and to some extent to type 2 and 5 digestates (Figure 2.2).

The organic fraction of municipal solid waste (OFMSW) is usually treated in large-scale municipal biogas facilities. OFMSW is obtained after a mechanical biological treatment (MBT) of mixed-collection urban waste. MBT of such a stream usually produces a fibrous cellulosic biodegradable material with a structured bulk capacity without needing a structuring co-substrate. MBT is often followed by a Dry-AD thermophilic digester. Another emerging configuration is Dry-AD of (source-separated) biowaste, which contains a significant contribution of green waste serving as a structuring material. In Figure 2.2, they are shown as being composed mostly of types 3 and 5 presenting low nutrient content that can be associated with the typical cellulosic material obtained from MBT. The restrictions and end-of-waste scenarios described for sludge digestate apply similarly to MBT digestate from mixed-collection waste.

Guilayn *et al.* (2019a)'s typology has been confirmed by two other studies. The first of these was a French project named Concept-Dig (Jimenez *et al.*, 2019) with the objective of proposing a tool to help with the characterization of post-treatment agrowaste digestate according to the feedstock type and the process involved (Jimenez *et al.*, 2019). The first part of the project was dedicated to recovering data from 72 farms operating digesters. Raw digestate and post-treated digestate quality was analysed. Similarly to Guilayn *et al.* (2019a), a statistical analysis was made. Applied to only agrowaste AD, the results revealed 8 groups associated with their main feedstock and showed a significant difference between ruminant slurry digestate (more fibrous, associated with high OM level and poor nutrient content) and non-ruminant slurry digestate (which contains more nutrients as illustrated by Figure 2.3). In between, some co-substrates such as biowaste (i.e., agroindustrial, coming from supermarkets or restaurants) and grease (agro-industrial wastes) impact the fertilizer potential of the digestate as shown in Figure 2.3. Thanks to this classification, different agricultural digestate types were proposed and some global characteristics in terms of amendment potential of N/P/K fertilizer were made, by typology.

The second study confirming the feedstock fingerprint in digestate quality was undertaken by Fernández-Dominguez *et al.* (2021). In this study, the authors go further into the OM quality of digestates by using indicators translating accessibility and complexity of OM with sequential extraction methods combined with a 3D fluorescence method. Both agricultural and urban digestates and their respective feedstocks were characterized. Statistical analysis identified four groups: (i) pig slurry; (ii) fibrous manure and OFMSW; (iii) pig slurry in co-digestion with fibrous wastes; and (iv) WWTP sludge digestates. Clustering revealed most digestates were classified together with their respective feedstock, meaning that OM accessibility and complexity from substrate act as a fingerprint after AD. This fingerprint is highlighted for each group of digestates/feedstocks in terms of more or less accessible OM fractions and more or less complex associated molecules. Another main result of this study was that the operational conditions of AD (i.e., temperature and HRT) were not significant enough to impact digestate OM quality. Indeed, these parameters are usually associated with the feedstock type.

In conclusion, feedstock type is the main factor controlling digestate quality. In the case of OM (the major digestate component), fibre-like feedstock enhances OM in the digestate because lignin-like compounds (e.g., manure, crop residues, straw) are not biodegradable , whereas biowaste-like feedstocks (e.g., OFMSW and biowaste itself) decrease OM in the digestate because their content is more biodegradable. With regard to nutrients, pig slurry, biowaste, lipidic agro-industrial effluents and sewage sludge increased fertilizing potential (N, P, K). However, all these results are related to raw digestates. If national or European regulations are considered, only some agricultural digestates can be considered as products (e.g.,

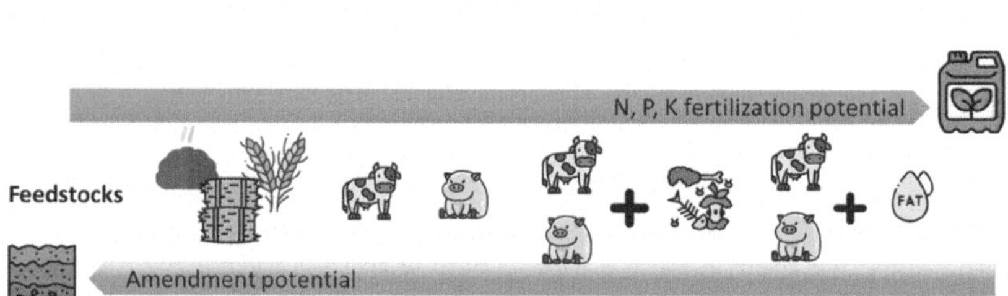

Figure 2.3 Schematic typology set-up in the Concept-Dig project.

in France). Guilayn *et al.* (2019a) compared Figure 2.2 with European regulations and showed that raw digestates are out of the range (i.e., not sufficiently dried, not containing enough nutrients, etc.). Therefore, post-treatment application on raw digestates can be a solution to meet these regulations.

2.2.2.2 *Post-treatments impact*

In this section, a focus is made on the most classical post-treatments found in AD plants: phase separation, storage, and composting.

2.2.2.2.1 Phase separation

Digestate phase separation is a key step for enhancing digestate management practices. Indeed, two products are obtained: (i) a liquid fraction (LF) containing nutrients and most of the moisture, and which is easily pumpable, and (ii) a solid fraction (SF) containing less moisture (making it easier to transport) and rich in OM and fiber (Guilayn *et al.*, 2019b). Phase separation impacts the quality of both these obtained digestates. The SF concentrates 30–90% of the OM into 5–50% of the initial mass and the LF concentrates most of the ammoniacal nitrogen and potassium, while phosphorus tends to follow the organic matter and is thus concentrated in the SF, depending on the phase separator efficiency, which is related to the technology used – as shown by Guilayn *et al.* (2019b). Figure 2.4 presents the mass distribution of several digestate composition

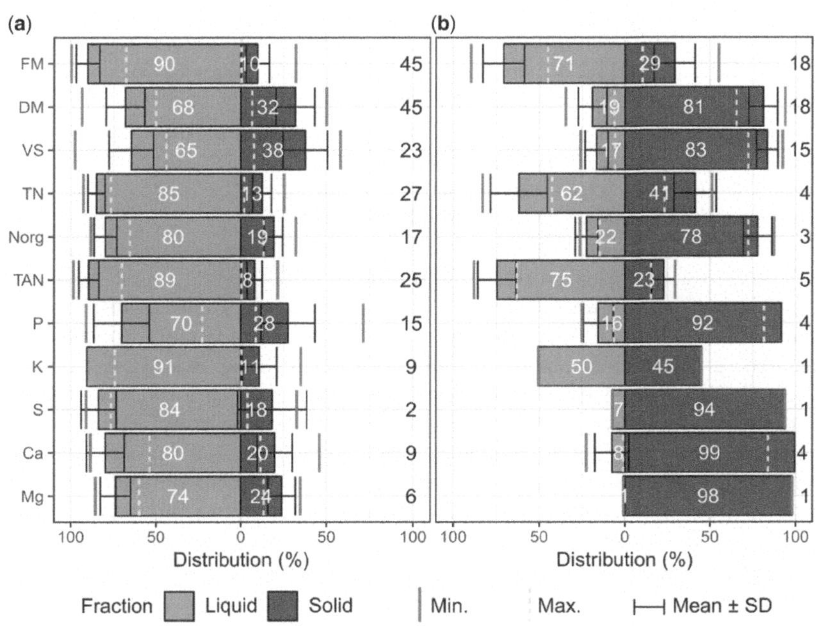

Figure 2.4 Mass distribution profiles of digestate mechanical separation according to low (a) and high (b) efficiency categories, according to Guilayn *et al.* (2019b).

parameters according to efficiency of the phase separator. The most commonly applied technologies are the screw press and decanting centrifuge but vibrating screens and filter presses are also reported (Akhiar *et al.*, 2021; Guilayn *et al.*, 2019b). According to Guilayn *et al.* (2019b), fibrous agricultural digestates tend to be separated by screw presses, which are perform poorly when compared to centrifuges.

The SF has a higher soil bulking capacity and a higher amendment potential than LF which has a higher fertilizer potential and a better infiltration capacity in soil. Thus, each product can be used for different agronomic services, that is for soil improvement and as a crop fertilizer.

The numbers in the right indicate the number of observations. FM: fresh matter. DM: dry matter. VS: volatile solids. TN: total nitrogen. Norg: organic nitrogen. TAN: total ammoniacal nitrogen. P: total phosphorus. K: total potassium. S: total sulfur. Ca: total calcium. Mg: total magnesium.

2.2.2.2.2 Composting

Composting of digestates is a developed technology in agricultural, urban and industrial AD plants. Composting is an aerobic process that decomposes the biodegradable fraction of the OM into CO_2 and microbial biomass during a period ranging from 3 weeks up to 1 to 2 months. Due to the favourable thermodynamics and aerobic conditions, higher variability of microorganisms (i.e., bacteria, actinomycetes, fungi) can be found and allow a higher OM decomposition than AD itself (Guilayn *et al.*, 2020). Two steps are observed during composting: fermentation and maturation. During fermentation, the aerobic condition induces an intensive self-heating process. Mesophilic and thermophilic bacteria are mostly involved. Temperatures can reach more than 70°C within the mounds of digestate inducing a hygienization process during the fermentation step. Temperature, exposure time and the number of turnings are usual regulatory parameters (example: temperatures of 55, 60 and 65°C are requested for a period of respectively 14, 6 and 3 days, Amlinger and Blytt (2013)). The maturation step then follows, involving actinomycete and fungal activity leading to lignocellulose degradation and the humification process. Composting converts ammonia to nitrates through nitrification. However, depending on the process operation, nitrates can be denitrified (under anoxic conditions) and lost as N_2, or ammonia can be volatilized as NH_3. According to Zeng *et al.* (2012), 2–43% of total N can be volatilized as ammonia and up to 76% including both nitrification and denitrification reactions.

Usually, composting is performed on SF digestates after phase separation with or without bulking agents such as green waste, wood chips, sawdust and recirculated compost grinding refuse (Epstein, 2011). Bulking agents are used to improve the biodegradability of OM by inducing a porosity and aeration enhancement and by increasing C/N ratios (25:1–30:1)

Maynaud *et al.* (2017) showed that digestate compost composition and properties lead to a high amendment potential. Indeed, the authors showed that composting decreases the biodegradability of SF digestates and increases molecule complexity.

2.2.2.2.3 Storage

AD is mainly a continuous process needing storage tanks for digestates (raw or LF) or storage areas for SF digestates in order to use them as organic fertilizers. However, storage conditions can lead to physico-chemical and biological reactions in the stored digestates, inducing quality impacts. Indeed, covered storage of liquid digestates prevents N loss whereas storage can lead to N loss of up to 80% (Girault *et al.*, 2017).

2.3 DIGESTATE USE IN LAND APPLICATION

The classical best use of digestates and by-products is their agronomic use in crop soils. Indeed, as previously mentioned, because of their stabilized OM and mineralized nutrients, digestates can be considered as organic amendments or organic fertilizers when properly handled and managed, as illustrated by Figure 2.5. Three types of fertilization can be performed using organic residues in crop soils: (i) chemical fertilizing (available NPK nutrients used at short-/mid-term to feed plants); (ii) biological fertilizing (labile OM able to feed the soil microorganisms in the short term); and (iii) physical fertilizing (given the amendment potential of stabilized OM to improve the soil structure). Different types of digestates have been demonstrated to be effective replacements for mineral fertilizers while reducing environmental impacts (Koszel & Lorencowicz, 2015; Panuccio *et al.*, 2019; Verdi *et al.*, 2019; Walsh *et al.*, 2012). Nevertheless, regulatory constraints, good management and spreading practices are necessary to control the potential environmental and health risks presented by chemical fertilizers, raw manure and sludge (Nkoa, 2014).

2.3.1 Chemical fertility

Mineral elements promote chemical fertility by feeding plants' nutritional needs. Nitrogen, phosphorus and potassium (N, P and K) are the main targeted macro-elements.

2.3.1.1 Nitrogen

The nitrogen fertilizing value of digestate depends on:

(i) **mineral nitrogen content (i.e., ammonia, nitrites/nitrates),** which itself depends on the feedstocks and post-treatments. More than half of the organic nitrogen contained in the raw digested inputs is found as ammonium quickly mobilized by plants, conferring a strong fertilizing potential. The value of this potential varies from one typology to another. Post-treatments also play a major role in the immediate availability of nitrogen, and digestates can be classified as follows, in terms of the short-term (one crop year) fertilizing value equivalent coefficient: Liquid digestates > Raw digestates > Solid digestates > Composted digestates and Dried digestates (Reibel & Leclerc, 2018).

(ii) **mineralization rate of the organic fraction of nitrogen in the digestates.** The mineralization kinetics depend on the type of feedstock and the environmental conditions (temperature and, to a large extent, soil

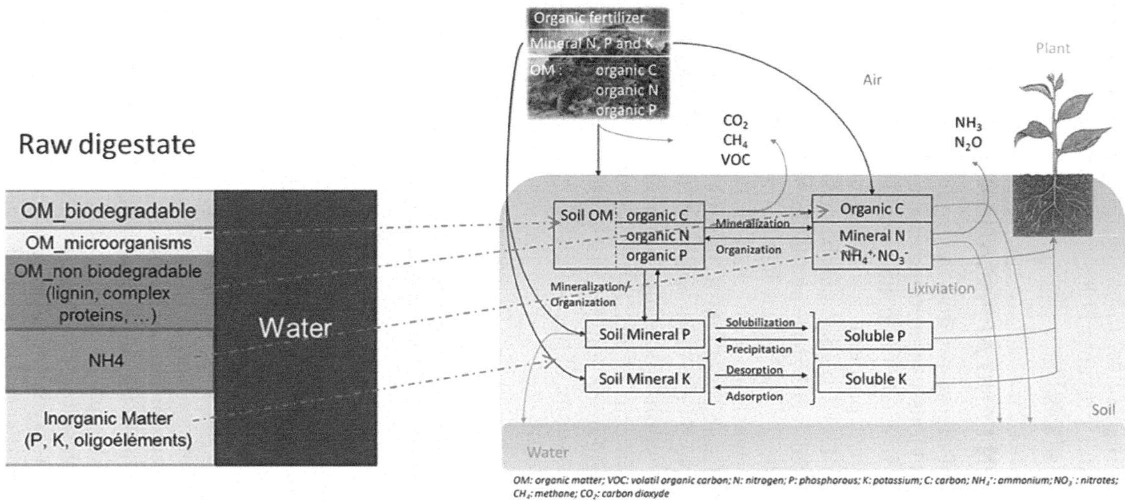

Figure 2.5 Adequation between digestate composition and soil/plants needs in terms of organic matter and nutrients (adapted from Houot *et al.*, 2014).

moisture). During OM mineralization of digestates, part of the carbon is used as an energy source for the soil microflora, and nitrogen is used as a nutrient. A phenomenon of reorganization of mineral nitrogen is often observed. Indeed, microorganisms are activated and OM is used as the energy source for growth. This requires a proportional amount of nitrogen, which is part of the composition of many essential molecules in cells. If the supplied OM does not contain enough N to meet this demand, microorganisms will immobilize the N from the soil solution for their growth (Reibel & Leclerc, 2018). For raw and liquid digestate with a C to N ratio below 15, studies report that the rate of net nitrogen mineralization is usually positive (Cavalli *et al.*, 2016; Jimenez *et al.*, 2020). On the other hand, for the solid phase of digestate or for digestate from fibrous feedstocks and C to N ratios >15, net immobilization of nitrogen can occur (Jimenez *et al.*, 2020). Indeed, the C to N ratio has been shown to harm mineralized N and its recovery in plant tissues (Jimenez *et al.*, 2020; Le Roux *et al.* (2021). It seems to be a good indicator for the mineralization potential of organic N and for the apparent recovery fraction of N by plants, as shown by Jimenez *et al.* (2020), who proposed a first classification according to digestate typology: (i) Group 1: fibrous digestates with high C:N ratio (>25), not suitable for N fertilization alone; (ii) Group 2: intermediary, protein-like and fibrous substrate digestates with C:N ratio of 13–15; (iii) Group 3: municipal solid wastes and biowates digestates with C:N ratio >15 on average, high ammonium and nitrate levels in the labile fraction and high levels of total N recovery; (iv) Group 4: protein-like digestates with a poor C:N ratio (<10) and high total N recovery.

(iii) **the phenomenon of ammonia volatilization and other N loss sources.** This phenomenon is also a function of the type of feedstock, post-treatments, storage, spreading and pedoclimatic conditions (Reibel & Leclerc, 2018). Digestate pH is about 7.5–8.5, promoting an equilibrium shift from NH_4^+ to volatile NH_3 (Guilayn *et al.*, 2020). Girault *et al.* (2017) showed that liquid digestate seems more sensitive to NH_3 emissions during storage compared to the same raw digestate with a loss that can range from 35–100%. Conversely, Holly *et al.* (2017), highlighted a reduction of NH_3 losses of around 30% for liquid digestate compared to raw digestate. The mechanism explaining this difference is not known, especially since the ammonia concentrations are similar for raw digestate and liquid digestate. A loss of up to 80% can be avoided in storage, in particular on digestate from livestock effluent by covering the digestate during storage. The loss of nitrogen during application is also very important and can be more difficult to avoid.

During spreading, Holly *et al.* (2017) observed only 5% ammonia emissions. However, other studies have shown 22% of ammonia loss during slurry digestate spreading (Reibel & Leclerc, 2018). The spreading conditions have a strong influence on the apparent recovery fraction of N by crops: a liquid digestate applied on the surface can lose from 35–100% of its ammonia (Houot *et al.*, 2014; Nyord *et al.*, 2008). Climatic conditions on the spreading day itself and on following days are a strong factor explaining the variability observed in certain field trials because this parameter greatly influences ammonia volatilization.

Leaching of nitrates is also another loss risk and depends mainly on pedoclimatic and agricultural conditions. Different organic nitrogen inputs (type, dose, period, method of supply and management) as well as land use and soil type are the factors explaining leaching. Svoboda *et al.* (2013) reported that the results relating to leaching induced by digestates are rare and contradictory; their own work showed that the digestion residues of cattle and pig slurry induce leaching similar to undigested slurries. According to Houot *et al.* (2014), the spreading period is also a factor that has been extensively studied and it has been shown that the most favourable period is spring rather than late summer or autumn (before the start of drainage), especially for digestates containing rapidly available nitrogen (therefore liquid and raw liquid digestates).

In anoxic zones, if soil porosity is not sufficient, denitrification can also occur leading to nitrate conversion into N_2. In addition, both nitrification and denitrification reactions can produce N_2O emissions. Several factors can influence these N_2O emissions: the pH of the soil, the temperature, the nitrogen content of the digestate, as well as the soil moisture, making it difficult to generalize about the results found during digestate tests, which can sometimes also be contradictory (Reibel & Leclerc, 2018).

At an equivalent dose of nitrogen, several authors have obtained similar crop yields by spreading digestates as from mineral fertilizer or pig slurry (Reibel & Leclerc, 2018). In some cases, the yields or the quality of the crops are higher with the digestate, while other studies report negative effects. Le Roux *et al.* (2021) presented an overview of the apparent N recovery fraction by plants after several types of digestates were spread in France. Although a negative correlation appears between this coefficient and the C to N ratio, extreme types of digestate are missing (i.e., very high and very low C/N ratios). SFs from digestates and digestate composts obtained lower crop yields than raw or liquid digestate, which contains more mineral N and lower C to N ratios.

In addition to the intrinsic properties of digestates, the efficiency of nitrogen in the field depends on several factors such as pedoclimatic conditions and spreading methods. The diversity of experimental conditions and pedoclimatic contexts makes it difficult to make comparisons between studies (Smith *et al.*, 2010).

2.3.1.2 Phosphorus

With regard to phosphorus fertilization potential, P is theoretically conserved in the digestate. Similarly to nitrogen, organic P can be converted to inorganic P, but in one study P was dominantly present as inorganic P (>80% of P total) in digestates (Regelink *et al.*, 2021). Unlike nitrogen, P is mainly found in a solid form in raw digestate, mostly converted to ortho-phosphates as salts (such as calcium phosphate and struvite) and/or adsorbed to particles. Zhang *et al.* (2012) observed that 70% of P is solid whereas 30% is found in interstitial fluid. Losses (more or less significant) can be observed due to accumulation in the reactor (Marcato *et al.*, 2008; Möller & Müller, 2012; Schievano *et al.*, 2011). According to (Reibel & Leclerc, 2018) the fertilizing value of P in the digestate depends first on the P content, varying according to the feedstock, with total P contents extremely variable (from 4 to more than 20 g P/kg DM). The highest levels are observed for digestates from wastewater treatment sludge, with up to 21 g P/kg DM. Digestates from pig slurry AD and animal by-products also have high levels of P, though slightly lower. In contrast, the lowest content is associated with digestates from bio-waste and green waste AD (Houot *et al.*, 2014). However, total P content is not a sufficient indicator of the P fertilizing value of a digestate, as Jimenez *et al.* (2020) and Regelink *et al.* (2021) have shown. Phosphorus speciation and the soil's physico-chemical conditions govern the apparent phosphorus recovery by plants. Indeed, Jimenez *et al.* (2020) showed that phosphorus speciation and post-treatments impact phosphorus chemical availability as well as phosphorus recovery by plants in a calcareous soil, where phosphorus is at lowest availability for plants (i.e., high pH value and carbonates levels).

Digestate post-treatments also have an impact on phosphorus speciation and fertilizing value. According to (Reibel & Leclerc, 2018), the following classification can be made (% DM): P liquid digestate from a screw press > raw digestate > P dried digestate > P digestate solid fraction separation by centrifuge > P digestate solid fraction separation by a screw press > composted digestate. Indeed, the liquid phase which contains orthophosphates retains a fertilizing value in P greater than the solid phase of the digestate or compost, with composting breaking down the concentration of P in the digestate. This classification makes it possible to estimate the phosphorus concentration in a digestate. However, in terms of raw matter, a different order is obtained, with the products richest in DM providing more P (in % raw matter): P dried digestate > P composted digestate > P digestate solid fraction separation by centrifuge > P digestate solid fraction separation by a screw press > raw digestate > liquid digestate.

Moreover, unlike for nitrogen, physico-chemical reactions govern the fate of P in soil. Ahmad *et al.* (2018) reported more than 80% of applied P is rapidly immobilized becoming unavailable for plants following the adsorption/precipitation processes or (re-)conversion into organic form. However, Jimenez *et al.* (2020) applied 9 digestates from different feedstocks for plant pot tests. The apparent recovery fraction of phosphorous in the analysed plant tissues were higher than using the chemical fertilizer. The result suggested that the P availability in those digestates was similar or better than in chemical fertilizers.

2.3.1.3 Potassium

Very little literature deals with this element, which is less important in the fertilization of crops. Potassium is found in dissolved form and therefore is more concentrated in the liquid phase after phase separation (about 80% of

K is found in the liquid phase). However, an excess of potassium can increase the risk of calcium deficiency for a plant since the calcium ion can compete with root uptake of other cations such as potassium. According to the results of field trials performed in France (Reibel & Leclerc, 2018) potassium is 100% available in most cases.

2.3.2 Biological and physical fertilization

With regard to biological and physical soil fertilization, digestates can provide stable OM to compensate soil humus mineralization as soil improvers or soil amendments. Depending on the nature and complexity of the molecules, OM in soils is crucial for feeding microbiological activity (biological fertility), providing bulk capacity, and for improving soil structure, cation exchange, pH buffering and water retention capacities among other effects (physical structure). The complete distinction between mineral fertilizers and soil improvers is not always consistent. Some materials as dried digestates are both rich in OM and nutrients, allowing them to act both as organic soil improvers and 'slow-release' fertilizers (Guilayn *et al.*, 2020). Moreover, mineral fertilizers can be used to enrich and balance nutrient content in organic soil improvers such as composts, providing readily available nutrients along with stable OM.

Voelkner *et al.* (2015) reported an improvement in water holding capacity after incorporating digestate in different sandy and clayey-loamy soils. This phenomenon can be explained, according to the authors, by the decreasing fungus/bacteria ratio. Indeed, strong development of a fungal community can cause a water repellent effect. However, it is mainly bacterial communities that seem to develop in the presence of digestate (Voelkner *et al.*, 2015). In addition, the stability of aggregates seems to be enhanced (Beck & Brandhuber, 2012; Erhart *et al.* 2014; Frøseth *et al.* 2014). For Cationic Exchange Capacity (CEC), the trend is not clear (Voelkner *et al.*, 2015). Regarding the potential negative effects, an accumulation of suspended matter and mono-valent cations (Na+, K+, NH4+, etc.) can have a negative effect on soil stability by causing a phenomenon of slaking crust in soil (Menasseri-Aubry *et al.* 2016). In the literature, the amending potential of digestates varies from 10–354 kg C/t of raw digestate, with the following classification: Composted digestate > Solid digestate > Raw digestate > Liquid digestate (Reibel & Leclerc, 2018). However, the monitoring of digestate uses for agricultural purposes is new, and above all concerns carbon storage and physical soil structure effects. Long term observations are still missing.

Few data exist in the literature regarding biological fertility (Bourgeteau-Karimi *et al.*, 2022; Wentzel *et al.*, 2015). Depending on the digestate type, soil nature, climate and agricultural practices, results reveal both positive and negative impacts. Some studies have shown that digestate spreading increases soil microbial activity (Abubaker *et al.*, 2012; Caracciolo *et al.*, 2015; García-Sánchez *et al.*, 2015; Hupfauf *et al.*, 2016; Odlare *et al.*, 2008; Walsh *et al.*, 2012) whereas others did not report significant effects in comparison with non-digested feedstocks (Coelho *et al.*, 2019; Wentzel *et al.*, 2015). However, all the studies agree that spreading of digestate induces a modification of the microbial community structure (García-Sánchez *et al.*, 2015; Gómez-Brandón *et al.*, 2016; Sapp *et al.*, 2015). Adding digestate to the soil promotes the bacterial population of the soil by increasing its growth (Walsh *et al.*, 2012) and its diversity (Sapp *et al.*, 2015). Conversely, fungi seem to not benefit from this practice, since adding digestate to the soil has no effect on the growth of the fungal population, and would even have a negative effect on the structure of this community (Voelkner *et al.*, 2015; Wentzel & Joergensen, 2016).

There is also little data available in the scientific literature concerning the impact of digestates on soil macrofauna. However, it seems that the supply of digestates induces an increase in biomass and the abundance of earthworms in agricultural soils (Koblenz *et al.*, 2015), though these results do not represent a consensus, since another study demonstrates the deleterious effect of AD digestate on earthworms (Ross *et al.*, 2017). To sum up, considering micro- or macroorganisms in the soil, the results available in the scientific literature are contradictory, with each study reporting effects specific to particular digestates, applied under unique experimental conditions, leading to non-generic conclusions. Rationalizing data seems crucial to considering all the variables which can impact soil biology. It is, therefore, necessary to acquire more knowledge to allow the identification of generic effects.

2.3.3 Risks and relative safety associated with land spreading of digestates

AD enhances N availability for plant growth, but it can also enhance the potential loss of ammonia by volatilization during storage (minimized if the storage is covered) and during spreading, compared to feedstocks as previously mentioned. This loss can be limited by using appropriate agricultural practices for digestates spreading under optimal climatic conditions (dry and cool temperatures) using tools such as drag hose spreader. Greenhouse gases can also be produced during storage, depending also on spreading and operational conditions. However, some studies have shown that AD decreases CH_4 and N_2O emissions compared with feedstock spreading (Holly *et al.*, 2017; Reibel & Leclerc, 2018).

Another limitation of digestate use is their metallic trace element content. AD tends to increase trace mineral concentrations as a result of the degradation of the organic matrix, with a clearer tendency of this for sewage sludge and OFMSW. For digested sewage sludge, speciation of metallic trace elements seems to evolve towards more stable forms and being less available during digestion (Reibel & Leclerc, 2018). Concerning organic micropollutants, AD modifies the concentrations of organic contaminants. For example, the anaerobic digestion of urban sludge has demonstrated a potential for degradation of polycyclic aromatic hydrocarbons, nonylphenol and polychlorobiphenyls (Reibel & Leclerc, 2018).

Jiang *et al.* (2020) reviewed the fate of pathogens during AD, concluding that some parameters have significant effects on pathogens, such as temperature (i.e., thermophilic conditions, and during hygienization) and high volatile fatty acids content, as well as high ammonia content; however, these two latter are inhibitors of AD. By contrast, others parameters such as feedstock type, HRT and reactor type (batch versus continuous) have no significant effect on pathogen content. Moreover, some pathogens are thermotolerant (spore forming, gram-negative bacteria) and resistant to 'cooking' temperature, whereas non-spore forming gram-positive bacteria can be inactivated. However, during storage, a complementary inactivation can occur because of unfavourable abiotic conditions (aeration, light) and biotic conditions (predation, competition). In particular, in the case of easily biodegradable organic matter which has been degraded during treatment, the phenomena of competition between microorganisms will occur (Maynaud *et al.*, 2020).

2.4 CONCLUSION AND FUTURE PERSPECTIVES

Land spreading is the most common strategy for application of agrowaste digestates. They have a high potential as soil amendments and fertilizers. Many parameters influence the agronomic value and environmental safety of digestates, with the two predominant factors being the nature of the feedstocks and the post-treatments influencing the nutrients and the concentration of elements, trace elements, organic micropollutants and the biochemical composition of the organic matter. Other variables also interfere positively or negatively with the agronomic value of digestates: some AD process conditions, soil characteristics (texture, biological activity, etc.), climatic context (temperature, humidity, etc.), cropping systems (crop succession, agricultural practices) and spreading methods (doses, frequency of application, incorporation into the soil, etc.). Therefore, considering the overall conditions together can lead the type of digestate being adapted to the agrosystem's needs. There is a need to reconcile all the available data to better analyze the effects and impacts associated with each type of digestate. Moreover, experimental data and conditions about biological, physical and chemical soil fertility associated with agricultural practices are needed to improve knowledge and lead to safe and environmentally friendly use of digestates.

However, according to Guilayn *et al.* (2020), several legal and technical bottlenecks exist for non-agricultural digestate providers, as follows:

(i) digestate spreading regulations are strict for non-source-separated AD feedstock digestates. Therefore, these digestates are considered as wastes and have to follow waste regulations requiring high administrative cost procedures for approval and spreading;
(ii) the production of non-agricultural digestates is often constant throughout the year, while agricultural needs are highly seasonal. Digestates thus must be stored for a long time until the growing season, or even transported to distant regions (Gong *et al.*, 2013);
(iii) digestate characteristics might be time-variable, depending on feedstock variations and process performance stability;
(iv) the production of digestate may exceed the capacity of the local available arable lands for receiving nutrients (Nkoa, 2014). Increasing land competition implies high and growing digestate transportation costs. Moreover, the number of large and centralized facilities exceeding local

nutrient spreading capacity is tending to grow since they present better economic feasibility. Despite the high fertilizer value of digestate, land spreading is not enough on its own to overcome the digestate challenge, especially for large-scale and/or non-agricultural digesters. Therefore, new ways of digestate valorization through upgrading technologies could be a future subject of interest.

REFERENCES

Abubaker J., Risberg K. and Pell M. (2012). Biogas residues as fertilisers – effects on wheat growth and soil microbial activities. *Applied Energy*, **99**, 126–134, https://doi.org/10.1016/j.apenergy.2012.04.050

Ahmad M., Ahmad M., El-Naggar A. H., Usman A. R., Abduljabbar A., Vithanage M., Elfaki J., Abdulelah A. F. and Al-Wabel M. I. (2018). Aging effects of organic and inorganic fertilizers on phosphorus fractionation in a calcareous sandy loam soil. *Pedosphere*, **28**, 873–883, https://doi.org/10.1016/S1002-0160(17)60363-1

Akhiar A., Guilayn F., Torrijos M., Battimelli A., Shamsuddin A. H. and Carrère H. (2021). Correlations between digestate composition and process conditions allow to better characterize the liquid fraction of full-scale digestates. *Energy*, **14**(4), 971, https://doi.org/10.3390/en14040971

Amlinger F. and Blytt L. D. (2013). How to Comply with the EU Animal By-Products Regulations at Composting and Anaerobic Digestion Plants. European Compost Network ECN e.V., Bochum, Germany.

Batstone D. J., Keller J., Angelidaki I., Kalyuzhnyi S. V., Pavlostathis S. G., Rozzi A., Sanders W. T. M., Siegrist H. and Vavilin V. A. (2002). Anaerobic Digestion Model No.1. (ADM1). IWA Scientific and Technical Report No. 13. IWA Publishing, London, UK. ISBN:1-900222-78-7.

Beck R. and Brandhuber R. (2012) Effekte der Gärrestdüngung auf Humusbilanz und Bodenstruktur-Zwischenbilanz. In: Bayerische Landesanstalt für Landwirtschaft (ed.) Schriftenreihe no. 11/2012: Düngung mit Biogasgärresten–effektiv-umweltfreundlich-bodenschonend. ISSN 1611-4159, pp. 49–58.

Brunetti G., Farrag K., Plaza C. and Senesi N. (2012). Advanced techniques for characterization of organic matter from anaerobically digested grapemarc distillery effluents and amended soils. *Environmental Monitoring and Assessment*, **184**(4), 2079–2089, https://doi.org/10.1007/s10661-011-2101-z

Caracciolo A. B., Bustamante M. A., Nogues I., Di Lenola M., Luprano M. L. and Grenni P. (2015). Changes in microbial community structure and functioning of a semiarid soil due to the use of anaerobic digestate derived composts and rosemary plants. *Geoderma*, **245**, 89–97, https://doi.org/10.1016/j.geoderma.2015.01.021

Cavalli D., Cabassi G., Borrelli L., Geromel G., Bechini L., Degano L. and Gallina P. M. (2016). Nitrogen fertilizer replacement value of undigested liquid cattle manure and digestates. *European Journal of Agronomy*, **73**, 34–41, https://doi.org/10.1016/j.eja.2015.10.007

Coelho J. J., Hennessy A., Casey I., Woodcock T. and Kennedy N. (2019). Responses of ryegrass, white clover, soil plant primary macronutrients and microbial abundance to application of anaerobic digestates, cattle slurry and inorganic N-fertiliser. *Applied Soil Ecology*, **144**, 112–122. 4 CSNM, 2019, https://doi.org/10.1016/j.apsoil.2019.07.011

Epstein E. (2011). Industrial Composting. Environmental Engineering and Facilities Management. Taylor and Francis Group, New York.

Erhart E., Siegl T., Bonell M., Unterfrauner H., Peticzka R., Ableidinger C., … and Hartl W. (2014). Fertilization with liquid digestate in organic farming-effects on humus balance, soil potassium contents and soil physical properties. EGU General Assembly Conference Abstracts, p. 4419.

Fernández-Domínguez D., Patureau D., Houot S., Sertillanges N., Zennaro B. and Jimenez J. (2021). Prediction of organic matter accessibility and complexity in anaerobic digestates. *Waste Management*, **136**, 132–142, https://doi.org/10.1016/j.wasman.2021.10.004

Frøseth R. B., Bakken A. K., Bleken M. A., Riley H., Pommeresche R., Thorup-Kristensen K. and Hansen S. (2014). Effects of green manure herbage management and its digestate from biogas production on barley yield, N recovery, soil structure and earthworm populations. *European Journal of Agronomy*, **52**, 90–102, https://doi.org/10.1016/j.eja.2013.10.006

García-Sánchez M., Siles J. A., Cajthaml T., García-Romera I., Tlustoš P. and Száková J. (2015). Effect of digestate and fly ash applications on soil functional properties and microbial communities. *European Journal of Soil Biology*, **71**, 1–12, https://doi.org/10.1016/j.ejsobi.2015.08.004

Girault R., Roux J. C., Mazoyer J., Nunes G., Auvinet N., Saint-Cast Leroux S. and Guiziou F. (2017). Impact des stratégies de post-traitement des digestats sur les émissions gazeuses à l'échelle de la filière. Recueil des Résumés JRI 2017.

Gómez-Brandón M., Juárez M. F.-D., Zangerle M. and Insam H. (2016). Effects of digestate on soil chemical and microbiological properties: a comparative study with compost and vermicompost. *Journal of Hazardous Materials*, **302**, 267–274, https://doi.org/10.1016/j.jhazmat.2015.09.067

Gong H., Yan Z., Liang K. Q., Jin Z. Y. and Wang K. J. (2013). Concentrating process of liquid digestate by disk tube-reverse osmosis system. *Desalination*, **326**, 30–36, https://doi.org/10.1016/j.desal.2013.07.010

Guilayn F., Jimenez J., Martel J. L., Rouez M., Crest M. and Patureau D. (2019a). First fertilizing-value typology of digestates: a decision-making tool for regulation. *Waste Management*, **86**, 67–79, https://doi.org/10.1016/j.wasman.2019.01.032

Guilayn F., Jimenez J., Rouez M., Crest M. and Patureau D. (2019b). Digestate mechanical separation: efficiency profiles based on anaerobic digestion feedstock and equipment choice. *Bioresource Technology*, **274**, 180–189, https://doi.org/10.1016/j.biortech.2018.11.090

Guilayn F., Rouez M., Crest M., Patureau D. and Jimenez J. (2020). Valorization of digestates from urban or centralized biogas plants: a critical review. *Reviews in Environmental Science and Biotechnology*, **19**(2), 419–462, https://doi.org/10.1007/s11157-020-09531-3

Holly M. A., Larson R. A., Powell J. M., Ruark M. D. and Aguirre-Villegas H. (2017). Greenhouse gas and ammonia emissions from digested and separated dairy manure during storage and after land application. *Agriculture, Ecosystems & Environment*, **239**, 410–419, https://doi.org/10.1016/j.agee.2017.02.007

Houot S., Pons M. N., Pradel M., Tibi A., Aubry C., Augusto L., … and Viguié C. (2014). Valorisation des matières fertilisantes d'origine résiduaire sur les sols à usage agricole ou forestier. Doctoral dissertation, Ministère de l'Agriculture.

Hupfauf S., Bachmann S., Juárez M. F.-D., Insam H. and Eichler-Löbermann B. (2016) Biogas digestates affect crop P uptake and soil microbial community composition. *Science of the Total Environment*, **542**, 1144–1154, https://doi.org/10.1016/j.scitotenv.2015.09.025

Jiang Y., Xie S. H., Dennehy C., Lawlor P. G., Hu Z. H., Wu G. X., … and Gardiner G. E. (2020). Inactivation of pathogens in anaerobic digestion systems for converting biowastes to bioenergy: a review. *Renewable and Sustainable Energy Reviews*, **120**, 109654, https://doi.org/10.1016/j.rser.2019.109654

Jimenez J., Nicolas S., Sabine H. and Patureau D. (2019). Caractérisation de la matière organique: outil d'aide à la compréhension des mécanismes de la digestion anaérobie et de la qualité du digestat. Recueil des Résumés JRI 2019.

Jimenez J., Grigatti M., Boanini E., Patureau D. and Bernet N. (2020). The impact of biogas digestate typology on nutrient recovery for plant growth: accessibility indicators for first fertilization prediction. *Waste Management*, **117**, 18–31, https://doi.org/10.1016/j.wasman.2020.07.052

Karimi B., Sadet-Bourgeteau S., Cannavacciuolo M., Chauvin C., Flamin C., Haumont A.... & Ranjard L. (2022). Impact of biogas digestates on soil microbiota in agriculture: a review. *Environmental Chemistry Letters*, 1–24, https://doi.org/10.1007/s10311-022-01451-8

Koblenz B., Tischer S., Rücknagel J. and Christen O. (2015). Influence of biogas digestate on density, biomass and community composition of earthworms. *Industrial Crops and Products*, **66**, 206–209.

Koszel M. and Lorencowicz E. (2015). Agricultural use of biogas digestate as a replacement fertilizers. *Agriculture and Agricultural Science Procedia*, **7**, 119–124, https://doi.org/10.1016/j.aaspro.2015.12.004

Le Roux C., Michaud, A., Bodèle, C., Burel, E., Carré, M., Lagrange, H., Levasseur, F., Moreira, M., Morvan, T., Stangret, V. (2021) "Synthèse des travaux sur la détermination des coefficients d'équivalent engrais azotés (Keq N) sur les digestats dans des dispositifs expérimentaux". COMIFER Conference. 24th–25th November 2021. Clermont-Ferrand. France.

Marcato C. E., Pinelli E., Pouech P., Winterton P. and Guiresse M. (2008). Particle size and metal distributions in anaerobically digested pig slurry. *Bioresource Technology*, **99**(7), 2340–2348, https://doi.org/10.1016/j.biortech.2007.05.013

Maynaud G., Druilhe C., Daumoin M., Jimenez J., Patureau D., Torrijos M., … and Wéry N. (2017). Characterisation of the biodegradability of post-treated digestates via the chemical accessibility and complexity of organic matter. *Bioresource Technology*, **231**, 65–74, https://doi.org/10.1016/j.biortech.2017.01.057

Maynaud G., Druilhe C., Jimenez J., Patureau D., Pourcher A. M. and Wéry N. (2020). Élimination de bactéries pathogènes dans les digestats agricoles par compétition avec les bactéries indigènes. *Environnement, Risques & Santé*, **19**(1), 35–40.

Mazzini S., Borgonovo G., Scaglioni L., Bedussi F., D'Imporzano G., Tambone F. and Adani F. (2020). Phosphorus speciation during anaerobic digestion and subsequent solid/liquid separation. *Science of the Total Environment*, **734**, 139284, https://doi.org/10.1016/j.scitotenv.2020.139284

Menasseri-Aubry S. (UMR INRA/AO sol agronomie et spatialisation). (2016). SPACE, Les enjeux agronomiques et sanitaires des digestats, 16th september 2016.

Möller K. and Müller T. (2012). Effects of anaerobic digestion on digestate nutrient availability and crop growth: a review. *Engineering in Life Sciences*, **12**, 242–257, https://doi.org/10.1002/elsc.201100085

Moscoviz R. and Jimenez J. (2021). Improving anaerobic digestion mass balance calculations through stoichiometry and usual substrate characterization. *Bioresource Technology*, **337**, 1–10, https://doi.org/10.1016/j.biortech.2021.125402

Nkoa R. (2014). Agricultural benefits and environmental risks of soil fertilization with anaerobic digestates: a review. *Agronomy for Sustainable Development*, **34**, 473–492, https://doi.org/10.1007/s13593-013-0196-z

Nyord T., Søgaard H. T., Hansen M. N. and Jensen L. S. (2008). Injection methods to reduce ammonia emission from volatile liquid fertilisers applied to growing crops. *Biosystems Engineering*, **100**, 235–244, https://doi.org/10.1016/j.biosystemseng.2008.01.013

Odlare M., Pell M. and Svensson K. (2008). Changes in soil chemical and microbiological properties during 4 years of application of various organic residues. *Waste Management*, **28**, 1246–1253, https://doi.org/10.1016/j.wasman.2007.06.005

Panuccio M. R., Papalia T., Attinà E., Giuffrè A. and Muscolo A. (2019). Use of digestate as an alternative to mineral fertilizer: effects on growth and crop quality. *Archives of Agronomy and Soil Science*, **65**(5), 700–711, https://doi.org/10.1080/03650340.2018.1520980

Regelink I. C., Egene C. E., Tack F. M. G. and Meers E. (2021). Speciation of P in solid organic fertilisers from digestate and biowaste. *Agronomy*, **11**, 2233, https://doi.org/10.3390/agronomy11112233

Reibel A. and Leclerc B. (2018). Valorisation agricole des digestats: Quels impacts sur les cultures, le sol et l'environnement? Revue de littérature. Méthanisation en Provence-Alpes-Côte d'Azur.

Ross C.-L., Wilken V., Krück S., Nielsen K., Sensel-Gunke K. and Ellmer F. (2017). Assessing the impact of soil amendments made of processed biowaste digestate on soil macrofauna using two different earthworm species. *Archives of Agronomy and Soil Science*, **63**, 1939–1950, https://doi.org/10.1080/03650340.2017.1316380

Sapp M., Harrison M., Hany U., Charlton A. and Thwaites R. (2015). Comparing the effect of digestate and chemical fertiliser on soil bacteria. *Applied Soil Ecology*, **86**, 1–9, https://doi.org/10.1016/j.apsoil.2014.10.004

Schievano A., D'Imporzano G., Salati S. and Adani F. (2011). On-field study of anaerobic digestion full-scale plants (part I): an on-field methodology to determine mass, carbon and nutrients balance. *Bioresource Technology*, **102**(17), 7737–7744, https://doi.org/10.1016/j.biortech.2011.06.006

Smith K. A., Jeffrey W. A., Metcalfe J. P., Sinclair A. H. and Williams J. R. (2010). Nutrient value of digestate from farm-based biogas plants. Treatment and Use of Organic Residues in Agriculture: Challenges and Opportunities toward Sustainable Management: Proceedings of the 14th Ramiran International Conference, Instituto Superior de Agronomia, Universidad Técnica de Lisboa Lisboa, Portugal.

Svoboda N., Taube F., Wienforth B., Kluß C., Kage H. and Herrmann A. (2013). Nitrogen leaching losses after biogas residue application to maize. *Soil and Tillage Research*, **130**, 69–80, https://doi.org/10.1016/j.still.2013.02.006

Verdi L., Kuikman P. J., Orlandini S., Mancini M., Napoli M. and Dalla Marta A. (2019). Does the use of digestate to replace mineral fertilizers have less emissions of N2O and NH3? *Agricultural and Forest Meteorology*, **269–270**, 112–118, https://doi.org/10.1016/j.agrformet.2019.02.004

Voelkner A., Ohl S., Holthusen D., Hartung E., Dörner J. and Horn R. (2015). Impact of mechanically pre-treated anaerobic digestates on soil properties. *Journal of Soil Science and Plant Nutrition*, **15**(4), 882–895.

Walsh J. J., Rousk J., Edwards-Jones G., Jones D. L. and Williams A. P. (2012). Fungal and bacterial growth following the application of slurry and anaerobic digestate of livestock manure to temperate pasture soils. *Biology and Fertility of Soils*, **48**, 889–897, https://doi.org/10.1007/s00374-012-0681-6

Wentzel S. and Joergensen R. G. (2016). Effects of biogas and raw slurries on grass growth and soil microbial indices. *Journal of Plant Nutrition and Soil Science*, **179**, 215–222, https://doi.org/10.1002/jpln.201400544

Wentzel S., Schmidt R., Piepho H.-P., Semmler-Busch U. and Joergensen R. G. (2015). Response of soil fertility indices to long-term application of biogas and raw slurry under organic farming. *Applied Soil Ecology*, **96**, 99–107, https://doi.org/10.1016/j.apsoil.2015.06.015

Zeng Y., de Guardia A., Daumoin M. and Benoist J.-C. (2012). Characterizing the transformation and transfer of nitrogen during the aerobic treatment of organic wastes and digestates. *Waste Management*, **32**, 2239–2247, https://doi.org/10.1016/j.wasman.2012.07.006

Zhang Y., Lashermes G., Houot S., Doublet J., Steyer J. P., Zhu Y. G., Barriuso E. and Garnier P. (2012). Modelling of organic matter dynamics during the composting process. *Waste Management*, **32**, 19–30, https://doi.org/10.1016/j.wasman.2011.09.008

doi: 10.2166/9781789062755_0045

Chapter 3

Valorization of digestate from anaerobic digestion of organic fraction of municipal solid waste

*Mai Sofia, Barampouti Elli Maria, Malamis Dimitris and Moustakas Konstantinos**

National Technical University of Athens, School of Chemical Engineering, Unit of Environmental Science & Technology, 9 Iroon Polytechniou Str., Zographou Campus, GR-15780 Athens, Greece
**Corresponding author e-mail: konmoust@central.ntua.gr*

ABSTRACT

Anaerobic digestion is one of the most widely applied treatment methods for the organic fraction of municipal solid waste (OFMSW). Within this work, the quantification of OFMSW that is anaerobically digested along with the annual production of digestate and biogas (current and potential) is carried out. It is estimated that globally around 440 $*10^6$ tn of digestate could be produced and thus the need for their valorisation is urgent. A detailed review of the digestate's physico-chemical characteristics is performed revealing the great fluctuations reported. Alternative valorisation pathways of OFMSW digestate are discussed in terms of their technical efficiency, end-product quality, economic viability and scale of current applications. Technoeconomic challenges are also pointed out concluding that upscaling of valorisation schemes is all scientific and technical efforts. The integration of anaerobic digestion of OFMSW with a proposed digestate valorisation strategy may close the resource loop and materialise the circular economy concept.

Keywords: Biorefinery, digestate, valorisation pathways, secondary products

3.1 INTRODUCTION

Municipal solid waste (MSW) is the waste collected and disposed of under a municipality's responsibility including streams such as food waste, plastics, glass, paper and metal. In some cases, demolition and construction waste along with household hazardous waste are also included (Speight, 2015). The major stream of MSW is the organic fraction of municipal solid waste (OFMSW), which consists of food waste and kitchen leftovers from households and the HORECA sector (food service and hotels) and the term 'food waste' or 'biowaste' is also used for this waste stream. The separation and collection of OFMSW could be either mechanical (MS-OFMSW) or at-source (SS-OFMSW) (Barampouti

Editor(s): Vinay Kumar Tyagi, Kaoutar Aboudi and Cigdem Eskicioglu

et al., 2019). In Europe, source-separation is promoted in the revised Waste Framework Directive that defines that food waste should be source-separated and either recycled or/and collected separately from the other waste streams by the end of 2023, aiming to boost secondary resources. In addition, 65% of MSW should be recycled by 2035. Furthermore, by 2027, compost will not be considered a recycling option. Thus, source separation of biowaste is imperative for a resource-efficient and circular treatment of biowaste. Currently, in Europe, around half of the OFMSW produced is separately collected while the rest remains in mixed MSW. These rates vary significantly from less than 10% in Cyprus, Spain, Portugal, Bosnia-Herzegovina and Turkey to over 80% in Slovenia and Austria, indicating that they can be further increased (Van der Linden & Almut, 2020).

The two most established processes for the treatment of OFMSW are anaerobic digestion (AD) and composting. Although the latter is more widely applied, the application of AD is rising, given its environmental benefits. The selection among the treatment processes is also dependent on the characteristics of OFMSW and the efficiency of the separation and collection scheme (Van der Linden & Almut, 2020). There are three main anaerobic system configurations (EPA, 2020):

- **Stand-alone digesters** that are mainly devoted to treat OFMSW but other organic waste streams could also be accepted (e.g., manure, agricultural residues and sludge).
- **On-farm digesters** that are mainly devoted to treat manure but other organic waste streams could also be accepted (e.g., food waste and agricultural residues).
- **Water resource recovery facility digesters** that are devoted to mainly treat excess sludge from aerobic wastewater treatment plants but may also accept food waste.

Figure 3.1 illustrates the geographical (on continent base) fractionation of MSW, pointing out the percentage of OFMSW in MSW and the respective fraction of OFMSW that is anaerobically treated. In the same figure, data on the current and potential annual production of digestate and biogas are also presented. The potential production data are estimated considering that all OFMSW is led to AD plants.

It is estimated that in the United States 63×10^6 tons of OFMSW were produced in 2018 accounting for almost 22% of MSW. Around 11×10^6 tons of OFMSW were anaerobically treated in 2016, 86% in stand-alone digesters, 12% in water resource recovery facility digesters and the rest 2% in on-farm digesters (Al Seadi *et al.*, 2013b).

Similarly in Europe (EU-28), 86×10^6 tons OFMSW were produced in 2017 accounting for 34% of MSW, while the existing infrastructure could annually treat 38×10^6 tons (55% in composting plants and 45% in AD plants) (Van der Linden & Almut, 2020). However, according to International Energy Agency (IEA, 2018), just 2.2 Mtoe biogas was produced from OFMSW in Europe, 3.8 Mtoe in the United States and 2.3 Mtoe in China.

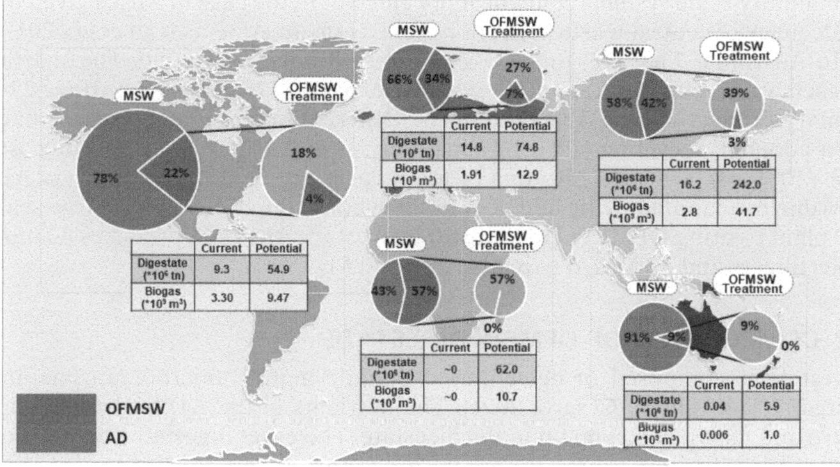

Figure 3.1 Fractionation of MSW on the continent basis, OFMSW production, AD fraction of OFMSW and current and potential annual production of the digestate and biogas.

In Asia, 278×10^6 tons of food waste were produced in 2014, 29.8% in China, 28.8% in South East Asia, 11.2% in Indonesia, 5.9% in Japan, 4.3% in Vietnam, 4.3% in the Philippines and 3.2% in Thailand. Even though there are established strategies and targets for the management of OFMSW, it is reported that this issue has been poorly addressed in Asia. The main management alternatives include: production of animal feedstuff, compost, biogas and digestate as fertilizer, burning and landfilling. In an effort to promote AD of food waste, 138 full-scale plants have been constructed in different cities over China. Most of them apply single-stage wet AD process. Yet, other configurations also exist. For instance, in Longgang, Changchuan and Suzhou a hydrothermal pretreatment step has been included leading to elevated efficiencies. Anaerobic treatment is also adopted in Hong Kong for the management of food waste as six AD plants of total daily capacity $1.3–1.5 \times 10^3$ tons. Full-scale AD plants also operate in Thailand and India. The operation of large-scale plants has indicated serious technoeconomic and logistics challenges. This, in some cases, led to a shift to smaller scale systems where the process is easier to control, the area occupied is smaller, the daily feedstock needs and transportation costs are lower (Joshi & Visvanathan, 2019).

On the other hand, the management of MSW in Africa is problematic since the legislative framework is not easily enforced, the economic issues are huge and the citizens awareness is low with a direct impact on the environment. The annual production of MSW in Africa was 125×10^6 tons (2012) with 65% being produced in sub-Saharan Africa, and is expected to double by 2025. Nevertheless, just 68×10^6 tons are collected while the rest are disposed in an uncontrolled manner. Fifty-seven percent of total MSW corresponds to OFMSW, mainly produced during the primary stages of the supply chain.

Community-scale systems could be more efficient in the case of Africa given the inadequacy of central waste collection and treatment systems (Lin *et al.*, 2013).

In Australia, 7.68×10^6 tons OFMSW were produced in 2020. The sharing of each value chain stage is 7% wholesale retail, 17% manufacturing, 22% primary production and 51% consumers. Different management schemes have been adopted; recycling (1.2×10^6 tons), recovery (2.9×10^6 tons) and disposal (3.2×10^6 tons). Only 40 150 tons of food waste are being processed via AD, 40% during manufacturing and 60% in the wholesale-retail stage. A large-scale AD plant treating OFMSW (137 tons/d) in Western Australia produces 60 m^3/d biofertilizers and 29 MW/d (Arcadis, 2019; FIAL, 2021).

3.2 COMPOSITION OF OFMSW DIGESTATE

Digestate is composed of decomposed organic matter, microorganisms and inorganic substances (Cesaro, 2021). The nutrients of the AD feedstock, both micro and macro, also end up in the digestate. The exact digestate composition is regulated by the substrate composition, the anaerobic system and the operational conditions applied (Barampouti *et al.*, 2020). Hence, it is imperative to characterize the AD raw material when the digestate's composition is of interest.

It has been acknowledged that OFMSW composition and production rates depend on its origin, eating habits, seasonality, sorting system along with socio-economic background. Additionally, the handling (transportation, storage) of OFMSW from households to final disposal/management strongly affects its composition. Barampouti *et al.* (2019) performed a detailed mapping of OFMSW composition, revealing the high fluctuations of the concentration of most components pointing out the suitability of this feedstock for bioprocessing given its composition (moisture (~70%), organic matter (over 70%), proteins (14% d.b.), carbohydrates (~50%), fat and oil (~13%) metals and macro/micro-nutrients).

These high fluctuations of OFMSW composition are directly reflected in the composition of OFMSW digestate. Composition profiles of digestate streams resulting from OFMSW reported in the literature have been tabularized including parameters and are further discussed below.

Tables 3.1–3.3 summarize the composition of stand-alone digesters, that process food waste as a sole substrate, on-farm digesters, that co-digest food waste with manure and agricultural residues and water resources recovery facility digesters that also process biosolids, respectively.

It is evident from Tables 3.1–3.3 that there are great fluctuations in all parameters apart from pH and carbon content within the same type of digester indicating that a general prediction of the composition of digestate cannot be made unless the feedstock's composition is specifically defined. The mean values along with the respective range of all parameters can be compared in Figure 3.2.

The mean TS concentration of stand-alone and on-farm digesters is over 10%, much higher than the respective value in the case of water resource recovery facility digesters. This fact is strongly related to the characteristics of the input

Table 3.1 Composition of digestate produced from stand-alone digesters that process food waste as a sole substrate.

Reference	TS	C	TN	N-NH_4	Norg	TP	TK	C/N	pH
	(%)	(% TS)	(% TS)	(% TS)	(% TKN)	(% TS)	(% TS)	(dimensionless)	(dimensionless)
Beggio *et al.* (2019)	8.8	38.3		8.2		0.7	3.6		7.7
Massaccesi *et al.* (2013)	44.9		1.6	0.1	92.4	1.3	1.5	8.3	6.4
Massaccesi *et al.* (2013)	43.7		1.5	0.1	92.7	3.1	0.5	29.8	8.4
Massaccesi *et al.* (2013)	50.2		1	0.1	88.3	5.4	1.5	37.5	8.9
Massaccesi *et al.* (2013)	37.3		1.8	0.2	86.4	4.5	0.9	28	8.7
Stoknes *et al.* (2016)	2.4		10	0.1		1	4		
Tampio *et al.* (2015)	6.7	38.6	11.6	6		2	4.4	3.3	8
Tampio *et al.* (2016)	6.8	39.5	12.8	6.6				3.1	8
Tampio *et al.* (2016)	7.88	32.9	9.9	2.2				3.3	7.6
Tampio *et al.* (2016)	2	34.2	23.6	19.6				1.5	8.3
Tampio *et al.* (2016)	3.2	32	14	9.9				2.3	8.3
Peng *et al.* (2022)		22.4	2.4					9.5	7.9
Peng *et al.* (2022)		45.6	1.4					32.6	7.7
Krishnasamy *et al.* (2014)	12	18.8	1	0.2	34.2	0.4	0.4	18.7	8.4
Svensson *et al.* (2018)	18.5	42.7	2.4					17.6	
Svensson *et al.* (2018)	13.4								3.9
Yu *et al.* (2021)	1.7		0.3						8.2
Yu *et al.* (2021)	2.8		0.3						7.8

(Continued)

Table 3.1 Composition of digestate produced from stand-alone digesters that process food waste as a sole substrate (*Continued*).

Reference	TS	C	TN	N-NH_4	Norg	TP	TK	C/N	pH
	(%)	(% TS)	(% TS)	(% TS)	(% TKN)	(% TS)	(% TS)	(dimensionless)	(dimensionless)
Yu *et al.* (2021)	2.6		0.2						7.9
Yu *et al.* (2021)	3.4		0.3						8
Chuka-ogwude *et al.* (2021)				0.4					8.1
Opatokun *et al.* (2015)		42	5.8						
Nicholson *et al.* (2017)	5.1		8	5.8	27.4				8.1
Nicholson *et al.* (2017)	6.1		5.4	3.9	27.6				8.4
Nicholson *et al.* (2017)	5.4		7.8	6.3	19.1				8.8
Nicholson *et al.* (2017)	4.4		6.9	6.2	10.0				8.7
Cheong *et al.* (2020)				0.5					7.8
Li *et al.* (2020)	11								7.9
Liu *et al.* (2020)		43.5	1.9						
Wang *et al.* (2020)	3.6		0.5	0.3		0.1			8.6
Zhang *et al.* (2022)	2.5	36.5							8
Opatokun *et al.* (2017)	9.8	42.1	5.8			2	0.6		8
Akarsu *et al.* (2019)		46	2.7						
Peng *et al.* (2022)	1.99–7.88	32.8–39.5	4.2	0.3		0.9	2.3	2.6	7.6–8.3
Peng *et al.* (2022)	0.72–51.2	34.6	1.3–12.4	0.17–2.75		0.2–0.9	0.6–1	1.3–29.8	8.3
Chen *et al.* (2019)	20								
Barzee *et al.* (2019)			0.3	0.3			0.2		

Table 3.2 Composition of digestate produced from on-farm digesters that co-digest food waste with manure and agricultural residues.

Reference	TS	C	TN	N-NH_4	Norg	TP	TK	C/N	pH
	(%)	(% TS)	(% TS)	(% TS)	(% TKN)	(% TS)	(% TS)	(dimensionless)	(dimensionless)
Abubaker *et al.* (2012)	6.1	31.1	12.9	8.68	33	1.47	2.62	7.1	7.9
Abubaker *et al.* (2012)	1.7	41.1	15.3	11.8	23	1.18	6.47	11	8
Abubaker *et al.* (2012)	5.9	40.6	8.98	5.59	37.7	0.68	6.27	12.1	8.7
Arab and McCartney (2017)	37	24.5	1.6	0.62		0.15	1.05		8.5
Grigatti *et al.* (2020)	34	30.2		0.025	26	0.51	0.53	18.9	8.9
Tambone *et al.* (2009)	5.9	40.4	6.5	3.33				6.7	
Abubaker *et al.* (2012)	3.6	37.7		6.34				3.4	
Tambone *et al.* (2010)	3.8	3.86	14.37	9.4	34.6			7.7	

Table 3.3 Composition of digestate produced from water resource recovery facility digesters that co-digest biosolids with food waste.

Reference	TS	C	TN	N-NH_4	Norg	TP	TK	C/N	pH
	(%)	(% TS)	(% TS)	(% TS)	(% TKN)	(% TS)	(% TS)	(dimensionless)	(dimensionless)
Di Maria *et al.* (2014)					55.4			2.7	7.6
Di Maria *et al.* (2014)					47.0			7.0	7.4
Di Maria *et al.* (2014)					58.4			7.4	7.3
Di Maria *et al.* (2014)					69.0			10.0	7.3
Di Maria *et al.* (2014)					75.0			13.3	7.1
Tampio *et al.* (2016)	3.4	39.5	6.4	5.0				6.1	7.6
do Carmo Precci Lopes *et al.* (2022)	3.4	17.1				5.6		9.3	

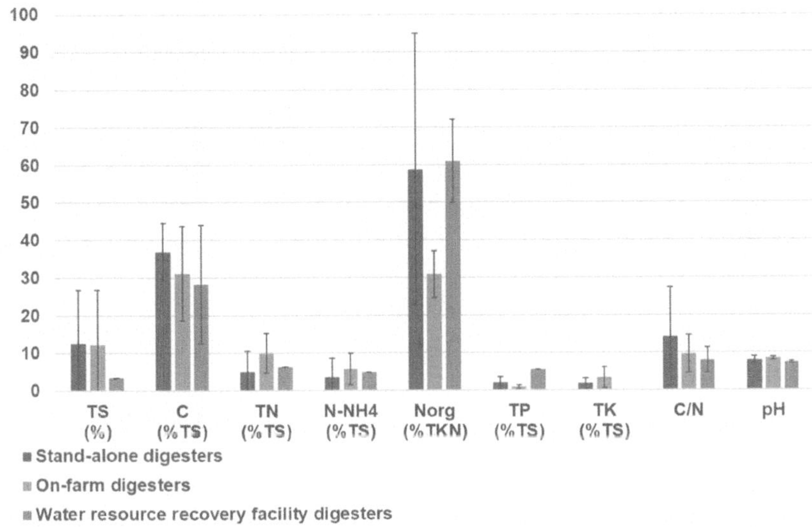

Figure 3.2 Mean composition of the digestate from stand-alone, on-farm and water resource recovery facility digesters.

digesters' substrate and the digestion operational conditions. Nevertheless, the carbon content (%TS) falls within the same range. The pH values for all cases ranged from 7.38 to 8.40. The most alkaline digestates are those resulting from on-farm digesters. This fact is in accordance with the highest ammoniacal concentration (5.72%TS) that these digestates present and the alkaline nature of manure (8–12). The highest phosphorous content was obtained for digestates from water resource recovery facility digesters, indicating the high phosphorous content of biosolids. The carbon-to-nitrogen ratio, a critical parameter for the short-term nitrogen availability, for on-farm and water resource recovery facility digesters ranges within 4.5–9.5, while the respective value of stand-alone digester presents much higher fluctuations (from 1.13 to 27.17).

As presented in Tables 3.1–3.3, the heavy metal content of manure and sewage sludge digestate is higher than that of the OFMSW digestate turning its land application a more sustainable option in economic and environmental terms (Lu & Xu, 2021).

3.3 MAPPING OF VALORIZATION PATHWAYS

In the context of integrated sustainable OFMSW management, AD and the resulting digestate valorization are integral components. Several valorization pathways have been reported in the literature and will be discussed in the following sections. In Figure 3.3, the alternative valorization pathways along with the respective end products are presented.

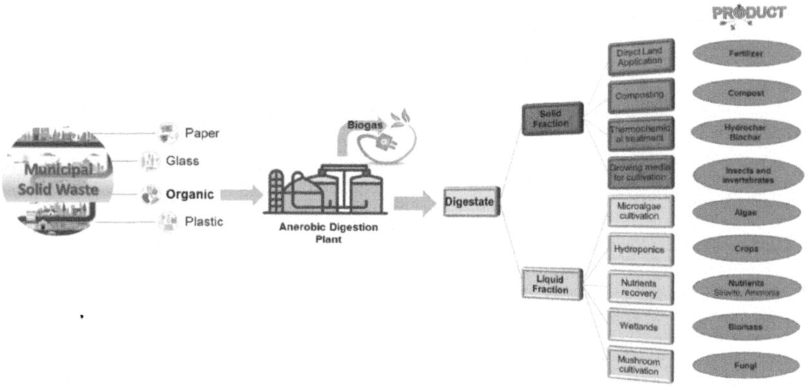

Figure 3.3 Alternative valorization pathways for the OFMSW digestate.

3.3.1 Dewatering

Digestate is usually separated into liquid and solid fractions. For the case of OFMSW digestate, around 90% is the liquid fraction whereas the rest 10% is the solid part. In mass terms, 87% of the raw OFMSW ends up in the digestate. This separation step reduces the transportation cost and facilitates downstream processing (Dutta *et al.*, 2021). Nevertheless, several issues, technical as well as economic, are raised during digestate dewatering. These may be attributed to various factors but the main obstacle is the poor dewaterability of the OFMSW digestate (Zhang *et al.*, 2022). The latter is influenced by the high calcium and sodium contents of digestate, which originate from bones, shells and salt of food (Liu *et al.*, 2017; Zhang *et al.*, 2022). It is also reported that the fluctuation of protein, carbohydrate and extracellular polymeric substances (EPS) contents of digestate significantly affects its dewaterability. The higher the soluble protein, the lower the dewaterability (Yu *et al.*, 2021). Another controlling factor is digestion time. Lü *et al.* (2015) suggested that the digestion times should range from 8 to 30 days (Lü *et al.*, 2015).

The poor dewaterability of the OFMSW digestate is also reflected on the average total solids in the solid fraction of the digestate (9% when using a belt press, 13% for a screw press and 22% for a centrifugation system), which is generally lower than the respective values of manure (24%) and sewage sludge (15–35%) digestates (Lü *et al.*, 2015).

The addition of flocculating agents (inorganic or polymeric compounds) prior to mechanical dewatering is usually applied and provides higher separation efficiencies (Yu *et al.*, 2021). Other technological schemes that could enhance solid–liquid separation are ozonation and thermal pretreatment (Zhang *et al.*, 2022).

3.3.2 Direct land application

Food waste digestate could stand as an effective biofertilizer, which provides significant amounts of nitrogen, reducing the requirements for nutrient additions from conventional fertilizers (such as ammonium nitrate). It also contains other useful macronutrients such as phosphorous in the form of phosphorus oxides, potassium in the form of potash, sulfur, magnesium and trace elements. Its nitrogen–phosphorous–potassium (NPK) ratio can be compared to the respective value of fossil fertilizers, which renders food waste digestate appropriate for agricultural applications (Dutta *et al.*, 2021).

Apart from the elevated nutrient content, food waste digestate use is advantageous due to its low metals and pathogens concentrations and to its elevated water-retention capacity. It is reported that the nitrogen and phosphorous contents of OFMSW digestates from on-farm digesters are easily accessible by plants and may act as slow-release fertilizers. Furthermore, the use of the solid fraction of the digestate may enhance the fertility of the soil by inhibiting pathogens growth (Dutta *et al.*, 2021).

Moreover, the elevated agronomic value of food waste digestate may be further improved by applying several pretreatment schemes such as autoclaving (Cesaro, 2021).

The digestate is considered suitable and safe for use in agriculture, gardening and pastures. For digestates, resulting from food waste, specific controls should be performed to ensure the safety of these products. To standardize the safe use of renewable fertilizers, a 'matrix' of renewable fertilizer (renewable fertilizer matrix) has been developed in consultation with agricultural safety and other stakeholders in the food chain. The necessity for pasteurization or not and the relative time spans are defined depending on the type of crop.

The digestate could be certified in accordance with the BSI PAS110 Directive, under the Biofertilizer Certification System (BCS), which is an independent quality assurance system, ensuring the market that a digestate is safe, stable and suitable for use. Certification schemes allow farmers and growers to upgrade existing standards or add new ones when needed. In general, when certified material is used, the involvement of other authorities (Special Environment Agency, SEPA or NRW) is not considered necessary. However, in the cases where a renewable fertilizer comes from food waste or other animal by-products, additional legal restrictions usually apply.

It is noted that, in May 2019, the European Parliament and the Council adopted rules related to the market of European Union fertilizing products (2016/0084 COD), while it would be applicable in June 2022. This regulation sets rules for all types of fertilizers, including organic, mineral, biostimulants and industrial by-products. This legislation promotes the market intrusion of waste-based fertilizers boosting the circular bioeconomy in Europe. Eleven component material categories (CMCs) are established placing the OFMSW digestate in the fifth CMC titled 'Digestate other than fresh crop digestate.' Ensuring the safety of the land application of these materials, limits for pathogens and metals are set (Cesaro, 2021).

When using renewable fertilizers resulting from the treatment of substrates such as food waste (or other permitted animal by-products), the prescribed standards must always be observed. These include:

- Prevent access to digestion stored inside the farm area before being applied to the soil.
- Observance of the minimum intervals for grazing and harvesting, as stipulated by the regulation of disposal of digestion.
- Maintain a record of the observed grazing and harvesting intervals.

In some cases, animal by-products such as manure can be applied directly to the soil without pre-treatment. The possibility of direct disposal of such by-products (ABP) allows their integration into digestive processes, without prior pasteurization. All other types of animal by-products require their pasteurization before being fed to AD processes. While there are animal by-products for which the processes must include an additional stage of decontamination and pasteurization, which usually takes place at 70°C for 1 h.

The whole digestate is hard to be stored and transported and thus be used as liquid fertilizer given its water content (Song *et al.*, 2021). Instead, the solid digestate fraction can be dispensed with conventional dispersers, while the whole digestion and the liquid fraction are best applied with specialized deposition equipment. This category includes tank dispensing and material injection systems, by injecting into strips on the ground, incorporating into the ground, or injecting into the surface. Infusion by incorporation into the soil can significantly reduce ammonia and nitrogen losses and increase the availability of plant nutrients.

Special software are available, such as MANNER-NPK and PLANET/PLANET Scotland, which are widely used to calculate the digestibility of nutrients available for cultivation. Users should aim to meet 50–60% of the total nitrogen crop requirements from digestate, while the remaining percentage should be replaced by synthetic nitrogen fertilizer. Complete coverage of nitrogen requirements by digestate alone is not recommended, as it may pose risks to the quality and yields of the crop. As a general rule, about 50% of the phosphates and 80% of the potassium contained in the digestate are available for cultivation during the time of application. These values are useful in calculating the nutrient requirements of the crop, especially for cases where the supplement of phosphates and potassium is required.

Digestate also provides significant amounts of sulfur and magnesium. As no data are available on the limits of the specific elements for the development of the crop, their introduction is carried out in such quantities, in order to maintain the normal reserves of the soil, in the specific components.

The use of digestate versus synthetic fertilizers can reduce the carbon dioxide footprint of farms that use it. Replacing them, in fact, with food waste digestate can reduce it by 20 kg CO_2eq/ton applied digestate or by 1 ton CO_2eq/ha.

The main drawbacks of direct land use of food waste digestate are the emissions of NH_3, CO_2 and N_2O (Cesaro, 2021). Given the high ammoniacal and organic contents of OFMSW digestate, up to 70% of nitrogen is expected to be lost during land use (Dutta *et al.*, 2021; Song *et al.*, 2021).

Nevertheless, Di Maria *et al.* (2020) at an LCA analysis reported that the direct land application increased the global warming effect, the human health and the freshwater eutrophication indicators, implying the necessity of a post-treatment step of the digestate (Lu & Xu, 2021).

The integration of digestion into the overall nutrient management design of a farm should aim to maximize the use of these nutrients. The nutrients provided by renewable fertilizers will help farmers optimize the yield of their crops, reducing the cost of cultivation, as well as its environmental impact.

3.3.3 Composting

Composting, as an aerobic biological stabilization process could potentially face the constraints posed by direct land application and thus is a widely used technological choice for the solid digestate (Dutta *et al.*, 2021; Song *et al.*, 2021).

The regulation PAS110 in the UK has set detailed microbiological limits rendering pasteurization or similar processes indispensable (Fuldauer *et al.*, 2018). These limits may be met by the application of composting since organic matter is stabilized and pathogens are removed during the thermophilic composting phase. Another merit of composting is the significant volume reduction that favors the economics and marketability of the final product (Fuldauer *et al.*, 2018; Lu & Xu, 2021). Apart from the UK, composting is widely applied elsewhere in Europe, as well as in the US contributing to sustainable agriculture (Chen *et al.*, 2021).

The composting performance is strongly dependent on the feedstock's composition, temperature, moisture, pH, aeration and mixing frequency (Lu & Xu, 2021). Figure 3.4 illustrates the main controlling parameters of composting, along with the targeted values, the respective characteristics of the OFMSW

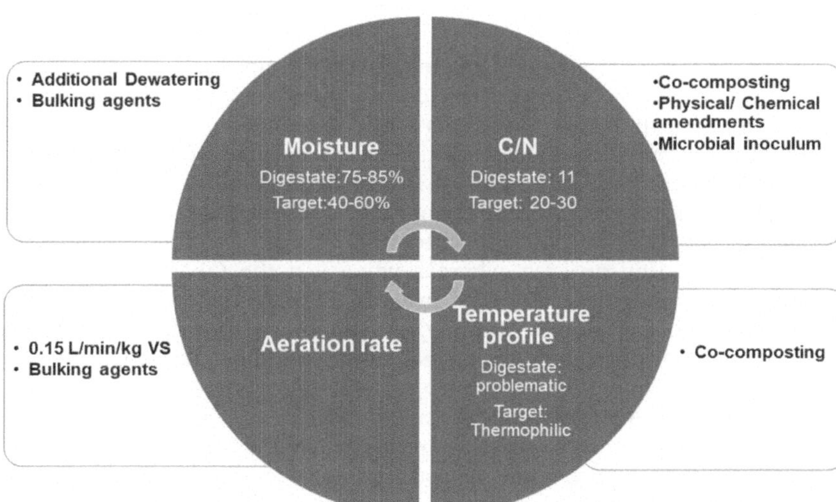

Figure 3.4 Composting of the OFMSW digestate; targeted values of controlling parameters, characteristics of the OFMSW digestate and corrective actions and approaches that can be adopted.

digestate and the corrective actions and approaches that can be adopted. Additionally, special care should be given to the greenhouse gas emissions, odors and heavy metal content (Chen *et al.*, 2021).

The initial moisture of solid OFMSW digestate (75–85%) cannot meet the optimum moisture for composting (40–60%) and thus an additional dehydration step is needed. Centrifugation has been studied for this purpose achieving water content lower than 70% (Lu & Xu, 2021). Another approach usually applied is co-composting of the OFMSW digestate with appropriate bulking materials that could control moisture, C/N ratio, ammonia emissions and accelerate the biological processes (Dutta *et al.*, 2021; Song *et al.*, 2021).

The C/N ratio is of high importance during composting and should range from 20 to 30. The OFMSW digestate is typically lower (Tables 3.1–3.3) mainly because of the high protein degradation rate (higher than the carbohydrates rate) and the subsequent elevated nitrogen concentrations (Dutta *et al.*, 2021; Lu & Xu, 2021). The ratio could be balanced by the use of bulking agents such as woodchips, sawdust, mature compost, and so on in various loadings (Manu *et al.*, 2021).

Apart from C/N adjustment, the challenges associated with NH_3 volatilization and N_2O emissions could be overcome by various mitigation strategies like the addition of physical/chemical compounds and inoculums that facilitate nitrification. Regarding physical additives, zeolite and biochar have been used with promising results since their elevated surface area and ion-exchange capacity could reduce the nitrogen losses via sorption phenomena. Regarding chemical additives, the use of metal salts for ammonia precipitation, acids for acidification or nitrification inhibitors have been reported. Another option is the acceleration of nitrification via proper microbial additives (Manu *et al.*, 2021).

The aeration rate is strongly related to the composting temperature. A balanced air feed is necessary. Sufficient air is essential for the exothermic microbial activity that kills pathogens while excess air cools down the pile reducing the composting rate. Usually, either aeration or temperature is controlled. A ventilation rate of 150 mL/min/kg is ideal for organic matter's stabilization with increased humic compounds concentration (Lu & Xu, 2021). Co-composting of OMSW digestate with MSW ensured the typical temperature profile of composting process, whereas when just the OFMSW digestate is composted, very short thermophilic periods (if any) are observed resulting in poor composting products (Manu *et al.*, 2021). Thus, for efficient OFMSW digestate composting, proper corrective actions should be adopted.

3.3.4 Thermo-chemical processing

Thermochemical processes, such as pyrolysis, hydrothermal carbonization and gasification have been widely used in view of digestates valorization (Cesaro, 2021).

Gasification must be carried out under oxygen sub-stoichiometric conditions and in temperatures between 600 and 1200°C. By transforming the digestate into syngas, rich in H_2 and CH_4, the overall energy gain is increased (Cesaro, 2021).

In order to examine the influence of energy yield on process economics, Singlitico *et al.* (2020) analyzed six different pathways for OFMSW digestate gasification. The researchers noted that integrating AD with the steam gasification of the solid fraction of digestate could considerably increase the renewable energy yield. In this case, the expenses were the lowest and the net present value was indeed the maximum. Furthermore, the methane produced from this process could substitute natural gas, minimizing the CO_2 emissions. Additionally, a drying step may be necessary in order to achieve the optimal solids concentration (>30%) for gasification (Cesaro, 2021).

Biochar synthesis via pyrolysis includes drying of the solid digestate and processing at 300–900°C in an oxygen-limited environment. This could also stand as an appealing alternative for the OFMSW digestate valorization (Dutta *et al.*, 2021). Biochar has been extensively studied because it is considered a carbon-negative substance. Only a few research studies, though, have looked into the possibility of using the OFMSW digestate for biochar production (Dutta *et al.*, 2021).

OFMSW digestate-based biochars can be used as a fertilizer, soil amendment or as a raw material for functionalized products (Cesaro, 2021; Dutta *et al.*, 2021). Such biochars could be further improved with special properties and utilized as advanced materials in various applications (e.g., wastewater treatment sector, soil amendment and catalysis). Biochar produced from the OFMSW digestate showed lower metals content than biochar generated from other fuel sources (e.g., sludge). Moreover, elements like nitrogen and sulfur could promote the synthesis of multifunctional groups with heteroatoms, resulting in biochars with interesting catalytic properties for environmental applications (Dutta *et al.*, 2021). Bio-oil and syngas are generated in addition to biochar. The digestate pyrolysis bio-oil can be utilized as a fuel for engines. Syngas may be used as a source of energy, or can be employed in AD to boost performance (Peng & Pivato, 2019). After the pyrolysis of OFMSW digestate at 500°C, the calorific value of the bio-oil was 13.5 MJ/kg, and of the syngas 13.0 MJ/kg (Peng & Pivato, 2019).

The OFMSW digestate may be a suitable feedstock for biochar synthesis because of its abundance and properties. The feedstock selected has a significant impact on biochar characteristics. The OFMSW digestate is advantageous, due to its high organic matter content that ensures large specific surface area and improved absorption properties of the biochar produced when suitable operational conditions are adopted (Dutta *et al.*, 2021). Additionally, it has more phosphorus and potassium than the original digestate as well as higher water-holding capacity, resulting in superior soil amendment performance (Cesaro, 2021).

The OFMSW digestate was pyrolyzed at 300–700°C in a study by Opatokun *et al.* (2017), yielding biochar with 34–45% carbon, 2–5% nitrogen and 36–60% ash and elevated phosphorous, potassium and other micronutrients contents. The ash and fixed carbon content of biochar rose with temperature, and the biochar's surface area reached up to 463 m^2/g at 800°C pyrolysis temperature (Dutta *et al.*, 2021). Similarly, Li *et al.* (2020) suggested a pyrolysis temperature of 500°C for OFMSW solid digestate resulting in biochar that can be used as a solid fuel, a biofertilizer or soil conditioner.

Yang *et al.* (2020) suggested the addition of the syngas produced from the OFMSW digestate into the AD reactor in order to maximize biogas production. The improvement was attributed not just to the stoichiometric conversion of syngas into CH_4, but also to the favorable effects on the AD of OFMSW (Cesaro, 2021).

Metals (Cu, Pb, and Zn) have been immobilized in industrial soil employing the OFMSW digestate biochar. Gusiatin *et al.* (2016) demonstrated that the pH, ash content and dissolved organic carbon content of biochar were the most important factors in metal immobilization efficiency (Gusiatin *et al.*, 2016). Unlike raw digestate, which alters soil microbial populations, biochar made from digestate immobilizes metals due to its high pH (around 9–10) and ion-exchange capacity. Biochars with elevated fixed carbon concentration (56–83%) can improve carbon sequestration and thus these biochars could be used as soil remediation agents enhancing plants' yields and soil fertility (Peng & Pivato, 2019).

Li *et al.* (2020) investigated a hydrothermal pretreatment step for the whole OFMSW digestate improving dewaterability and as the next step, biogas and biochar production from the liquid and solid fractions, respectively (Li *et al.*, 2020). Mass and energy balances revealed that this treatment system is promising in terms of recovery of either energy or nutrients (Cesaro, 2021).

Hydrothermal carbonization (HTC), a thermochemical process, has also attracted attention as a prospective digestate valorization approach (Cesaro, 2021). It is a thermal treatment carried out at lower temperatures (170–300°C) with self-generated pressure (up to 10 MPa) in aqueous media. It produces three products: a gaseous product mostly composed of CO_2, a liquid product comprising water-soluble molecules, and a solid product (hydrochar) of advanced functionalities (Cesaro, 2021; Dutta *et al.*, 2021). An appealing feature of HTC is that it may take place in an aqueous medium, obviating the need for drying which is highly energy consuming and is a prerequisite for the other thermochemical processes (gasification, pyrolysis) (Cesaro, 2021; Dutta *et al.*, 2021). The moisture content of the OFMSW digestate makes it an excellent raw material for HTC producing hydrochars with increased carbon sequestration suitable for a wide range of uses (Cesaro, 2021).

A study on the HTC of the OFMSW digestate at 140–200°C revealed that digestate mass and volume were reduced by 37% and 69%, respectively, while hydrophobicity was enhanced. The produced hydrochar was deemed a stable and ecofriendly product that may be used for a spectrum of applications (Zhang *et al.*, 2021).

Akarsu *et al.* (2019) examined how HTC factors affected the yields and characteristics of hydrochar manufactured from OFMSW digestate. The latter resulted in a higher hydrochar yield when the working temperature and treatment interval were raised from 175°C to 200°C and 15 to 30 minutes (Cesaro, 2021).

Bernardo *et al.* (2020) investigated the effect of the OFMSW digestate pH (3–8.3) on hydrochar properties. The drop in pH had no consequence on hydrochar features, but it resulted in porous products with increased ash, reduced surface area, but equivalent sorption capacity. The HTC has also been studied for the digestate of mechanically sorted OFMSW demonstrating an increase in carbon content (Cesaro, 2021; Luque Di Salvo *et al.*, 2020).

Previous research suggests that using hydrochar in agriculture optimizes soil characteristics, enhances carbon sequestration, increases plant development, and boosts soil microbial populations. Additionally, hydrochar could be used as functional material with a wide spectrum of applications such as advanced wastewater treatment, medicines and dyes. Nevertheless, to fully investigate the possibility of OFMSW digestate-derived hydrochar, further investigation is necessary (Dutta *et al.*, 2021).

Furthermore, recent research has investigated the combustion properties and calorific content of hydrochar in view of its use as solid fuel. The yield of OFMSW digestate-derived hydrochar varied from 43% to 58%, and it had presented a superior combustion reactivity than lignite. The high ash and low carbon contents of OFMSW digestate, on the other hand, might be a problem, causing slagging and fouling during burning, restricting its use as a renewable fuel. In the study of Parmar and Ross (2019), the hydrochar produced from food and garden waste digestate had a carbon content of 27–32%, an ash content of 53–59%, and a high heating value of 15 MJ/kg. The properties of this hydrochar were similar to MSW-derived digestates but inferior to agricultural residue digestates. Intermediate water-soluble products of HTC process were identified and included carboxylic acids and furans and should either be recycled within the process of further valorization (Cao *et al.*, 2021; Dutta *et al.*, 2021).

3.3.5 Growing media for cultivation

Food waste digestate could also stand as an appropriate medium for various cultivation species (Cesaro, 2021).

Both fractions of the OFMSW digestate could be valorized. The liquid fraction could support the mushroom growth by its fertilizing properties (elevated nitrogen content), while the solid fraction could be utilized as a growing medium for invertebrates and insects. These applications have been verified by experiments conducted with the effluents of conventional separation technologies (Cesaro, 2021).

Mushroom cultivation is a technological scheme that achieves both food production and waste management, by converting regionally available biowaste streams in mushroom protein (Jasinska *et al.*, 2016; O'Brien *et al.*, 2019). The characteristics of food waste digestate favour mushroom growth; high fibre and nutrient content (mainly nitrogen) and elevated pH. O'Brien *et al.* (2019) observed high mushroom yields when solid digestate from an on-farm digester was used for the cultivation of *Pleurotus ostreatus*. The highest mushroom yield (1.15 kg mushrooms/kg dry substrate) was achieved at 35% digestate concentration while doubling the digestate concentration, mushroom growth was inhibited because of the elevated salt concentration of OFMSW (Cesaro, 2021; O'Brien *et al.*, 2019). In the study of Jasinska *et al.* (2016), mixtures of OFMSW with commercial compost were evaluated for *Agaricus arvensis* cultivation. Although higher yields were achieved with commercial products, the mushroom yield increased as the digestate content increased. In another study, the positive effect of food waste digestate in agar media cultures for mycelium growth was also confirmed. The fastest and the best growth was observed for *Coprinus comatus* (Jasińska *et al.*, 2017). Food waste digestate was also successfully applied to the cultivation of *Agaricus subrufescens* and *Agaricus bisporus*, substituting poultry manure in straw composts without reducing the mushroom yields or composition. The only issue raised was the presence of some compact digestate lumps that need to be reduced with appropriate mixing and dilution (Stoknes *et al.*, 2013).

On the other hand, given the urgent need for alternative protein sources, insects reared on digestate seem as a promising option (Spranghers *et al.*, 2017). Girotto and Cossu (2019) described the integration of invertebrate growing processes into biorefineries that target to simultaneously manage biowaste and produce marketable end-products such as biofuels and proteins. Digestates could be used as biowaste in the aforementioned process (Girotto & Cossu, 2019). In this context, Spranghers *et al.* (2017) studied food waste digestate, among other substrates, for the growth of *Hermetia illucens* larvae which is a typical black soldier fly. These flies presented elevated ash but a lower lipid content than the respective ones that had derived from the unfermented substrate (Cesaro, 2021; Spranghers *et al.*, 2017). Furthermore, the calcium content was over the recommendations for pigs and poultry although the phosphorous content was in compliance. Al Seadi *et al.* (2013b) also reported that earthworms produced via vermicomposting of the OFMSW digestate can be used to feed chickens (Al Seadi *et al.*, 2013a; Guilayn *et al.*, 2020). Similarly, Krishnasamy *et al.* (2014) reported that vermicomposting of food waste digestate with sawdust at a ratio 70:30 achieved high earthworm after 75 days providing a 95% pathogens reduction (Krishnasamy *et al.*, 2014). Even though vermicomposting is a promising valorization technique for OMSW digestate, its high ammonia content could inhibit earthworm growth (Guilayn *et al.*, 2020).

3.3.6 Microalgae cultivation

Microalgae cultivation could be integrated with the AD process valorizing both carbon dioxide of biogas and the liquid fraction of the digestate. The carbon dioxide is photosynthetically converted into algal biomass and the nutrient content of the digestate could effectively support algal growth (Cesaro, 2021; Dutta *et al.*, 2021). This valorization approach could enhance the economic viability of algae cultivation (Peng & Pivato, 2019) since energy (Cesaro, 2021) and bio-based products are produced. Apart from electricity, valuable lipids, amino acids, carbohydrates, proteins, minerals and vitamins, pigments, phenols and phytosterols are generated from the microalgal biomass that may stand as secondary resources for feed, or other bio-chemical blocks (Dutta *et al.*, 2021; Koutra *et al.*, 2018; Kumar *et al.*, 2021; Peng & Pivato, 2019). *Chlorella PY-ZU1*, *C. vulgaris*, *C. pyrenoidosa*, *C. pyrenoidosa*, *Tetradesmus obliquus*, *Botryococcus braunii*, *N. oleoabundans* and *S. obliquus* are among the microalgae species that have been efficiently cultivated on the liquid fraction of the OFMSW digestate (Chuka-ogwude *et al.*, 2021; Dutta *et al.*, 2021; Massa *et al.*, 2017).

The critical factor for the utilization of the liquid fraction of the OFMSW digestate on microalgae cultivation is the selection of suitable dosage and dilution in order to avoid nutrient imbalances (Dutta *et al.*, 2021; Kaur *et al.*, 2020). Other important parameters that should be taken into consideration are light requirements, turbidity, and ammonia inhibition (Dutta *et al.*, 2021). More specifically, dilution is applied in order to reduce the suspended solids concentration given that this method is the least expensive and most efficient for effective algal growth (Kaur *et al.*, 2020). In addition, in order to avoid ammonia toxicity, the respective composition should be 20–200 mg NH_4^+/L. Thus, given the high ammonium content of food waste digestates (Tables 3.1–3.3), they should be diluted (Peng & Pivato, 2019). In microalgae cultivation, photo-limitation increases linearly with culture depth and biomass content. The more the dark zones, the lower the productivity (Chuka-ogwude *et al.*, 2021). Turbidity may also impede the effective light penetration that can in turn lead to reduced biomass yields (Chuka-ogwude *et al.*, 2021; Peng & Pivato, 2019).

The C/N ratio is another critical factor that should be taken into account. The optimal ratio for microalgae growth is 4–8 (Chong *et al.*, 2022). Thus, food waste digestates seem favorable (Tables 3.1–3.3) for microalgae cultivation. A high C/N ratio implies fast absorption of nitrogen by methanogens and thus higher biomass production (Chong *et al.*, 2022). On the other hand, the low phosphorous to nitrogen (P/N) ratio is a drawback of food waste digestate, since it does not meet the phosphorous requirements for algal growth (7–16) (Chong *et al.*, 2022). The imbalance in the ratios of P/N and C/N could be faced by the combination of different feedstock materials with suitable characteristics (Chong *et al.*, 2022).

McDowell *et al.* (2020) studied the valorization of OFMSW digestate into microalgae aiming at protein production. It was proved that the levels of crude protein and calcium were higher when OFSMW digestate was applied as a medium in contrast to the synthetic fraction. Similarly, Mayers *et al.* (2017) investigated the utilization of the liquid fraction of the OFMSW digestate as a nutrient source (N and P) for *Nannochloropsis* sp. cultivation. It was concluded that the diluted liquid digestate may provide up to 16% and 100% of the algae phosphorous and nitrogen demand, respectively (Cesaro, 2021; Dutta *et al.*, 2021). In this way, the environmental impact of phosphorous and nitrogen is mitigated in comparison with artificial fertilizers (Dutta *et al.*, 2021). However, the selection of the specific microalgae strain has a significant impact on its growth even on the same digestate (Cesaro, 2021).

In this context, Yu *et al.* (2017) tested 10 microalgal species for cultivation on the diluted liquid fraction of the OFMSW digestate. Among them, just four were grown while two of them (*Chlorella* SDEC-18 and *Scenedesmus* SDEC-8), presented high tolerance and 70% higher lipid content. The latter implies their suitability for the production of biodiesel (Cesaro, 2021). Massa *et al.* (2017) highlighted that the low phosphorous content could be the limiting factor for algal growth (Massa *et al.*, 2017).

Furthermore, microalgae cultivation could achieve simultaneously municipal wastewater treatment (Dutta *et al.*, 2021) with high chemical oxygen demand (COD), nitrogen and phosphorous removals and biodiesel production. This was proved when *Scenedesmus bijuga* was used as a microalgal strain and the liquid fraction of OFMSW digestate, diluted with municipal wastewater (1:20) as growth media, achieving 1.49 g/L microalgae production and 0.016 g/L/d lipid productivity (Cesaro, 2021; Dutta *et al.*, 2021; Shin *et al.*, 2015). In a semi-continuous system, more elevated productivity was achieved when the liquid fraction of the OFMSW digestate was diluted at a 1:2 ratio in contrast to when the digestate was diluted at a 1:5 (Chen *et al.*, 2016).

In another study, food waste digestate was ozonated and then was cultivated with *Chlorella* PY-ZU1. Microalgae growth was increased when 2 g/gC ozone was applied since the compounds that could be assimilated by the algae were increased. Apart from the algae productivity, their possible uses are also influenced by the cultivation systems (Cesaro, 2021; Cheng *et al.*, 2016).

Moreover, Torres-Franco *et al.* suggested that low-intensity red LEDs are a promising alternative for the enhancement of microalgae-based ponds treating digestates, mainly for sunlight complementation, aiming for lower area requirements (Torres-Franco *et al.*, 2021).

Despite the encouraging yields, the capital cost of algal systems is exorbitantly high, whereas the revenues from algal biomass are almost minimal. Thus, these microalgae systems could stand as a viable alternative, only if the respective yields are increased. Furthermore, the legislative constraints related to the marketing of algal products from digestate should be tackled (Fuldauer *et al.*, 2018).

3.3.7 Hydroponics

The liquid fraction of OFMSW digestate could also be valorized in hydroponic cultivation systems in order to provide nutrients and boost plant yields (Cesaro, 2021; Guilayn *et al.*, 2020). In these cultivation systems, plants are not grown on soil but in water (Stoknes *et al.*, 2016). This way, soil quality issues are partially resolved and nutrient losses are minimized via recirculation (Sheets *et al.*, 2015). Nevertheless, the effect of microbial contamination during hydroponic cultivation, which is vital in order to ensure that vegetables and fruits are safe for human consumption were studied (Sheets *et al.*, 2015). Spread of pathogens is favored from the high nitrogen content of the liquid fraction of the OFMSW digestate, especially when recirculation is applied. Regular microbiological analytical tests should be conducted in order to ensure food safety (Fuldauer *et al.*, 2018).

Krishnasamy *et al.* (2012) cultivated silverbeet on the liquid fraction of the OFMSW digestate. A dilution of 80% presented an elevated growth rate, better nutrients' utilization and 95% COD removal after 50 days. Indeed, the performance of liquid digestate was superior to commercial solutions (Cesaro, 2021; Krishnasamy *et al.*, 2012; Sheets *et al.*, 2015). It is worth mentioning that dilution is beneficial since the high ammonia and salts concentrations inhibit growth of plants. Additionally, the high COD concentration causes low oxygen levels at plants' roots (Song *et al.*, 2021).

Stoknes *et al.* (2016) developed an integrated system that utilizes the energy produced from AD, the carbon dioxide of biogas and the digestate. An innovative energy-autonomous greenhouse was proposed in which plants were cultivated on the digestate (Cesaro, 2021).

Fuldauer *et al.* (2018) also studied the hydroponics valorization pathway and suggested the use of a dewatering step via a sifter in order to recycle nutrients loss leading to a more viable process. It was also reported that although hydroponics is an odor-proof process, double door systems, granular activated carbon filters, extractor fans and so on could be used in order to phase possible odor issues. Nevertheless, operational costs in hydroponic systems are elevated compared to the revenues from crop sales. Thus, it was suggested that vertical hydroponics or other modifications could improve the economics of this valorization option (Fuldauer *et al.*, 2018).

3.3.8 Nutrients recovery

The liquid fraction of the digestate has high ammonium and phosphate concentrations and thus could serve as a feedstock for nutrient recovery. Technological options in this direction are membrane-based purification, ammonia stripping, struvite precipitation and thermal distillation (Orner *et al.*, 2021).

Utilizing biofertilizers, like struvite ($MgNH_4PO_4{\cdot}6H_2O$), which naturally occurs as crystals when the $Mg{:}NH_4{:}PO_4$ ratio exceeds 1:1:1, presents an interesting option for chemical or inorganic fertilization. In contrast to other fertilizers, struvite has lower water solubility, enhancing its output by reducing unrestricted nutrient loss in the soil. Struvite also improves impervious environmental conditions (Parra-Orobio *et al.*, 2021).

Plants treating industrial and municipal wastewater can commercially exploit the precipitation of struvite for fertilizer production. Struvite can also be recovered from the digestate via precipitation; incorporating magnesium, ammoniacal nitrogen and phosphates in the form of $MgNH_4PO_4{\cdot}6H_2O$ (Orner *et al.*, 2021).

More specifically, treatment of liquid digestate, deriving from the digestion of agricultural residues, manure and fish processing waste, in a struvite precipitation process achieved removal efficiencies of 44% for COD, 89% for ammonium (NH_4^+) and 60% for phosphates (PO_4^{3-}). Although 89% of the digestate's ammoniacal nitrogen was recovered as struvite, an external phosphorous source was needed (Peng & Pivato, 2019).

In this direction, some other researchers have also analyzed the production of struvite utilizing AD processes. Kjerstadius *et al.* (2015), studying food waste co-digestion, concluded that 229 mg struvite/kg substrate can be recovered. Similarly, Wid and Ayut (2019) managed to extract 136 g struvite/kg substrate. In these studies, the analysis was carried out in single-phase reactors. Vinardell *et al.* (2021) also examined struvite production from the OFMSW digestate in a single-phase configuration. Parra-Orobio *et al.* (2021) reported that a yield of 75.60 g struvite/kg could be recovered in the case of two-stage AD, while 11.64 g struvite/kg were produced when single-stage anaerobic process was applied. These yields were in accordance with the study of Wid and Ayut (2019), which reported an optimum yield of 136 g struvite/kg food waste, pointing out the privileged effect of the feedstock type and composition, along with the digester's configuration. Thus, it is evident that food waste could stand as an efficient source of struvite (Parra-Orobio *et al.*, 2021).

Liquid nutrient recovery procedures may provide different rates of phosphorous and nitrogen recovery. The most productive recovery method should be considered struvite precipitation, recovering 75% and 10% of phosphorous and nitrogen, respectively. Supplementary nitrogen content could be further removed by adopting technological alternatives such as thermal distillation and ammonia stripping (Orner *et al.*, 2021).

Ammonia stripping includes transferring of aqueous ammonia ($NH_{3(aq)}$) into the gas phase at elevated pH and its absorption back into an acidic liquid phase to stabilize it as ammoniacal nitrogen. This process includes the use of acidic media (e.g., H_2SO_4 or HNO_3) that finally led to the recovery of market fertilizing end-products such as ammonium nitrate or ammonium sulfate (Vaneeckhaute *et al.*, 2017). This method has also been seen in commercial and industrial exploitation extracting up to 90% of NH_4 (Orner *et al.*, 2021).

Thermal distillation first uses thermal energy to induce evaporation of $NH_{3(aq)}$ and then continues to utilize condensation. Paired with acid, this process can generate ammonium sulfate or ammonium phosphate and other liquid fertilizers (Tun *et al.*, 2016). Also, it has managed to recover 95% NH_4 in the digestate during its experimental phase (7500 gal/d) and has extensive commercial application (Orner *et al.*, 2021). During the thermal or vacuum evaporation process, pH of the liquid digestate must be decreased in order to eliminate the possibility of ammonia stripping and also to minimize ammonia concentrations in the condensate. Nearly 98% of total nitrogen remains in the concentrate solution when the digestate from on-farm digesters was processed via a vacuum evaporation process, while 2.5% of TKN remained in the condensate accounting for 94.4% of input in mass terms. The practical applications of vacuum evaporation and ammonia stripping were compared by Li *et al.* (2016). It was concluded that in the case of ammonia stripping with $Ca(OH)_2$, low ammoniacal nitrogen (137 mg/L) was achieved, while in the case of vacuum evaporation much higher concentrations (2998 mg/L) were observed, implying a high plant nutrient availability. The key factor for both of these methods was the initial pH found in the digestate (Peng & Pivato, 2019).

Alternatively, two other methods that treat sustainably liquid digestate providing sufficient results are membrane purification and surface scraped heat exchange. These provide concentrated liquors with high values of secondary nutrients (phosphorous and nitrogen) in the form of liquid fertilizer, while the condensate could be used as reclaimed water given its quality characteristics. Despite the high nutrients' concentrations in the concentrate, its use for land applications is limited by the high salinity. Similarly, membrane purification of digestate combined with aerobic treatment could provide a solid solution for digestate management because in this way we can avoid external addition of carbon sources (Peng & Pivato, 2019).

3.3.9 Wetland

Wetlands and reed beds have been widely investigated for the processing of waste sludge. Constructed wetlands could stand as a treatment option mainly of the liquid fraction of OFMSW digestate achieving high biological oxygen demand and nitrogen removals due to the nitrification and denitrification that take place (Peng & Pivato, 2019). Very few studies have been reported on this subject. Among them, Ediviani *et al.* (2018a, 2018b) investigated the possibility of using ornamental aquatic macrophytes to remove ammoniacal nitrogen from the liquid fraction of food waste digestate in a horizontal subsurface flow wetland. Additionally, *Canna indica* presented the highest nitrogen uptake improving significantly the effluent quality along with the environmental aesthetics (Ediviani *et al.*, 2018a).

3.4 TECHNO-ECONOMIC CHALLENGES

The valorization of OFMSW digestate is a prerequisite for the successful implementation of AD processes in line with the principles of circular bioeconomy.

Its conventional application as fertilizer cannot stand as an environmentally and/or economically viable option. The revenues from the fertilizing product usually do not counterbalance the processing cost. In addition, environmental concerns are raised regarding nitrogen leaching and the lack of arable land. In this context, sustainable alternative options are sought. In order to efficiently identify these pathways, one should consider whether a dewatering step shall be included. The different moisture content of the substrates (whole digestate, liquid digestate and solid digestate) indicates different valorization strategies. In any case, the valorization of food waste digestate should be incorporated within a biorefinery framework. The valorization of food waste digestate within a biorefinery would strongly support its development by reducing the raw materials cost and by encouraging the valorization of underexploited feedstocks. Furthermore, analysis of energy and mass flows along with life-cycle assessment shall enlighten the integrated process. This way all resources will be fully utilized and secondary pollution will be minimized.

Although the land application could promote the closed loops of carbon and nitrogen utilization, it also poses the threat of nitrogen losses.

Thermochemical treatment may be applied to either whole digestates or their solid fractions aiming to recover energy carriers and value-added products such as biochar, hydrochar and so on. Biochar could stand as a full-scale industrial product. Yet, cases of food waste digestate biochar marketing were not identified. Although thermochemical processes are well-established, they are highly energy-consuming impeding their viability for capacities lower than a specific threshold.

The use of food waste digestate for microalgae cultivation is another appealing alternative that encompasses the quintessence of circular economy principles. However, this technology is of low technological readiness level and it should overcome the barriers of low growth rates, ammonia inhibition and light penetration.

Hydroponic cultivation also seems as a promising solution compromising the conventional fertilizing applications with the innovative cultivating approaches. A possible contamination leakage has to be ruled out by selecting either specific plant species or operational conditions.

Even though some of the valorization strategies have been investigated for several years, there still lacks a systematic approach to navigating proper digestate management. Another aspect that should also be taken into consideration is the legislative framework when the end-products from food waste digestate valorization are intended for the production of food or feed. Specific regulations have to be defined for the safe utilization of these products in order to gain the confidence of all involved stakeholders. Furthermore, market studies have to be set up in order to ensure the economic viability of the processes. These two aspects have to be efficiently addressed prior to up-scaling.

In any case, the up-scaling of the aforementioned valorization solutions is worth all scientific and technical efforts.

3.5 CONCLUDING REMARKS

This study focused on the presentation of all valorization solutions of the digestate resulting from the AD of the organic fraction of MSW. Given the new legislative framework along with the need to comply with the circular bioeconomy strategy, alternative valorization solutions should be sought since the conventional use as a fertilizer is no longer viable. In this context, food waste digestate could be converted to value-added products or be used as a nutrient solution or growing medium. Thus, the integration of the AD of OFMSW with a proposed valorization strategy may close the resource loop and materialize the circular economy concept.

REFERENCES

Abubaker J., Risberg K. and Pell M. (2012). Biogas residues as fertilisers – effects on wheat growth and soil microbial activities. *Applied Energy*, **99**, 126–134, https://doi.org/10.1016/j.apenergy.2012.04.050

Akarsu K., Duman G., Yilmazer A., Keskin T., Azbar N. and Yanik J. (2019). Sustainable valorization of food wastes into solid fuel by hydrothermal carbonization. *Bioresource Technology*, **292**, 121959, https://doi.org/10.1016/j.biortech.2019.121959

Al Seadi T., Drosg B., Fuchs W., Rutz D. and Janssen R. (2013a). 12 – biogas digestate quality and utilization A2 – Wellinger. In: The Biogas Handbook, A. Wellinger, J. Murphy, D. Baxter (eds.), Woodhead Publishing Series in Energy, Woodhead Publishing, United Kingdom, 267–301.

Al Seadi T., Owen N., Hellström H. and Kang H. (2013b). *Source Separation of MSW.*

Arab G. and McCartney D. (2017). Benefits to decomposition rates when using digestate as compost co-feedstock: part I – focus on physicochemical parameters. *Waste Management*, **68**, 74–84, https://doi.org/10.1016/j.wasman.2017.07.018

Arcadis. (2019). National food waste baseline executive summary. 9.

Barampouti E. M., Mai S., Malamis D., Moustakas K. and Loizidou M. (2019). Liquid biofuels from the organic fraction of municipal solid waste: a review. *Renewable and Sustainable Energy Reviews*, **110**, 298–314, https://doi.org/10.1016/j.rser.2019.04.005

Barampouti E. M., Mai S., Malamis D., Moustakas K. and Loizidou M. (2020). Exploring technological alternatives of nutrient recovery from digestate as a secondary resource. *Renewable and Sustainable Energy Reviews*, **134**, 110379, https://doi.org/10.1016/j.rser.2020.110379

Barzee T. J., Edalati A., El-Mashad H., Wang D., Scow K. and Zhang R. (2019). Digestate biofertilizers support similar or higher tomato yields and quality than mineral fertilizer in a subsurface drip fertigation system. *Frontiers in Sustainable Food Systems*, **3**, 1–13, https://doi.org/10.3389/fsufs.2019.00058

Beggio G., Schievano A., Bonato T., Hennebert P. and Pivato A. (2019). Statistical analysis for the quality assessment of digestates from separately collected organic fraction of municipal solid waste (OFMSW) and agro-industrial feedstock. Should input feedstock to anaerobic digestion determine the legal status of digestate? *Waste Management*, **87**, 546–558, https://doi.org/10.1016/j.wasman.2019.02.040

Bernardo M., Correa C. R., Ringelspacher Y., Becker G. C., Lapa N., Fonseca I., Esteves I.A.A.C. and Kruse A. (2020). Porous carbons derived from hydrothermally treated biogas digestate. *Waste Management*, **105**, 170–179, https://doi.org/10.1016/j.wasman.2020.02.011

Cao Z., Hülsemann B., Wüst D., Oechsner H., Lautenbach A. and Kruse A. (2021). Effect of residence time during hydrothermal carbonization of biogas digestate on the combustion characteristics of hydrochar and the biogas production of process water. *Bioresource Technology*, **333**, 125110, https://doi.org/10.1016/j.biortech.2021.125110

Cesaro A. (2021). The valorization of the anaerobic digestate from the organic fractions of municipal solid waste: challenges and perspectives. *Journal of Environmental Management*, **280**, 111742, https://doi.org/10.1016/j.jenvman.2020.111742

Chen R., Liu Y. and Liao W. (2016) Using an environmentally friendly process combining electrocoagulation and algal cultivation to treat high-strength wastewater. *Algal Research*, **16**, 330–337, https://doi.org/10.1016/j.algal.2016.03.032

Chen H., Osman A. I., Mangwandi C. and Rooney D. (2019). Upcycling food waste digestate for energy and heavy metal remediation applications. *Resources, Conservation and Recycling: X*, **3**, 100015.

Chen T., Qiu X., Feng H., Yin J. and Shen D. (2021) Solid digestate disposal strategies to reduce the environmental impact and energy consumption of food waste-based biogas systems. *Bioresource Technology*, **325**, 124706, https://doi.org/10.1016/j.biortech.2021.124706

Cheng J., Ye Q., Xu J., Yang Z., Zhou J. and Cen K. (2016). Improving pollutants removal by microalgae Chlorella PY-ZU1 with 15% CO_2 from undiluted anaerobic digestion effluent of food wastes with ozonation pretreatment. *Bioresource Technology*, **216**, 273–279, https://doi.org/10.1016/j.biortech.2016.05.069

Cheong J. C., Lee J. T. E., Lim J. W., Song S., Tan J. K. N., Chiam Z. Y., Yap K. Y., Lim E. Y., Zhang J., Tan H. T. W. and Tong Y. W. (2020). Closing the food waste loop: food waste anaerobic digestate as fertilizer for the cultivation of the leafy vegetable, xiao bai cai (*Brassica rapa*). *Science of the Total Environment*, **715**, 136789, https://doi.org/10.1016/j.scitotenv.2020.136789

Chong C. C., Cheng Y. W., Ishak S., Lam M. K., Lim J. W., Tan I. S., Show P. L. and Lee K. T. (2022). Anaerobic digestate as a low-cost nutrient source for sustainable microalgae cultivation: a way forward through waste valorization approach. *Science of the Total Environment*, **803**, 150070, https://doi.org/10.1016/j.scitotenv.2021.150070

Chuka-ogwude D., Ogbonna J. C. and Moheimani N. R. (2021). Depth optimization of inclined thin layer photobioreactor for efficient microalgae cultivation in high turbidity digestate. *Algal Research*, **60**, 102509, https://doi.org/10.1016/j.algal.2021.102509

Di Maria F., Sordi A., Cirulli G., Gigliotti G., Massaccesi L. and Cucina M. (2014). Co-treatment of fruit and vegetable waste in sludge digesters. An analysis of the relationship among bio-methane generation, process stability and digestate phytotoxicity. *Waste Management*, **34**, 1603–1608, https://doi.org/10.1016/j.wasman.2014.05.017

Di Maria F., Sisani F., El-Hoz M. and Mersky R. L. (2020). How collection efficiency and legal constraints on digestate management can affect the effectiveness of anaerobic digestion of bio-waste: an analysis of the Italian context in a life cycle perspective. *Science of the Total Environment*, **726**, 138555, https://doi.org/10.1016/j.scitotenv.2020.138555

do Carmo Precci Lopes A., Ebner C., Gerke F., Wehner M., Robra S., Hupfauf S., and Bockreis A. (2022). Residual municipal solid waste as co-substrate at wastewater treatment plants: an assessment of methane yield, dewatering potential and microbial diversity. *Science of the Total Environment*, **804**, 149936, https://doi.org/10.1016/j.scitotenv.2021.149936

Dutta S., He M., Xiong X. and Tsang D. C. W. (2021). Sustainable management and recycling of food waste anaerobic digestate: a review. *Bioresource Technology*, **341**, 125915, https://doi.org/10.1016/j.biortech.2021.125915

Ediviani W., Priadi C. R. and Moersidik S. S. (2018a). Nutrient uptake from liquid digestate using ornamental aquatic macrophytes (*Canna indica*, *Iris pseudacorus*, *Typha latifolia*) in a constructed wetland system. *Journal of Physics: Conference Series*, **1022**, 012052.

Ediviani W., Priadi C. R. and Moersidik S. S. (2018b). Nutrient recovery from anaerobic digester liquid effluent using ornamental aquatic macrophytes. IOP Conference Series: Earth and Environmental Science.

EPA. (2020). Wasted Food Measurement Methodology Scoping Memo.

FIAL. (2021). National Food Waste Strategy Feasibility Study – Final Report.

Fuldauer L. I., Parker B. M., Yaman R. and Borrion A. (2018). Managing anaerobic digestate from food waste in the urban environment: evaluating the feasibility from an interdisciplinary perspective. *Journal of Cleaner Production*, **185**, 929–940, https://doi.org/10.1016/j.jclepro.2018.03.045

Girotto F. and Cossu R. (2019). Role of animals in waste management with a focus on invertebrates' biorefinery: an overview. *Environmental Development*, **32**, 100454, https://doi.org/10.1016/j.envdev.2019.08.001

Grigatti M., Barbanti L., Hassan M. U. and Ciavatta C. (2020). Fertilizing potential and CO_2 emissions following the utilization of fresh and composted food-waste anaerobic digestates. *Science of the Total Environment*, **698**, 134198, https://doi.org/10.1016/j.scitotenv.2019.134198

Guilayn F., Rouez M., Crest M., Patureau D. and Jimenez J. (2020). Valorization of digestates from urban or centralized biogas plants: a critical review. *Reviews in Environmental Science and Biotechnology*, **19**, 419–462, https://doi.org/10.1007/s11157-020-09531-3

Gusiatin Z. M., Kurkowski R., Brym S. and Wiśniewski D. (2016). Properties of biochars from conventional and alternative feedstocks and their suitability for metal immobilization in industrial soil. *Environmental Science and Pollution Research*, **23**, 21249–21261, https://doi.org/10.1007/s11356-016-7335-4

International Energy Agency. (2018). Biogas production by region and by feedstock type.

Jasinska A. J., Wojciechowska E., Krzesinski W., Spizewski T., Stokens K. and Krajewska K. (2016). Mushroom cultivation on substrates with addition of anaerobically digested food waste. *Acta Horticulturae*, **1123**, 199–205, https://doi.org/10.17660/ActaHortic.2016.1123.28

Jasińska A., Dawidowicz L., Siwulski M. and Kilinowski P. (2017). Growth of mycelium of different edible and medicinal mushrooms on medium supplemented with digestate from AD biogas plant. *Notulae Botanicae Horti Agrobotanici Cluj-Napoca*, **45**, 498–506, https://doi.org/10.15835/nbha45210954

Joshi P. and Visvanathan C. (2019). Sustainable management practices of food waste in Asia: technological and policy drivers. *Journal of Environmental Management*, **247**, 538–550, https://doi.org/10.1016/j.jenvman.2019.06.079

Kaur G., Wong J. W. C., Kumar R., Patria R. D., Bhardwaj A., Uisan K., and Johnravindar D. (2020). Value addition of anaerobic digestate from biowaste: thinking beyond agriculture. *Current Sustainable/Renewable Energy Reports*, **7**, 48–55, https://doi.org/10.1007/s40518-020-00148-2

Kjerstadius H., Haghighatafshar S. and Davidsson A. (2015). Potential for nutrient recovery and biogas production from blackwater, food waste and greywater in urban source control systems. *Environmental Technology (United Kingdom)*, **36**, 1707–1720.

Koutra E., Economou C. N., Tsafrakidou P. and Kornaros M. (2018). Bio-based products from microalgae cultivated in digestates. *Trends in Biotechnology*, **36**, 819–833, https://doi.org/10.1016/j.tibtech.2018.02.015

Krishnasamy K., Nair J. and Bäuml B. (2012). Hydroponic system for the treatment of anaerobic liquid. *Water Science and Technology*, **65**, 1164–1171, https://doi.org/10.2166/wst.2012.031

Krishnasamy K., Nair J. and Bell R. (2014). Evaluation of anaerobic digestate as a substrate for vermicomposting. *International Journal of Environment and Waste Management*, **14**, 149–164, https://doi.org/10.1504/IJEWM.2014.064084

Kumar V., Jaiswal K. K., Tomar M. S., Rajput V., Upadhyay S., Nanda M., Vlaskin M. S., Kumar S. and Kurbatova A. (2021). Production of high value-added biomolecules by microalgae cultivation in wastewater from anaerobic digestates of food waste: a review. *Biomass Conversion and Biorefinery*. https://doi.org/10.1007/s13399-021-01906-y

Li X., Guo J., Dong R., Ahring B. K. and Zhang W. (2016). Properties of plant nutrient: comparison of two nutrient recovery techniques using liquid fraction of digestate from anaerobic digester treating pig manure. *Science of the Total Environment*, **544**, 774–781, https://doi.org/10.1016/j.scitotenv.2015.11.172

Li C., Li J., Pan L., Zhu X., Xie S., Yu G., Wang Y., Pan X., Zhu G. and Angelidaki I. (2020). Treatment of digestate residues for energy recovery and biochar production: from lab to pilot-scale verification. *Journal of Cleaner Production*, **265**, 121852, https://doi.org/10.1016/j.jclepro.2020.121852

Lin C. S. K., Pfaltzgraff L. A., Herrero-Davila L., Mubofu E. B., Abderrahim S., Clark J. H., Koutinas A., Kopsahelis N., Stamatelatou K., Dickson F., Thankappan S., Mohamed Z., Brocklesby R. and Luque R. (2013). Food waste as a valuable resource for the production of chemicals, materials and fuels. Current situation and global perspective. *Energy and Environmental Science*, **6**, 426–464, https://doi.org/10.1039/c2ee23440h

Liu L., Yan H., Tan W. and Zhu G. (2017). Influence of Na+, K+, Mg2+, Ca2+, and Fe3+on filterability and settleability of drilling sludge. *Chinese Journal of Chemical Engineering*, **25**, 658–664, https://doi.org/10.1016/j.cjche.2016.10.008

Liu J., Huang S., Chen K., Wang T., Mei M. and Li J. (2020) Preparation of biochar from food waste digestate: pyrolysis behavior and product properties. *Bioresource Technology*, **302**, 122841, https://doi.org/10.1016/j.biortech.2020.122841

Lu J. and Xu S. (2021). Post-treatment of food waste digestate towards land application: a review. *Journal of Cleaner Production*, **303**, 127033, https://doi.org/10.1016/j.jclepro.2021.127033

Lü F., Zhou Q., Wu D., Wang T., Shao L. and He P. (2015). Dewaterability of anaerobic digestate from food waste: relationship with extracellular polymeric substances. *Chemical Engineering Journal*, **262**, 932–938, https://doi.org/10.1016/j.cej.2014.10.051

Luque Di Salvo J., De Luca G., Cipollina A. and Micale G. (2020). Effect of ion exchange capacity and water uptake on hydroxide transport in PSU-TMA membranes: a DFT and molecular dynamics study. *Journal of Membrane Science*, **599**, 117837, https://doi.org/10.1016/j.memsci.2020.117837

Manu M. K., Li D., Liwen L., Jun Z., Varjani S. and Wong J. W. C. (2021). A review on nitrogen dynamics and mitigation strategies of food waste digestate composting. *Bioresource Technology*, **334**, 125032, https://doi.org/10.1016/j.biortech.2021.125032

Massa M., Buono S., Langellotti A. L., Castaldo L., Martello A., Paduano A., Sacchi R. and Fogliano V. (2017). Evaluation of anaerobic digestates from different feedstocks as growth media for *Tetradesmus obliquus*, *Botryococcus braunii*, *Phaeodactylum tricornutum* and *Arthrospira maxima*. *New Biotechnology*, **36**, 8–16, https://doi.org/10.1016/j.nbt.2016.12.007

Massaccesi L., Sordi A., Micale C., Cucina M., Zadra C., Di Maria F., and Gigliotti G. (2013). Chemical characterisation of percolate and digestate during the hybrid solid anaerobic digestion batch process. *Process Biochemistry*, **48**, 1361–1367, https://doi.org/10.1016/j.procbio.2013.06.026

Mayers J. J., Ekman Nilsson A., Albers E. and Flynn K. J. (2017). Nutrients from anaerobic digestion effluents for cultivation of the microalga *Nannochloropsis* sp. – impact on growth, biochemical composition and the potential for cost and environmental impact savings. *Algal Research*, **26**, 275–286, https://doi.org/10.1016/j.algal.2017.08.007

McDowell D., Dick J. T., Eagling L., Julius M., Sheldrake G. N., Theodoridou K., and Walsh P. J. (2020). Recycling nutrients from anaerobic digestates for the cultivation of *Phaeodactylum tricornutum*: a feasibility study. *Algal Research*, **48**, 101893, https://doi.org/10.1016/j.algal.2020.101893

Nicholson F., Bhogal A., Cardenas L., Chadwick D., Misselbrook T., Rollett A., Taylor M., Thorman R. and Williams J. (2017). Nitrogen losses to the environment following food-based digestate and compost applications to agricultural land. *Environmental Pollution*, **228**, 504–516, https://doi.org/10.1016/j.envpol.2017.05.023

O'Brien B. J., Milligan E., Carver J. and Roy E. D. (2019). Integrating anaerobic co-digestion of dairy manure and food waste with cultivation of edible mushrooms for nutrient recovery. *Bioresource Technology*, **285**, 121312, https://doi.org/10.1016/j.biortech.2019.121312

Opatokun S. A., Strezov V. and Kan T. (2015). Product based evaluation of pyrolysis of food waste and its digestate. *Energy*, **92**, 349–354, https://doi.org/10.1016/j.energy.2015.02.098

Opatokun S. A., Yousef L. F. and Strezov V. (2017). Agronomic assessment of pyrolysed food waste digestate for sandy soil management. *Journal of Environmental Management*, **187**, 24–30, https://doi.org/10.1016/j.jenvman.2016.11.030

Orner K. D., Smith S. J., Breunig H. M., Scown C. D. and Nelson K. L. (2021). Fertilizer demand and potential supply through nutrient recovery from organic waste digestate in California. *Water Research*, **206**, 117717, https://doi.org/10.1016/j.watres.2021.117717

Parmar K. R. and Ross A. B. (2019). Integration of hydrothermal carbonisation with anaerobic digestion; opportunities for valorisation of digestate. *Energies*, **12**, 1586, https://doi.org/10.3390/en12091586

Parra-Orobio B. A., Pérez-Vidal A. and Torres-Lozada P. (2021). Potential production of struvite from the anaerobic digestion of food waste: analysis in one-phase and two-phase configurations. *Water Science and Technology*, **84**, 1048–1058, https://doi.org/10.2166/wst.2021.299

Peng W. and Pivato A. (2019). Sustainable management of digestate from the organic fraction of municipal solid waste and food waste under the concepts of back to earth alternatives and circular economy. *Waste and Biomass Valorization*, **10**, 465–481, https://doi.org/10.1007/s12649-017-0071-2

Peng W., Zhang H., Lü F., Shao L. and He P. (2022). From food waste and its digestate to nitrogen self-doped char and methane-rich syngas: evolution of pyrolysis products during autogenic pressure carbonization. *Journal of Hazardous Materials*, **424**, 127249, https://doi.org/10.1016/jjhazmat.2021.127249.

Sheets J. P., Yang L., Ge X., Wang Z. and Li Y. (2015). Beyond land application: emerging technologies for the treatment and reuse of anaerobically digested agricultural and food waste. *Waste Management*, **44**, 94–115, https://doi.org/10.1016/j.wasman.2015.07.037

Shin D. Y., Cho H. U., Utomo J. C., Choi Y. N., Xu X. and Park J. M. (2015). Biodiesel production from *Scenedesmus bijuga* grown in anaerobically digested food wastewater effluent. *Bioresource Technology*, **184**, 215–221, https://doi.org/10.1016/j.biortech.2014.10.090

Singlitico A., Dussan K., O'Shea R., Wall D., Goggins J., Murphy J. D., and Monaghan R. F. D. (2020). Can thermal energy recovery from digestate make renewable gas from household waste more cost effective? A case study for the Republic of Ireland. *Journal of Cleaner Production*, **261**, 121198, https://doi.org/10.1016/j.jclepro.2020.121198

Song S., Lim J. W., Lee J. T. E., Cheong J. C., Hoy S. H., Hu Q., Tan J. K. N., Chiam Z., Arora S., Lum T. Q. H., Lim E. Y., Wang C. H., Tan H. T. W. and Tong Y. W. (2021). Food-waste anaerobic digestate as a fertilizer: the agronomic properties of untreated digestate and biochar-filtered digestate residue. *Waste Management*, **136**, 143–152, https://doi.org/10.1016/j.wasman.2021.10.011

Speight J. G. (2015). Waste gasification for synthetic liquid fuel production, In: Woodhead Publishing Series in Energy, Gasification for Synthetic Fuel Production, R. Luque, J. G. Speight (eds.), Woodhead Publishing, 2015, 277–301.

Spranghers T., Ottoboni M., Klootwijk C., Ovyn A., Deboosere S., De Meulenaer B., Michiels J., Eeckhout M., De Clercq P. and De Smet S. (2017). Nutritional composition of black soldier fly (*Hermetia illucens*) prepupae reared on different organic waste substrates. *Journal of the Science of Food and Agriculture*, **97**, 2594–2600, https://doi.org/10.1002/jsfa.8081

Stoknes K., Beyer D. M. and Norgaard E. (2013). Anaerobically digested food waste in compost for *Agaricus bisporus* and *Agaricus subrufescens* and its effect on mushroom productivity. *Journal of the Science of Food and Agriculture*, **93**, 2188–2200, https://doi.org/10.1002/jsfa.6026

Stoknes K., Scholwin F., Krzesiński W., Wojciechowska E. and Jasińska A. (2016). Efficiency of a novel 'food to waste to food' system including anaerobic digestion of food waste and cultivation of vegetables on digestate in a bubble-insulated greenhouse. *Waste Management*, **56**, 466–476, https://doi.org/10.1016/j.wasman.2016.06.027

Svensson K., Kjørlaug O., Higgins M. J., Linjordet R. and Horn S. J. (2018). Post-anaerobic digestion thermal hydrolysis of sewage sludge and food waste: effect on methane yields, dewaterability and solids reduction. *Water Research*, **132**, 158–166, https://doi.org/10.1016/j.watres.2018.01.008

Tambone F., Genevini P., D'Imporzano G. and Adani F. (2009). Assessing amendment properties of digestate by studying the organic matter composition and the degree of biological stability during the anaerobic digestion of the organic fraction of MSW. *Bioresource Technology*, **100**, 3140–3142, https://doi.org/10.1016/j.biortech.2009.02.012

Tambone F., Scaglia B., D'Imporzano G., Schievano A., Orzi V., Salati S., and Adani F. (2010). Assessing amendment and fertilizing properties of digestates from anaerobic digestion through a comparative study with digested sludge and compost. *Chemosphere*, **81**, 577–583, https://doi.org/10.1016/j.chemosphere.2010.08.034

Tampio E., Ervasti S. and Rintala J. (2015). Characteristics and agronomic usability of digestates from laboratory digesters treating food waste and autoclaved food waste. *Journal of Cleaner Production*, **94**, 86–92, https://doi.org/10.1016/j.jclepro.2015.01.086

Tampio E., Salo T. and Rintala J. (2016). Agronomic characteristics of five different urban waste digestates. *Journal of Environmental Management*, **169**, 293–302, https://doi.org/10.1016/j.jenvman.2016.01.001

Torres-Franco A. F., Silva G., Freitas M. P., Passos F., Mota Filho C. R. and Figueredo C. C. (2021). Effect of digestate loading rates on microalgae-based treatment under low LED light intensity. *Environmental Technology (United Kingdom)*. https://doi.org/10.1080/09593330.2021.1914178

Tun L. L., Jeong D., Jeong S., Cho K., Lee S. and Bae H. (2016). Dewatering of source-separated human urine for nitrogen recovery by membrane distillation. *Journal of Membrane Science*, **512**, 13–20, https://doi.org/10.1016/j.memsci.2016.04.004

Van der Linden A. and Almut R. (2020). Bio-waste in Europe-turning challenges into opportunities. European Environment Agency, Luxembourg.

Vaneeckhaute C., Lebuf V., Michels E., Belia E., Vanrolleghem P. A., Tack F. M. G., and Meers E. (2017) Nutrient recovery from digestate: systematic technology review and product classification. *Waste and Biomass Valorization*, **8**, 21–40, https://doi.org/10.1007/s12649-016-9642-x

Vinardell S., Astals S., Koch K., Mata-Alvarez J. and Dosta J. (2021). Co-digestion of sewage sludge and food waste in a wastewater treatment plant based on mainstream anaerobic membrane bioreactor technology: a techno-economic evaluation. *Bioresource Technology*, **330**, 124978, https://doi.org/10.1016/j.biortech.2021.124978

Wang X., Wang W., Zhou B., Xu M., Wu Z., Liang J., and Zhou L. (2020). Improving solid–liquid separation performance of anaerobic digestate from food waste by thermally activated persulfate oxidation. *Journal of Hazardous Materials*, **398**, 122989, https://doi.org/10.1016/j.jhazmat.2020.122989

Wid N. and Ayut L. F. (2019). Anaerobic digestion of organic waste in UMS campus for resource recovery and waste reduction. In: Green Engineering for Campus Sustainability, A. Yaser (ed.), Springer, Singapore, pp. 133–143.

Yang Z., Liu Y., Zhang J., Mao K., Kurbonova M., Liu G., Zhang R. and Wang W. (2020). Improvement of biofuel recovery from food waste by integration of anaerobic digestion, digestate pyrolysis and syngas biomethanation under mesophilic and thermophilic conditions. *Journal of Cleaner Production*, **256**, 120594, https://doi.org/10.1016/j.jclepro.2020.120594

Yu Z., Song M., Pei H., Han F., Jiang L. and Hou Q. (2017). The growth characteristics and biodiesel production of ten algae strains cultivated in anaerobically digested effluent from kitchen waste. *Algal Research*, **24**, 265–275, https://doi.org/10.1016/j.algal.2017.04.010

Yu Y., Li P., Zhang J., Li J. and Yu R. (2021). Comprehensive insights into the organic fractions on solid–liquid separation performance of anaerobic digestates from food waste. *Science of the Total Environment*, **800**, 149608, https://doi.org/10.1016/j.scitotenv.2021.149608

Zhang C., Shao M., Wu H., Wang N., Chen Q. and Xu Q. (2021). Management and valorization of digestate from food waste via hydrothermal. *Resources, Conservation and Recycling*, **171**, 105639, https://doi.org/10.1016/j.resconrec.2021.105639

Zhang C., Shao M., Wu H., Wang N., Wang X., Wang Q., and Xu Q. (2022). Mechanism insights into hydrothermal dewatering of food waste digestate for products valorization. *Science of the Total Environment*, **804**, 150145, https://doi.org/10.1016/j.scitotenv.2021.150145

doi: 10.2166/9781789062755_0075

Chapter 4

Management of digestate from anaerobic digestion of municipal sludge

*Umme Sharmeen Hyder, Farokhlaqa Kakar, Frances Okoye and Elsayed Elbeshbishy**

Civil Engineering, Ryerson University, 350 Victoria St. Toronto, Ontario, M5B 2K3, Canada
**Corresponding author e-mail: elsayed.elbeshbishy@ryerson.ca*

ABSTRACT

A large quantity of anaerobically digested sludge or digestate produced from municipal wastewater treatment processes is a worldwide concern. Digestate contains a lot of water which increases its volume. Consequently, transportation and disposal become challenging. The dewatering process helps to reduce the water content of the digestate by solid–liquid separation. Dewatering digestate is critical for lowering the digestate volume. Reduction of moisture content in the dewatered cake can be achieved through a conditioning process before dewatering. The application of organic polymers or inorganic chemicals as conditioners is a standard practice in municipal wastewater treatment plants. However, researchers are investing time and efforts to enhance the dewatering process through various chemicals, oxidants, and skeleton builders. Advanced dewatering techniques such as ultrasonication or electro-dewatering, and so on, are also applied to improve the dewatering of the digestate. The efficiency of digestate dewatering depends on the sources and several characteristics of sludges. Capillary suction time, filtration resistance, cake solid contents are key parameters considered in sludge dewatering indices. Advanced dewatering methods such as electro-dewatering of oxidation often help attain a required level of biosolids disinfection. The chapter describes a comprehensive summary of municipal digestate dewatering processes, conditioning methods with mechanisms, advanced dewatering techniques, and disinfection of biosolids and per biosolids regulations.

Keywords: Digestate management, anaerobic digestion, municipal sludge, resource recovery

4.1 INTRODUCTION

As wastewater directives become more stringent for environmental protection, the volume of sludge produced from municipal water resource recovery facilities (WRRF) and wastewater treatment plants (WWTP) increases substantially. Municipal sludge is the primary waste product of WRRFs, requiring proper

Editor(s): Vinay Kumar Tyagi, Kaoutar Aboudi and Cigdem Eskicioglu

stabilization before disposal or reuse. Anaerobic digestion (AD) of sludge has emerged as a well-established, cost-effective, and attractive technology for municipal sludge processing. Municipal sludge has properties that are well-suited to the digestion process, such as its moisture content, the potential for biogas production, and nutrient recovery (Logan & Visvanathan, 2019). Biogas that is generated from municipal sludge can provide 50% of the energy requirements of WRRFs (Gude, 2015). The by-product of this biogas production process is the digestate which contains up to 90% of the nutrients in the original feed. This digestate can constitute a significant economic and legal concern to the WRRF. While the nutrient content in the digestate allows for its current use in the land application, the presence of heavy metals and emerging contaminants can constrain its use. This chapter explores the current technological approaches, regulations, and innovative research into managing municipal sludge digestate.

Indiscriminate management of digestate can be detrimental to the environment through nutrient release/runoff and volatile gas emissions, including ammonia and methane (Barampouti *et al.*, 2020). Existing and projected regulatory frameworks ensure safe digestate standards for disposal and reuse applications that protect public and environmental health. These standards regulate digestate characteristics, including odor, organic matter, degree of fermentation, pathogen presence (Siebielec *et al.*, 2018). Careful consideration of digestate management shows that related regulations are also impacted, including those implemented to address climate change, renewable energy, and fertilizer applications.

The digestate management technology can be broken down into practices that are implemented pre-digestion, where the focus is on the feedstock to improve digestion efficiency, and post-digestion, which is targeted toward the digestate volume reduction and quality improvement. The end-use applications dictate the extent of post-digestion management. Physical, chemical, and biological techniques have been explored for contaminant removal from the digestate. There is the potential that the current land application of the digestate becomes even more restricted. Upcoming research into digestate management has considered such applications as combined digestate treatment and microalgal cultivation, the digestate use as a solid fuel or co-combustion, and production of volatile fatty acids as a green carbon source (Gude, 2015). Presently, great attention is given to biogas yield during digestion and not as much to improve the quality of the digestate. An integrated approach toward biogas production and digestate quality needs to be implemented for progress toward a circular economy. Such practices will enhance the value of the digestate as a resource by providing opportunities for innovative use, meeting strict regulations that protect the environment, and potentially reducing costs associated with an application.

4.2 SOURCES AND CHARACTERISTICS OF ANAEROBIC DIGESTATE

The feedstock into the anaerobic digester of a WRRF is typically primary sludge, secondary sludge, or a mixture of both classes of sludge. Some facilities accept sludge or organic matter from other WRRFs, food or agriculture industries to improve biogas production in the digesters. Primary sludge contains grit, grease, scum, and sediments that are removed from municipal wastewater coming into the facility. Primary sludge tends to have more bioavailable organic matter. Secondary sludge also referred to as waste-activated sludge, comprises microorganisms from the biological treatment stage. Secondary sludge has fewer contaminants as most have been removed in the primary treatment, lower carbon, and high nutrient content. Before AD, primary and secondary sludge is mixed and thickened to improve digestion efficiency. The mixing stage has the practical benefit of reducing complexity in sludge management within the facility and enhancing biogas yield compared to sole secondary sludge digestion (Gherghel *et al.*, 2019).

The properties of the digestate resulting from municipal sludge are due to the properties of the municipal sludge itself and the operation of the digesters. The optimum pH for AD lies between 6.5 and 7.2. However, the extent to which the digestion has proceeded will cause volatile fatty acids to accumulate, reducing the pH or the production of ammonia and carbon dioxide, which will increase the pH. The temperature at which digestion is conducted will affect the population of pathogens that remain in the digestate (Appels *et al.*, 2008). Thermophilic conditions can achieve up to 3-log removal of pathogens. Up to 90% of the macro- and micro-nutrients that are found in the municipal sludge end up in the digestate. Examples of the contents of municipal sludge digestates are presented in Table 4.1. The inert nature of heavy metals and other inorganic contaminants results in their retention and concentration within the digester. However, the nutrients become more bioavailable after digestion. Organic nitrogen is converted into ammonia, while some of the phosphorous is converted into ortho-phosphorous in the digestate. In these forms, the nutrients are more accessible to plants if the digestate was to be used as fertilizer. A wide range of total solids (TS) concentrations is provided in Table 4.1 as the digestate is often thickened after digestion for more accessible transportation or in preparation of reuse applications.

4.3 FACTORS AFFECTING ANAEROBIC DIGESTATE CHARACTERISTICS

The physical, chemical or biological characteristics of anaerobic digestate depend on the feed sludge type, source water composition, and stabilization process. Biological stabilization processes such as AD produce a large quantity of digested sludge. Anaerobic digestate contains more than 90% water and needs to be dewatered for volume reduction (Mowla *et al.*, 2013; Yin *et al.*, 2004). The solid content of the digestate improves by about 20–30% that undergoes dewatering processes (Novak, 2006; Wu *et al.*, 2020). Dewatered sludge is called biosolids. The characteristics of biosolids depend on the conditioning mechanism and dewatering technology. Dewatering of anaerobic digestate minimizes the volume of sludge and improves handling properties. The need for disposal or utilization of biosolids makes it necessary to identify the factors that affect digestate characteristics. Solid

Table 4.1 Characteristics of typical municipal sludge digestate.

Ref.	pH	TS	TCOD	TN	TP	As	Cu	Mo	Ni	Zn	Pb
Rékási *et al.* (2019)	6.62	18.4% W/W		39 400 mg/kg	32 509 mg/kg	3.38 mg/kg	296 mg/kg	5.27 mg/kg	19.3 mg/kg	795 mg/kg	
Zuliani *et al.* (2016)	7.9	28 g/l	18 g/l	2750 mg/l, NH_3 53%	480 mg/l OrthoP 6.4%						
Hosseini Koupaie *et al.* (2019)	8.0	15.7 g/l	17.1 g/l	1300 mg/l NH_3 54%	1400 mg/l						
Pecorini *et al.* (2020)	8.12	20%					433 mg/kg		33.7 mg/kg	847 mg/kg	61.5 mg/kg
Cristina *et al.* (2019)	7.5–7.7	4.4–4.8%		7.4%, NH_3 21%	4.2% –5.8%	2.1–2.8 mg/kg	340–357 mg/kg		120–163 mg/kg	650–918 mg/kg	70–92 mg/kg

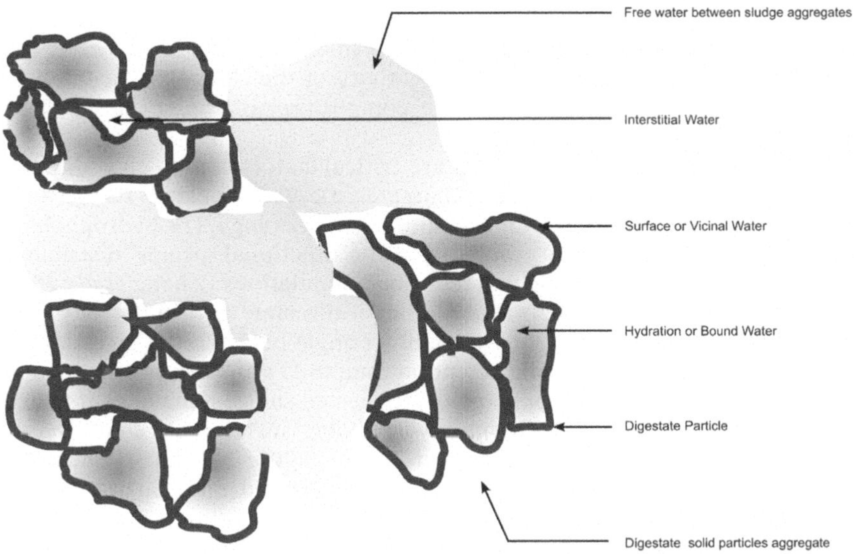

Figure 4.1 Different forms of water in biosolids. Adapted from Tsang and Vesilind (1990).

concentrations such as TS, volatile solids, suspended solids, whether the source wastewater is industrial or residence sludge, type of sludge are essential parameters influencing digestate characteristics. The digestate solids with a density of 1.0 g/cm^3 for organic and 2.3 g/cm^3 for inorganic solids affect the settling features and centrifugal properties (Casallas-Ojeda *et al.*, 2021). Sludge particle size has the highest effect on filterability and dewaterability.

Sludge with more super colloidal particles less than 1 μm is very difficult to dewater because it binds the filtration media and increases the resistance to flow (Qi *et al.*, 2017). Free water is the surrounding water to solids' particles – interstitial water trapped inside the sludge flocs by the capillary force. The vicinal or surface water firmly adhered particle surface, fibers, adhesive strength, and was impossible to remove by any mechanical force. Intercellular water or hydration water which is chemically bound can only be removed by thermal energy. Figure 4.1 demonstrates different forms of water in biosolids. Free water has the least binding energy of 0.28 kJ/kg, whereas interstitial water was the most acute binding energy of more than 5 kJ/kg (Kopp & Dichtl, 2000). The digestate sludge flocs consist of small channels and pores through which water flows inside the floc. For a 100 μm aggregate, the porosity of a floc ranges from 0.66 to 0.99 (Chakraborti, 2005). Only the entrapped interstitial water is involved in an advective flow through the sludge pores. It is directly removed by dewatering processes, whereas compact flocs with more porosity yield a greater force on the surrounding fluid than a loose floc. Polymer flocculated sludge yields a more compact sludge floc. Polymer flocculated floc exerts 12% advection flow, whereas advection flow is 53% for activated sludge (Tsou *et al.*, 2002).

Sludge flocs are negatively charged, and the negative surface charge is exerted by the sludge solids, polymers, and microbes (Morgan *et al.*, 1990). Carbohydrates (C) contribute to the negativity of the sludge surface, but the overall surface charge depends on the composition of the surface polymer (Sanin *et al.*, 2011).

Surface charge and hydrophobicity are critical factors of municipal digestate dewaterability. These impact the characteristics of digested sludge flocs, flocculation, settleable, dewaterability, and sludge rheology. The hydrophobicity of the digestate solid surface results from the functional groups' distribution variation caused by the protein and microbial populations (Zhang *et al.*, 2020). The contact angle is directly proportional to digestate hydrophobicity (Liao *et al.*, 2001; Zhang *et al.*, 2020). The contact angle between the digestate and water increased by 13° at a solid retention time (SRT) of 15 days; high contact angle resulted in higher hydrophobicity in digested sludge (Zhang *et al.*, 2020). Negative surface free charges imply hydrophobic surfaces, whereas positive values indicate hydrophilic surfaces (Gotoh *et al.*, 2003; O'Connell *et al.*, 2009; van Oss, 2003). Anaerobically digested sludge exerted more hydrophobicity with the negative surface free energy of −172 MJ/m^2 than waste activated sludge 416 MJ/m^2 (Zhang *et al.* 2020). The digestate with higher hydrophobicity will dewater easily (Wu *et al.*, 2017a, 2017b).

Activated sludge has extracellular polymer substances (EPS) that hold microbial colonies, which carries a net negative charge within a pH range of 5.0–7.0, which has a 7% contribution to the hydrophobicity (Busch & Stumm, 1968; Urbain *et al.*, 1993). The hydrophobicity for activated sludge increases at higher SRT and decreases with age of the sludge. Increased sludge age encounters starvation and a lack of oxygen that lowers the cell hydrophobicity (Palmgren *et al.*, 1998). The digestate hydrophobicity increases at high SRT (Liss *et al.*, 2002; Van den Broeck *et al.*, 2012).

Digested sludge has fertilized value containing macronutrients of NPK: 3.3–2.5–0.3% (Sanin *et al.*, 2011). Digested sludge resulting from industrial wastewater sources may contain heavy metals such as Al, As, Fe, Pb, Zn, Mn, Co, Cr (Su *et al.*, 2019). To keep the metal contaminants low in municipal digested sludge to 0.2–10 mg/kg dry solids, separating the industrial input is recommended than extracting metal from biosolids (Garg *et al.*, 2021).

4.4 DIGESTATE MANAGEMENT STRATEGIES

4.4.1 Pre-digestion treatment

4.4.1.1 Thickening

The thickening process improves solid sludge content and reduces the volume. It increases the concentration of solids in wastewater for subsequent operations. Thickening before AD minimizes the volume and improves the efficiency of the digestion and dewatering process. Thickened sludge helps to reduce anaerobic digestor volume and cost because anaerobic digesters are designed based on solid confinement. Suspension thickening involves settling suspended particles and then compressing settled solids under the influence of gravitational force.

The theory behind settling is that the suspended particles' settling rate should be faster than the hydraulic rise of the liquid overflow. Often high molecular weight polymers are used to form floc and increase particle settling rates. Particle drag coefficient and interparticle interference often impact particle settling rate. Sedimentation mechanics involves gravitational and hydrodynamic force, which ultimately influences settling. Settling rate is a function of solid concentration, settling velocity, fluid velocity, particle size, shape, and so on. The thickening ability of sludge is measured by the sludge volume index (SVI) (Brandt *et al.*, 2017).

$$\text{SVI} = \frac{\text{ml of sludge} \times 1000}{\text{suspended solids concentration, mg/l}}$$

Deflocculation, pin settling, and filamentous bacterial growth, often cause bulking in activated sludge. Increased C, organic N, and sulfide (S^{2-}) also cause sludge bulking (Spinosa & Vesilind, 2007). Sulfide promotes the development of filamentous bacteria, which makes sludges challenging to settle (Sanin *et al.*, 2011). Gravity thickening, gravity belt thickening, rotary drum thickening, and dissolved air floatation thickening are the popular thickening processes applied in municipal sludge treatment plants. Gravity thickening of sludge improves the solid concentration and decreases the volume by eliminating free water.

4.4.1.2 Sludge pre-treatment

Sludge can be pre-treated before AD to increase the solids loading, increase volatile solid reduction, increase biogas production, and produce class A biosolids. Pre-treatment results in increased hydrolysis. Pre-treatment methods can be chemical, physical, electrical, or thermal. The pre-treatment processes are applied to the sludges produced from secondary treatment processes as they cannot be digested easily in anaerobic conditions. Pre-treatment of these sludges is accomplished through mechanical shear, pressure drop, ultrasonic waves, or applying heat, chemicals, or enzymes.

The pre-treatment process's effect has a significant impact on sludge post-treatment processes. The result of the pre-treatment process is AD, enhanced dewatering, and pathogen reduction. Pre-treatment causes the disintegration of activated sludge, which is difficult to degrade due to its microorganism.

The ideal temperature for the heat-treatment process ranges from 40 to 180°C (Barjenbruch *et al.*, 2000; Kepp *et al.*, 2000). Carbohydrates and sludge fatty acids are readily biodegradable, but the proteins are not. Protein cell walls prevent it from hydrolysis. The protein cell wall can be destroyed by a low temperature of 60–100°C heat treatment, making the proteins accessible to biological degradation (Zhang *et al.*, 2018). Anaerobic digestate from a batch digestor was pre-treated by low-temperature heat treatment and dewatered using a centrifuge. The solid content of dewatered biosolids improved by 4.6% compared to the unpretreated digestate (Zhang *et al.* 2018).

Chemical pre-treatment is done by adding chemicals to enhance solubilization and requires low energy (Bordeleau & Droste, 2011). On the contrary, chemicals incur an extra cost, might cause harm to beneficial microbes, and cause corrosion to equipment. Chemical hydrolysis often forms intermediate composites that are non-biodegradable due to the increasing chemical dose (Neyens *et al.*, 2003). Chemical pre-treatment can be categorized into two subcategories alkaline and oxidative. Acids, alkali, ozone, hydrogen peroxides are often used for chemical pre-treatment (Tanaka *et al.*, 1997). These chemicals convert difficult-to-degrade compounds into readily degradable compounds. Alkali or water-soluble ionic salts such as calcium or magnesium hydroxide, sodium or potassium hydroxide are used for alkaline pre-treatment (Kim *et al.*, 2003). If the alkali dose is high, that might adversely affect the performance of anaerobic microbes (Mouneimne *et al.*, 2003). The thermal alkaline technique combines the thermal and low-dose alkali pre-treatment to avoid chemical overdose. The thermal alkaline process reduces the high-temperature requirement than thermal treatment to produce the same degree of solubilization (Carrère *et al.*, 2010). Oxidative pre-treatment that uses ozone or hydrogen peroxide (H_2O_2) is known as advanced oxidation processes (AOPs), which enhance sludge's dewaterability (Mustranta & Viikari, 1993). The oxidative pre-treatment process disintegrates the sludge structure by producing hydroxyl radicals (OH°). Hydrogen peroxide (H_2O_2) is an excellent oxidizer, but the oxidation power can be enhanced with a combination of thermal, ultrasound, ultraviolet, or other techniques (Eskicioglu *et al.*, 2008). The use of enzymes increases degradation by 10%, especially for treating primary sludges with high lignocellulose content (Dursun *et al.*, 2006).

Ultrasounds, mills, homogenizers are used for mechanical pre-treatment. Mechanical energy is applied in the form of pressure, translational or rotational energy. Mechanical energy causes tension and deformation in the disintegration's solid structure, resulting in cell disruption. Short retention time and increased gas production during the digestion process have resulted from the mechanical pre-treatment process.

The freezing and thawing change the sludge floc formation into a more dense structure. The freezing and thawing will reduce the bound water content and improve dewaterability.

Anaerobic sludge that underwent the pre-treatment process and disintegrated needs less polymer or flocculants than non-pre-treated sludge. Furthermore, it results in a lower solid content in dewatered sludge (Carrère *et al.*, 2010).

4.4.2 Post-digestion treatment

The sludge produced after AD possesses a challenge for final application and disposal due to its high content of water, organic matter, pathogen, and toxic contaminant. Several post digestion technologies are employed separately or integrated to overcome these challenges. Post-digestion technologies enhance dewatering, reduce and stabilize solid and pathogen content, and remove toxic metals. Some promising post digestion technologies include alkali stabilization, physicochemical iron-based oxidation, heat treatment, or further aerobic or AD of digested sludge. Post-digestion treatment technologies boost anaerobic digestate management in a more profitable and environmentally sustainable way.

4.4.2.1 Conditioning

The conditioning process, employed before thickening or dewatering, enhances the solid–liquid separation with an effort to convert digestate sludge into biosolids.

4.4.2.1.1 Conditioning mechanism

Conditioning methods can alter digestate sludge properties. One such method is chemical addition or chemical conditioning. The other is physical conditioning that uses temperature to change sludge's physical properties.

The digestate is a suspension of colloidal particles held apart by the repulsion of the same charged surfaces. Because of aversion, it lacks affinity between the particle surfaces, resulting in water restoration within the sludge floc structure and difficult water removal.

The use of chemical conditioners implies the charge neutralization concept in conditioning. Digestate sludge particles are negatively charged. The addition of an adequate amount of adsorptive and high molecular weight cationic chemical conditioners eliminates the charge electrostatic repulsion and allows coagulation. Coagulation involves destabilizing the sludge elements, decreasing the repulsive electrostatic forces between the components (López-Maldonado *et al.* 2014). With gentle stirring coagulated particles get flocculated (Guan *et al.*, 2016; Wei *et al.*, 2018). A gentle stirring disperses the conditioning chemical through the sludge during conditioning, which brings the particles and colloidal suspension together.

When added to the digestate sludge, high molecular weight cationic polyelectrolytes allow adsorbing more particles, thus forming a bridging structure that increases with time as the number of polymer bridges included grows (Bhosale & Berg, 2012). Crosslink between colloid particles and polymers occurs via covalent bonds such as esterification, addition, reaction, and bridge formation, crosslink through ionic bonds or combined a covalent or ionic bond (Hou *et al.*, 2012). Figure 4.2 demonstrates the digestate particle agglomeration through charge neutralization and flocculation of particles through adsorption on polymer and bridging mechanism.

Even though the sludge particles are anionic, a charge neutralization mechanism is not required for wastewater treatment technologies. Cationic, anionic, or non-ionic conditioners, even inorganic chemicals such as iron or aluminum hydroxide or a combination of both, are used for digestate conditioning.

4.4.2.1.2 Factors affecting conditioning

Type and properties of sludge, type, and properties of conditioner, solution physical properties establish the efficacy of the conditioning process. Molecular weight, charge types are essential factors for good conditioning. Organic and inorganic conditioner performance is greatly influenced by the solution's physical chemistry, pH, and alkalinity. pH affects the dissociation and solubility of the inorganic conditioners and the ionization of the organic polymer functional groups. If the pH value is lower than the acid dissociation

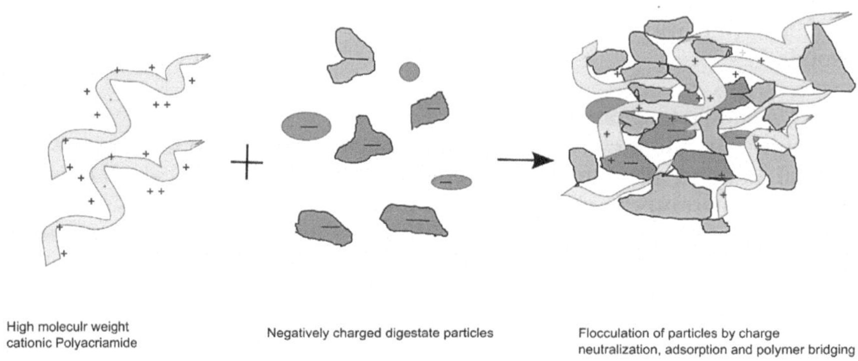

Figure 4.2 Polymer bridging and flocculation.

constant (p*K*a value), the ionization group of cationic electrolyte exhibits a cationic and stretched nature. Being stretched out is essential for polymers to make the bridging mechanism effective. The anionic or negatively charged polymers are most valuable when the sludge pH is higher than 8.0. Cationic or positively charged polymers are most useful at sludges with pH values below or slightly above 7.0 (McFarland, 2001). Alkalinity affects the dose and performance of inorganic chemicals. The higher the alkalinity, the higher is the inorganic chemical dose. The properties of digestate solids, such as particle size, the surface charge, surface area, solid concentration, and so on, influence the conditioner dose and efficacy.

If the solid concentration is low, the interaction between solids is also at a low level. Conversely, with increased solid concentration, interaction increases; therefore, excess coagulants are required to remove the external charge. Suppose the particle's surface charge increases, the conditioner dose increases. The digestate sludge particles retain hydration water on the surface and create a buffer that stops close contact with the particles (Shao *et al.*, 2009). Several organic conditioners such as polymers and inorganic conditioners like ferric chloride or lime are used to overcome these hydration effects (McFarland, 2001). These chemicals reduce or eliminate the repulsive force, which allows the solid particles to come closer and flocculate, facilitating water removal. Smaller particle size means a larger specific surface area resulting in higher negative surface charge and electrical repulsion between particles and higher frictional resistance to the water movement (Carrasco & Gao, 2019; Olböter & Vogelpohl, 1993). If the surface area is more, hydration tends to increase chemical demand and resistance to dewater. The conditioners' application helps particles combine either by charge neutralization or bridging mechanism. This accumulation forms large aggregates, increases particle size, is easy to settle, and dewater. The compressibility of the digestate causes deformation of the solid particles and the reduction of void spaces. When the void spaces reduce, the compressed portion reduces the water movement, lowering dewatering efficiency.

Conversely, if the sludge can be conditioned, porosity increases in the flocculant sludge matrix facilitate water movement and removal during the dewatering process (US EPA, 1987). A lower mixing rate disperses polymer well in the digestate. However, excessive mixing breaks the polymer flocs, leading to revered conditioning and breaking the long polymer chain, making them less effective.

4.4.2.1.3 Chemical conditioning

Inorganic chemicals such as $FeCl_3$, $Ca(OH)_2$, $FeSO_4$, and organic polymers are mainly used for chemical conditioning (Sri Bala Kameswari *et al.*, 2013). Organic polymer or polyelectrolytes create high molecular weight long-chain positive charges once dissociated in water. Polyelectrolytes worked in two mechanisms: they suppressed the surface charge of the solid particle and neutralized them. Secondly, they adsorb the particle, bridge the gap between particles, and create a floc matrix – molecular weight significantly impacts conditioning (Wu *et al.*, 2021). High molecular weight cationic polymers adsorb negatively charged particles only on the outside of the floc and form abridge. This adsorbed layer creates an electrostatic repulsion and steric hindrance, thereby preventing other polymer molecules from neutralizing and generating a large amount of intra floc water.

On the contrary, low molecular weight polymers penetrate the flocs, aggregate and neutralize the inner part of the floc, and lower water retention inside the floc (Du *et al.*, 2017). Polymer conditioning in conjunction with iron salts shows better dewaterability and lesser chemical consumption (Chitikela & Dentel, 1998; Novak, 2006). As the polymer dose increases, sludge dewaterability and filterability increase (Murujew *et al.*, 2020). The polymer dose has to be an optimum amount because overdosing causes disaggregation and redispersion of the flocs, decreasing filterability and dewaterability (Murujew *et al.*, 2020). The optimum dose for inorganic chemicals is much higher than the organic polymer. A study on digestate dewaterability testing showed that 2.7–2.9 kg/t polymer is required when added in conjunction with $FeCl_3$ to obtain an adequate level of dewaterability. However, to attain an equal level of dewaterability, at least 81 kg/t $FeCl_3$ is needed, which is 30 times more than the necessary organic polymer dose (Krishnamurthy & Viraraghavan, 2005). Table 4.2 describes a comparison of various chemical and physical conditioning methods, mechanisms, chemical doses, and energy requirements.

4.4.2.1.4 Physical conditioning

Physical conditioning can save the cost of chemical use but is not as efficient as polymer conditioning. Physical conditioning can be done by high temperature and pressure, freeze/thaw conditioning, and application of high energy such as electrical, mechanical, or sonic energy (Amrullah *et al.*, 2019; Liu *et al.*, 2020; Menon *et al.*, 2020; Sri Bala Kameswari *et al.*, 2013; Yang *et al.*, 2019; Zhang *et al.*, 2018). A combined high temperature (180–240°C) and pressure (20–27 bar) for 30 min breaks the cell wall and releases bound and interstitial water (Amrullah *et al.*, 2019; Barber *et al.*, 2012; Scherzinger & Kaltschmitt, 2021;

Table 4.2 Comparative analysis of various conditioning methods for enhancement of dewatering.

Technology	Mechanism	Special requirements	Chemical dosage/Energy requirement	References
Chemical conditioning				
Coagulation and flocculation	Agglomeration of digestate solid particles	Slow speed stirring	0.1–5 wt% dry chemical	Ayol *et al.* (2004), López-Maldonado *et al.* (2014), Wei *et al.* (2018)
Acid or alkali treatment	Regulate surface charge; cell wall destruction to release intracellular water	Anti-corrosion equipment	5–10 wt% dry acid or alkali	MacDonald *et al.* (2018), Shao *et al.* (2012), Wang *et al.* (2018)
Advanced oxidation	Lysis of water-holding EPS Cell rupture to release intracellular water	pH control; multi-step chemical doses	20–50 wt% dry oxidant	Ferrentino *et al.* (2020), Liu *et al.* (2020a, 2020b), Neyens and Baeyens (2003a, 2003b)
Enzymatic treatment	Degradation of EPS	pH and temperature control	1–5 wt% dry chemical	Chitikela and Dentel (1998), Dursun *et al.* (2006), Yu *et al.* (2008)
Physical conditioning				
Thermal treatment	Breakage of gel structure releases cell water content at 60–180°C for 10–30 min	The pressure bearing heat exchanger or vessel	Hydrothermal or thermal hydrolysis	Abelleira *et al.* (2012), Amrullah *et al.* (2019), Wu *et al.* (2020), Zhang *et al.* (2018)
Freeze and thaw	Formation of homogenous ice-crystal; cell swelling, floc structure changed to release intracellular water	Sludge is frozen at –15°C at a certain amount of time, then thawed at room temperature		Diak and Örmeci (2017), Dilek Sanin *et al.* (1994)
Non-chemical additive addition	Filter aids act as skeleton builders to improve permeability resists compression	Agitation and application of highly porous inert minerals fly ash, lime gypsum, or carbonaceous materials coal, wood chips, lignite, and so on	30–50 wt% dry matter	McFarland (2001), Sanin *et al.* (2011)
Electrical treatment	Electrophoresis, electromigration, electrolysis, electroosmosis	A special device, anti-corrosion polar plates	Electrodewatering Biosolid moisture content below 50%	Gao *et al.* (2021), Lv *et al.* (2019), Mahmoud *et al.* (2011), Masihi and Badalians Gholikandi (2018), Wu *et al.* (2020)
Sonication	Disintegration of colloidal bio floc, cell destruction for releasing intracellular water	A specialized device, challenging to scale up	600–800 kJ/kg dry matter is optimum sonication energy	Boruszko (2020), Cella *et al.* (2016), Pham *et al.* (2010), Yin *et al.* (2004)

Spinosa & Vesilind, 2007). Heat conditioning helps release all the bound water and pathogen destruction by no or low polymer addition. The Cambi process digested sludge heated by direct application of steam at 165°C and 6 bar for 30 min increases dewaterability by 60–80% (Barber *et al.*, 2017).

If the digestate is allowed to freeze naturally, it dehydrates the sludge by freezing the water associated with it, while freezing particles push away impurities and particulate matter. As the ice thaws, particles retain their new size, lower in density. As the freezing rate is prolonged, the interstitial water gets crystalline to squeeze into compact granules. As the ice thawed, it dewaters easily and achieves 50–70% solid concentration (Diak & Örmeci, 2017; Spinosa & Vesilind, 2007). Freezing and thawing function of freezing temperature, time, and duration meets the class B pathogen criterion by US EPA as it can significantly reduce the number of fecal coliform and other indicator microbes (Dilek Sanin *et al.*, 1994; US EPA, 1987).

4.4.2.2 Dewatering

Dewatering mainly reduces water content in the digestate and improves handling properties. In municipalities, dewatering of the digestate is primarily done by applying mechanical energy by using belt filter press centrifuges. Eighty percent of water in the digestate is bound to water which cannot be separated with conventional dewatering processes such as centrifuge dewatering or filter press due to strong bonding between water molecules with solid surface (Mowla *et al.*, 2013). The limiting factors for municipal digestate dewatering are settleability, filterability, and bound water content. Settleability of the digestate means low sedimentation rate of the colloidal particles; filterability represents compressibility of the solids (Mowla *et al.*, 2013). Bound water content chemically bound to the solid particles constitutes EPS (Mowla *et al.*, 2013). The pre-treatment methods before digestion or conditioning after digestion can improve dewatering.

4.4.2.2.1 Dewatering technologies

The sludge particle is surrounded by surface water and contains chemically held hydration water inside it and tightly held vicinal water to the surface of the particles. There are also trapped water within the floc which is the interstitial water. There is a critical difference between vicinal and interstitial water. When the physical confinement is removed, the interstitial water moves freely. However, vicinal water remains adhered to the solid surface.

The free water is easiest to remove from the digestate by thickening, drainage, filtration, or mechanical dewatering. The application of mechanical energy destroys the floc structure or compresses the sludge floc to remove the interstitial water. Application of mechanical energy cannot remove vicinal water and hydration water. In contrast, only thermal energy can remove vicinal water and is difficult to remove by mechanical dewatering unless conditioned with chemicals (Wu *et al.*, 2020). Activated sludges containing large internal surfaces consist of biopolymer, microbial cells, accumulated debris containing a high amount of non-removing water, and settle solid concentration.

The dewatering mechanism involves filtration and expression phases. Filtration removes water surrounding the particles, and expression involves squeezing out water from the particles. Several tests used water released during the filtration phase to characterize dewatering. Still, the water removed during the expression phase determines the solid content of the dewatered cake.

4.4.2.2.2 Sludge properties that affect dewatering

Organic and inorganic matter constitute the complex gelatinous structure of digested sludge. Studies examined various aspects such as particle size, flocs structure, EPS, and so on influence the dewatering characteristics of digestate (Mahmoud *et al.*, 2011; Zhou *et al.*, 2014). Bound water is the critical parameter for solid–liquid separation of sludge and biosolids. Mechanical force can remove free water (Vaxelaire & Cézac, 2004). Bound water can be of three types: interstitial, vicinal, and hydration water. However, the presence of bound water or hydration water, which is 3–8% of the total water content of sludge, makes sludges difficult to dewater but not always less bound water results in better dewaterability (Colin & Gazbar, 1995; Katsiris & Kouzeli-Katsiri, 1987; Liao *et al.*, 2000).

Microbes produce a polysaccharide (PS) from the particles' cell wall, called extracellular polysaccharides (EPS). Sludge flocs stratification was investigated and described as four layers, such as the pellet, tightly bound EPS (TB-EPS), loosely bound EPS (LB-EPS), and slime from the inner to the outer part (Yu *et al.*, 2008). Pellet and EPS form a capsule. EPS is tightly bound to the cell wall and loosely attached to slime. Mixed culture microbes sometimes cause protein (PN), nucleic acid, or lipids other than PS. Other than PS the rest of the components are intracellular. Extracellular PS is formed mainly from cell lysis and enzyme entrapment and consists of 40–90% of EPS (Dai *et al.*, 2017).

Researchers reported that soluble PNs remain mainly in the pellet up to 61% and TB-EPS (10%), while PS was dispersed equally in all the layers of EPS (Cao *et al.*, 2018). Figure 4.3, adapted from Wang *et al.* (2020) demonstrates EPS stratification of waste-activated sludge. PN is shifted from the pellet and TB-EPS fractions to the slime fraction (Li & Yang, 2007). Conversely, PS had no noteworthy transference trends (Carrasco & Gao, 2019; Shao *et al.*, 2009). Polysaccharide is easily degraded during AD, leaving PN content abundant in digested sludge slime EPS (Zhang *et al.*, 2020). PN has amino acid groups which retain 98% water in the pellet, making PN in the pellet to be hydrophilic (Cetin & Erdincler, 2004). Since PN possesses a high water-holding capacity; therefore, decreasing protein portion in sludge EPS could increase sludge dewaterability (Cetin & Erdincler, 2004; Jin *et al.*, 2004). During AD PN moves from pellet to slime fraction, but the overall PN is reduced in the slime fraction due to decreases in EPS content in the digestate. Digestates contain 10 mg EPS/suspended solids (SS) compared to 21 mg EPS/SS in raw sludge (Houghton *et al.*, 2000). EPS yield decreased by 15–43% in municipal and C content reduced to 4–28% in EPS of the digestate (Houghton *et al.*, 2000). Since PN is difficult to degrade than C or PS, PN remains abundant in slime EPS in the digestate, but a decrease in the overall EPS content results in the reduced amount of PN and PN/PS in slime EPS reduced in the digestate (Houghton

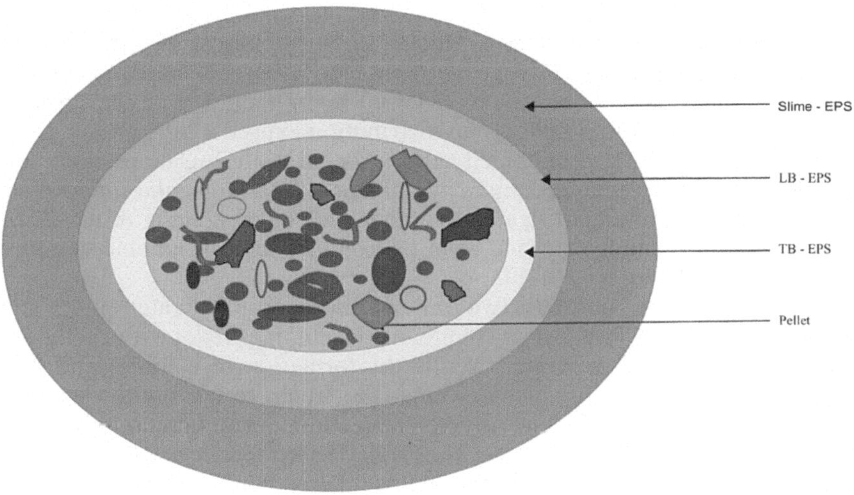

Figure 4.3 EPS stratification of waste-activated sludge. Adapted from Wang *et al.* (2020).

et al., 2000; Yu *et al.*, 2008). A decrease in the PN content and PN/PS in slime EPS of the digestate enhances dewaterability (Houghton *et al.*, 2000; Zhou *et al.*, 2014).

The surface charge affects the dewatering properties of digested sludge. PN content is also correlated to surface charge (Liao *et al.*, 2001; Morgan *et al.*, 1990). Protein (PN) contains a positive charge (Wu *et al.*, 2021; Zhang *et al.*, 2020). PN binds with anions such as carboxyl and phosphates', thus decreasing the overall surface charge (Dai *et al.*, 2017; Morgan *et al.*, 1990). Surface charge is inversely proportional to digestate hydrophobicity (Liao *et al.*, 2000). Suppose the surface charge is decreased, hydrophobicity increases (Zhang *et al.*, 2020). Increase in hydrophobicity improves the dewatering of the digestate (Zhang *et al.*, 2020).

The molecular function of PN is classified as binding, catalytic, and transporter activities. Among these three functions, catalytic activity and binding contribute by 51% and 40%, respectively, and dominate during sludge hydrolysis (Zhang *et al.*, 2020). The catalytic activity of PN in the EPS provides energy to the heterographic bacteria, which accelerates cell lysis and the destruction of EPS structure (Zhang *et al.*, 2015). This improves acidification during AD and the destruction of bonding between organic fractions with a water molecule (García-Bernet *et al.*, 2011). The cell lysis caused by PN catalytic activity releases mechanically bound water to free water, improving dewaterability (Kopp & Dichtl, 2000; Zhang *et al.*, 2019). PN related to the binding activity influences hydrophobicity and dewaterability. Cation and anion binding functional PNs are released during various stages of AD. Anion binding groups interact selectively and non-covalently with charged atoms or net negatively charged atom groups (Zhang *et al.*, 2020). Cation binding groups interact selectively and non-covalently with cations, charged particles, or net positively charged atom groups (Zhang *et al.*, 2020). Since anion binding, PN contains more free amino acid groups, which is the main reason for the sludge surface hydrophilicity (Cetin & Erdincler, 2004). Removal of anion binding proteins indicated increased sludge hydrophobicity, thereby improved digestate dewaterability (Zhang *et al.*, 2020). Surface hydrophobicity of the digestate increases due to the functional group distribution caused by PN and microbeś distribution. Surface hydrophobicity influences dewatering more than total hydrophobicity. Surface hydrophobicity influences intermolecular interactions, such as binding with small ligands or other molecules (Zhang *et al.*, 2020). Transformation of sludge surface from hydrophilic to hydrophobic enhances dewaterability (Wu *et al.*, 2017b).

EPS also contains microbes. Two or more microbes in activated sludge stick together by their EPS and form a colony. As the colonies react with each other, a larger structure floc is formed. While the formation of colonies and flocs, divalent cations (Mg^{2+}), trivalent cations (Fe^{3+}) help to bind with negatively charged digested sludge particles by neutralization, bridging, and polymer gel formation (Liao *et al.*, 2001; Zhou *et al.*, 2014). Cations reduce a negative surface charge, facilitate coagulation, improve flocculation and settling (Mikkelsen, 2002; Urbain *et al.*, 1993). Microbes and bacteria in EPS thus help flocculation contribute to the surface charge and settling properties (Houghton *et al.*, 2000).

Divalent cations enhance sludge dewaterability. Divalent cations Ca^{2+} and Mg^{2+} bind with a negatively charged chemical group (Eriksson & Alm, 1991). Divalent cations act like bridging agents in the EPS matrix (Nguyen *et al.*, 2007). Ca^{2+} has higher amounts and has higher binding capacity than Mg^{2+}. Therefore Ca^{2+} is a vital role player in floc strength and breakage (Guan *et al.*, 2012; Jin *et al.*, 2004; Park *et al.*, 2013). Activated sludge treatment plants, if monovalent cations are more than divalent cations, result in poor dewatering and affluent properties (Neyens & Baeyens, 2003a, 2003b).

Particle size is another important property that affects sludge or biosolids dewatering. Studies to explore the impact of particle size on sludge dewaterability show that smaller particles negatively affect dewatering characteristics (Carrasco & Gao, 2019; Olböter & Vogelpohl, 1993; Shao *et al.*, 2009). Thermal and ultrasound treatment generated more super colloidal particles with 1–100 μm size particles and significantly impacted dewaterability. They can blind filtration media and sludge cakes. The porosity of the cake and the filter media is further affected by blinding or pore blockage when particles are smaller than 40 μm (Novak *et al.*, 1988). Due to the high compressibility, an increase in mechanical stress reduces porosity in sludge. A decrease in the porosity increases the resistance to flow, thereby hampering free water flow through the cake and the filter (Qi *et al.*, 2011). Filterability of digestate decreases if particle size decreases; moreover, dewaterability of thermophilic anaerobically digested sludge is higher than mesophilic digested sludge as thermophilic digestate contains fewer fine particles (Karr & Keinath, 1978).

Suppose the flocs' strength and size can be increased, the filterability and compactibility of the municipal digestate can be improved as well (Cetin & Erdincler, 2004). An inverse relationship between bound water and floc size was established. Bound water content decreased linearly with increasing floc size of 20–100 μm (Liao *et al.*, 2001).

Floc size and floc size distribution are two important physical properties that influence dewatering. Larger flocs create an inter floc channel to release water efficiently. Studies show that the floc size distribution of digested sludge contains fewer small flocs. Floc size and floc size distribution are the results of dynamic equilibrium while transformation and formation of aggregates. Research shows the image sizing system of 1000–2000 flocs ranges between 8.3 and 614.0 μm equivalent spherical diameter, indicating that larger floc size and percent volume of large flocs has smaller capillary suction time (CST) (Spinosa & Vesilind, 2007).

4.4.2.2.3 Dewatering rates

The dewatering rate can be measured by filtration. During filtration, sludge cake is formed by the accumulation of small particles. The surface friction of the particles creates a resistance to the flow of water. The filtration rate is measured as SRF, which is inversely proportional to the cake permeability. SRF can be measured with the following equation (Christensen, 1983; Spinosa & Vesilind, 2007).

$$r = \frac{2PA^2b}{\mu w}$$

where $r = \text{SRF}\,(\text{cm/g})$, $P = \text{Applied vacuum}\left(\frac{\text{g}}{\text{cm sec}^2.}\right)$, $A = $ Area of the filter paper (cm^2), $b = $ slope of $\frac{\text{T1}}{\text{V1}}$ vs. V1 plot (s/cm^6), $\mu = $ viscosity of the filtrate $\left(\frac{\text{g}}{\text{cm s}}\right)$, $w = $ weight of dry solids per unit volume of filtrate (g/cm^3).

If the amount of small particles is high in compressible slurries, they cause binding of the cake, which significantly reduces the rate of filtration. Anaerobically digested waste-activated sludge exhibits a lower filtration rate due to a substantial amount of released microbes in the solution from the floc.

4.4.2.2.4 Effects of shear

The effect of shear on sludge depends on the intensity and time of shear. The product of shear time (t) and intensity (G) will influence the rate of polymer and the dose of polymer. Research studies reported G_t for filter press ranges from 8000 to 40 000 (Spinosa & Vesilind, 2007). Several factors such as polymer demand, polymer mixing, feeding, and pressing cycle of the filter press equipment to depend on G_t. Conditioning polymer dose is proportional to G_t. Sludges with high sodium concentrations are very sensitive to shear and break the flocs in activated sludge as the mixing or shear time increases. Digested sludges sensitive to shear should be dewatered using low shear devices such as belt filter presses (Spinosa & Vesilind, 2007).

CST measurement is an estimate of the shear ability and the ability for reflocculation. The shear strength required for digestate centrifuging is very high. Sludge viscosity can create a relationship between shear stress and shear rate. An effluent or thin sludge rheological exhibit is Newtonian. If the fluid is Newtonian, shear stress is directly proportional to the shear rate, where viscosity is the proportionality constant. Still, municipal sludges are non-Newtonian means shear rate or velocity gradient is not directly proportional to shear stress. The viscosity of the sludge is a factor of temperature, solid content, particle size distribution. Measuring the viscosity of centrate or filtrate can help determine the optimum polymer dose. When municipal sludge is conditioned with less than optimum polymer dose, filtrate or centrate viscosity is similar to water. But with increased polymer dose, the filtrate or centrate viscosity is proportional to the conditioner dose. The excess polymer not adsorbed by the sludge remains in the water phase. Therefore, optimum polymer dose can be determined only by measuring the filtrate or centrate viscosity, noting the polymer dose at which the viscosity will start increasing again. In the case or use of inorganic chemicals, filtrate viscosity remains the same as water viscosity.

Viscosity also represents the particle interaction magnitude in a suspension. Sludge viscosity exhibits a clear positive correlation with bound water content and CST (Niu *et al.*, 2013). The study result shows an increasing trend on digestate CST and bound water content with an increased viscosity from 3.8 to 11.0 mPa s (Jin *et al.*, 2004). If sludge flocs experience high shear, flocs are broken, small particles are released, and increase the resistance to filtration as the small particles clog the filter cake (Shao *et al.*, 2009). Compressibility and porosity are crucial rheological properties of dewatered cake which impact filtration's water removal efficiency. Lower compressibility means higher permeability (Qi *et al.*, 2011). Low compressibility accompanies the porous structure of dewatered sludge, inhibits the deformation, and promotes water removal (Qi *et al.*, 2011).

4.4.2.2.5 Characterization of dewatering

Dewatering is often characterized by the dry solid content of the biosolids. Popular dewatering indices are specific resistance to filtration (SRF) and capillary suction time (CST). Filterability and dewaterability are parameters to measure how easily the sludge releases water. SRF and CST can measure filterability and dewaterability. CST is a rapid test that depends on gravity describes the water discharge velocity of the digested sludge. CST apparatus can calculate the time required for the water portion of the conditioned sludge to travel to a fixed distance of 1 cm in a filter paper. Capillary suction of a standard and thick filter paper draws out water from a sample of conditioned sludge. The time required by the water to travel 1 cm is the CST in seconds. CST depends on TS. A short CST less than 20 s indicates dewaterability readily, whereas a long CST expresses poor dewaterability (To *et al.*, 2016). SRF test measures the resistance of the sludge to release water through a porous media under applied vacuum or pressure (Christensen, 1983). The higher the SRF, the more difficult it is to dewater biosolids. Sludge with low SRF values of 10^{10}–$10^{12.5}$ min/kg is easy to dewater, $10^{12.5}$–$10^{14.5}$ min/kg on an average is dewaterable, while high SRF values of $10^{14.5}$–10^{15} min/kg indicates the difficulty to dewater (To *et al.*, 2016).

4.4.2.3 Several dewatering processes

Centrifuge, belt filter press, and chamber filter press are the major dewatering processes used for the dewatering of municipal digestate. Vacuum filtration is mainly used for laboratory analysis based on filtration and cake formation.

The working press of the belt press depends on the digestate distribution and the number of rollers used for pressing. Three operating steps for the belt press include free drainage, wedge zone dewatering, and expression zone dewatering (Wang *et al.*, 2021). Most of the water was removed from the first two phases with appropriate polymer doses. The centrifuge is the most widely used dewatering equipment in municipal sludge treatment facilities. Centrifuge performance mainly depends on the feed rate, torque, polymer dose, and centrifugal force (Christensen *et al.*, 2015). The centrifuge operates on three principles. The first is that clarification means separating solids from the liquid, consolidating the solids, and transmitting the solids to the emission point. The centrifuge's overall performance depends on the performance of these three operations on water removal and solids capture in cake solids. Cake solid concentration and centrate quality increase dramatically with the increase in torque. Table 4.3 describes a comparison among several dewatering processes.

Several technologies such as electroosmosis are combined with mechanical dewatering to improve the dewatering performance. An applicant of electrical current through sludge cake during dewatering increases water removal. The possible explanation could be the applied current moves the particle double-layer toward an opposite electrical charge that causes further water removal (Wacławek *et al.*, 2016). Whatever be the dewatering process, researchers have shown that filter cake solid content can be doubled by applying 20 V across the cake during dewatering (Cai *et al.*, 2019). Vicinal water retained by sludge particles remains in the sludge cake and cannot be removed by pressure filtration. Higher voltages in electroosmosis cause the vicinal water to be released, resulting in increased water removal (Gao *et al.*, 2021).

4.5 DISINFECTION OF BIOSOLIDS

The best possible way to apply biosolids is to use them in landfills. Biosolids can be a good source of nitrogen and phosphorus. At the same time, biosolids' application poses a risk on the land application because of heavy metals and pathogens such as viruses, bacteria, protozoa, helminth, and parasitic eggs. Infectious pathogens present in biosolids can contaminate the land and crops. Different stabilization processes such as AD can significantly reduce the pathogenicity. These are called the processes to significantly reduce pathogens (PSRP), but intestinal parasites or parasitic eggs remain unaffected by these stabilization processes (Boczek, 2019). Further stabilization is required for complete inactivation. These are called processes to reduce further pathogens (PFRP) (Boczek, 2019). These are aimed to reduce the level of the pathogen but not achieve complete disinfection.

US Environmental Protection Agency (US EPA) published the standards for applying and disposing of municipal digestate and biosolids called USEPA 1993:40 CFR part 503 subpart D. This standard regulates the specific level of pathogens in biosolids before disposal. US EPA regulations classify biosolids as class A and class B and indicate the density of pathogens in biosolids as pathogen numbers per unit mass. Class A biosolids are ready to use wanting of any additional restrictions. Biosolids have fecal coliforms, salmonella, enteric

Table 4.3 Comparison of different dewatering processes (Metcalf & Eddy, Inc., 2003; Spinosa & Vesilind, 2007; Turovskiĭ & Mathai, 2006).

Dewatering equipment	Cake solids (%)		Solids capture efficiency (%)		Advantages	Disadvantages
	Range	Typical	Range	Typical		
Centrifuge	15–35	24	85–98	92	• Produces dry biosolids cake • Clean apparatus and minimum odor problem • Less space required • Quick start-up and shut down • Simple set up • It does not need constant operator attention • Small capital cost to capacity ratio	• Requires grit removal, digestate sludge, and sludge grinding • Relatively high suspended solid concentration in the centrate • Elevated operating cost • Needs skilled maintenance personnel
Vacuum filter	15–30	20	90–98	95	Skilled personnel not required to operate and maintenance Continuous operation Low maintenance requirement	Maximum strength required per unit of dewatered biosolids Constant operator attention required Vacuum pumps create noise
Belt filter press	18–30	23	90–98	95	Comparatively low capital and operating cost Easy operation, maintenance, and shutdown	Low levels of dry solid obtained Sensitive to digested sludge property Delicate to polymer type, dose, and rate Significant consumption of wash water
Pressure filter press	20–50	38	90–98	95	The high cake solids content Efficient capture of solids Proper dewatering equipment for sludges difficult to dewater	Batch process High operating cost, high labor cost Skilled maintenance workers required Inorganic chemicals require a conditioning process; these chemicals produce further solid
Drying bed and drying lagoons	25–30				No chemical consumption, low cost and energy consumption minimal capital cost, applicable where there is available land, organic matter is further stabilized in lagoons, not susceptible to sludge variations Retain better solid content than other mechanical methods	Large area required Stabilized sludge is required Design requires to consider climate effect Sludge cake removal is labor-intensive Odor potential
Electro-dewatering	25–50				Automatic dewatering, good results for difficult sludge and biosolids, better odor improvement and pathogen killing	Batch operation Moderate-to-high capital cost New technology electricity-dependent Difficult to predict performance without bench-scale testing

viruses, and viable helminth below the detectable level by the PFRP processes or heat-treatment processes to meet class A biosolids criterion (Boczek, 2019; EPA, 2020). Class A biosolids will have fecal coliform density <1000 MPN/g dry solids, salmonella density <3 MPN/4 g dry solids, enteric viruses <1 PFU/4 g dry solids, viable helminth ova <1 ova/4 g dry solids (EPA, 2020; US EPA, 2015). Biosolids are detectable but reduced by PSRP processes to levels that do not threaten humans, animals, or crops. Class B biosolids are designated when fecal coliform limits are less than 2.0×10^6 MPN/g TS or less than 2.0×10^6 CFU/g TS (EPA, 2020; Rhodes *et al.*, 2015; US EPA, 2015).

Active disinfection involves destroying microbes' cell structure to make it a non-viable or replication mechanism. Heat or alkali treatment processes or biological processes reduce the vector attraction reduction (VAR) potential. These processes break down 38% of volatile solids (Metcalf & Eddy *et al.*, 2014). They also reduce existing nutrients that can instigate microbial activities or initiate odor-producing potential (Adams *et al.*, 2008; Wéry *et al.*, 2008). Microbes lose the replication capability if the cell structure is broken, but chemical or alkaline treatments of at least pH 12.0 stop the cell multiplication process (Wang *et al.*, 2021). If the chemicals are removed or the physical condition changes, microbes that survived might multiply again.

To meet the pathogen and VAR requirements, the US EPA defines PSRP and PFRP as treatment methods before land application of biosolids. PSRP decreases but does not eliminate the chances of pathogen contamination. Aerobic digestion, AD, drying, composting, and lime stabilization can be named as popular PSRP processes. Aerobic digestion biosolids are mixed with air or oxygen at 20°C for 40 days or 56°C for 60 days (McFarland, 2001). In drying, biosolids are dried in sand beds or basins for three months. In AD, biosolids are digested in the absence of air for 35–55°C for 15 days (Flemming *et al.*, 2017). While composting, the temperature of the biosolids is raised at 40°C for five days or 55°C for 4 h and kept in a vessel, static aerated pile, or windrow composting (Wéry *et al.*, 2008). For lime stabilization biosolids, pH is raised at pH 12 using lime and maintained a reaction time of 2 h (Huang *et al.*, 2018).

While PSRP has potential for pathogen transmission, PFRP reduces pathogen below the detectable levels and eliminates the pathogen-associated restriction to land application. PFRP processes recommended for biosolids disinfection include heat drying, heat-treatment, composting, thermophilic aerobic digestion, beta ray irradiation, gamma-ray irradiation, pasteurization, and lime treatment.

Composting is a biological treatment done in a vessel or static aerated pile at 55°C for three days or windrow composting 55–75°C for 15 days (Spinosa & Vesilind, 2007). Temperature 55–75°C is high enough to destroy most of the pathogen. Direct or indirect hot gases are applied to biosolids at 80°C that kills pathogens and reduces the volume by 10% (Spinosa & Vesilind, 2007). Thermal heat treatment differs from heat drying in that the temperature is 180°C for 30 min (Barber *et al.*, 2012; Zhang *et al.*, 2018). The thermophilic aerobic digestion liquid digested sludge was treated at 55°C for 10 days in the presence of air (Wacławek *et al.* 2016). Biosolids can be radiated by direct ionization of particles generated in accelerators are called beta rays or indirect electromagnetic ionizing radiation of radionuclides or isotopes called gamma rays. Radiation interrupts cell division. Viruses are more resistant to radiation than bacteria due to their smaller size. The influence of radiation depends on their development stages. Zygotes are more resistant to radiation than morula or blastula stages of eggs. Dry media requires more radiation than liquid sludges. Beta rays of 1 Mega radiation (Mrad) at ambient temperature and gamma rays of 1.0 Mrad from certain isotopes Cobalt-60 or Cesium-135 at room temperature can radiate biosolids (Metcalf & Eddy *et al.*, 2014). Pasteurization is heating sludge at 70°C and retaining the elevated temperature for 30 min (Spinosa & Vesilind, 2007). Digested sludge or biosolids can be heated by hot water heat exchange, direct steam injection, and so on. The combined AD and pasteurization and storage SRT of 6 days can eliminate pathogen regrowth probabilities (Brisolara & Bourgeois, 2019). The quick lime (CaO) or hydrated lime $Ca(OH)_2$ are utilized to increase the pH of the biosolids to 11.0 or higher in lime treatment to kill pathogens (Brisolara & Bourgeois, 2019). The application of lime also, when reacted with water, generates heat about 70°C. Studies show that a pH value above 11.5 with 5–7 kg/m^3 sludge suspension kills *E. coli* and *Salmonella* spp (Metcalf & Eddy *et al.*, 2014). Destruction of a pathogen is a function of time along with raising pH. Gram-negative bacteria, parasites like thin shell ova are destroyed quickly, whereas tuberculosis causing *Mycobacterium* and thick shell ova can withstand lime for a long time (Metcalf & Eddy *et al.*, 2014).

4.6 STORAGE AND HANDLING

Dewatered municipal sludge cake or biosolids contain 15–40% solids that do not flow by gravity in a pipe (Fitzmorris *et al.*, 2009). Biosolids need to be transported by mechanical conveyors. Sometimes dewatered cake from a centrifuge drop directly into a collectible container and then moved to storage or the final disposal site. Often reciprocating piston hydraulic pumps or progressive cavity pumps are used to pump dewatered cake. Pumps offer better odor control because the pump is traveled through an enclosed pipe. Dewatered cake can be a pump with 1400–7000 kPa pressure (Spinosa & Vesilind, 2007). Belt conveyors or screw conveyors are often used to transport dewatered and dried municipal biosolids. Belt conveyors do horizontal conveying and have a width of not more than 40 inches (Spinosa & Vesilind, 2007). Free-flowing dried biosolids in the form of pellets are not recommended to convey through a conveyer belt inclined to an angle of more than 20° (Spinosa & Vesilind, 2007). The conveyor belts should have sidewalls or cover to convey dried palletized biosolids if required. Screw conveyors can convey dried biosolids very well and are mounted on a shaft or without a shaft. Screw conveyors are installed horizontally and inclined at less than 45° (Spinosa & Vesilind, 2007).

Vertical bins or hoppers are used to store biosolids. Multiple hoppers with a 400 m^3 capacity are typically utilized for biosolids storage in municipal wastewater treatment plants (Turovskiĭ & Mathai, 2006). Hoppers are covered and have blowers, scrubbers, and ductworks to control orders. At the bottom of the hoppers, biosolids are discharged through screw augers. Sometimes biosolids are stored in open or enclosed concrete pads.

4.7 APPLICATIONS OF BIOSOLIDS

Dewatered municipal biosolids can be used in landfills or land. Biosolids can be applied to specific lands such as agricultural, forest, dedicated disposal sites, or disturbed land. For the land application of biosolids, several factors such as sunlight, soil microbes, desiccants combine to inactivate residual pathogen and other toxic organic matters. Trace metals of biosolids are trapped in the soil matrix. Nutrients present in the biosolids are taken up by plants or converted into biomass. Biosolids have to meet at least class B requirements for pathogen and vector attraction to meet the eligibility for agricultural land application, but to be applied to the lawn or home gardens or to be sold or given away in bags; biosolids must meet class A criterion (Boczek, 2019; EPA, 2020; Rhodes *et al.*, 2015). Biosolids can be applied to the land in the same way as animal manure is applied. Biosolids with a solids percentage of 20–30 can be spread over land using tractors (Lowman *et al.*, 2013). Biosolids regulation 40 CFR Part 503 covers landfilling of biosolids. Anaerobically digested sludges are reduced in volume by centrifuge dewatering oven palletized for better transportation. Leachate control is required for the disposal of biosolids in dedicated sanitary landfills. Often the sanitary landfill is compacted with a tractor or roller and wrapped with 14 inches layers of clean soil. New solid wastes are used as a covering that minimizes odor or flies.

Applying biosolids to agricultural land use is advantageous because the organic matter enhances soil composition, moisture, water-holding capacity, infiltration of water, and aeration of the soil. The macro-nutrients such as potassium, nitrogen, phosphorus, and micronutrients such as iron, manganese, copper, zinc, and so on, support plant growth and reduce the chemical fertilizer requirements (Möller & Müller, 2012).

4.8 RESOURCE AND ENERGY RECOVERY OPTIONS FROM BIOSOLIDS

There is about five times more energy in the wastewater than is needed for the treatment, with a total energy potential of 851 trillion BTU/year (Radke, 2017). For decades, researchers have conducted energy recovery from wastewater, developing many methods that have led to renaming of certain WWTPs to water resource recovery facilities (Pecorini *et al.*, 2020). As seen in Figure 4.4, the by-product of each method is significantly different. The quality of biosolids, the energy produced from biosolids, and public perception and regulation are the main drivers for the choice of recovery options and the potential of energy recovery from wastewater biosolids (NACWA). Biosolids are not used in their full potential using the conventional methods despite partial nutrient and energy recovery. Furthermore, the presence of pathogens, hormones, antibiotics, heavy metals, and persistent organic pollutants in the sludge challenges the continuation of the current disposal methods. On the other hand, thermochemical methods such as incineration, pyrolysis and gasification

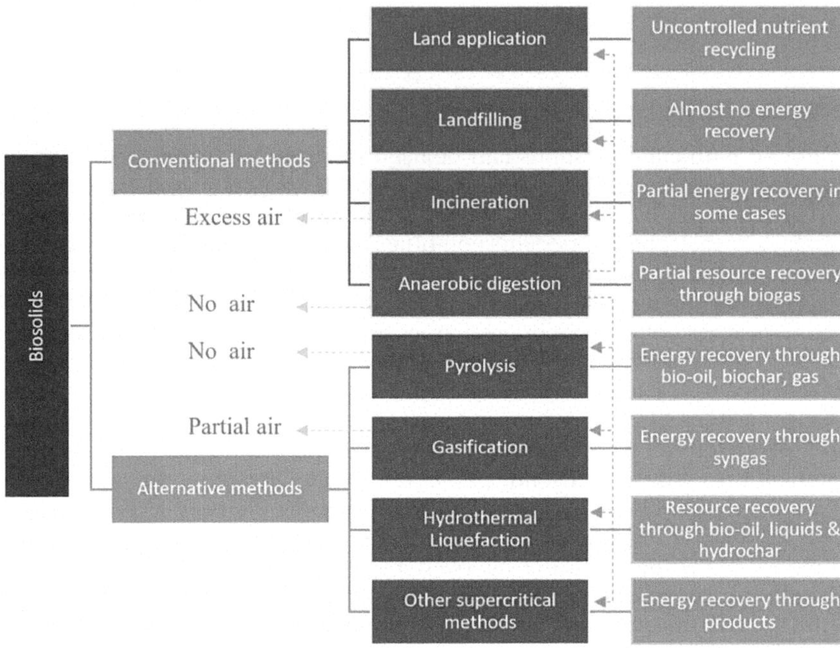

Figure 4.4 Conventional and alternative resource recovery methods for biosolids and their energy and nutrient recovery extent. Dashed lines indicate the digestate of the AD used for other processes. Adapted from Bora *et al.* (2020).

at relatively high pressure and temperature produce residues with high heat value (Sanin *et al.*, 2011).

Land application is one of the most established and prominent methods of sludge disposal. Since this method does not require any special equipment except the transportation and spreading vehicle and potentially returns the nutrients to the soil, it has been a popular choice for wastewater treatment plants. However, the challenge is that the nutrient ratio in the digestate is always the same, while the excess amount of nutrients causes eutrophication and over-fertilization. In addition, concerns have been raised about the presence of harmful substances such as pathogens, heavy metals, and polycyclic aromatic hydrocarbons (PAH) at different concentrations in the land-applied sludge (The Water Resource Foundation, 1975).

Combustion or incineration is the best established and most commonly used technology for converting biosolids into heat. Biomass fuel is burnt or oxidized in excess air during combustion to produce heat. The first combustion stage involves the evolution of combustible vapors from the biomass, which burns like flames. The hot combustion gases are sometimes used directly for product drying, but usually, they are passed through a heat exchanger to produce hot air, hot water, or steam. The most significant advantage of incineration is the reduction in the volume up to 70%, the destruction of pathogens and toxic organic compounds owing to the high operating temperatures, and the odor reduction (Al Seadi *et al.*, 2012).

Pyrolysis thermally decomposes the volatile components of biosolids in a temperature range of 200–800°C in the absence of the air. The by-products are biochar, a charcoal that can sequester carbon, and increase plant growth, py-gas used as a fuel, and a light non-aqueous liquid called pyrolysis oil (py-oil) that is used as a liquid fuel after conditioning and an aqueous pyrolysis liquid (APL) that with no known use up to date (Liu *et al.*, 2020a, 2020b). Advantages of biosolids pyrolysis include volume reduction, pathogen and emerging contaminant concentrations minimization, and a value-added soil amendment (Bora *et al.*, 2020).

Gasification needs higher temperature input from 700 to 1400°C with little air. Both volatile components and non-volatile carbon char from pyrolysis are converted into syngas during gasification. During gasification heat (generating power and process heat), high-value fuels and chemicals, including synthetic gas (syngas), renewable methane, hydrogen, and synthetic liquid fuels, are produced (NACWA, 2010). However, high water content, ash content, tar formation tendency, and contaminants such as sulfur and nitrogen are the main challenges associated with gasification (Kumar *et al.*, 2017). Furthermore, given that the moisture content of less than 35% is required for the gasification, considerable energy is needed to dry the biosolids (Radke, 2017).

4.9 REGULATIONS FOR ANAEROBIC DIGESTATE MANAGEMENT

The European Union (EU) for the circular economy and bioeconomy strategy is the foundation of sustainable development for resource efficiency. AD is one of the leading technologies contributing to achieving these goals. Worldwide AD has been used on a small and large scale with different capacities but has a vast difference in the adoption method. Policy drivers, socio-economic barriers, existing infrastructure, and technology availability and reliability are among the main factors impacting AD's implication around the world. Consequently, many regulations have been in place to manage the digestate from AD in different regions.

Policies that mainly impact the anaerobic digestate management are the comprehensive agricultural and waste management policies. Since serious environmental issues and health problems can arise from the discharge of the digestate from biosolids through air and water, facilities and farmers must meet specific requirements from regulatory agencies. These regulations include but are not limited to the digestate quality before disposal (e.g., class A or class B) and the level of toxic material. Further, regulations mandating the diversion of organic materials from landfills raise the potential application of the AD and digestate for land application and energy recovery.

Recently, the use of digestate as a fertilizer in agriculture has been regulated by the EU Fertilizer Regulation, promoting the utilization of the digestate from recycled biowaste, not including the digestate from the wastewater sludge. Preliminary studies on EU Fertilizer Regulation claim higher toxicity by the existence of more pronounced traces of organic pollutants than those derived from source separation. Furthermore, the issue associated with heavy metals, pharmaceutical residues, and personal care product residues is mentioned. The sludge digestate is currently directed to landfills or waste-energy plants.

Nonetheless, regulators have initiated many incentives for the growth of AD technology. These incentive programs include feed-in tariffs (FiTs), credits for carbon reductions, credits for renewable energy, credits for renewable transportation fuel, credits for nutrient load reduction, payments for producing renewable heat, and tax exemptions. These incentives have been introduced as the traditional agricultural business practices managing the digestate which cannot recover the high capital and long-term maintenance and operation costs.

4.10 RECENT DEVELOPMENTS

Ten mega electron volt (MeV) electronic beam is highly efficient at disinfecting anaerobic municipal digestate. A recent study shows that biosolids treated with 10 MeV electronic beams do not show pathogen regrowth in municipal digestate. Since chlorine dioxide and iron compounds have disinfection ability, 10 MeV electronic beam combined with ferrate enhances biosolids pathogen disinfection and stabilization (Pillai & Reimers, 2010). Compounds like volatile organic sulfur compounds (VOSC), nitrogenous compounds such as ammonia, volatile fatty acids, odorous volatile aromatic compounds (OVAC) are the source of odor in ADB and biosolids cake (Gao *et al.*, 2017; Gruchlik *et al.*, 2017). A lab-scale and a full-scale study were conducted to identify the cause of biosolids odor and odor reduction in biosolid cake. Researchers investigated the effect of cation addition, the impact of AD parameters such as mixing, detention time, temperature on biosolids cake odor production. The study results demonstrate a strong correlation between total volatile organic sulfur (TVOSC) concentrations on anaerobic digestate and dewatered cake odor. TVOSC increased with the iron content of raw sludge. During centrifugation, shear force releases iron-bound biopolymer, which causes odor in the dewatered cake. Centrifuge parameters such as torque and bowl speed affect the emission of TVOSC from the dewatered biosolid cake. Increased SRT during mesophilic AD resulted in decreased TVOSC release. Several advanced digestion processes, such as the microsludge cell lysis process during digestion and advanced oxidation post digestion or alum conditioning before dewatering, were employed to lower the odor of biosolids (Adams *et al.*, 2008). A combination of ferric chloride and poly aluminum chlorohydrate at a concentration of 4000 and 4500 mg/kg TS and anaerobic digestor yields biogas with a 93% reduction in volatile sulfur compounds, 82% improved fecal coliform deactivation, and improved dewaterability (Akgul *et al.*, 2017). Anaerobically digested municipal biosolids treated with Fenton's reagent improved dewaterability to a great extent. Biosolids treated with Fe^{2+}: H_2O_2 in 1:10 ratio 36 mM Fe^{2+} and 360 mM H_2O_2 lead to SRF and CST reduction by 99.8% and 98.8%, respectively, with a value 1.04×10^{12} min/kg and 19 s indicating good dewaterability ( ).

4.11 CONCLUSION

The performance of dewatering of municipal anaerobic digestate is challenging to predict. The efficiency of the dewatering process depends on sources of sludge, sludge particle characteristics, dewatering process adopted, dewatering device performances, and so on. Dewatering of the digestate is a costly process. Therefore, the enhancement of dewatering process performance has been a concern to modern researchers. The conditioning process as a post-treatment technology is adopted before the dewatering process. The use of polymer is expensive and, excess polymers can pose an ecotoxicological threat to the aquatic environment (Hamaide *et al.*, 2014). Therefore, other divalent or trivalent cations are added during conditioning, reducing the polymer demand. Biosolids regulations and biosolids reuse potential create the necessity for disinfection of biosolids. Advanced conditioning methods such as electro dewatering, peroxide, or persulfate oxidation are often introduced to improve the quality of biosolids.

REFERENCES

Abelleira J., Pérez-Elvira S. I., Portela J. R., Sánchez-Oneto J. and Nebot E. (2012). Advanced thermal hydrolysis: optimization of a novel thermochemical process to aid sewage sludge treatment. *Environmental Science and Technology*, **46**, 6158–6166, https://doi.org/10.1021/es204203y

Adams G. M., Witherspoon J., Erdal Z., Forbes B., McEwen D., Hargreaves R., Higgins M. and Novak J. (2008). Biosolids Processing Modifications for Cake Odor Reduction (Phase 3 of Identifying and Controlling the Municipal Wastewater Environment). IWA Publishing, UK. https://doi.org/10.2166/9781780403694

Akgul D., Abbott T. and Eskicioglu C. (2017). Assessing iron and aluminum-based coagulants for odour and pathogen reductions in sludge digesters and enhanced digestate dewaterability. *Science of the Total Environment*, **598**, 881–888, https://doi.org/10.1016/j.scitotenv.2017.04.141

Al Seadi T., Lukehurst C., Al Saedi T., Lukehurst C., Al Seadi T. and Lukehurst C. (2012). Quality management of digestate from biogas plants used as fertiliser. IEA Bioenergy, Task 40.

Amrullah A., Paksung N. and Matsumura Y. (2019). Cell structure destruction and its kinetics during hydrothermal treatment of sewage sludge. *Korean Journal of Chemical Engineering*, **36**, 433–438, https://doi.org/10.1007/s11814-018-0212-8

Appels L., Baeyens J., Degrève J. and Dewil R. (2008). Principles and potential of the anaerobic digestion of waste-activated sludge. *Progress in Energy and Combustion Science*, **34**, 755–781, https://doi.org/10.1016/j.pecs.2008.06.002

Ayol A., Dentel S. K. and Filibeli A. (2004). Use of drainability and filterability simulations for evaluation of oxidative treatment and polymer conditioning of sludge. *Water Science and Technology*, **50**, 9–16, https://doi.org/10.2166/wst.2004.0521

Barampouti E. M., Mai S., Malamis D., Moustakas K. and Loizidou M. (2020). Exploring technological alternatives of nutrient recovery from digestate as a secondary resource. *Renewable and Sustainable Energy Reviews*, **134**, 110379, https://doi.org/10.1016/j.rser.2020.110379

Barber B., Lancaster R. and Kleiven H. (2012). Thermal hydrolysis: the missing ingredient for better biosolids? *Water & Wastewater International; Houston*, **27**, 36–38.

Barber B., Nilsen P. J. and Christy P. (2017). Cambi solidstream®: thermal hydrolysis as a pre-treatment for dewatering to further reduce operating costs. Presented at the Water Environment Federation Technical Exhibition and Conference 2017, WEFTEC 2017, pp. 919–932.

Barjenbruch M., Hoffmann H., Kopplow O. and Tränckner J. (2000). Minimizing of foaming in digesters by pre-treatment of the surplus-sludge. *Water Science and Technology*, **42**, 235–241, https://doi.org/10.2166/wst.2000.0215

Bhosale P. S. and Berg J. C. (2012). The dynamics of polymer bridge formation and disruption and its effect on the bulk rheology of suspensions. *Langmuir*, **28**, 16807–16811, https://doi.org/10.1021/la3039297

Boczek L. (2019). EPA 40 CFR Part 503 Regulations Biosolids 101: Pathogen and Vector Attraction Reduction Regulations. Presented at the Office of Water Webinar series on Biosolids, NA, Ohio.

Bora R. R., Richardson R. E. and You F. (2020). Resource recovery and waste-to-energy from wastewater sludge via thermochemical conversion technologies in support of circular economy: a comprehensive review. *BMC Chemical Engineering*, **2**, 1–16.

Bordeleau ÉL and Droste R. L. (2011). Comprehensive review and compilation of pretreatments for mesophilic and thermophilic anaerobic digestion. *Water Science and Technology*, **63**, 291–296, https://doi.org/10.2166/wst.2011.052

Boruszko D. (2020). Influence of sonication on changes in micro-and macro-elements content and availability of organic matter in stabilized sewage sludge used for agricultural purposes. *Desalination and Water Treatment*, **199**, 57–65, https://doi.org/10.5004/dwt.2020.25848

Brandt M. J., Johnson K. M., Elphinston A. J. and Ratnayaka D. D. (2017). Chapter 13 – energy use, sustainability and waste treatment. In: Twort's Water Supply (7th ed.), M. J. Brandt, K. M. Johnson, A. J. Elphinston and D. D. Ratnayaka (eds.), Butterworth-Heinemann, Boston, pp. 553–580.

Brisolara K. F. and Bourgeois J. (2019). Biosolids and sludge management. *Water Environment Research*, **91**, 1168–1176, https://doi.org/10.1002/wer.1212

Busch P. L. and Stumm W. (1968). Chemical interactions in the aggregation of bacteria bioflocculation in waste treatment.

Cai M., Wang Q., Wells G., Dionysiou D. D., Song Z., Jin M., Hu J., Ho S.-H., Xiao R. and Wei Z. (2019). Improving dewaterability and filterability of waste activated sludge by electrochemical Fenton pretreatment. *Chemical Engineering Journal*, **362**, 525–536, https://doi.org/10.1016/j.cej.2019.01.047

Cao B., Zhang W., Du Y., Wang R., Usher S. P., Scales P. J. and Wang D. (2018). Compartmentalization of extracellular polymeric substances (EPS) solubilization and cake microstructure in relation to wastewater sludge dewatering behavior assisted by horizontal electric field: effect of operating conditions. *Water Research*, **130**, 363–375, https://doi.org/10.1016/j.watres.2017.11.060

Carrasco M. and Gao W. (2019). Sludge particle size and correlation with soluble organic matter and conditioning characteristics after freezing treatments. *Water, Air, & Soil Pollution*, **230**, 54, https://doi.org/10.1007/s11270-019-4107-z

Carrère H., Dumas C., Battimelli A., Batstone D. J., Delgenès J. P., Steyer J. P. and Ferrer I. (2010). Pretreatment methods to improve sludge anaerobic degradability: a review. *Journal of Hazardous Materials*, **183**, 1–15, https://doi.org/10.1016/j.jhazmat.2010.06.129

Casallas-Ojeda M. R., Marmolejo-Rebellón L. F. and Torres-Lozada P. (2021). Identification of factors and variables that influence the anaerobic digestion of municipal biowaste and food waste. *Waste and Biomass Valorization*, **12**, 2889–2904, https://doi.org/10.1007/s12649-020-01150-x

Cella M. A., Akgul D. and Eskicioglu C. (2016). Assessment of microbial viability in municipal sludge following ultrasound and microwave pretreatments and resulting impacts on the efficiency of anaerobic sludge digestion. *Applied Microbiology and Biotechnology*, **100**, 2855–2868, https://doi.org/10.1007/s00253-015-7139-3

Cetin S. and Erdincler A. (2004). The role of carbohydrate and protein parts of extracellular polymeric substances on the dewaterability of biological sludges. *Water Science and Technology*, **50**, 49–56, https://doi.org/10.2166/wst.2004.0532

Chakraborti R. (2005). Effects of Floc Size and Shape in Particle Aggregation. pp. 95–120.

Chitikela S. and Dentel S. K. (1998). Dual-chemical conditioning and dewatering of anaerobically digested biosolids: laboratory evaluations. *Water Environment Research*, **70**, 1062–1069, https://doi.org/10.2175/106143098X123408

Christensen G. L. (1983). Units for specific resistance. *Journal (Water Pollution Control Federation)*, **55**, 417–419.

Christensen M. L., Keiding K., Nielsen P. H. and Jørgensen M. K. (2015). Dewatering in biological wastewater treatment: a review. Water research. *Special Issue on Sludge Research*, **82**, 14–24.

Colin F. and Gazbar S. (1995). Distribution of water in sludges in relation to their mechanical dewatering. *Water Research*, **29**, 2000–2005, https://doi.org/10.1016/0043-1354(94)00274-B

Cristina G., Camelin E., Pugliese M., Tommasi T. and Fino D. (2019). Evaluation of anaerobic digestates from sewage sludge as a potential solution for improvement of soil fertility. *Waste Management*, **99**, 122–134, https://doi.org/10.1016/j.wasman.2019.08.018

Dai Y., Huang S., Liang J., Zhang S., Sun S., Tang B. and Xu Q. (2017). Role of organic compounds from different EPS fractions and their effect on sludge dewaterability by combining anaerobically mesophilic digestion pre-treatment and Fenton's reagent/lime. *Chemical Engineering Journal*, **321**, 123–138, https://doi.org/10.1016/j.cej.2017.03.107

Diak J. and Örmeci B. (2017). Ferrrate(VI) and freeze-thaw treatment for oxidation of hormones and inactivation of fecal coliforms in sludge. *Water Science and Technology*, **75**, 1625–1632, https://doi.org/10.2166/wst.2017.021

Dilek Sanin F., Aarne Vesilind P. and James Martel C. (1994). Pathogen reduction capabilities of freeze/thaw sludge conditioning. *Water Research*, **28**, 2393–2398, https://doi.org/10.1016/0043-1354(94)90055-8

Du Y., Cao B., Zhang W., Yang P., Xu Q., Wang D. and Shen X. (2017). Improvement of wastewater sludge dewatering properties using polymeric aluminum-silicon complex flocculants conditioning: importance of aluminum/silicon ratio. *Colloids and Surfaces A: Physicochemical and Engineering Aspects*, **530**, 134–145, https://doi.org/10.1016/j.colsurfa.2017.07.042

Dursun D., Turkmen M., Abu-Orf M. and Dentel S. K. (2006). Enhanced sludge conditioning by enzyme pre-treatment: comparison of laboratory and pilot scale dewatering results. *Water Science and Technology*, **54**, 33–41, https://doi.org/10.2166/wst.2006.544

EPA, U.S. (2020). Standards for the Use or Disposal of Sewage Sludge. United States Environmental Protection Agency, Office of Research and Development, Washington DC.

Eriksson L. and Alm B. (1991). Study of flocculation mechanisms by observing effects of a complexing agent on activated sludge properties. *Water Science and Technology*, **24**, 21–28, https://doi.org/10.2166/wst.1991.0180

Eskicioglu C., Prorot A., Marin J., Droste R. L. and Kennedy K. J. (2008). Synergetic pretreatment of sewage sludge by microwave irradiation in presence of H_2O_2

for enhanced anaerobic digestion. *Water Research*, **42**, 4674–4682, https://doi.org/10.1016/j.watres.2008.08.010

Ferrentino R., Merzari F. and Andreottola G. (2020). Optimisation of Fe^{2+}/H_2O_2 ratio in Fenton process to increase dewaterability and solubilisation of sludge. *Environmental Technology (United Kingdom)*, **41**, 2946–2954.

Fitzmorris K. B., Sarmiento F. and O'Callaghan P. (2009). Biosolids and sludge management. *Water Environment Research*, **81**, 1376–1393, https://doi.org/10.2175/106143009X12445568399730

Flemming C. A., Simhon A. and Odumeru J. A. (2017). Pathogen characterization of fresh and stored mesophilic anaerobically digested biosolids. *Water Environment Research*, **89**, 2031–2042, https://doi.org/10.2175/106143017X14839994522704

Fontmorin J.-M. and Sillanpää M. (2017). Dewatering and removal of metals from urban anaerobically digested sludge by Fenton's oxidation. *Environmental Technology*, **38**, 495–505, https://doi.org/10.1080/09593330.2016.1199598

Gao H., Zhang W., Song Z., Yang X., Yang L., Cao M., Wang D. and Liao G. (2017). Transfer behavior of odorous pollutants in wastewater sludge system under typical chemical conditioning processes for dewaterability enhancement. *Scientific Reports*, **7**, 1–12, https://doi.org/10.1038/s41598-016-0028-x

Gao S., Wang Y., Zhang D., Fan X., Guo Y., Li E. and Zheng H. (2021). Insight to peroxone-Fe(III) joint conditioning-horizontal electro-dewatering process on water reduction in activated sludge: performance and mechanisms. *Journal of Hazardous Materials*, **402**, 123441, https://doi.org/10.1016/j.jhazmat.2020.123441

García-Bernet D., Buffière P., Latrille E., Steyer J.-P. and Escudié R. (2011). Water distribution in biowastes and digestates of dry anaerobic digestion technology. *Chemical Engineering Journal*, **172**, 924–928, https://doi.org/10.1016/j.cej.2011.07.003

Garg S., Ahammed M. M. and Shaikh I. (2021). Heavy metal fractionation in aerobic and anaerobic sewage sludge. *Lecture Notes in Civil Engineering*, **104**, 11–19, https://doi.org/10.1007/978-981-15-8151-9_2

Gherghel A., Teodosiu C. and De Gisi S. (2019). A review on wastewater sludge valorisation and its challenges in the context of circular economy. *Journal of Cleaner Production*, **228**, 244–263, https://doi.org/10.1016/j.jclepro.2019.04.240

Gotoh K., Nakata Y., Tagawa M. and Tagawa M. (2003). Wettability of ultraviolet excimer-exposed PE, PI and PTFE films determined by the contact angle measurements. *Colloids and Surfaces A: Physicochemical and Engineering Aspects*, **224**, 165–173, https://doi.org/10.1016/S0927-7757(03)00263-2

Gruchlik Y., Fouché L., Joll C. A. and Heitz A. (2017). Use of alum for odor reduction in sludge and biosolids from different wastewater treatment processes. *Water Environment Research*, **89**, 2103–2112, https://doi.org/10.2175/106143017X15054988926406

Guan B., Yu J., Fu H., Guo M. and Xu X. (2012). Improvement of activated sludge dewaterability by mild thermal treatment in $CaCl_2$ solution. *Water Research*, **46**, 425–432, https://doi.org/10.1016/j.watres.2011.11.014

Guan Q., Tang M., Zheng H., Teng H., Tang X. and Liao Y. (2016). Investigation of sludge conditioning performance and mechanism by examining the effect of charge density on cationic polyacrylamide microstructure. *Desalination and Water Treatment*, **57**, 12988–12997, https://doi.org/10.1080/19443994.2015.1057530

Gude V. G. (2015). Energy positive wastewater treatment and sludge management. *Edorium Journal of Waste Management*, **1**, 10–15.

Hamaide T., Deterre R. and Feller J.-F. (2014). Environmental Impact of Polymers. ISTE Ltd and John Wiley & Sons, Inc., UK, USA.

Hosseini Koupaie E., Azizi A., Bazyar Lakeh A. A., Hafez H. and Elbeshbishy E. (2019). Comparison of liquid and dewatered digestate as inoculum for anaerobic digestion of organic solid wastes. *Waste Management*, **87**, 228–236, https://doi.org/10.1016/j.wasman.2019.02.014

Hou H., Di Vona M. L. and Knauth P. (2012). Building bridges: crosslinking of sulfonated aromatic polymers – A review. *Journal of Membrane Science*, **423–424**, 113–127, https://doi.org/10.1016/j.memsci.2012.07.038

Houghton J. I., Quarmby J. and Stephenson T. (2000). The impact of digestion on sludge dewaterability. *Process Safety and Environmental Protection*, **78**, 153–159, https://doi.org/10.1205/095758200530547

Huang X., Dong W., Wang H. and Feng Y. (2018). Role of acid/alkali-treatment in primary sludge anaerobic fermentation: insights into microbial community structure, functional shifts and metabolic output by high-throughput sequencing. *Bioresource Technology*, **249**, 943–952, https://doi.org/10.1016/j.biortech.2017.10.104

Jin B., Wilén B.-M. and Lant P. (2004). Impacts of morphological, physical and chemical properties of sludge flocs on dewaterability of activated sludge. *Chemical Engineering Journal*, **98**, 115–126, https://doi.org/10.1016/j.cej.2003.05.002

Karr, P. R. and Keinath T. M. (1978). Influence of particle size on sludge dewaterability. *Water Pollution Control Federation*, **50**, 1911–1930.

Katsiris N. and Kouzeli-Katsiri A. (1987). Bound water content of biological sludges in relation to filtration and dewatering. *Water Research*, **21**, 1319–1327, https://doi.org/10.1016/0043-1354(87)90004-2

Kepp U., Machenbach I., Weisz N. and Solheim O. E. (2000). Enhanced stabilisation of sewage sludge through thermal hydrolysis – three years of experience with full scale plant. *Water Science and Technology*, **42**, 89–96, https://doi.org/10.2166/wst.2000.0178

Kim J., Park C., Kim T.-H., Lee M., Kim S., Kim S.-W. and Lee J. (2003). Effects of various pretreatments for enhanced anaerobic digestion with waste activated sludge. *Journal of Bioscience and Bioengineering*, **95**, 271–275, https://doi.org/10.1016/S1389-1723(03)80028-2

Kopp J. and Dichtl N. (2000). The influence of free water content on sewage sludge dewatering. In: Chemical Water and Wastewater Treatment VI, H. H. Hahn, E. Hoffmann and H. Ødegaard (eds.), Springer, Berlin, Heidelberg, pp. 347–356.

Krishnamurthy S. and Viraraghavan T. (2005). Chemical conditioning for dewatering municipal wastewater sludges. *Energy Sources*, **27**, 113–122, https://doi.org/10.1080/00908310490448154

Kumar V., Chopra A. K. and Kumar A. (2017). A review on sewage sludge (biosolids) a resource for sustainable agriculture. *Archives of Agriculture and Environmental Science*, **2**, 340–347, https://doi.org/10.26832/24566632.2017.020417

Li X. Y. and Yang S. F. (2007). Influence of loosely bound extracellular polymeric substances (EPS) on the flocculation, sedimentation and dewaterability of activated sludge. *Water Research*, **41**, 1022–1030, https://doi.org/10.1016/j.watres.2006.06.037

Liao B. Q., Allen D., Droppo I., Leppard G. and Liss S. N. (2000). Bound water content of activated sludge and its relationship to solids retention time, floc structure, and surface properties. *Water Environment Research*, **72**, 722–730, https://doi.org/10.2175/106143000X138346

Liao B. Q., Allen D. G., Droppo I. G., Leppard G. G. and Liss S. N. (2001). Surface properties of sludge and their role in bioflocculation and settleability. *Water Research*, **35**, 339–350, https://doi.org/10.1016/S0043-1354(00)00277-3

Liss S., Liao B. Q., Droppo I., Allen D. and Leppard G. (2002). Effect of solids retention time on floc structure. *Water Science and Technology*, **46**, 431–438, https://doi.org/10.2166/wst.2002.0512

Liu C., Chen X. and Cai H. (2020a). Aerobically digested sludge conditioning by Fe^{2+}/citrate chelated-Fe^{2+} activated peroxymonosulfate oxidation. *Chemical Engineering Journal*, **400**, 125954, https://doi.org/10.1016/j.cej.2020.125954

Liu Z., Mayer B. K., Venkiteshwaran K., Seyedi S., Raju A. S. K., Zitomer D. and McNamara P. J. (2020b). The state of technologies and research for energy recovery from municipal wastewater sludge and biosolids. *Current Opinion in Environmental Science and Health*, **14**, 31–36, https://doi.org/10.1016/j.coesh.2019.12.004

Logan M. and Visvanathan C. (2019). Management strategies for anaerobic digestate of organic fraction of municipal solid waste: current status and future prospects. *Waste Management & Research: The Journal for a Sustainable Circular Economy*, **37**, 27–39, https://doi.org/10.1177/0734242X18816793

López-Maldonado E. A., Oropeza-Guzman M. T., Jurado-Baizaval J. L. and Ochoa-Terán A. (2014). Coagulation–flocculation mechanisms in wastewater treatment plants through zeta potential measurements. *Journal of Hazardous Materials*, **279**, 1–10, https://doi.org/10.1016/j.jhazmat.2014.06.025

Lowman A., McDonald M. A., Wing S. and Muhammad N. (2013). Land application of treated sewage sludge: community health and environmental justice. *Environmental Health Perspectives*, **121**, 537–542, https://doi.org/10.1289/ehp.1205470

Lv H., Liu D., Zhang Y., Yuan D., Wang F., Yang J., Wu X., Zhang W. and Dai X. (2019). Effects of temperature variation on wastewater sludge electro-dewatering. *Journal of Cleaner Production*, **214**, 873–880, https://doi.org/10.1016/j.jclepro.2019.01.033

MacDonald B. A., Oakes K. D. and Adams M. (2018). Molecular disruption through acid injection into waste activated sludge – a feasibility study to improve the economics of sludge dewatering. *Journal of Cleaner Production*, **176**, 966–975, https://doi.org/10.1016/j.jclepro.2017.12.014

Mahmoud A., Olivier J., Vaxelaire J. and Hoadley A. F. A. (2011). Electro-dewatering of wastewater sludge: influence of the operating conditions and their interactions effects. *Water Research*, **45**, 2795–2810, https://doi.org/10.1016/j.watres.2011.02.029

Masihi H. and Badalians Gholikandi G. (2018). Employing electrochemical-Fenton process for conditioning and dewatering of anaerobically digested sludge: a novel approach. *Water Research*, **144**, 373–382, https://doi.org/10.1016/j.watres.2018.07.054

McFarland, M. J. (2001). Ch 3: Biosolids and sludge Processing. In Biosolids Engineering. McGraw-Hill Education, New York, USA, pp. 1–248.

Menon U., Suresh N., George G., Ealias A. M. and Saravanakumar M.P. (2020). A study on combined effect of Fenton and Free Nitrous Acid treatment on sludge dewaterability with ultrasonic assistance: Preliminary investigation on improved calorific value. *Chemical Engineering Journal*, **382**, 123035. https://doi.org/10.1016/j.cej.2019.123035

Metcalf & Eddy, Inc. (2003). Wastewater Engineering: Treatment and Reuse. Fourth edition/revised by George Tchobanoglous, Franklin L. Burton, H. David Stensel. McGraw-Hill, Boston.

Metcalf & Eddy, Tchobanoglous G., Stensel H. D., Tsuchihashi R., Burton F. L., Abu-Orf M., Bowden G. and Pfrang W. (2014). Wastewater engineering: treatment and resource recovery.

Mikkelsen L. (2002). Physico-chemical characteristics of full scale sewage sludges with implications to dewatering. *Water Research*, **36**, 2451–2462, https://doi.org/10.1016/S0043-1354(01)00477-8

Möller K. and Müller T. (2012). Effects of anaerobic digestion on digestate nutrient availability and crop growth: a review. *Engineering in Life Sciences*, **12**, 242–257, https://doi.org/10.1002/elsc.201100085

Morgan J. W., Forster C. F. and Evison L. (1990). A comparative study of the nature of biopolymers extracted from anaerobic and activated sludges. *Water Research*, **24**, 743–750, https://doi.org/10.1016/0043-1354(90)90030-A

Mouneimne A. H., Carrère H., Bernet N. and Delgenès J. P. (2003). Effect of saponification on the anaerobic digestion of solid fatty residues. *Bioresource Technology*, **90**, 89–94, https://doi.org/10.1016/S0960-8524(03)00091-9

Mowla D., Tran H. N. and Allen D. G. (2013). A review of the properties of biosludge and its relevance to enhanced dewatering processes. *Biomass and Bioenergy*, **58**, 365–378, https://doi.org/10.1016/j.biombioe.2013.09.002

Murujew O., Geoffroy J., Fournie E., Socionovo Gioacchini E., Wilson A., Vale P., Jefferson B. and Pidou M. (2020). The impact of polymer selection and dose on the incorporation of ballasting agents onto wastewater aggregates. *Water Research*, **170**, 115346, https://doi.org/10.1016/j.watres.2019.115346

Mustranta A. and Viikari L. (1993). Dewatering of activated sludge by an oxidative treatment. *Water Science and Technology*, **28**, 213–221, https://doi.org/10.2166/wst.1993.0051

NACWA (2010). Renewable Energy Resources: Banking on Biosolids 23.

Neyens E. and Baeyens J. (2003a). A review of thermal sludge pre-treatment processes to improve dewaterability. *Journal of Hazardous Materials*, **98**, 51–67, https://doi.org/10.1016/S0304-3894(02)00320-5

Neyens E. and Baeyens J. (2003b). A review of classic Fenton's peroxidation as an advanced oxidation technique. *Journal of Hazardous Materials*, **98**, 33–50, https://doi.org/10.1016/S0304-3894(02)00282-0

Neyens E., Baeyens J. and Creemers C. (2003). Alkaline thermal sludge hydrolysis. *Journal of Hazardous Materials*, **97**, 295–314, https://doi.org/10.1016/S0304-3894(02)00286-8

Nguyen T. P., Hankins N. P. and Hilal N. (2007). A comparative study of the flocculation behaviour and final properties of synthetic and activated sludge in wastewater treatment. Desalination. *EuroMed*, **2006**(204), 277–295.

Niu M., Zhang W., Wang D., Chen Y. and Chen R. (2013). Correlation of physicochemical properties and sludge dewaterability under chemical conditioning using inorganic coagulants. *Bioresource Technology*, **144**, 337–343, https://doi.org/10.1016/j.biortech.2013.06.126

Novak, J. T., Goodman, G. L., Pariroo, A., & Huang, J.-C. (1988). The blinding of sludges during filtration. *Journal of Water Pollution Control Federation*, **60**, 206–214.

Novak J. T. (2006). Dewatering of sewage sludge. *Drying Technology*, **24**, 1257–1262, https://doi.org/10.1080/07373930600840419

O'Connell C., Sherlock R., Ball M. D., Aszalós-Kiss B., Prendergast U. and Glynn T. J. (2009). Investigation of the hydrophobic recovery of various polymeric biomaterials after 172 nm UV treatment using contact angle, surface free energy and XPS measurements. *Applied Surface Science*, **255**, 4405–4413, https://doi.org/10.1016/j.apsusc.2008.11.034

Olböter L. and Vogelpohl A. (1993). Influence of particle size distribution on the dewatering of organic sludges. *Water Science and Technology*, **28**, 149–157, https://doi.org/10.2166/wst.1993.0037

Palmgren R., Jorand F., Nielsen P. H. and Block J. C. (1998). Influence of oxygen limitation on the cell surface properties of bacteria from activated sludge. *Water Science and Technology*, **37**, 349–352, https://doi.org/10.2166/wst.1998.0663

Park M., Tsai S.-L. and Chen W. (2013). Microbial biosensors: engineered microorganisms as the sensing machinery. *Sensors*, **13**, 5777–5795, https://doi.org/10.3390/s130505777

Pecorini I., Peruzzi E., Albini E., Doni S., Macci C., Masciandaro G. and Iannelli R. (2020). Evaluation of MSW compost and digestate mixtures for a circular economy application. *Sustainability (Switzerland)*, **12**, 3042. https://doi.org/10.3390/su12073042.

Pham T. T. H., Brar S. K., Tyagi R. D. and Surampalli R. Y. (2010). Influence of ultrasonication and Fenton oxidation pre-treatment on rheological characteristics of wastewater sludge. *Ultrasonics Sonochemistry*, **17**, 38–45, https://doi.org/10.1016/j.ultsonch.2009.06.007

Pillai S. D. and Reimers R. S. (2010). Disinfecting and Stabilizing Biosolids Using E-Beam and Chemical Oxidants. IWA Publishing, UK. https://doi.org/10.2166/9781843392996.

Qi Y., Thapa K. B. and Hoadley A. F. A. (2011). Application of filtration aids for improving sludge dewatering properties – a review. *Chemical Engineering Journal*, **171**, 373–384, https://doi.org/10.1016/j.cej.2011.04.060

Qi L., Wu S., Cheng J. and Hu Y. (2017). The effects of physicochemical properties of sludge on dewaterability under chemical conditioning with amphoteric polymer. *Journal of Polymers and the Environment*, **25**, 1262–1272, https://doi.org/10.1007/s10924-016-0884-2

Radke C. H. (2017). Advancing Resource Recovery from Biosolids and Biosolids as a Resource Christine H. Radke, PMP Research Program Director.

Rékási M., Mazsu N., Draskovits E., Bernhardt B., Szabó A., Rivier P. A., Farkas C., Borsányi B., Pirkó B., Molnár S., Kátay G. and Uzinger N. (2019). Comparing the agrochemical properties of compost and vermicomposts produced from municipal sewage sludge digestate. *Bioresource Technology*, **291**, 121861, https://doi.org/10.1016/j.biortech.2019.121861

Rhodes E. R., Boczek L. A., Ware M. W., McKay M., Hoelle J. M., Schoen M. and Villegas E. N. (2015). Determining pathogen and indicator levels in class B municipal organic residuals used for land application. *Journal of Environmental Quality*, **44**, 265–274, https://doi.org/10.2134/jeq2014.04.0142

Sanin F. D., Clarkson W. W., Vesilind P. A. and Vesilind P. A. (2011). Sludge Engineering: the Treatment and Disposal of Wastewater Sludges, New ed., DEStech Publications, Inc, Lancaster, PA.

Scherzinger M. and Kaltschmitt M. (2021). Thermal pre-treatment options to enhance anaerobic digestibility – a review. *Renewable and Sustainable Energy Reviews*, **137**, 110627, https://doi.org/10.1016/j.rser.2020.110627

Shao L., He P., Yu G. and He P. (2009). Effect of proteins, polysaccharides, and particle sizes on sludge dewaterability. *Journal of Environmental Sciences*, **21**, 83–88, https://doi.org/10.1016/S1001-0742(09)60015-2

Shao L., Wang X., Xu H. and He P. (2012). Enhanced anaerobic digestion and sludge dewaterability by alkaline pretreatment and its mechanism. *Journal of Environmental Sciences*, **24**, 1731–1738, https://doi.org/10.1016/S1001-0742(11)61031-0

Siebielec G., Siebielec S. and Lipski D. (2018). Long-term impact of sewage sludge, digestate and mineral fertilizers on plant yield and soil biological activity. *Journal of Cleaner Production*, **187**, 372–379, https://doi.org/10.1016/j.jclepro.2018.03.245

Spinosa, L. and Vesilind P.A. (2007). Sludge into Biosolids - Processing, Disposal, Utilization. IWA Publishing, UK. https://doi.org/10.2166/9781780402215

Sri Bala Kameswari K., Kalyanaraman C., Sudharsan Varma V., Porselvam S. and Thanasekaran K. (2013). Significance of chemical conditioning to improve dewaterability of biosludge generated from tanneries. *Clean Technologies and Environmental Policy*, **15**, 945–953, https://doi.org/10.1007/s10098-012-0561-7

Su M., Kong L., Liao C., Chen D. and Shih K. (2019). 22 – Stabilization of cadmium in industrial sludge–generation of crystalline products. In: Industrial and Municipal

Sludge, M. N. V. Prasad, P. J. de C. Favas, M. Vithanage and S. V. Mohan (eds.), Butterworth-Heinemann, Oxford, UK, pp. 503–524. https://doi.org/10.1016/B978-0-12-815907-1.00022-2

Tanaka S., Kobayashi T., Kamiyama K. and Signey Bildan M. L. N. (1997). Effects of thermochemical pretreatment on the anaerobic digestion of waste activated sludge. Water science and technology. *Water Quality and Pollution Control in Asia*, **35**, 209–215.

The Water Resource Foundation (1975). Resource Recovery.

To V. H. P., Nguyen T. V., Vigneswaran S. and Ngo H. H. (2016). A review on sludge dewatering indices. *Water Science and Technology; London*, **74**, 1–16.

Tsang K. R. and Vesilind P. A. (1990). Moisture distribution in sludges. *Water Science and Technology*, **22**, 135–142, https://doi.org/10.2166/wst.1990.0108

Tsou G. W., Wu R. M., Yen P. S., Lee D. J. and Peng X. F. (2002). Advective flow and floc permeability. *Journal of Colloid and Interface Science*, **250**, 400–408, https://doi.org/10.1006/jcis.2002.8317

Turovskii I. S. and Mathai P. K. (2006). Wastewater Sludge Processing. Wiley-Interscience, Hoboken, NJ.

Urbain V., Block J. C. and Manem J. (1993). Bioflocculation in activated sludge: an analytic approach. *Water Research*, **27**, 829–838, https://doi.org/10.1016/0043-1354(93)90147-A

US EPA (1987). Design manual: dewatering municipal wastewater sludges., EPA/625/1;87/014. U.S. Environmental Protection Agency, Office of Research and Development, Center for Environmental Research Information, Cincinnati, OH.

US EPA, O. (2015). Biosolids Laws and Regulations [WWW Document]. US EPA.

Van den Broeck R., Van Dierdonck J., Nijskens P., Dotremont C., Krzeminski P., van der Graaf J. H. J. M., van Lier J. B., Van Impe J. F. M. and Smets I. Y. (2012). The influence of solids retention time on activated sludge bioflocculation and membrane fouling in a membrane bioreactor (MBR). *Journal of Membrane Science*, **401–402**, 48–55, https://doi.org/10.1016/j.memsci.2012.01.028

Van Oss C. J. (2003). Long-range and short-range mechanisms of hydrophobic attraction and hydrophilic repulsion in specific and aspecific interactions. *Journal of Molecular Recognition*, **16**, 177–190, https://doi.org/10.1002/jmr.618

Vaxelaire J. and Cézac P. (2004). Moisture distribution in activated sludges: a review. *Water Research*, **38**, 2215–2230, https://doi.org/10.1016/j.watres.2004.02.021

Wacławek S., Grübel K., Dennis P., Vinod V. T. P. and Černík M. (2016). A novel approach for simultaneous improvement of dewaterability, post-digestion liquor properties and toluene removal from anaerobically digested sludge. *Chemical Engineering Journal*, **291**, 192–198, https://doi.org/10.1016/j.cej.2016.01.103

Wang T., Xu B., Zhang X., Yang Q., Xu B. and Yang P. (2018). Enhanced Biogas Production and Dewaterability from Sewage Sludge with Alkaline Pretreatment at Mesophilic and Thermophilic Temperatures. 3 Biotech 229.

Wang W., Yan Y., Zhao Y., Shi Q. and Wang Y. (2020). Characterization of stratified EPS and their role in the initial adhesion of anammox consortia. *Water Research*, **169**, 115223, https://doi.org/10.1016/j.watres.2019.115223

Wang Z., Liu T., Duan H., Song Y., Lu X., Hu S., Yuan Z., Batstone D. and Zheng M. (2021). Post-treatment options for anaerobically digested sludge: current status and future prospect. *Water Research*, **205**, 117665, https://doi.org/10.1016/j.watres.2021.117665

Wei H., Gao B., Ren J., Li A. and Yang H. (2018). Coagulation/flocculation in dewatering of sludge: a review. *Water Research*, **143**, 608–631, https://doi.org/10.1016/j.watres.2018.07.029

Wéry N., Lhoutellier C., Ducray F., Delgenès J.-J. and Godon J.-J. (2008). Behaviour of pathogenic and indicator bacteria during urban wastewater treatment and sludge composting, as revealed by quantitative PCR. *Water Research*, **42**, 53–62, https://doi.org/10.1016/j.watres.2007.06.048

Wu B., Ni B.-J., Horvat K., Song L., Chai X., Dai X. and Mahajan D. (2017a). Occurrence state and molecular structure analysis of extracellular proteins with implications on the dewaterability of waste-activated sludge. *Environmental Science & Technology*, **51**, 9235–9243, https://doi.org/10.1021/acs.est.7b02861

Wu B., Su L., Song L., Dai X. and Chai X. (2017b). Exploring the potential of iTRAQ proteomics for tracking the transformation of extracellular proteins from enzyme-disintegrated waste activated sludge. *Bioresource Technology*, **225**, 75–83, https://doi.org/10.1016/j.biortech.2016.11.041

Wu B., Dai X. and Chai X. (2020). Critical review on dewatering of sewage sludge: influential mechanism, conditioning technologies and implications to sludge re-utilizations. *Water Research*, **180**, 115912, https://doi.org/10.1016/j.watres.2020.115912

Wu W., Ma J., Xu J. and Wang Z. (2021). Mechanistic insights into chemical conditioning by polyacrylamide with different charge densities and its impacts on sludge dewaterability. *Chemical Engineering Journal*, **410**, 128425, https://doi.org/10.1016/j.cej.2021.128425

Yang M. B. H., Bai Y., Liu R., Hu C. and Qu J. (2019). Using activated peroxymonosulfate by electrochemically generated FeII for conditioning and dewatering of anaerobically digested sludge. *Chemical Engineering Journal*, **391**, 123–603. https://doi.org/10.1016/j.cej.2019.123603.

Yin X., Han P., Lu X. and Wang Y. (2004). A review on the dewaterability of bio-sludge and ultrasound pretreatment. *Ultrasonics Sonochemistry*, **11**, 337–348, https://doi.org/10.1016/j.ultsonch.2004.02.005

Yu G.-H., He P.-J., Shao L.-M. and Zhu Y.-S. (2008). Extracellular proteins, polysaccharides and enzymes impact on sludge aerobic digestion after ultrasonic pretreatment. *Water Research*, **42**, 1925–1934, https://doi.org/10.1016/j.watres.2007.11.022

Zhang P., Guo J.-S., Shen Y., Yan P., Chen Y.-P., Wang H., Yang J.-X., Fang F. and Li C. (2015). Microbial communities, extracellular proteomics and polysaccharides: A comparative investigation on biofilm and suspended sludge. *Bioresource Technology*, **190**, 21–28, https://doi.org/10.1016/j.biortech.2015.04.058

Zhang J., Li N., Dai X., Tao W., Jenkinson I. R. and Li Z. (2018). Enhanced dewaterability of sludge during anaerobic digestion with thermal hydrolysis pretreatment: new insights through structure evolution. *Water Research*, **131**, 177–185, https://doi.org/10.1016/j.watres.2017.12.042

Zhang W., Dong B. and Dai X. (2019). Mechanism analysis to improve sludge dewaterability during anaerobic digestion based on moisture distribution. *Chemosphere*, **227**, 247–255, https://doi.org/10.1016/j.chemosphere.2019.03.150

Zhang W., Dai X., Dong B. and Dai L. (2020). New insights into the effect of sludge proteins on the hydrophilic/hydrophobic properties that improve sludge dewaterability during anaerobic digestion. *Water Research*, **173**, 115503, https://doi.org/10.1016/j.watres.2020.115503

Zhou J., Zheng G., Zhang X. and Zhou L. (2014). Influences of extracellular polymeric substances on the dewaterability of sewage sludge during bioleaching. *PLoS ONE*, **9**, 102–688.

Zuliani L., Frison N., Jelic A., Fatone F., Bolzonella D. and Ballottari M. (2016). Microalgae cultivation on anaerobic digestate of municipal wastewater, sewage sludge and agro-waste. *International Journal of Molecular Sciences*, **17**, 1692, https://doi.org/10.3390/ijms17101692

doi: 10.2166/9781789062755_0111

Chapter 5

Valorization of anaerobic digestate from microalgal treatment

*Ahmed Tawfik**

National Research Centre, Water Pollution Research Department, 12622 Dokki, Cairo, Egypt
**Corresponding author e-mail: prof.tawfik.nrc@gmail.com*

ABSTRACT

Microalgae are a promising low-cost substrate for bioenergy production via the anaerobic digestion (AD) process. However, they have strong cell walls that prevent the easy dissolution of their components. Low biomass productivity, unbalanced carbon-to-nitrogen ratio, recalcitrant cellular components, and the high cost of cultivation, harvesting, and dewatering, make the AD of algae not economically attractive for the valorization of this biomass. Moreover, algal cells contain ammonia, aluminium, sulphide, sodium and potassium, which usually affect the bioenergy production via AD. Anaerobic co-digestion of algal biomass with complementary co-substrates could be a suitable strategy to valorize algae successfully. The synergistic effect of anaerobic co-digestion of microalgae with other biodegradable co-substrates is mainly attributed to the presence of essential trace elements and micronutrients leading to a nutrient balance. To attain zero bio-waste, the digestate from AD of algal biomass could be reused for agricultural purposes due to its richness in nutrients and minerals.

Keywords: Microalgae, anaerobic co-digestion, digestate, bioenergy, inhibition, reuse

5.1 INTRODUCTION

A major portion of wastewater is treated with algal pond systems due to the low-cost technology and the production of useful biomass (Allam *et al.*, 2016; Bassuney & Tawfik, 2017). The excess sludge and biomass from algal pond systems are very rich in carbohydrates, lipids and proteins, which could be useful for bioenergy production through the anaerobic digestion (AD) process (Osama *et al.*, 2020, 2021; Tawfik *et al.*, 2022a). However, the biodegradable content of microalgae (i.e. carbohydrates and lipids) is quite low. Moreover, it has a strong cell wall preventing the easy dissolution of these components. In general, diatom and green algae are more resistant to AD due to their rigid cell walls composed of silica and highly polymerized cellulose/polymers, respectively. Euglenophyta and Cyanobacteria are more readily biodegradable

Editor(s): Vinay Kumar Tyagi, Kaoutar Aboudi and Cigdem Eskicioglu

due to the absence of rigid cell walls (Santos-Ballardo *et al.*, 2016). *Tribonema* cell walls contain cellulose II based on 1,6-linked glucan, 1,3 and 1,4-linked xylose, 1,3-linked rhamnose and, mostly, 1,3; 1,4; and 1,6-linked glucose (Torruella *et al.*, 2018), which render them a certain degree of resistance to AD.

Most microalgae species used in AD are of *Scenedesmus* and *Chlorella*. Nevertheless, low biomass productivity, unbalanced carbon-to-nitrogen ratio, recalcitrant cellular components and the high cost of cultivation, harvesting, and dewatering make the AD process economically not attractive for the valorization of algal biomass (Ali *et al.*, 2021). The synergistic effect of anaerobic co-digestion of microalgae with other biodegradable co-substrates is an interesting strategy and is mainly attributed to the presence of essential trace elements and micronutrients (Soltan *et al.*, 2019).

The review of Lee and Lee (2019) discussed the nutrient-rich anaerobic digestate for the cultivation of microalgae and biogas upgrading. Moreover, the cultivation of microalgae in the digestate is rather challenging due to its high ammonia concentration and low C/N ratio (Bonetta *et al.*, 2014; Koutra *et al.*, 2021; Xia & Murphy, 2016). Monocultures and/or co-cultures of microalgae and bacteria (Yu *et al.*, 2019) and fungi species (Kumsiri *et al.*, 2018), were examined for the valorization of various digestate types (Massa *et al.*, 2017; Tawfik *et al.*, 2021a). Different feeding strategies, dilution ratios and pre-treatment methods have been tested to optimize the quality production of this biomass (Massa *et al.*, 2017; Romero-Villegas *et al.*, 2018; Zhu *et al.*, 2019). The determination of algal biomass composition depends mainly on the cultivated species and the imposed conditions to identify the subsequent steps towards the biorefinery approach. *Chlorella vulgaris* has a rich protein content of 58% and is commonly employed for wastewater treatment at pilot and full-scale applications (Stiles *et al.*, 2018). The harvested microalgae biomass could be utilized for the production of nutraceuticals, agrochemicals, forage and biofuels (Safi *et al.*, 2014; Sakarika & Kornaros, 2019). Moreover, the algal biomass extracts provided phytochemicals (alkaloids, phenols, flavonoids, glycosides, and tannins).

The digestate of swine manure was utilized by *Chlorella PY-ZU1* and supplemented with 15% CO_2 and phosphorus, resulting in biomass productivity of 4.81 g/L, with a composition of 13% carbohydrates, 19.5% lipids and 42% proteins (Cheng *et al.*, 2015). The *C. vulgaris* was cultured on the digestate from the pulp and paper industry and resulted in algal biomass of 2.91 gVS/L, with complete removal of ammonia and phosphates (Tao *et al.*, 2017). Co-cultivation of *C. vulgaris* and aerobic bacteria on the digestate from swine manure resulted in biomass concentration of 2.3 g/L and removal efficiency of 97.2% for COD, 94% for total nitrogen (TN) and 99.7% for total phosphorus (Wang *et al.*, 2016, 2019). Supplementation of digestate from municipal wastewater plant with 2 g/L of glycerol-enhanced mixotrophic growth of *C. vulgaris*, reaching up to 64 mg/L and lipid productivity of 20.4 mg/L (Ge *et al.*, 2018).

The aim of this chapter is to: (1) Assess algal biomass solely and as co-substrate for bioenergy production through AD, as well as the use of the resulting digestate as a soil conditioner. (2) Study the effect of algal cells composition (ammonia, aluminium, sulphide, sodium and potassium on biogas production via AD. (3)

Investigate the interaction between algal aggregate sludge and co-substrate, emphasizing hydrolytic enzyme activities by anaerobes. (4) Finally, to examine the cultivation of algal cells for the valorization of the digestate.

5.2 HARVESTING EXCESS MICROALGAE

The utilization of algal biomass for bioenergy production, treatment of anaerobic digestate, co-digestion process, soil fertility, and the production of value-added by-products are certainly minimizing the cost and closing the loop cycle for bio-waste (Shi *et al.*, 2019).

The harvesting of algal cells is a consuming energy process, but it is important for the accomplishment of the utilization of this biomass (Figure 5.1).

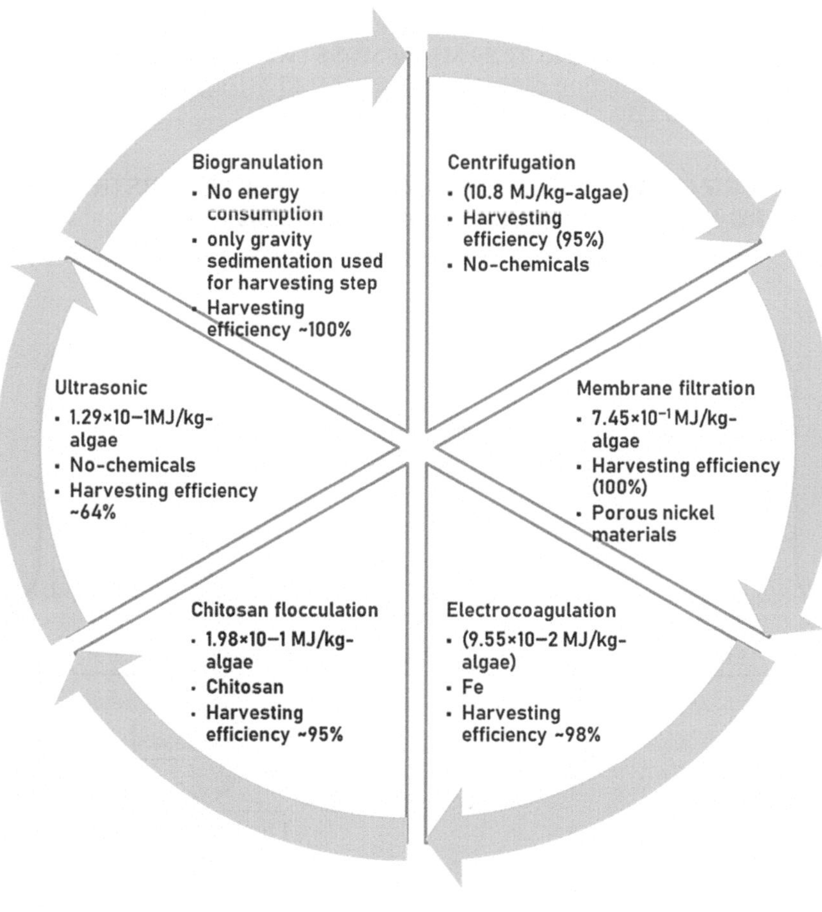

Figure 5.1 Different techniques for the harvesting of excess microalgae.

Algal biomass harvesting by centrifugation method consumes electricity of 10.8 MJ/kg-algae, which is considered high (Brentner *et al.*, 2011). The electrocoagulation process requires lower consumption of electricity; however, its downstream has shown to be affected by the presence of metals in the algal biomass (Richardson & Johnson, 2015). The chitosan flocculation technology harvested more than 95% which is mainly proportional to the initial algal biomass concentration with a relatively low energy consumption of 1.98×10^{-1} MJ/kg-algae output. However, the technology application consumes huge quantities of chemicals, which negatively affects its implementation at full-scale plants (Beach *et al.*, 2012; Billuri *et al.*, 2015). A lower algal biomass harvesting capacity of 64% is achieved by ultrasonic technology. Nevertheless, the advantage of ultrasonic technology is the low consumption of electricity, chemicals and materials, as well as the possibility of the separation of lipids and proteins by standing acoustic waves. On the other hand, bio-granulation technology is quite new for algal biomass harvesting, and it consists of harvesting by gravity sedimentation, where the heating value of the harvesting process ranged from 13.33 to 22.79 MJ/kg-MLSS (Wang *et al.*, 2020). However, the electricity consumption of this technique is 9.45 MJ/kg-algae, which is still being higher than other technologies (Shi *et al.*, 2019).

5.3 VALORIZATION OF ALGAL BIOMASS AS A SOLE SUBSTRATE FOR BIOENERGY PRODUCTION AND SOIL CONDITIONING

Figure 5.2 shows the bioenergy and algal chain productivity from wastewater.

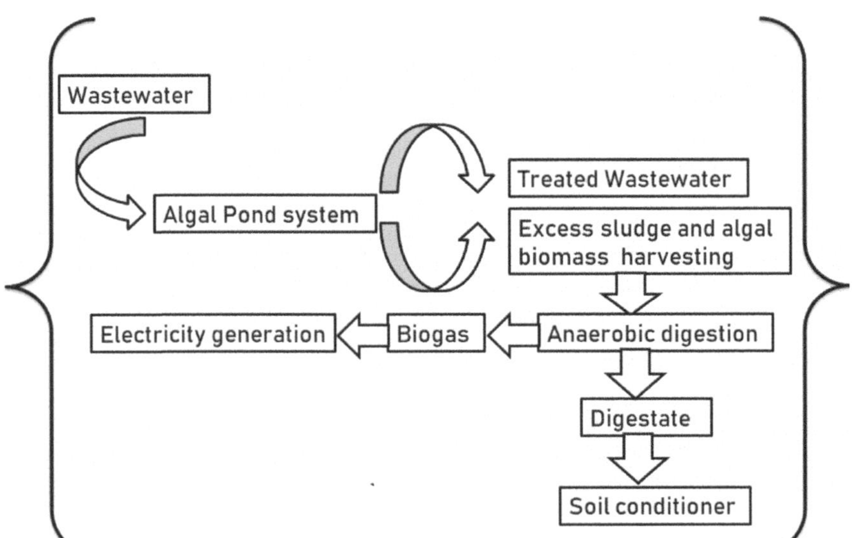

Figure 5.2 Bioenergy and algal chain productivity from wastewater.

Diatoms and green algae are presumably resistant to AD due to the presence of rigid cell walls containing silica and cellulose/polymers. Cyanobacteria and Euglenophyta are easily biodegradable due to their non-rigid cell walls (Santos-Ballardo *et al.*, 2016). Filamentous yellow-green algae (*Tribonema*) has cell walls containing 1,3 and 1,4-linked xylose, cellulose II based on 1,6-linked glucan, 1,3; 1,4; and 1,6-linked glucose and 1,3-linked rhamnose (Torruella *et al.*, 2018), which could hinder their degree of resistance in the AD process.

AD of *C. vulgaris* provided a methane yield (MY) of 158 mL CH_4/gVS at a C/N ratio of 6 and a temperature of 19°C (Zhang, *et al.*, 2019). Microalgal biomass grown in wastewater was harvested and anaerobically digested by anaerobes resulting in biomethane productivity of 264–287 mLCH_4/gVS (Solé-Bundó *et al.*, 2017a). The MY from AD of the microalgae *Chlorella* sp. grown on pretreated swine manure amounted to 250 mL/gVS (Wang *et al.*, 2016). In another study, AD of blue algal biomass was assessed in two-phased continuous anaerobic reactors, operating at organic loading rates (OLRs) of 2.0, 4.0, 6.0, and 8.00 gVS/L d (Zhong *et al.*, 2013). The authors reported that MY increased at increasing the OLR from 2.0 to 6.00 gVS/L d and dropped at OLR of 8.0 gVS/L d, due to ammonium accumulation and the system inhibition. The MY was optimized by Zhong *et al.* (2012) at a level of 160 CH_4/gVS at OLR of 6.0 gVS/L d and the AD of blue algal sludge cells achieved a MY of 201 mL/gVS at 35°C.

AD of flocculated algae inhibited biogas production by 40% at a high concentration of 200 mg aluminium/L (Sukias & Craggs, 2011). In the study carried out by Tsapekos *et al.* (2018), AD of microalgae harvested from wastewater treatment plants achieved a MY of 313 ± 13 N mL/gVS, which was found to correspond to 77% of the estimated energy output. MY of 98.4 mL CH_4/gCOD was harvested from thermophilic AD of *Scenedesmus* spp. in a continuous stirred-tank reactor (Greses *et al.*, 2017). Zamalloa *et al.* (2012) found that the biogas production rate from AD of *Scenedesmus* in a CSTR is 1.3-fold higher at a temperature of 55°C than at 35°C. In another study performed by Greses *et al.* (2017), the MY and biodegradability of 164 mLCH_4/gCOD and 46.8% were achieved from thermophilic AD of *Scenedesmus* spp. in CSTR at OLR of 0.3 gCOD/L d and SRT of 70 d, respectively.

5.4 INHIBITORY CONDITIONS IN AD OF ALGAL BIOMASS

Physico-chemical characteristics of the harvested and excess algal cells determine their efficiency for bioenergy production. The algal cells contain ammonia, aluminium, sulphide, sodium and potassium, which affect bioenergy production via AD process (Figure 5.3).

5.4.1 Ammonia concentration

Ammonia-nitrogen is mainly produced during the anaerobic degradation of proteins and urea. Ammonium ion (NH_4^+) and free ammonia (NH_3) are the two most common forms of the present inorganic nitrogen. Free ammonia is the most toxic element due to its capability to pass through the bacterial cell membrane, causing severe proton imbalance and deficiency of potassium (Chen

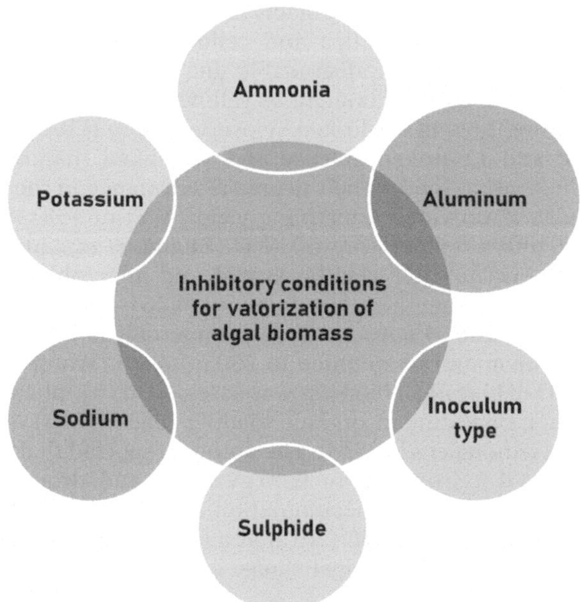

Figure 5.3 Inhibitory conditions for valorization of algal biomass.

et al., 2008). The NH_3 concentration mainly depends on pH, temperature, and total ammonia concentration (Hansen *et al.*, 1998). An increase in temperature would have a positive effect on the anaerobic microbial growth rate, but it may generate high free ammonia concentrations. An increase in pH provided a higher toxicity level due to a shift to a higher ratio from NH_3 to NH_4^+. Ammonia concentrations <200 mg/L are necessary for anaerobes to carry out their activities and supply essential nutrient sources for the microorganisms (Liu & Sung, 2002). Free ammonia (NH_3) of 560–568 mg/L caused the inhibition of 50% of methanogens at a pH value of 7.6 and thermophilic temperature conditions (Sung & Liu, 2003). Surprisingly, the acetogenic population is highly tolerant compared to the methanogens. NH_3 of 4051–5734 mg/L slightly affected the acidogens but the activity of methanogens was reduced by 56.1% (Chen *et al.*, 2008). Under thermophilic conditions, methanogens could tolerate ammonia concentrations up to 2 g/L and complete inhibition occurs at a concentration of 10 g-NH_4-N/L (Sung & Liu, 2003). The high nitrogen content and low C/N ratio of the algal biomass represents the most barriers in the biomethanization process. Ammonium inhibition occurred due to the protein degradation under anaerobic conditions, causing toxicity and lower biogas production (Sukias & Craggs, 2011).

Total ammonia nitrogen (TAN) could be toxic and inhibitory to methanogens at a range of 3000–15 000 mg/L and pH values of 7.2–7.6 (Inglesby & Fisher, 2012). The ammonium toxicity is acute and critical when increasing pH levels in the fermentation medium, mainly due to a shift to high free ammonia (NH_3) fraction (Appels *et al.*, 2008). *C. vulgaris* cultivated on the anaerobic digestate was declined with an increase in N and P concentration. Ammonium (NH_4-N) concentration of 500 mg/L in the feedstock of *C. vulgaris* highly caused toxicity and declined the growing efficiency (Uggetti *et al.*, 2014; Xia & Murphy, 2016). Similarly, NH_4-N concentration exceeding 402.5 mg/L causes a decline in the growth of *C. vulgaris* due to ammonium accumulation in the cultivated cells (Fernandes *et al.*, 2020). However, MY from AD of blue algae was slightly reduced from 113 to 108 $mLCH_4/gVS$ when increasing the level of TAN from 995 to 3959 mg/L (Zhong *et al.*, 2013). The level of 3959 mg/L for TAN is toxic not only to hydrolytic-fermentative bacteria but also to methanogens (Appels *et al.*, 2008).

As previously mentioned, the increase in TAN in the algal feedstock is due to proteins degradation. Therefore, anaerobic co-digestion of algal biomass with other substrates is a good option to mitigate the ammonium inhibition effect and balance of the C/N ratio for methanogens. *Chlorella* spp. and *Scenedesmus* spp. were supplemented with food grease and waste paper to balance the C/N ratio of the feedstock, which positively affects the biomethanization performance avoiding excessive ammonia toxicity and inhibition (Park & Li, 2012; Yen & Brune, 2007). Co-digestion of a mixture of primary sludge (75%) and microalgae 25% (VS-based) increased methane productivity by 65% and mitigate the inhibition effect of ammonia toxicity compared to mono-digestion of microalgae under mesophilic conditions (Solé-Bundó *et al.*, 2019a). In the same line, Olsson *et al.* (2018) found that supplying algae to sewage sludge at a ratio of 37/63% (w/w) improved methane production.

5.4.2 Aluminium concentration

Classical chemical coagulation of algae by aluminium sulphate (alum) is commonly used to harvest the biomass from culture ponds and remove the suspended particles from the treated effluent to produce a good effluent quality. However, the presence of alum in the algal sludge represents a big barrier for AD and bioenergy production (Sukias & Craggs, 2011). The methanogenesis and biogas production was reduced by 40% in the AD of algal biomass coagulated with an aluminium concentration of 1600 g/m^3. Thus, the physical binding of aggregates alum flocs is strong and reduces the hydrolytic enzymes accessibility into the biomass, which inhibits the methanogenic activities. Alum inhibition of algal biomass AD is mainly due to binding of algal cells/substrate that is proteins, phosphate immobilization, and the toxicity by the alum concentration (George *et al.*, 1995).

5.4.3 Inoculum type

The bacterial seed is necessary to start up the AD of algal cells to accelerate the process and minimize the period of lag phase. The inoculum type affects the activities of methanogens and subsequently biogas production. The inoculum harvested from piggery psychrophilic anaerobic pond was better than the inoculum sludge harvested from mesophilic digester of municipal wastewater for harvesting energy from algal biomass sludge (Sukias & Craggs, 2011). The use of this inoculum also mitigates the inhibition effect and ammonium toxicity. Thus, the inoculum harvested from anaerobic pond treating piggery wastewater is recommended to avoid ammonia inhibition during algal biomass digestion.

5.4.4 Sulphide

Sulphate ions (SO_4^{2-}) commonly existed in biomass and biowaste. Furthermore, SO_4^{2-} is mainly used as an electron acceptor and reduced by sulphate reducing bacteria (SRB) into sulphide (Chen *et al.*, 2008). The inhibition of SRB occurred due to the competition for substrates and the toxicity of sulphide. SRB can metabolise substrates, such as long-chain fatty acids (LCFA), aromatic compounds, alcohols and organic acids. SBR compete with the acetogenic and/or methanogenic archaea for by-products such as acetate, butyrate, propionate, and H_2 during the AD process, which results in an inhibition of the process.

The ratio COD/SO_4^{2-} controls the competition of SBR and other anaerobes for substrate minimizing the sulphide toxicity. The degradation of propionate mainly occurred by the high competition of SRB and the acetogens for ethanol and butyrate.

Temperature minimizes the competition between hydrogenotrophic methanogens and SRB when the latter is dominant under mesophilic conditions while methanogens prefer thermophilic temperatures (Jiang *et al.*, 2019).

Non-dissociated hydrogen sulphide (H_2S) is toxic for sulphate reducers and methanogens, because it diffuses freely through the cell membrane of microorganisms, causing protein denaturation and interfering with assimilatory sulphur metabolism (Chen *et al.*, 2008). Thus, the inhibition of microorganisms occurred at 0.002–0.003 mole/l H_2S or 0.003–0.006 mole/L total S (Rueter *et al.*, 1994). However, the sulphide concentration of 150 mg/l has shown to be acceptable for methanogens (O'Flaherty *et al.*, 1998).

5.4.5 Sodium

A low sodium concentration of the feedstock is essential for methanogens due to its importance for the oxidation of nicotinamide adenine dinucleotide hydride (NADH) and adenosine triphosphate (ATP) formation. High sodium concentrations could inhibit the activity of the microorganisms and interfere strongly with their metabolism (Chen *et al.*, 2008; Feijoo *et al.*, 1995; Jiang *et al.*, 2019). Hydrogenotrophic methanogens growth occurs at a sodium concentration of 350 mg/L and their moderate inhibitory effects started at a level of 3500 and 5500 mgNa^+/L. The strong inhibition of methanogenic archaea occurred at a sodium concentration of 8800 mg/L under mesophilic digestion (Chen *et al.*, 2008). Furthermore, 50% inhibition occurred for acetic acid, propionic acid, and n-butyric acid by VFA-degrading bacteria at sodium concentrations of 7000, 10 500, and 19 000 mg/l, respectively (Feijoo *et al.*, 1995). Acetate-utilizing bacteria are more susceptible to high sodium concentration compared to H_2/CO_2-utilising microorganisms and propionate-utilizing bacteria (Liu & Boone, 1991). Reducing sodium ions toxicity to methanogens could be obtained with the addition of calcium (339 mg/L) and potassium (326 mg/L) into the fermentation medium.

5.4.6 Potassium

The passive potassium ions influx occurs to neutralize the membrane potential at high concentrations in the feedstock during AD (Chen *et al.*, 2008). Potassium concentration below 400 mg/L is acceptable for mesophilic and thermophilic anaerobic conditions. Nevertheless, higher levels of potassium cause an inhibitory effect, particularly to the thermophilic anaerobes (Kugelman & Mccarty, 1965). The presence of calcium, magnesium, sodium and ammonium in the feedstock could mitigate the inhibition effect of a high concentration of potassium (Soltan *et al.*, 2017). The optimum sodium, calcium and sodium were reported to be 564, 837 and 379 mg/L (Feijoo *et al.*, 1995).

5.5 MITIGATION OF THE INHIBITORY EFFECT FOR BIOENERGY PRODUCTION VIA ANAEROBIC CO-DIGESTION OF ALGAL BIOMASS WITH ORGANIC WASTE

Microalgae are CO_2 consuming microorganisms and have a positive impact on minimizing greenhouse gas (GHGs) emissions (Farghaly & Tawfik, 2017; Tawfik *et al.*, 2022b; Wazeri *et al.*, 2018). Moreover, they could be utilized as a co-substrate for the AD process to enhance biogas production. Microalgal biomass was utilized as a co-substrate for the enhancement of biogas production from pig manure (Solé-Bundó *et al.*, 2019b). MY was improved by 47% for co-digestion of *C. vulgaris* and potato processing waste at a mixing ratio of 1:3 (Zhang *et al.*, 2019). The addition of *C. vulgaris* creates a balance of C/N ratio for methanogenesis, which improved their bio-metabolism activities. Moreover, the carbohydrates content was increased in the feedstock due to the addition of algal biomass. Supplementation of *C. vulgaris* into the co-digestion process reduced the ammonia toxicity and the formation of free ammonium nitrogen that negatively affects the biomethanization process. This was not the case for anaerobic co-digestion of wheat straw with mixed microalgae, where MY only increased by a value of 5–9% (Solé-Bundó *et al.*, 2017b). This was mainly due to the abundance of lignocellulosic materials in both substrates, which needs specific conditions for the accomplishment of the biodegradation process. Co-digestion of mixed microalgae and food waste increased the MY by 4.99-fold as compared to mono-digestion of microalgae (Zhen *et al.*, 2016). MY was improved by 13–28% in the co-digestion of swine manure and *Chlorella* sp. at the C/N ratio of 35/1 (Wang *et al.*, 2016).

In another attempt, algae biomass (33%, 50%, and 67%) was anaerobically co-digested with fat, oil, and grease waste and the result in MY was 0.54 LCH_4/g VS d. Furthermore, it has been shown that lipids degradation contributed 68–83% of the methane potential productivity (Park & Li, 2012). Continuous AD of blue algae was examined with corn straw in continuous digesters operating at a hydraulic retention time (HRT) of 10 days and 35°C (Zhong *et al.*, 2013). The C/N ratio for the co-digestion process was optimized at 20:1 and

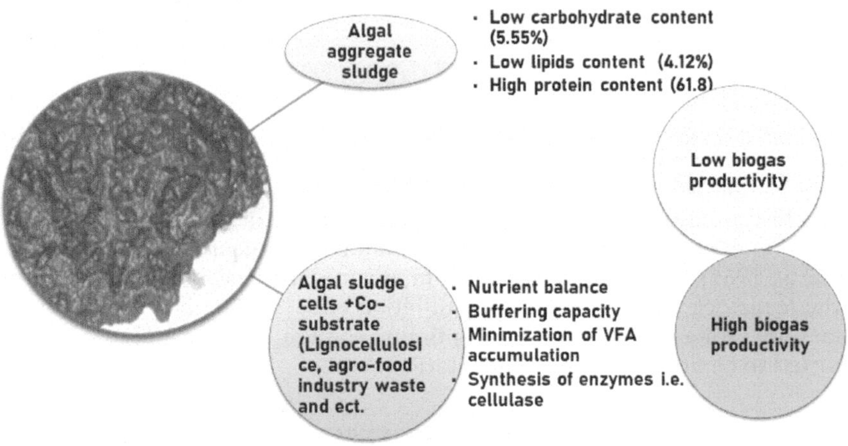

Figure 5.4 Mitigation of the inhibition effect for bioenergy productivity via anaerobic co-digestion of algal biomass with negative organic value waste.

OLR of 6.0 gVS/L d. MY and methane production rate were 234 mL CH_4/gVS and 1404 mL CH_4/L d, respectively, and volatile solids degradation was 63%. Zhong *et al.* (2012) found that anaerobic co-digestion of *Microcystis* spp. with corn straw at 35°C at an initial concentration of 20 gVS/L and C/N ratio of 20:1 achieved MY of 325 mL/gVS. The addition of algal sludge in the co-digestion process creates nutrient balance for anaerobic microorganisms and increases the buffering capacity, avoiding the occurrence of the acidification process. Supplying corn straw in the fermentation of algal sludge biomass increases cellulase activity, which positively affects the methane production rate.

Figure 5.4 shows the positive effect of co-substrates addition to microalgae AD. Cellulases are secreted by hydrolytic microorganisms (i.e., *Clostridium themocellum*) during the fermentation of cellulosic materials (Suto & Tomita, 2001). A mixture of algal species (*Nannochloropsis limnetica*) and piggery slurry (60:40 v/v) was co-digested for biogas production (Tsapekos *et al.*, 2018). Higher biogas production of 23% (equivalent to 355 ± 27 N mL/gVS) was achieved for co-digestion compared to mono-digestion of livestock manure. Similarly, *Chlorella* spp. and primary sludge were anaerobically co-digested in a continuous membrane bioreactor under thermophilic conditions resulting in an MY of 302 mL/gVS at OLR of 0.17–0.5 gCOD/L.d (Serna-García *et al.*, 2020). In another study by Serna-García *et al.* (2021), mesophilic anaerobic co-digestion of *Chlorella* spp. and primary sludge in membrane bioreactor at an SRT of 70 d and HRT of 30 d achieved a MY of 370 mLCH_4/gVS.

Enzymes play a key factor in the AD process. The deficiency of carbohydrate content in the algal sludge and the cell walls minimize the energy yield and productivity (Bakr *et al.*, 2021; Osama *et al.*, 2019). Co-digestion of algal sludge with another substrate enhances the biodegradation activity due to an increase of the hydrolytic enzymes in the fermentation medium (Figure 5.5)

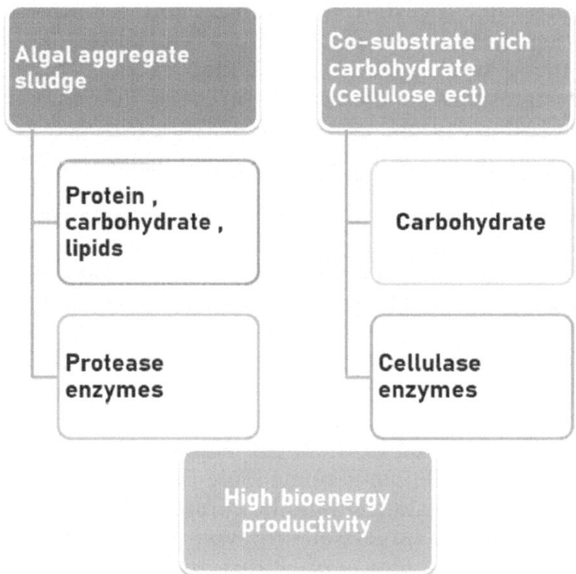

Figure 5.5 Interaction between algal aggregate sludge and co-substrate showing hydrolytic enzyme activities by anaerobes.

(Abdullahi *et al.*, 2020; Nasr *et al.*, 2021; Tawfik *et al.*, 2021b). Hydrolytic bacteria synthesize the cellulase enzyme from the degradation of waste rich in complex sugars, which is utilized to break down the cell walls of the algal aggregates. The latter release protease enzymes by hydrolytic anaerobes for the degradation of proteins (i.e., lignocellulosic wastes). Suto and Tomita (2001) found that the addition of cellulosic rich waste to the anaerobic digester highly induce cellulase excretion by anaerobes (i.e., *Clostridium thermocellum*).

5.6 VALORIZATION OF THE DIGESTATE FROM ANAEROBIC DIGESTION OF MICROALGAE FOR AGRICULTURAL USES

Harvesting the algal biomass after the treating process is regularly carried out. The utilization of the harvested algal biomass by the AD process for bioenergy production is an economic approach. To attain zero biowaste, the digestate from the AD of algal biomass could be reused for agricultural purposes due to its richness with nutrients and minerals. The pH of the digestate was found to be slightly alkaline at a level of 7.0, which was attributed to the higher ammonium concentration in the fermentation medium due to protein degradation (Passos & Ferrer, 2015). Those pH values are compatible and comply with reuse.

The electrical conductivity (EC) of the digestate was 5.9–8.2 dS/m (Alburquerque *et al.*, 2012). The resulted digestate was stabilized due to the presence of volatile fatty acids at a level of 10 mg COD-eq/L (Di Maria *et al.*, 2014). Anaerobic digestate is rich in mineral fertilizers (Nkoa, 2014). The main nutrient in the algal digestates was nitrogen (80 g/kg TS), which is useful for the plants' growth (Solé-Bundó *et al.*, 2017a). Fortunately, microalgae digestates provide the typical nitrogen values that are required for fertility of soils (Alburquerque *et al.*, 2012). Thus, nitrogen values are much higher than those obtained for the digestate resulting from AD of sewage sludge (36–40 g/kg TS) (Di Maria *et al.*, 2014; Di Mauro *et al.*, 2019; Gell *et al.*, 2011). The NH_4^+-N/TKN ratio of the digestate resulting from algal biomass varied from 30.9 to 33.8%, which could be used as a soil amendment rather than chemical fertilizer (Teglia *et al.*, 2011). Phosphorous (P) and potassium (K^+) were detected in the algal biomass digestate at levels of 3.2–3.9 g P/kg TS and 2.2–5.2 g K/kgTS, respectively (Solé-Bundó *et al.*, 2017b). These values are higher than those obtained from AD of sewage sludge where the digestate contained only 2.2–3.0 gK^+/kgTS and 3.2–3.8 g P/kg TS, respectively, for both nutrients (Di Maria *et al.*, 2014; Gell *et al.*, 2011; Tambone *et al.*, 2010). Ca^{2+}, Mg^{2+} and Na^+ were detected in the algal biomass digestate, which are also necessary nutrients for soil amendments (Solé-Bundó *et al.*, 2017a). However, the salt content of the digestate should be carefully analysed before applying the algal digestates to the target soils to avoid their salinization due to the presence of Na^+ (Daliakopoulos *et al.*, 2016).

5.7 CONCLUDING REMARKS

Microalgae are CO_2 consuming microorganisms and have a positive impact on minimizing GHGs emissions. Moreover, they could be utilized as a co-substrate in the AD process to enhance biogas production from organic wastes due to their richness in nutrients, carbohydrates, proteins and lipids, which are useful elements for anaerobes. In addition, the algal aggregates release protease enzymes in AD contributing to the degradation of protein-rich and lignocellulosic wastes. The algal biomass digestate is rich in Ca^{2+}, Mg^{2+} and Na^+, which are necessary elements for soil amendments. Nevertheless, the salts content of the digestate should be carefully analysed to avoid soil salinization due to the presence of Na^+.

ACKNOWLEDGEMENTS

This study was supported by the Science, Technology & Innovation Funding Authority (STIFA)-Egypt, under the Basic Science and Pollution Control grant (ID: 26271 and 41591), Academy of Scientific Research and Technology (ASRT)-Egypt (Code: Call no. 2/2019/ASRT-Nexus), and bilateral research project between ASRT-Egypt and National Natural Science Foundation of China. The author is grateful to the National Research Centre-Egypt for partially supporting this research (Project ID: 12030202).

REFERENCES

Abdullahi K., Elreedy A., Fujii M., Ibrahim M. G. and Tawfik A. (2020). Robustness of anaerobes exposed to cyanuric acid contaminated wastewater and achieving efficient removal via optimized co-digestion scheme. *Journal of Advanced Research*, **24**, 211–222, [online] https://doi.org/10.1016/j.jare.2020.02.006

Alburquerque J. A., de la Fuente C., Ferrer-Costa A., Carrasco L., Cegarra J., Abad M. and Bernal M. P. (2012). Assessment of the fertiliser potential of digestates from farm and agroindustrial residues. *Biomass and Bioenergy*, **40**, 181–189, https://doi.org/10.1016/j.biombioe.2012.02.018

Ali M., Elreedy A., Fujii M. and Tawfik A. (2021). Co-metabolism based adaptation of anaerobes to phenolic saline wastewater in a two-phase reactor enables efficient treatment and bioenergy recovery. *Energy Conversion and Management*, **247**, 114722, [online] https://doi.org/10.1016/j.enconman.2021.114722

Allam A., Tawfik A., El-Saadi A. and Negm A. (2016). Potentials of using duckweed (*Lemna gibba*) for treatment of drainage water for reuse in irrigation purposes. *Desalination and Water Treatment*, **57**(1), 459–467.

Appels L., Baeyens J., Degrève J. and Dewil R. (2008). Principles and potential of the anaerobic digestion of waste-activated sludge. *Progress in Energy and Combustion Science*, **34**(6), 755–781, https://doi.org/10.1016/j.pecs.2008.06.002

Bakr M. H., Nasr M., Ashmawy M. and Tawfik A. (2021). Predictive performance of auto-aerated immobilized biomass reactor treating anaerobic effluent of cardboard wastewater enriched with bronopol (2-bromo-2-nitropropan-1,3-diol) via artificial neural network. *Environmental Technology and Innovation*, **21**, 101327, [online] https://doi.org/10.1016/j.eti.2020.101327

Bassuney D. and Tawfik A. (2017). Baffled duckweed pond system for treatment of agricultural drainage water containing pharmaceuticals. *International Journal of Phytoremediation*, **19**(8), 774–780, [online] https://doi.org/10.1080/15226514.2017.1284756

Beach E. S., Eckelman M. J., Cui Z., Brentner L. and Zimmerman J. B. (2012). Preferential technological and life cycle environmental performance of chitosan flocculation for harvesting of the green algae *Neochloris oleoabundans*. *Bioresource Technology*, **121**, 445–449, [online] https://doi.org/10.1016/j.biortech.2012.06.012

Billuri M., Bonner J. S., Fuller C. B. and Islam M. S. (2015). Impact of natural cationic polymers on charge and clarification of microalgae suspensions. *Environmental Engineering Science*, **32**(3), 212–221, https://doi.org/10.1089/ees.2014.0301

Bonetta S., Bonetta S., Ferretti E., Fezia G., Gilli G. and Carraro E. (2014). Agricultural reuse of the digestate from anaerobic co-digestion of organic waste: microbiological contamination, metal hazards and fertilizing performance. *Water, Air, and Soil Pollution*, **225**(8), 1–11, https://doi.org/10.1007/s11270-014-2046-2

Brentner L. B., Eckelman M. J. and Zimmerman J. B. (2011). Combinatorial life cycle assessment to inform process design of industrial production of algal biodiesel. *Environmental Science and Technology*, **45**(16), 7060–7067, https://doi.org/10.1021/es2006995

Chen Y., Cheng J. J. and Creamer K. S. (2008). Inhibition of anaerobic digestion process: a review. *Bioresource Technology*, **99**(10), 4044–4064, https://doi.org/10.1016/j.biortech.2007.01.057

Cheng J., Xu J., Huang Y., Li Y., Zhou J. and Cen K. (2015). Growth optimisation of microalga mutant at high CO_2 concentration to purify undiluted anaerobic digestion effluent of swine manure. *Bioresource Technology*, **177**, 240–246, [online] https://doi.org/10.1016/j.biortech.2014.11.099

Daliakopoulos I. N., Tsanis I. K., Koutroulis A., Kourgialas N. N., Varouchakis A. E., Karatzas G. P. and Ritsema C. J. (2016). The threat of soil salinity: a European scale review. *Science of the Total Environment*, **573**, 727–739, [online] https://doi.org/10.1016/j.scitotenv.2016.08.177

Di Maria F., Sordi A., Cirulli G., Gigliotti G., Massaccesi L. and Cucina M. (2014). Co-treatment of fruit and vegetable waste in sludge digesters. An analysis of the relationship among bio-methane generation, process stability and digestate phytotoxicity. *Waste Management*, **34**(9), 1603–1608, [online] https://doi.org/10.1016/j.wasman.2014.05.017

Di Mauro M. D., Fava G., Spampinato M., Aleo D., Melilli B., Saita M. G., Centonze G., Maggiore R. and D'antona N. (2019). Polyphenolic fraction from olive mill wastewater: scale-up and in vitro studies for ophthalmic nutraceutical applications. *Antioxidants*, **8**(10), 1–21.

Farghaly A. and Tawfik A. (2017). Simultaneous hydrogen and methane production through multi-phase anaerobic digestion of paperboard mill wastewater under different operating conditions. *Applied Biochemistry and Biotechnology*, **181**(1), 142–156, [online] https://doi.org/10.1007/s12010-016-2204-7

Feijoo G., Soto M., Méndez R. and Lema J. M. (1995). Sodium inhibition in the anaerobic digestion process: antagonism and adaptation phenomena. *Enzyme and Microbial Technology*, **17**(2), 180–188, https://doi.org/10.1016/0141-0229(94)00011-F

Fernandes F., Silkina A., Fuentes-Grünewald C., Wood E. E., Ndovela V. L. S., Oatley-Radcliffe D. L., Lovitt R. W. and Llewellyn C. A. (2020). Valorising nutrient-rich digestate: dilution, settlement and membrane filtration processing for optimisation as a waste-based media for microalgal cultivation. *Waste Management*, **118**, 197–208, [online] https://doi.org/10.1016/j.wasman.2020.08.037

Ge S., Qiu S., Tremblay D., Viner K., Champagne P. and Jessop P. G. (2018). Centrate wastewater treatment with *Chlorella vulgaris*: simultaneous enhancement of nutrient removal, biomass and lipid production. *Chemical Engineering Journal*, **342**, 310–320, [online] https://doi.org/10.1016/j.cej.2018.02.058

Gell K., van Groenigen J. W. and Cayuela M. L. (2011). Residues of bioenergy production chains as soil amendments: immediate and temporal phytotoxicity. *Journal of Hazardous Materials*, **186**(2–3), 2017–2025, [online] https://doi.org/10.1016/j.jhazmat.2010.12.105

George D. B., Berk S. G., Adams V. D., Ting R. S., Roberts R. O., Parks L. H. and Lott R. C. (1995). Toxicity of alum sludge extracts to a freshwater alga, protozoan, fish, and marine bacterium. *Archives of Environmental Contamination and Toxicology*, **29**(2), 149–158, https://doi.org/10.1007/BF00212964

Greses S., Gaby J. C., Aguado D., Ferrer J., Seco A. and Horn S. J. (2017). Microbial community characterization during anaerobic digestion of *Scenedesmus* spp. under mesophilic and thermophilic conditions. *Algal Research*, **27**, 121–130, [online] https://doi.org/10.1016/j.algal.2017.09.002

Hansen K. H., Angelidaki I. and Ahring B. K. (1998). Anaerobic digestion of swine manure: inhibition by ammonia. *Water Research*, **32**(1), 5–12, https://doi.org/10.1016/S0043-1354(97)00201-7

Inglesby A. E. and Fisher A. C. (2012). Enhanced methane yields from anaerobic digestion of *Arthrospira maxima* biomass in an advanced flow-through reactor with an integrated recirculation loop microbial fuel cell. *Energy and Environmental Science*, **5**(7), 7996–8006, https://doi.org/10.1039/c2ee21659k

Jiang Y., McAdam E., Zhang Y., Heaven S., Banks C. and Longhurst P. (2019). Ammonia inhibition and toxicity in anaerobic digestion: a critical review. *Journal of Water Process Engineering*, **32**, 100899, https://doi.org/10.1016/j.jwpe.2019.100899

Koutra E., Mastropetros S. G., Ali S. S., Tsigkou K. and Kornaros M. (2021). Assessing the potential of *Chlorella vulgaris* for valorization of liquid digestates from agro-industrial and municipal organic wastes in a biorefinery approach. *Journal of Cleaner Production*, **280**, 124352, [online] https://doi.org/10.1016/j.jclepro.2020.124352

Kugelman I. J. and Mccarty P. L. (1965). Cation toxicity and stimulation waste treatment in anaerobic. *Journal (Water Pollution Control Federation)*, **37**(1), 97–116.

Kumsiri B., Pekkoh J., Pathom-Aree W., Lumyong S. and Pumas C. (2018). Synergistic effect of co-culture of microalga and actinomycete in diluted chicken manure digestate for lipid production. *Algal Research*, **33**, 239–247, [online] https://doi.org/10.1016/j.algal.2018.05.020

Lee, Y.-J. and Lee, D.-J. (2019). Impact of adding metal nanoparticles on anaerobic digestion performance – A review. *Bioresource Technology*, **292**, 121926.

Liu Y. and Boone D. R. (1991). Effects of salinity on methanogenic decomposition. *Bioresource Technology*, **35**(3), 271–273, https://doi.org/10.1016/0960-8524(91)90124-3

Liu T. and Sung S. (2002). Ammonia inhibition on thermophilic aceticlastic methanogens. *Water Science and Technology: A Journal of the International Association on Water Pollution Research*, **45**(10), 113–120, https://doi.org/10.2166/wst.2002.0304

Massa M., Buono S., Langellotti A. L., Castaldo L., Martello A., Paduano A., Sacchi R. and Fogliano V. (2017). Evaluation of anaerobic digestates from different feedstocks as growth media for *Tetradesmus obliquus*, *Botryococcus braunii*, *Phaeodactylum tricornutum* and *Arthrospira maxima*. *New Biotechnology*, **36**, 8–16, [online] https://doi.org/10.1016/j.nbt.2016.12.007

Nasr M., Tawfik A., Awad H. M., Galal A., El-Qelish M., Abdul Qyyum M., Mumtaz Ali Khan M., Rehan M., Nizami A.-S. and Lee M. (2021). Dual production of hydrogen and biochar from industrial effluent containing phenolic compounds. *Fuel*, **301**, 121087, [online] https://doi.org/10.1016/j.fuel.2021.121087

Nkoa R. (2014). Agricultural benefits and environmental risks of soil fertilization with anaerobic digestates: a review. *Agronomy for Sustainable Development*, **34**(2), 473–492, https://doi.org/10.1007/s13593-013-0196-z

O'Flaherty V., Mahony T., O'Kennedy R. and Colleran E. (1998). Effect of pH on growth kinetics and sulphide toxicity thresholds of a range of methanogenic, syntrophic and sulphate-reducing bacteria. *Process Biochemistry*, **33**(5), 555–569, https://doi.org/10.1016/S0032-9592(98)00018-1

Olsson J., Schwede S., Nehrenheim E. and Thorin E. (2018). Microalgae as biological treatment for municipal wastewater – effects on the sludge handling in a treatment plant. *Water Science and Technology*, **78**(3), 644–654, https://doi.org/10.2166/wst.2018.334

Osama R., Ibrahim M. G., Fujiia M. and Tawfik A. (2019). Potentials of duckweed (*Lemna gibba*) for treatment of 1,4-dioxane containing wastewater using duckweed multi-ponds system. Energy Procedia, **35**, 101179.

Osama R., Awad H. M., Ibrahim M. G. and Tawfik A. (2020). Mechanistic and economic assessment of polyester wastewater treatment via baffled duckweed pond. *Journal of Water Process Engineering*, **35**, 101179, https://doi.org/10.1016/j.jwpe.2020.101179

Osama R., Awad H. M., Zha S., Meng F. and Tawfik A. (2021). Greenhouse gases emissions from duckweed pond system treating polyester resin wastewater containing 1,4-dioxane and heavy metals. *Ecotoxicology and Environmental Safety*, **207**, 111253, [online] https://doi.org/10.1016/j.ecoenv.2020.111253

Park S. and Li Y. (2012). Evaluation of methane production and macronutrient degradation in the anaerobic co-digestion of algae biomass residue and lipid waste. *Bioresource Technology*, **111**, 42–48, [online] https://doi.org/10.1016/j.biortech.2012.01.160

Passos F. and Ferrer I. (2015). Influence of hydrothermal pretreatment on microalgal biomass anaerobic digestion and bioenergy production. *Water Research*, **68**, 364–373, [online] https://doi.org/10.1016/j.watres.2014.10.015

Richardson J. W. and Johnson M. D. (2015). Financial feasibility analysis of NAABB developed technologies. *Algal Research*, **10**, 16–24, [online] https://doi.org/10.1016/j.algal.2015.03.020

Romero-Villegas G. I., Fiamengo M., Acién Fernández F. G. and Molina Grima E. (2018). Utilization of centrate for the outdoor production of marine microalgae at pilot-scale in flat-panel photobioreactors. *Journal of Biotechnology*, **284**, 102–114, [online] https://doi.org/10.1016/j.jbiotec.2018.08.006

Rueter P., Rabus R., Wilkest H., Aeckersberg F., Rainey F. A., Jannasch H. W. and Widdel F. (1994). Anaerobic oxidation of hydrocarbons in crude oil by new types of sulphate-reducing bacteria. *Nature*, **372**(6505), 455–458, https://doi.org/10.1038/372455a0

Safi C., Zebib B., Merah O., Pontalier P. Y. and Vaca-Garcia C. (2014). Morphology, composition, production, processing and applications of *Chlorella vulgaris*: a review. *Renewable and Sustainable Energy Reviews*, **35**, 265–278, https://doi.org/10.1016/j.rser.2014.04.007

Sakarika M. and Kornaros M. (2019). *Chlorella vulgaris* as a green biofuel factory: comparison between biodiesel, biogas and combustible biomass production. *Bioresource Technology*, **273**, 237–243, [online] https://doi.org/10.1016/j.biortech.2018.11.017

Santos-Ballardo D. U., Rossi S., Reyes-Moreno C. and Valdez-Ortiz A. (2016). Microalgae potential as a biogas source: current status, restraints and future trends. *Reviews in Environmental Science and Biotechnology*, **15**(2), 243–264, https://doi.org/10.1007/s11157-016-9392-z

Serna-García R., Borrás L., Bouzas A. and Seco A. (2020). Insights into the biological process performance and microbial diversity during thermophilic microalgae co-digestion in an anaerobic membrane bioreactor (AnMBR). *Algal Research*, **50**, 101981, [online] https://doi.org/10.1016/j.algal.2020.101981

Serna-García R., Ruiz-Barriga P., Noriega-Hevia G., Serralta J., Pachés M. and Bouzas A. (2021). Maximising resource recovery from wastewater grown microalgae and primary sludge in an anaerobic membrane co-digestion pilot plant coupled to a composting process. *Journal of Environmental Management*, **281**, 111890, https://doi.org/10.1016/j.jenvman.2020.111890

Shi R., Handler R. M. and Shonnard D. R. (2019). Life cycle assessment of novel technologies for algae harvesting and oil extraction in the renewable diesel pathway. *Algal Research*, **37**, 248–259, [online] https://doi.org/10.1016/j.algal.2018.12.005

Solé-Bundó M., Cucina M., Folch M., Tàpias J., Gigliotti G., Garfí M. and Ferrer I. (2017a). Assessing the agricultural reuse of the digestate from microalgae anaerobic digestion and co-digestion with sewage sludge. *Science of the Total Environment*, **586**, 1–9, https://doi.org/10.1016/j.scitotenv.2017.02.006

Solé-Bundó M., Eskicioglu C., Garfí M., Carrère H. and Ferrer I. (2017b). Anaerobic co-digestion of microalgal biomass and wheat straw with and without thermo-alkaline pretreatment. *Bioresource Technology*, **237**, 89–98, [online] https://doi.org/10.1016/j.biortech.2017.03.151

Solé-Bundó M., Garfí M., Matamoros V. and Ferrer I. (2019a). Co-digestion of microalgae and primary sludge: effect on biogas production and microcontaminants removal. *Science of the Total Environment*, **660**, 974–981, [online] https://doi.org/10.1016/j.scitotenv.2019.01.011

Solé-Bundó M., Passos F., Romero-Güiza M. S., Ferrer I. and Astals S. (2019b). Co-digestion strategies to enhance microalgae anaerobic digestion: a review. *Renewable and Sustainable Energy Reviews*, **112**, 471–482, [online] https://doi.org/10.1016/j.rser.2019.05.036

Soltan M., Elsamadony M. and Tawfik A. (2017). Biological hydrogen promotion via integrated fermentation of complex agro-industrial wastes. *Applied Energy*, **185**, 929–938, [online] https://doi.org/10.1016/j.apenergy.2016.10.002

Soltan M., Elsamadony M., Mostafa A., Awad H. and Tawfik A. (2019). Nutrients balance for hydrogen potential upgrading from fruit and vegetable peels via fermentation process. *Journal of Environmental Management*, **242**, 384–393, https://doi.org/10.1016/j.jenvman.2019.04.066

Stiles W. A. V., Styles D., Chapman S. P., Esteves S., Bywater A., Melville L., Silkina A., Lupatsch I., Fuentes Grünewald C., Lovitt R., Chaloner T., Bull A., Morris C. and Llewellyn C. A. (2018). Using microalgae in the circular economy to valorise anaerobic digestate: challenges and opportunities. *Bioresource Technology*, **267**, 732–742, [online] https://doi.org/10.1016/j.biortech.2018.07.100

Sukias J. P. S. and Craggs R. J. (2011). Digestion of wastewater pond microalgae and potential inhibition by alum and ammoniacal-N. *Water Science and Technology*, **63**(5), 835–840, https://doi.org/10.2166/wst.2011.101

Sung S. and Liu T. (2003). Ammonia inhibition on thermophilic anaerobic digestion. *Chemosphere*, **53**(1), 43–52, https://doi.org/10.1016/S0045-6535(03)00434-X

Suto M. and Tomita F. (2001). Induction and catabolite repression mechanisms of cellulase in fungi. *Journal of Bioscience and Bioengineering*, **92**(4), 305–311, https://doi.org/10.1016/S1389-1723(01)80231-0

Tambone F., Scaglia B., D'Imporzano G., Schievano A., Orzi V., Salati S. and Adani F. (2010). Assessing amendment and fertilizing properties of digestates from anaerobic digestion through a comparative study with digested sludge and compost. *Chemosphere*, **81**(5), 577–583, [online] https://doi.org/10.1016/j.chemosphere.2010.08.034

Tao R., Kinnunen V., Praveenkumar R., Lakaniemi A. M. and Rintala J. A. (2017). Comparison of *Scenedesmus acuminatus* and *Chlorella vulgaris* cultivation in liquid digestates from anaerobic digestion of pulp and paper industry and municipal wastewater treatment sludge. *Journal of Applied Phycology*, **29**(6), 2845–2856, https://doi.org/10.1007/s10811-017-1175-6

Tawfik A., Hassan Gamal K., Awad H., Hassan M., Rojas P., Sanz J. L., Elsamadony M., Pant D. and Fujii M. (2021a). Strengthen "the sustainable farm" concept via efficacious conversion of farm wastes into methane. *Bioresource Technology*, **341**, 125838, https://doi.org/10.1016/j.biortech.2021.125838

Tawfik A., Nasr M., Galal A., El-Qelish M., Yu Z., Hassan M. A., Salah H. A., Hasanin M. S., Meng F., Bokhari A., Qyyum M. A. and Lee M. (2021b). Fermentation-based nanoparticle systems for sustainable conversion of black-liquor into biohydrogen. *Journal of Cleaner Production*, **309**, 127349, [online] https://doi.org/10.1016/j.jclepro.2021.127349

Tawfik A., Hasanan K., Abdullah M., Badr O. A., Awad H. M., Elsamadony M., El-Dissouky A., Qyyum M. A. and Nizami A.-S. (2022a). Graphene enhanced detoxification of wastewater rich 4-nitrophenol in multistage anaerobic reactor followed by baffled high-rate algal pond. *Journal of Hazardous Materials*, **424**(PA), 127395, [online] https://doi.org/10.1016/j.jhazmat.2021.127395

Tawfik A., Niaz H., Qadeer K., Qyyum M. A., Liu J. J. and Lee M. (2022b). Valorization of algal cells for biomass and bioenergy production from wastewater: sustainable strategies, challenges, and techno-economic limitations. *Renewable and Sustainable Energy Reviews*, **157**, 112024, [online] https://doi.org/10.1016/j.rser.2021.112024

Teglia C., Tremier A. and Martel J. L. (2011). Characterization of solid digestates: part 2, assessment of the quality and suitability for composting of six digested products. *Waste and Biomass Valorization*, **2**(2), 113–126, https://doi.org/10.1007/s12649-010-9059-x

Torruella G., Grau-Bové X., Moreira D., Karpov S. A., Burns J. A., Sebé-Pedrós A., Völcker E. and López-García P. (2018). Global transcriptome analysis of the aphelid *Paraphelidium tribonemae* supports the phagotrophic origin of fungi. *Communications Biology*, **1**(1), 1–11, https://doi.org/10.1038/s42003-018-0235-z

Tsapekos P., Kougias P. G., Alvarado-Morales M., Kovalovszki A., Corbière M. and Angelidaki I. (2018). Energy recovery from wastewater microalgae through anaerobic digestion process: methane potential, continuous reactor operation and modelling aspects. *Biochemical Engineering Journal*, **139**, 1–7, [online] https://doi.org/10.1016/j.bej.2018.08.004

Uggetti E., Sialve B., Latrille E. and Steyer J. P. (2014). Anaerobic digestate as substrate for microalgae culture: the role of ammonium concentration on the microalgae productivity. *Bioresource Technology*, **152**, 437–443, [online] https://doi.org/10.1016/j.biortech.2013.11.036

Wang M., Lee E., Zhang Q. and Ergas S. J. (2016). Anaerobic co-digestion of swine manure and microalgae *Chlorella* sp.: experimental studies and energy analysis. *Bioenergy Research*, **9**(4), 1204–1215, https://doi.org/10.1007/s12155-016-9769-4

Wang L., Addy M., Liu J., Nekich C., Zhang R., Peng P., Cheng Y., Cobb K., Liu Y., Wang H. and Ruan R. (2019). Integrated process for anaerobically digested swine manure treatment. *Bioresource Technology*, **273**, 506–514, [online] https://doi.org/10.1016/j.biortech.2018.11.050

Wang Q., Shen Q., Wang J., Zhang Y., Zhang Z., Lei Z., Shimizu K. and Lee D. J. (2020). Fast cultivation and harvesting of oil-producing microalgae *Ankistrodesmus falcatus* var. *acicularis* fed with anaerobic digestion liquor via biogranulation in addition to nutrients removal. *Science of the Total Environment*, **741**, 140183, [online] https://doi.org/10.1016/j.scitotenv.2020.140183

Wazeri A., Elsamadony M., Roux S. L., Peu P. and Tawfik A. (2018). Potentials of using mixed culture bacteria incorporated with sodium bicarbonate for hydrogen production from water hyacinth. *Bioresource Technology*, **263**, 365–374.

Xia A. and Murphy J. D. (2016). Microalgal cultivation in treating liquid digestate from biogas systems. *Trends in Biotechnology*, **34**(4), 264–275, [online] https://doi.org/10.1016/j.tibtech.2015.12.010

Yen H. W. and Brune D. E. (2007). Anaerobic co-digestion of algal sludge and waste paper to produce methane. *Bioresource Technology*, **98**(1), 130–134, https://doi.org/10.1016/j.biortech.2005.11.010

Yu H., Kim J. and Lee C. (2019). Potential of mixed-culture microalgae enriched from aerobic and anaerobic sludges for nutrient removal and biomass production from anaerobic effluents. *Bioresource Technology*, **280**, 325–336, [online] https://doi.org/10.1016/j.biortech.2019.02.054

Zamalloa C., Boon N. and Verstraete W. (2012). Anaerobic digestibility of *Scenedesmus obliquus* and *Phaeodactylum tricornutum* under mesophilic and thermophilic conditions. *Applied Energy*, **92**, 733–738, [online] https://doi.org/10.1016/j.apenergy.2011.08.017

Zhang Y., Caldwell G. S., Zealand A. M. and Sallis P. J. (2019). Anaerobic co-digestion of microalgae *Chlorella vulgaris* and potato processing waste: effect of mixing ratio, waste type and substrate to inoculum ratio. *Biochemical Engineering Journal*, **143**, 91–100, [online] https://doi.org/10.1016/j.bej.2018.12.021

Zhen G., Lu X., Kobayashi T., Kumar G. and Xu K. (2016). Anaerobic co-digestion on improving methane production from mixed microalgae (*Scenedesmus* sp., *Chlorella* sp.) and food waste: kinetic modeling and synergistic impact evaluation. *Chemical Engineering Journal*, **299**, 332–341, [online] https://doi.org/10.1016/j.cej.2016.04.118

Zhong W., Zhang Z., Luo Y., Qiao W., Xiao M. and Zhang M. (2012). Biogas productivity by co-digesting Taihu blue algae with corn straw as an external carbon source. *Bioresource Technology*, **114**, 281–286, [online] https://doi.org/10.1016/j.biortech.2012.02.111

Zhong W., Chi L., Luo Y., Zhang Z., Zhang Z. and Wu W. M. (2013). Enhanced methane production from Taihu Lake blue algae by anaerobic co-digestion with corn straw in continuous feed digesters. *Bioresource Technology*, **134**, 264–270, [online] https://doi.org/10.1016/j.biortech.2013.02.060

Zhu S., Feng S., Xu Z., Qin L., Shang C., Feng P., Wang Z. and Yuan Z. (2019). Cultivation of *Chlorella vulgaris* on unsterilized dairy-derived liquid digestate for simultaneous biofuels feedstock production and pollutant removal. *Bioresource Technology*, **285**, 121353, [online] https://doi.org/10.1016/j.biortech.2019.121353

IWA PUBLISHING

doi: 10.2166/9781789062755_0131

Chapter 6

Nutrient recovery from anaerobic digestate

Minh T. Vu[1], Luong N. Nguyen[1], Johir A. H. Mohammed[1], Jakub Zdarta[2], M. Mofijur[1], Nirenkumar Pathak[1] and Long D. Nghiem[1]*

[1]Centre for Technology in Water and Wastewater, School of Civil and Environmental Engineering, University of Technology Sydney, Sydney, NSW 2007, Australia
[2]Institute of Materials Science and Engineering, Faculty of Mechanical Engineering and Management, Poznan University of Technology, Poznan, Poland
**Corresponding author e-mail: luongngoc.nguyen@uts.edu.au*

ABSTRACT

Anaerobic digestate or digested sludge that is produced from the digestion of organic waste streams by microorganisms in the absence of free oxygen is a concentrated source of nutrients (i.e., nitrogen and phosphorus). Given increasing concerns regarding limited natural resources in line with a circular economy perspective, there has been a paradigm shift from nutrient removal to recovery from this stream. In this chapter, the profile of nutrients in different types of digestate derived from a range of feedstock is meticulously characterised. A variety of conventional and emerging technologies to recover nutrients from anaerobic digestate are also comprehensively reviewed. In particular, this study briefly presents a techno-economic analysis and an environmental impact assessment of nutrient recovery from anaerobic digestate.

Keywords: Anaerobic digestion, digestate, nutrient recovery, circular economy, solid–liquid separation

6.1 INTRODUCTION

Anaerobic digestion is extensively applied to treat organic wastes such as sewage sludge, food waste, and agricultural waste and produce energy in the form of biogas (Selvaraj *et al.*, 2021). It has become an indispensable technology in the emerging green energy economy given its advantages over other technologies (i.e., cost-effective organic wastes treatment for wastewater and solid waste streams and energy alternative in rural areas through the digestion of agro-wastes and/or energy crops) (Campos *et al.*, 2019). Anaerobic digestion is a mature technology applied in many full-scale plants with a digester capacity ranging from a few

Editor(s): Vinay Kumar Tyagi, Kaoutar Aboudi and Cigdem Eskicioglu

hundred to thousands of cubic metres (Selvaraj *et al.*, 2021). In a metropolitan area, anaerobic digestion plants are used to treat sewage sludge and organic fraction of municipal solid wastes (Hanum *et al.*, 2019), whereas their rural counterparts on-farm are employed to co-treat animal manure and other organic residues (e.g., straw and other agricultural wastes) (Campos *et al.*, 2019).

Besides biogas, anaerobic digestion of organic wastes produces digestate known as digested sludge. Digestate is a mixture of biodegraded organic matter, water, anaerobic biological biomass, heavy metals, and nutrients (Barampouti *et al.*, 2020). During anaerobic digestion, hydrolysis occurs and facilitates the release of nutrients from the organic matter to the bulk liquid (Amini *et al.*, 2017; Ma *et al.*, 2018). Organic nitrogen is converted into ammonium ions and organic phosphorus is hydrolysed to soluble phosphorus (Nguyen *et al.*, 2021). Thus, the digestate is very rich in nutrients. The concentration of nutrients depends on the three key factors of the anaerobic digestion process including types of feedstock, digester designs and operating conditions, and digestate management methods (Sheets *et al.*, 2015).

The digestate released from anaerobic digestion is both a problem and an opportunity. The current practice of using digestate directly as fertilisers for land application causes a number of environmental and public health concerns, such as odour, excess nutrients in soils, which have potential for eutrophication in, water bodies due to agricultural run-off, and pathogenic problems (Vaneeckhaute *et al.*, 2017; Zhao & Liu, 2019). Moreover, in wastewater treatment plants (WWTPs), the circulation of the liquid fraction of digestate (i.e., a nutrient-concentrated stream) to the head of work for further treatment causes the gradual accumulation of nutrients in the treatment systems, decreased treatment efficiency, excess nutrients in the effluent, and struvite blockage (Ansari *et al.*, 2016; Vu *et al.*, 2021b). Nonetheless, high nitrogen and phosphorus contents in the digestate make it an ideal candidate for nutrient recovery for fertiliser production and other industrial applications. Nutrient recovery from the digestate is a win-win solution. It can address the environmental concerns, compensate for the alarming shortage of global minable phosphate rock reserves as the main source of current phosphorus, and prevent detrimental impacts of digestate on current WWTPs.

This chapter aims to have an overall picture of nutrient characteristics of anaerobic digestate in order to highlight its potential for nutrient recovery. Current practice for digestate management is discussed. The chapter provides an update on the technologies used for anaerobic digestate valorisation via nutrient recovery. The maturity of each technology for nutrient recovery from the digestate is indicated in this study. In addition, a brief techno-economic and environmental assessment to evaluate the significance of nutrient recovery from the digestate to society is an important aspect of this chapter.

6.2 NUTRIENT CONCENTRATIONS IN ANAEROBIC DIGESTATE

The composition of the digestate is regulated by the types of feedstock used for digestion and operating conditions (e.g., anaerobic digestion configurations, temperature, retention time, and pressure). The pH range of the digestate varies from neutral to alkaline. A typical digestate includes anaerobic microorganisms, water, digested organic matter, and nutrients (e.g., nitrogen (N), phosphorus (P), potassium (K), calcium (Ca), magnesium (Mg), and heavy metals) (Table 6.1). Although the digestate can be produced from various feedstocks (e.g., pig slurry, cattle and chicken manure, agricultural waste, food waste, and sewage sludge), the composition of the final digestion products is similar (Vaneeckhaute *et al.*, 2017). Due to only transformation among nutrient forms of feedstock during anaerobic digestion, digestate has normally the same content of nutrients as feedstock (Barampouti *et al.*, 2020; Sheets *et al.*, 2015; Xie *et al.*, 2017).

Nitrogen in the digestate is derived from the nitrogen-bounded proteins and its content highly depends on the type of the feedstock. The concentration of total nitrogen could account for up to 24% dry biomass (Barampouti *et al.*, 2020). The digestate using protein-rich sources (e.g., food waste) as feedstock has 40% total nitrogen content higher than that from manure (Menardo *et al.*, 2011; Tiwary *et al.*, 2015). During hydrolysis and fermentation of organic matter in anaerobic digestion, organically bound nitrogen is released as ammoniacal nitrogen (i.e., ammonium (NH_4^+) and free ammonia (NH_3)) which is mainly present in digestate (Christensen *et al.*, 2009). Indeed, total ammoniacal nitrogen accounts for 35–81% total nitrogen (Christensen *et al.*, 2009; Fouda *et al.*, 2013; Sheets *et al.*, 2015). The concentration of ammonium nitrogen (NH_4^+-N) in the digestate is reported in the range of 0.8–5 g/L (Limoli *et al.*, 2016; Tao & Ukwuani, 2015; Vu *et al.*, 2022).

The total input phosphorus content in the digestate is independent of the anaerobic digestion process but governed by the feedstock composition (Selvaraj *et al.*, 2021; Vaneeckhaute *et al.*, 2017). Due to microorganisms' metabolism under anaerobic conditions, organically bound phosphorus present in the feedstock is transformed to soluble orthophosphate (Salehiyoun *et al.*, 2020). Phosphorus content can reach 1.2 g/L in the digestate (Shi *et al.*, 2018). The anaerobic conditions are favourable for the precipitation of orthophosphate in the digestate under the forms of calcium phosphate ($Ca_3(PO_4)_2$) and magnesium phosphate ($Mg_3(PO_4)_2$) (Christensen *et al.*, 2009). It is estimated that approximately 90% of the total phosphorus in the digestate is within these precipitates, while the dissolved fraction only accounts for around 10% (Christensen *et al.*, 2009).

6.3 TECHNOLOGIES FOR NUTRIENT RECOVERY FROM ANAEROBIC DIGESTATE

6.3.1 Solid–liquid separation of the digestate

High organic matter and nutrient contents of the digestate make it possible for direct land application as a simple and economical method to recover nutrients. However, given the environmental and social concerns (e.g., odour, ammonia volatilisation, heavy metal pollution, biological soil contamination due to the presence of pathogenic microorganisms, and leaching of nutrients which leads to eutrophication and algal bloom), the use of digestate for direct land

Table 6.1 Nutrient characteristics of unprocessed digestate from various feedstock.

Parameters	Food waste	Sewage sludge	Agricultural feedstock	Cattle manure and slurry	Co-digestion organic matter (i.e., sewage sludge and vegetable waste or manure and industrial waste)
pH	7.9–8.3	7.5	7.5–8.4	7.3–8.89	5.6–8.3
Dry matter (DM) (%)	1.4–7.88	1.9–2.97	6.41–24	22–9.2	1.83–34.2
Organic matter (% DM)	38–73	48.4–73.6	69–77	23.9–81.5	23.9–81
Total carbon (% DM)	0.44–45.2	4.38	1.92–40	0.59–5.07	0.41–11.6
Total nitrogen (% DM)	0.06–15.7	0.005–0.378	0.14–8.8	0.05–0.62	0.12–5.04
Ammoniacal nitrogen (% DM)	0.05–10.8	–	0.04–4.5	0.16–2.358	0.0476–0.1987
Total phosphorus (% DM)	0.008–1.6	0.04	0.058–6.6	0.034–0.616	0.01–1.001
Total potassium (% DM)	0.03–8.1	0.00019	0.324–10	0.03–1.273	0.03–2.52
Total magnesium (% DM)	0.079–5.2	–	0.041–3.1	0.013–0.166	0.006–0.26
Total calcium (% DM)	0.014–1	–	0.077–3.1	0.044–0.846	0.01–1.56
Total sulphur (% DM)	0.01–1	–	0.01–0.041	0.008–0.048	0.004–0.096
References	(Peng & Pivato, 2019; Selvaraj *et al.*, 2021; Sheets *et al.*, 2015)	(Barampouti *et al.*, 2020; Salehiyoun *et al.*, 2020; Xie *et al.*, 2017)	(Barampouti *et al.*, 2020; Sheets *et al.*, 2015; Tambone *et al.*, 2017; Tiwary *et al.*, 2015; Walsh *et al.*, 2018)	(Huttunen *et al.*, 2014; Selvaraj *et al.*, 2021; Zhang *et al.*, 2020)	(Huttunen *et al.*, 2014; Tambone *et al.*, 2017; Tampio *et al.*, 2016; Yasar *et al.*, 2017)

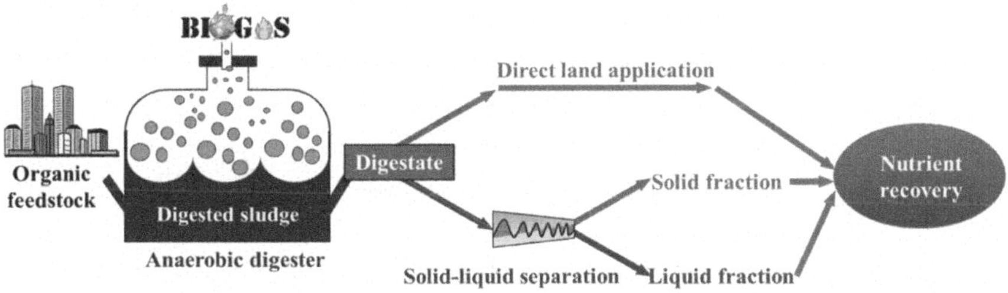

Figure 6.1 Overview of a current digestate processing approach.

application is limited and strictly regulated in most nations (Vaneeckhaute *et al.*, 2017). In addition, the storage of digestate can result in the emission of greenhouse gases (e.g., CH_4 and N_2O) (Selvaraj *et al.*, 2021).

Solid–liquid mechanical separation is extensively applied as an initial stage to valorise digestate in practice (Figure 6.1). The primary purpose is to reduce the volume of digestate and thus transportation costs as well as to ensure effective management of anaerobic digestate including handling, storage, and disposal. Two fractions (i.e., solids and liquids) obtained from dewatering digestate are processed in different ways to maximise their potential in nutrient recovery.

The solid fraction called biosolids contains 60–80% of the digestate dry matter, only 20–25% of the initial nitrogen, and almost all phosphorus of the digestate (40–90%) (Barampouti *et al.*, 2020; Campos *et al.*, 2019; Limoli *et al.*, 2016). However, most nutrients are present in organically bound compounds (Vaneeckhaute *et al.*, 2017). Thus, the possibility to recover nutrients from the solid fraction is limited. This biosolids fraction can be directly applied as an organic fertiliser in agriculture or can be used as a precursor for a variety of value-added materials (e.g., fuel, pyrochar, carbon-based materials, and nanocellulose) (Monlau *et al.*, 2015a, 2015b; Selvaraj *et al.*, 2021). The nutrient recovery from the solid fraction can be performed through composting, thermal treatment, drying, and biodrying (Figure 6.2).

In contrast, the liquid fraction is characterised by lower dry matter (3.3–6.6% of the digestate dry matter), phosphorus, and organic carbon contents but is rich in ammonium nitrogen (40–80% of the initial nitrogen) (Ledda *et al.*, 2013; Tambone *et al.*, 2017; Vaneeckhaute *et al.*, 2017). The liquid fraction with very high concentrations of soluble nutrients (i.e., 1100 mg/L NH_3-N and 500 mg/L PO_4^{3-}) is favourable for subsequent recovery (Vu *et al.*, 2019; Vu *et al.*, 2020). The possible technologies for nutrient recovery from the liquid fraction include ammonia stripping, chemical precipitation, adsorption and ion exchange, phycoremediation, membrane-based processes, and evaporation (Figure 6.2) (Barampouti *et al.*, 2020; Campos *et al.*, 2019; Selvaraj *et al.*, 2021; Vaneeckhaute *et al.*, 2017).

Mechanical separation methods using a screw press, centrifuge, and screening drum press are extensively applied in full-scale applications (Barampouti *et al.*, 2020). Screw presses are usually selected for the treatment of fibrous substrates

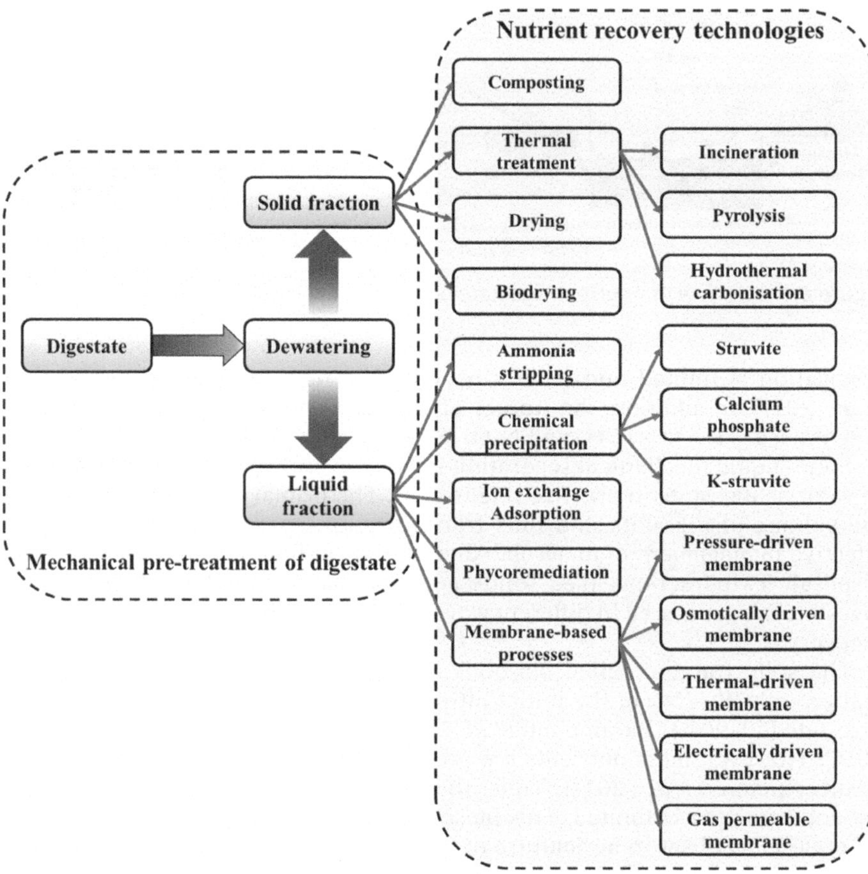

Figure 6.2 Overview of the latest technologies available for nutrient recovery from the digestate.

while centrifuge is adopted for non-fibrous feedstock. In general, a screw press is cheaper than a centrifuge due to less energy consumption (Barampouti *et al.*, 2020; Zarebska *et al.*, 2015). In practice, flocculants/coagulants such as $Al_2(SO_4)_3$, $FeCl_3$, $Fe_2(SO_4)_3$, $Ca(OH)_2$, and cationic polymers are mixed with digestate prior to the mechanical separation to improve separation efficiency for further improvement in subsequent nutrient recovery (Meixner *et al.*, 2015). It is reported that the addition of precipitating additives ($FeCl_3$ and $CaCO_3$) increased the separation efficiency of the suspended solids from 46% to 75% (Meixner *et al.*, 2015). Chen *et al.* (2021) used a template copolymer of acryloxy trimethylammonium chloride and acrylamide for dewatering sludge from a WWTP and achieved a filter biosolids cake moisture content of 74.7% at the polymer dosage of 40 mg/L.

6.3.2 Nutrient recovery from liquid fraction of the digestate

6.3.2.1 Ammonia stripping

Nitrogen recovery from the liquid fraction of the digestate using a stripping method is implemented through the physical conversion of ammoniacal nitrogen from the aqueous phase (NH_4^+) to the gas phase (NH_3 gas) at high pH and the presence of air or stream flows (Kinidi *et al.*, 2018; Limoli *et al.*, 2016; Yuan *et al.*, 2016). Ammonia stripping is already a mature technology (Vaneeckhaute *et al.*, 2017). Examples of commercial stripping processes for digested sludge and manure treatment include AMFER (Colsen, Netherland) and ANAStrip (GNS, Germany) (Vaneeckhaute *et al.*, 2017). These systems could remove up to 90% ammonia from the digestate (Vaneeckhaute *et al.*, 2017). This method is derived from equation (6.1).

$$NH_4^+ + OH^- \leftrightarrow H_2O + NH_3(g) \tag{6.1}$$

Ammonia stripping is usually performed using a packed-bed tower (Barampouti *et al.*, 2020; Fuchs & Drosg, 2013; Sengupta *et al.*, 2015). This configuration allows for an increased mass transfer area. The gas is then transferred to an air scrubber to allow for the adsorption of NH_3 gas from the gas to a liquid phase using sulphuric acid (H_2SO_4) in order to form and recover a concentrated solution of ammonium sulphate ($(NH_4)_2SO_4$) as an end-product (Vaneeckhaute *et al.*, 2017). Ammonium sulphate can be used as a fertiliser for agricultural production. Scaling problems are common to the ammonia stripping packed-bed towers (Barampouti *et al.*, 2020). A calcium oxide softening step is normally performed to remove magnesium, calcium, and carbonates prior to stripping in order to prevent the formation of scaling (Barampouti *et al.*, 2020).

The stripping efficiency is governed by a wide range of factors, such as ammonium content of substrate, hydraulic loading, packing density, retention time, ratio of air to liquid volume, air and liquid flowrates, but mostly pH and temperature (Kinidi *et al.*, 2018; Sengupta *et al.*, 2015; Shi *et al.*, 2018; Yuan *et al.*, 2016). Typically, high pH, temperature, air flowrate, and ammonium concentration would be beneficial to the conversion of ammonium towards NH_3 gas (Shi *et al.*, 2018). It is indicated that ammonia recovery efficiency increases from 80 to 92% as pH increases from 8 to 11 (Sengupta *et al.*, 2015). In practice, lime and caustic soda are usually used to increase pH before the stripping process (Sengupta *et al.*, 2015). However, at temperature over 80°C, ammonia can be fully recovered regardless of pH of the solution (Bonmatí & Flotats, 2003). The removal efficiency decreases significantly as air temperature decreases. For example, the stripping efficiency decreases from 95 to 75% as temperature decreases from 20 to 10°C (Sengupta *et al.*, 2015).

6.3.2.2 Chemical precipitation

Chemical precipitation is one of the most widely used technologies for nutrient recovery from the digestate (Puyol *et al.*, 2017; Shi *et al.*, 2018). The precipitation of calcium phosphate allows for the recovery of only phosphorus while struvite precipitation is a simple method for the simultaneous recovery of nitrogen and phosphorus. These precipitates are subsequently settled and harvested by gravity, crystallisation, or removed by filtration, whereby nutrient recovery is achieved. These precipitates can be used as fertilisers for agricultural production. This technology has been worldwide implemented at full-scale applications for nutrient recovery from products of the anaerobic digestion process (Table 6.2). The formation of phosphorous precipitates can be expressed in equations (6.2) and (6.3).

$$10Ca^{2+} + 6PO_4{}^{3-} + 2OH^- \rightarrow Ca_{10}(PO_4)_6(OH)_2 \downarrow \tag{6.2}$$

$$Mg^{2+} + NH^{4+} + H_nPO_4{}^{3-n} + 6H_2O \rightarrow MgNH_4PO_4.6H_2O \downarrow + nH^+, \text{ where } n - 0,1 \text{ or } 2 \tag{6.3}$$

In struvite precipitation, the addition of magnesium and caustic soda (NaOH) into the digestate is required for the occurrence of the reaction. The optimum pH for struvite precipitation is in the range of 8.5–9.5 (Daneshgar *et al.*, 2018; Günther *et al.*, 2018; Shi *et al.*, 2018). Due to the continuous release of protons during the struvite formation, the continuous supply of alkali is needed to maintain the pH in a suitable range for struvite precipitation. In addition to pH, the molar ratio of magnesium to ammonium to phosphate significantly affects the struvite formation efficiency and its purity. It is reported that this ratio of 1.6:0.6:1.0 is favourable for struvite formation (Sengupta *et al.*, 2015). In the

Table 6.2 Current commercial full-scale technologies for phosphorus recovery from the digestate and its liquid fraction.

Input materials	Process	Developer/location	Recovery principles	Product
Liquid fraction	Ostara Pearl	Ostara, Canada	Precipitation/ crystallisation	Struvite
Liquid fraction	MagPrex™	Centrisys/CNP, USA	Precipitation/ crystallisation	Struvite
Liquid fraction	Crystalactor™	Royal HaskoningDHV, Netherland	Precipitation/ crystallisation	CaP
Liquid fraction	Phosnix	Unitikia Ltd., Japan	Precipitation/ crystallisation	Struvite
Liquid fraction	Seaborne Gifhorn	Seaborne Environmental Research Laboratory, Germany	Precipitation/ crystallisation	Struvite
Liquid fraction/ digestate	ANPHOS®	Colsen, Netherland	Precipitation/ crystallisation	Struvite
Liquid fraction/ digestate	PHOSPAQ™	Paques, Netherland	Precipitation/ crystallisation	Struvite
Liquid fraction/ digestate	NuReSys	Nutrient Recovery Systems, Belgium	Precipitation/ crystallisation	Struvite

digestate, ammonium and phosphate are abundant while magnesium content is insufficient to ensure an effective reaction (Kataki *et al.*, 2016). For this reason, magnesium sourced from $MgCl_2$, $MgSO_4$, MgO, and $Mg(OH)_2$ is added to the digestate (Lee *et al.*, 2003; Quintana *et al.*, 2005). These requirements of alkali and magnesium supplementation predominantly contribute to the costs of struvite precipitation (i.e., approximately 75% of overall production costs) (Barampouti *et al.*, 2020). In this regard, the use of low-cost magnesium sources (e.g., industrial grade MgO, seawater, or bittern) can make this technology economically viable (Kataki *et al.*, 2016; Lee *et al.*, 2003). It is reported that struvite recovery efficiency can reach 85–97% (Sengupta *et al.*, 2015).

Phosphorus can be recovered from the digestate as calcium phosphate (Barampouti *et al.*, 2020; Song *et al.*, 2002; Tran *et al.*, 2014; Yeoman *et al.*, 1988). Instead of adding magnesium to form struvite, calcium sourced from lime is supplemented to the digestate to induce calcium phosphate formation. Hence, the amount of $Ca(OH)_2$ required will regulate the operational costs of this precipitation process. The pH range of 8–11 is favourable for the calcium phosphate precipitation (Yeoman *et al.*, 1988). CO_2-stripping may be performed prior to precipitation to prevent the competing formation of calcium carbonate. A removal efficiency of 50–90% P has been reported in the literature (Vaneeckhaute *et al.*, 2017).

6.3.2.3 Ion exchange and adsorption

Nutrient recovery using ion exchange and adsorption is based on the ion-selective retention ability of active materials (Barampouti *et al.*, 2020; Bunce *et al.*, 2018). This retention is driven by intermolecular and ionic forces towards adsorption and ion exchange, respectively. Ion exchange and adsorption are usually performed in column beds filled up with solid sorbents. A number of materials can be used as sorbents (e.g., locally sourced sands and gravel, furnace slag, fly ash, chemically modified clay, zeolites, resins, and biochar) (Barampouti *et al.*, 2020; Vaneeckhaute *et al.*, 2017; Vu *et al.*, 2021a). After the adsorption capacity of sorbents is saturated, regeneration is usually performed using nitric acid and sodium chloride. These saturated materials can be either directly applied onto agricultural fields as fertilisers or integrated with another process (e.g., air stripping and precipitation) for nutrient recovery from the nutrient-rich stream obtained from regeneration.

One of the most frequently used sorbents or ion exchangers for nutrient removal and recovery is natural zeolites. In nature, there are over 50 different types of naturally occurring zeolites with different traits, but most of them are generally aluminium silicate minerals with high cation-exchange capacity and high ammonium selective properties (Barampouti *et al.*, 2020; Hedström, 2001). Of these zeolites, clinoptilolite is the most popular for practice due to its low-cost availability (Vaneeckhaute *et al.*, 2017). Due to the exceptional properties of this material, it can be used for both ammonium and phosphate removal in different regimes. The recovery of ammonium from wastewater using zeolite is based on the replacement of ammonium ions with cations on its surface (i.e., Na^+, K^+, Ca^{2+}, etc.), while the presence of calcium, aluminium, and silicon in the composition of zeolite is beneficial for phosphorus removal via adsorption.

The adsorption capacity for ammonium is reported to be 19 g/kg for natural zeolite and 21 g/kg for Na-zeolite with the removal efficiencies ranging between 71 and 91% (Barampouti *et al.*, 2020). These figures for phosphorus are much lower (Barampouti *et al.*, 2020; Shi *et al.*, 2018).

The nutrient recovery efficiency by ion exchange and adsorption is governed by sorbent properties, physicochemical characteristics of the liquid fraction of the digestate, and operating conditions (Hedström, 2001; Vaneeckhaute *et al.*, 2017). Sorbents with high surface area, high porosity, and more reactive sites are beneficial for the nutrient recovery process. In addition, the key characteristics of this liquid fraction, such as initial suspended solid and nutrient concentrations, temperature, pH, and the presence of other competitive ions need to be taken into consideration when applying this technology. Fouling of the sorbent bed due to the high suspended solid in the liquid fraction of the digestate and the competition of other ions (e.g., Mg^{2+}, Ca^{2+}, Cl^-, SO_4^{2-}, and CO_3^{2-}) in this stream are challenges of this approach.

Although ion exchange and adsorption are robust techniques with high selectivity for ammonium and phosphate removal, their applicability in practice for the treatment of the liquid fraction of the digestate still need to be demonstrated. To date, the application of these techniques in nutrient recovery has been mostly implemented in lab-scale (Vaneeckhaute *et al.*, 2017). Further investigation into the process efficiency at pilot scale is necessary. The costs of these technologies are subject to the sorbent materials used, pre-treatment of the influent, and methods and frequency of regeneration. It is reported that the chemical cost of adsorption using zeolite is higher than struvite precipitation and NH_3 stripping. However, low labour cost for operation and low energy consumption are strengths of adsorption and ion-exchange techniques (Barampouti *et al.*, 2020; Zarebska *et al.*, 2015).

6.3.2.4 Membrane-based technologies

A number of pressure-driven membrane processes, such as microfiltration (MF), ultrafiltration (UF), nanofiltration (NF), and reverse osmosis (RO) can be applied to recover nutrients from the liquid fraction of the digestate. In MF and UF, porous membrane can retain most suspended solids, but not dissolved compounds (e.g., NH_4^+-N and PO_4^{3-}). Thus, the permeate is rich in ammonium and phosphate. In order to further purify the permeate, NF and RO are used. NF can achieve 5–23% NH_4^+ and 97–98% PO_4^{3-} (Barampouti *et al.*, 2020). In RO, the separation efficiency of these ions can achieve 99–100% (Barampouti *et al.*, 2020). It is reported that RO could concentrate NH_4^+-N to 7–10 g/L at the retention efficiency of 99.9% (Shi *et al.*, 2018). Nutrients can be recovered from these ammonium and phosphate concentrated streams by using ammonia stripping or precipitation as discussed in the earlier sections. However, using a single pressure-driven membrane process for nutrient recovery from the digestate liquid has yet to prove to be a feasible approach in large scale due to significant chemical and energy requirements for operation and membrane fouling.

To overcome drawbacks of the pressure-driven membrane processes (i.e., severe membrane fouling and energy consumption), non-pressure membrane technologies (i.e., forward osmosis (FO), membrane distillation (MD), electrodialysis (ED) have been recently used for nutrient recovery from wastewater. The overall principle is based on the enrichment of nutrients in wastewater due to the solute excellent retention of the membranes followed by chemical precipitation (e.g., calcium phosphate and struvite) or ammonia stripping.

Recently, FO has been demonstrated as a robust separation platform used to pre-concentrate the liquid fraction of the digestate for subsequent nutrient recovery (Ansari *et al.*, 2016; Vu *et al.*, 2018, 2019, 2021b). In the FO process, only water is transported through a semipermeable membrane from a feed solution (i.e., digestate liquid) to a draw solution (i.e., a high salinity solution) due to the difference in osmotic pressure of these two solutions, which leads to increased nutrient contents. Different from the pressure-driven membrane technologies, fouling layers in FO are not compacted and reversible due to the absence of hydraulic pressure (Holloway *et al.*, 2007; Vu *et al.*, 2019). For this reason, FO is usually recommended to be applied for the filtration of high-strength wastewater (e.g., the liquid fraction of the digestate). Several studies demonstrated the feasibility of using FO to concentrate the digestate liquid stream with high enrichment factors (Xie *et al.*, 2014). Xie *et al.* (2014) could achieve a concentration factor of 5 when using FO driven by $MgCl_2$ draw solution for pre-concentration of the digestate liquid stream.

Unlike FO, the enrichment of nutrients (ammonium and phosphate) for subsequent recovery using ED is an electrically driven membrane process (Thompson Brewster *et al.*, 2017; Ward *et al.*, 2018). In this process, the transportation of nutrients in the bulk feed is oriented by an electrical field produced by a cathode and an anode. Under the direct current, cations (i.e., NH_4^+-N) move towards the cathode, while anions (PO_4^{3-}) move to the anode (Barampouti *et al.*, 2020). The ED process is equipped with ion exchange membranes including cation-selective exchange membranes, anion-selective exchange membranes, and bipolar membranes (Thompson Brewster *et al.*, 2017; Ward *et al.*, 2018). It is reported that NH_4^+-N content in the digestate liquid could be concentrated to 16–21 g/L using ED (Barampouti *et al.*, 2020; Shi *et al.*, 2018). This figure is even much higher than that obtained from RO. ED has been usually applied to recover nutrients from the digestate in the lab-scale. There is only one ED pilot scale in the Netherlands that reported recovery of ammonium and potassium carbonates from the digestate (Barampouti *et al.*, 2020).

Nutrient recovery through the application of MD is a promising option for the liquid fraction of the digestate (Barampouti *et al.*, 2020; Shi *et al.*, 2018). The separation and enrichment of nutrients using MD is a thermally driven process. In MD, the volatile components (i.e., NH_3 and water) in the digestate liquid are converted to their gaseous phases before penetrating through the MD membrane to the permeate side, whereas non-volatile components (e.g., PO_4^{3-}) are retained in the feed side of the membrane. The concentrated phosphate stream in the feed side is recovered using precipitation or adsorption. On the permeate side, acidic solutions (e.g., HCl or H_2SO_4) are applied to absorb NH_3 gas for direct fertiliser production. It is reported that at $pH > 9.68$ and 45°C, ammonium in the digestate liquid could be concentrated in the permeate up to 18.3 g/L (Shi *et al.*, 2018).

Hydrophobic gas-permeable membrane technology is another option for ammonia recovery (Hasanoğlu *et al.*, 2010; Nagy *et al.*, 2019). In this system, the ammonium–ammonia balance shifts towards ammonia gas under an alkaline environment. Ammonia gas is transported through the membrane due to the difference in NH_3 concentration between the two sides of the membrane (Nagy *et al.*, 2019). This gradient remains constant due to the continuous reaction of NH_3 gas with sulphuric acid to form ammonium sulphate on the other side of the membrane. Hasanoğlu *et al.* (2010) successfully demonstrated the feasibility of using a macroporous hydrophobic membrane for ammonia recovery as $(NH_4)_2SO_4$.

6.3.2.5 Phycoremediation

Microalgae-based treatment has recently emerged as a cost-effective and environmentally friendly method for nutrient recovery from wastewater (Bunce *et al.*, 2018; Selvaraj *et al.*, 2021; Vaneeckhaute *et al.*, 2017; Vu *et al.*, 2022). Microalgae are photoautotrophic microorganisms that can be capable of using inorganic carbon (i.e., CO_2), light energy, and nutrients to produce biomass via its metabolism. Microalgae biomass is then harvested, whereby nutrients are extracted from the aqueous solution. Microalgae biomass can be directly applied to the field or used as feedstock for other processes (e.g., biochemical and biofuel productions) (Selvaraj *et al.*, 2021). The abundance of nutrients in the liquid fraction of the digestate is beneficial for microalgae cultivation. Once the nutrient usage alone accounts for 50% total cost of microalgae cultivation, using the digestate liquid phase is expected to reduce significantly the cost of cultivation (Selvaraj *et al.*, 2021).

Microalgae can be cultivated photoautotrophically, heterotrophically, and mixotrophically, which is dependent on the carbon source for their growth (Selvaraj *et al.*, 2021; Vu *et al.*, 2020). In autotrophic growth, microalgae produce the needed organic matter and energy by using inorganic carbon (e.g., CO_2) as a carbon source and sunlight as an energy source. In heterotrophic cultivation, microalgae use organic carbon for their metabolism under dark conditions. In the presence of both organic and inorganic carbon, microalgae can be capable of using these carbon sources for their growth under the mixotrophic mechanism. Thus, the fraction liquid of the digestate with plenty of organic and inorganic carbon can promote the mixotrophic growth of microalgae.

Indeed, several studies have been conducted using microalgae-based treatments in photobioreactors for simultaneous nutrient removal and recovery and biomass production from the digestate. Many types of microalgae can grow well in the digestate liquid stream, including freshwater microalgae (e.g., *Chlorella pyrenoidosa*, *Chlorella sorokiniana*, *Scenedesmus obliquus*, *Tetradesmus obliquus*, *Botryococcus braunii*) and marine microalgae (e.g., *Nannochloropsis salina* and *Phaeodactylum tricornutum*) (Selvaraj *et al.*, 2021). Sayedin *et al.* (2020) demonstrated that *Chlorella sorokiniana* could grow well and remove 95% and 78% of nitrogen and phosphorus, respectively from anaerobic digestate. However, the microalgae-based technology normally requires high hydraulic retention time, thus a high footprint due to the slow

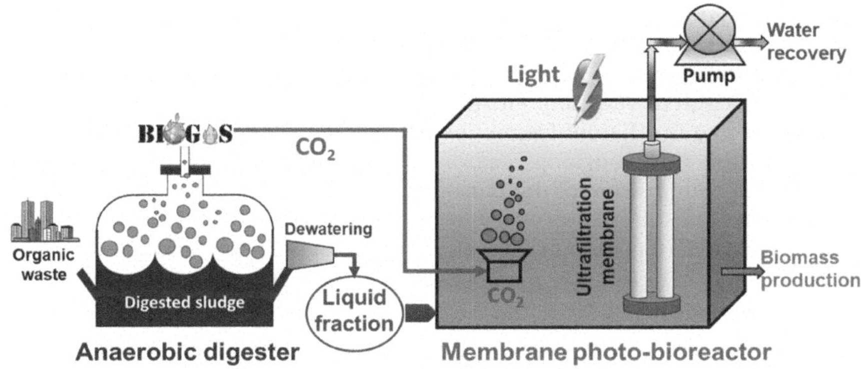

Figure 6.3 New approach for recycling of the liquid fraction of the digestate for microalgal biomass production and nutrient recovery.

kinetics of microalgae growth. Consequently, it has been rarely commercially applied for removing nutrients from wastewater (Vu *et al.*, 2022).

The aforementioned challenge can be addressed by integrating the microalgae system with a membrane separation (e.g., UF) to form a membrane photobioreactor (MPR). In the MPR, a high microalgal biomass concentration resulting from the excellent retention of the membrane can reduce the hydraulic retention time for the assimilation process, thereby decreasing the footprint required. Vu *et al.* (2022) have demonstrated the feasibility of using the digestate liquid stream to grow *Chlorella vulgaris* for nutrient recovery and biomass production. In this study, the novel developed MPR allowed for continuous cultivation of microalgae with 40% higher biomass concentration compared to that in conventional MPR. A graphical representation of the recycling of the liquid fraction of the digestate and biogas produced from anaerobic digestion for microalgal biomass production and nutrient recovery is presented in Figure 6.3.

6.3.3 Nutrient recovery from solid fraction of the digestate

6.3.3.1 Composting

Composting is a treatment method used to stabilise the digestate and obtain valuable and marketable end-products for agriculture (Barampouti *et al.*, 2020; Bustamante *et al.*, 2013). Due to the inherent characteristics of the digestate (e.g., odour, high humidity, high content of volatile fatty acid, and the presence of a wide range of pathogens), its direct use for agricultural soils is usually restricted (Selvaraj *et al.*, 2021). Composting can address these issues to valorise the digestate. The low ratio of carbon to nitrogen and biodegradability plus high humidity are the main challenges during composting the solid fraction of the digestate (Wang *et al.*, 2008). Mixing the digestate solids with bulking materials (e.g., straw and wood chips) in order to ensure optimal C/N ratio and aerobic conditions is applied in practice to facilitate the composting efficiency (Barampouti *et al.*, 2020).

6.3.3.2 Biodrying

Basically, the biodrying principle is similar to composting but focuses on eliminating the moisture of the digestate solid fraction and maintaining its high carbon content (Barampouti *et al.*, 2020; Wang *et al.*, 2008). In other words, biodrying is different from composting regarding the objectives of each process. Within the biodrying reactor, the thermal energy generated during the aerobic decomposition of the biosolids is combined with excess aeration to dry the waste. Phosphorus recovery from dried digestate using coupling of low pressure wet oxidation and nanofiltration, wet chemical extraction and struvite formation, and metallurgical phosphorus recovery is readily commercialised (Barampouti *et al.*, 2020).

6.3.3.3 Drying

Thermal drying is an effective biosolids management. This technology requires a relatively small footprint and has a high degree of disinfection, thus is becoming increasingly implemented by municipalities or private equities. The heating energy used for this process can be compensated by the waste heat of the anaerobic digestion process. Granule- or pellet-shaped, thermally dried biosolids can lead to their improved fertilising traits, reduced transportation costs, and facilitated land applications. Milorganite, for example, is a pelletised biosolids product that has been produced and marketed across the US for several decades (Selvaraj *et al.*, 2021). The typical biosolid drying systems that have been well established and extensively used on farms in Europe include belt dryers, mechanical drum dryers, and solar drying systems (Barampouti *et al.*, 2020; Fuchs & Drosg, 2013). One of the drawbacks of thermal drying is an energy-intensive requirement (Barampouti *et al.*, 2020). However, this can be less prohibitive as the cost of transportation is increasing over time.

6.3.3.4 Thermo-chemical treatment

Nutrient recovery from biosolids can be facilitated through thermochemical treatments such as combustion, pyrolysis, and hydrothermal carbonisation (Syed-Hassan *et al.*, 2017). The ashes produced from the combustion process contain 20–25% P_2O_5 (Barampouti *et al.*, 2020). Phosphorus recovery by acid leaching of biosolids ashes obtained by the combustion process has been reported with 78% P recovery efficiency (Shi *et al.*, 2018). Biochar and hydrochar produced from pyrolysis and hydrothermal carbonisation of biosolids have demonstrated their high nutrient capacity with improved fertilising characteristics (Barampouti *et al.*, 2020; Vaneeckhaute *et al.*, 2017). Pyrolysis can completely retain the phosphorus content of manure digestate. Biochars showed better nutrient recover performance than incineration ashes.

6.4 A TECHNO-ECONOMIC AND ENVIRONMENTAL IMPACT ASSESSMENT OF NUTRIENT RECOVERY FROM THE DIGESTATE

The economic viability of nutrient recovery from digestate is governed by initial capital expenditure, technologies applied, operating expenditures (i.e., chemical and energy consumption, pumping, human resource, etc.), and transportation costs (Campos *et al.*, 2019; Vaneeckhaute *et al.*, 2017). The digestate treatment technologies (e.g., solid–liquid separation, membrane filtration, bio drying, thermal drying, etc.) that allow for the significant reduction of volume can reduce transportation costs. In general, the proceeds from the sales of the recovered fertilisers from waste streams would not compensate for production costs, which makes nutrient recovery economically unfeasible (Campos *et al.*, 2019). Moreover, the recovered fertilisers only have a local market due to administrative restrictions and less competitive prices compared to commercial fertilisers. However, the economical balance should also take into account costs related to the environmental impacts caused by nutrient disposals. For example, nitrogen recovery from wastewater can help avoid the nitrogen pollution cost in Europe, which is between 70 and 100 billion euros per year (Sutton *et al.*, 2011). In addition, the implementation of a phosphorus recovery system in wastewater treatment plants could prevent the occurrence of scaling blockage caused by struvite. For instance, the operating costs of the Amsterdam West WWTP were reduced by 0.5 million euros/year after the installation of a struvite crystalliser (Campos *et al.*, 2019). In most cases, the nutrient recovery driver is subject to the legislation of each country rather than the economic feasibility of the process.

6.5 CONCLUSIONS

The chapter provides a brief overview of the current management of digestate produced from anaerobic digestion of organic matter before an attempt is made to link this current practice to possible strategies used to recover nutrients from this stream. The digestate is comprehensively characterised regarding the nutrient composition in order to highlight the potential for valorisation. In addition, this work attempts to systematically review up-to-date technologies available for enhancing the value of the digestate so that nutrients can be recovered efficiently and economically. A commercial application status of each technology for nutrient recovery is presented. The techno-economic aspects of nutrient recovery from anaerobic digestate are briefly discussed in consideration of the environmental impacts.

REFERENCES

Amini A., Aponte-Morales V., Wang M., Dilbeck M., Lahav O., Zhang Q., Cunningham J. A. and Ergas S. J. (2017). Cost-effective treatment of swine wastes through recovery of energy and nutrients. *Waste Management*, **69**, 508–517, https://doi.org/10.1016/j.wasman.2017.08.041

Ansari A. J., Hai F. I., Price W. E. and Nghiem L. D. (2016). Phosphorus recovery from digested sludge centrate using seawater-driven forward osmosis. *Separation and Purification Technology*, **163**, 1–7, https://doi.org/10.1016/j.seppur.2016.02.031

Barampouti E. M., Mai S., Malamis D., Moustakas K. and Loizidou M. (2020). Exploring technological alternatives of nutrient recovery from digestate as a secondary resource. *Renewable and Sustainable Energy Reviews*, **134**, 110379, https://doi.org/10.1016/j.rser.2020.110379

Bonmatí A. and Flotats X. (2003). Air stripping of ammonia from pig slurry: characterisation and feasibility as a pre- or post-treatment to mesophilic anaerobic digestion. *Waste Management*, **23**(3), 261–272, https://doi.org/10.1016/S0956-053X(02)00144-7

Bunce J. T., Ndam E., Ofiteru I. D., Moore A. and Graham D. W. (2018). A review of phosphorus removal technologies and their applicability to small-scale domestic wastewater treatment systems. *Frontiers in Environmental Science*, **6**(8).

Bustamante M. A., Restrepo A. P., Alburquerque J. A., Pérez-Murcia M. D., Paredes C., Moral R. and Bernal M. P. (2013). Recycling of anaerobic digestates by composting: effect of the bulking agent used. *Journal of Cleaner Production*, **47**, 61–69, https://doi.org/10.1016/j.jclepro.2012.07.018

Campos J. L., Crutchik D., Franchi Ó, Pavissich J. P., Belmonte M., Pedrouso A., Mosquera-Corral A. and del Río V. (2019). Nitrogen and phosphorus recovery from anaerobically pretreated agro-food wastes: a review. *Frontiers in Sustainable Food Systems*, **2**, 91.

Chen Y., Li X., Zizeng W., Feng L., Xie J., Lin Z., Xu Z., Liu B., Li X. and Zheng H. (2021). Research on a new cationic polyacrylamide (CPAM) with a cationic microblock structure and its enhanced effect on sludge condition and dewatering. *Environmental Science and Pollution Research*, **28**(37), 51865–51878, https://doi.org/10.1007/s11356-021-14325-3

Christensen M. L., Hjorth M. and Keiding K. (2009). Characterization of pig slurry with reference to flocculation and separation. *Water Research*, **43**(3), 773–783, https://doi.org/10.1016/j.watres.2008.11.010

Daneshgar S., Buttafava A., Capsoni D., Callegari A. and Capodaglio G. A. (2018). Impact of pH and ionic molar ratios on phosphorous forms precipitation and recovery from different wastewater sludges. *Resources*, **7**(4), 71, https://doi.org/10.3390/resources7040071

Fouda S., von Tucher S., Lichti F. and Schmidhalter U. (2013). Nitrogen availability of various biogas residues applied to ryegrass. *Journal of Plant Nutrition and Soil Science*, **176**(4), 572–584, https://doi.org/10.1002/jpln.201100233

Fuchs W. and Drosg B. (2013). Assessment of the state of the art of technologies for the processing of digestate residue from anaerobic digesters. *Water Science and Technology*, **67**(9), 1984–1993, https://doi.org/10.2166/wst.2013.075

Günther S., Grunert M. and Müller S. (2018). Overview of recent advances in phosphorus recovery for fertilizer production. *Engineering in Life Sciences*, **18**(7), 434–439, https://doi.org/10.1002/elsc.201700171

Hanum F., Yuan L. C., Kamahara H., Aziz H. A., Atsuta Y., Yamada T. and Daimon H. (2019). Treatment of sewage sludge using anaerobic digestion in Malaysia: current state and challenges. *Frontiers in Energy Research*, **7**(19).

Hasanoğlu A., Romero J., Pérez B. and Plaza A. (2010). Ammonia removal from wastewater streams through membrane contactors: experimental and theoretical analysis of operation parameters and configuration. *Chemical Engineering Journal*, **160**(2), 530–537, https://doi.org/10.1016/j.cej.2010.03.064

Hedström A. (2001). Ion exchange of ammonium in zeolites: a literature review. *Journal of Environmental Engineering*, **127**(8), 673–681, https://doi.org/10.1061/(ASCE)0733-9372(2001)127:8(673)

Holloway R. W., Childress A. E., Dennett K. E. and Cath T. Y. (2007). Forward osmosis for concentration of anaerobic digester centrate. *Water Research*, **41**(17), 4005–4014, https://doi.org/10.1016/j.watres.2007.05.054

Huttunen S., Manninen K. and Leskinen P. (2014). Combining biogas LCA reviews with stakeholder interviews to analyse life cycle impacts at a practical level. *Journal of Cleaner Production*, **80**, 5–16, https://doi.org/10.1016/j.jclepro.2014.05.081

Kataki S., West H., Clarke M. and Baruah D. C. (2016). Phosphorus recovery as struvite: recent concerns for use of seed, alternative Mg source, nitrogen conservation and fertilizer potential. *Resources, Conservation and Recycling*, **107**, 142–156, https://doi.org/10.1016/j.resconrec.2015.12.009

Kinidi L., Tan I. A. W., Abdul Wahab N. B., Tamrin K. F. B., Hipolito C. N. and Salleh S. F. (2018). Recent development in ammonia stripping process for industrial wastewater treatment. *International Journal of Chemical Engineering*, **2018**, 1–14, https://doi.org/10.1155/2018/3181087

Ledda C., Schievano A., Salati S. and Adani F. (2013). Nitrogen and water recovery from animal slurries by a new integrated ultrafiltration, reverse osmosis and cold stripping process: a case study. *Water Research*, **47**(16), 6157–6166, https://doi.org/10.1016/j.watres.2013.07.037

Lee S. I., Weon S. Y., Lee C. W. and Koopman B. (2003). Removal of nitrogen and phosphate from wastewater by addition of bittern. *Chemosphere*, **51**(4), 265–271, https://doi.org/10.1016/S0045-6535(02)00807-X

Limoli A., Langone M. and Andreottola G. (2016). Ammonia removal from raw manure digestate by means of a turbulent mixing stripping process. *Journal of Environmental Management*, **176**, 1–10, https://doi.org/10.1016/j.jenvman.2016.03.007

Ma H., Guo Y., Qin Y. and Li Y.-Y. (2018). Nutrient recovery technologies integrated with energy recovery by waste biomass anaerobic digestion. *Bioresource Technology*, **269**, 520–531, https://doi.org/10.1016/j.biortech.2018.08.114

Meixner K., Fuchs W., Valkova T., Svardal K., Loderer C., Neureiter M., Bochmann G. and Drosg B. (2015). Effect of precipitating agents on centrifugation and ultrafiltration performance of thin stillage digestate. *Separation and Purification Technology*, **145**, 154–160, https://doi.org/10.1016/j.seppur.2015.03.003

Menardo S., Gioelli F. and Balsari P. (2011). The methane yield of digestate: effect of organic loading rate, hydraulic retention time, and plant feeding. *Bioresource Technology*, **102**(3), 2348–2351, https://doi.org/10.1016/j.biortech.2010.10.094

Monlau F., Sambusiti C., Antoniou N., Zabaniotou A., Solhy A. and Barakat A. (2015a). Pyrochars from bioenergy residue as novel bio-adsorbents for lignocellulosic hydrolysate detoxification. *Bioresource Technology*, **187**, 379–386, https://doi.org/10.1016/j.biortech.2015.03.137

Monlau F., Sambusiti C., Ficara E., Aboulkas A., Barakat A. and Carrère H. (2015b). New opportunities for agricultural digestate valorization: current situation and perspectives. *Energy & Environmental Science*, **8**(9), 2600–2621, https://doi.org/10.1039/C5EE01633A

Nagy J., Kaljunen J. and Toth A. J. (2019). Nitrogen recovery from wastewater and human urine with hydrophobic gas separation membrane: experiments and modelling. *Chemical Papers*, **73**(8), 1903–1915, https://doi.org/10.1007/s11696-019-00740-x

Nguyen L. N., Kumar J., Vu M. T., Mohammed J. A. H., Pathak N., Commault A. S., Sutherland D., Zdarta J., Tyagi V. K. and Nghiem L. D. (2021). Biomethane production from anaerobic co-digestion at wastewater treatment plants: A critical review on development and innovations in biogas upgrading techniques. *Science of the Total Environment*, **765**, 142753, https://doi.org/10.1016/j.scitotenv.2020.142753

Peng W. and Pivato A. (2019). Sustainable management of digestate from the organic fraction of municipal solid waste and food waste under the concepts of back to earth alternatives and circular economy. *Waste and Biomass Valorization*, **10**(2), 465–481, https://doi.org/10.1007/s12649-017-0071-2

Puyol D., Batstone D. J., Hülsen T., Astals S., Peces M. and Krömer J. O. (2017). Resource recovery from wastewater by biological technologies: opportunities, challenges, and prospects. *Frontiers in Microbiology*, **7**, 2106, https://doi.org/10.3389/fmicb.2016.02106

Quintana M., Sánchez E., Colmenarejo M. F., Barrera J., García G. and Borja R. (2005). Kinetics of phosphorus removal and struvite formation by the utilization of by-product of magnesium oxide production. *Chemical Engineering Journal*, **111**(1), 45–52, https://doi.org/10.1016/j.cej.2005.05.005

Salehiyoun A. R., Di Maria F., Sharifi M., Norouzi O., Zilouei H. and Aghbashlo M. (2020). Anaerobic co-digestion of sewage sludge and slaughterhouse waste in existing wastewater digesters. *Renewable Energy*, **145**, 2503–2509, https://doi.org/10.1016/j.renene.2019.08.001

Sayedin F., Kermanshahi-pour A., He Q. S., Tibbetts S. M., Lalonde C. G. E. and Brar S. K. (2020). Microalgae cultivation in thin stillage anaerobic digestate for nutrient recovery and bioproduct production. *Algal Research*, **47**, 101867, https://doi.org/10.1016/j.algal.2020.101867

Selvaraj P. S., Periasamy K., Suganya K., Ramadass K., Muthusamy S., Ramesh P., Bush R., Vincent S. G. T. and Palanisami T. (2021). Novel resources recovery from anaerobic digestates: current trends and future perspectives. *Critical Reviews in Environmental Science and Technology*, **52**, 1–85.

Sengupta S., Nawaz T. and Beaudry J. (2015). Nitrogen and phosphorus recovery from wastewater. *Current Pollution Reports*, **1**(3), 155–166, https://doi.org/10.1007/s40726-015-0013-1

Sheets J. P., Yang L., Ge X., Wang Z. and Li Y. (2015). Beyond land application: emerging technologies for the treatment and reuse of anaerobically digested agricultural and food waste. *Waste Management*, **44**, 94–115, https://doi.org/10.1016/j.wasman.2015.07.037

Shi L., Simplicio W. S., Wu G., Hu Z., Hu H. and Zhan X. (2018). Nutrient recovery from digestate of anaerobic digestion of livestock manure: a review. *Current Pollution Reports*, **4**(2), 74–83, https://doi.org/10.1007/s40726-018-0082-z

Song Y., Hahn H. H. and Hoffmann E. (2002). The effect of carbonate on the precipitation of calcium phosphate. *Environmental Technology*, **23**(2), 207–215, https://doi.org/10.1080/09593332508618427

Sutton M. A., Oenema O., Erisman J. W., Leip A., van Grinsven H. and Winiwarter W. (2011). Too much of a good thing. *Nature*, **472**(7342), 159–161, https://doi.org/10.1038/472159a

Syed-Hassan S. S. A., Wang Y., Hu S., Su S. and Xiang J. (2017). Thermochemical processing of sewage sludge to energy and fuel: fundamentals, challenges and considerations. *Renewable and Sustainable Energy Reviews*, **80**, 888–913, https://doi.org/10.1016/j.rser.2017.05.262

Tambone F., Orzi V., D'Imporzano G. and Adani F. (2017). Solid and liquid fractionation of digestate: mass balance, chemical characterization, and agronomic and environmental value. *Bioresource Technology*, **243**, 1251–1256, https://doi.org/10.1016/j.biortech.2017.07.130

Tampio E., Salo T. and Rintala J. (2016). Agronomic characteristics of five different urban waste digestates. *Journal of Environmental Management*, **169**, 293–302, https://doi.org/10.1016/j.jenvman.2016.01.001

Tao W. and Ukwuani A. T. (2015). Coupling thermal stripping and acid absorption for ammonia recovery from dairy manure: ammonia volatilization kinetics and effects of temperature, pH and dissolved solids content. *Chemical Engineering Journal*, **280**, 188–196, https://doi.org/10.1016/j.cej.2015.05.119

Thompson Brewster E., Ward A. J., Mehta C. M., Radjenovic J. and Batstone D. J. (2017). Predicting scale formation during electrodialytic nutrient recovery. *Water Research*, **110**, 202–210, https://doi.org/10.1016/j.watres.2016.11.063

Tiwary A., Williams I. D., Pant D. C. and Kishore V. V. N. (2015). Emerging perspectives on environmental burden minimisation initiatives from anaerobic digestion technologies for community scale biomass valorisation. *Renewable and Sustainable Energy Reviews*, **42**, 883–901, https://doi.org/10.1016/j.rser.2014.10.052

Tran A. T. K., Zhang Y., De Corte D., Hannes J.-B., Ye W., Mondal P., Jullok N., Meesschaert B., Pinoy L. and Van der Bruggen B. (2014). P-recovery as calcium phosphate from wastewater using an integrated selectrodialysis/crystallization process. *Journal of Cleaner Production*, **77**, 140–151, https://doi.org/10.1016/j.jclepro.2014.01.069

Vaneeckhaute C., Lebuf V., Michels E., Belia E., Vanrolleghem P. A., Tack F. M. G. and Meers E. (2017). Nutrient recovery from digestate: systematic technology review and product classification. *Waste and Biomass Valorization*, **8**(1), 21–40, https://doi.org/10.1007/s12649-016-9642-x

Vu M. T., Ansari A. J., Hai F. I. and Nghiem L. D. (2018). Performance of a seawater-driven forward osmosis process for pre-concentrating digested sludge centrate: organic enrichment and membrane fouling. *Environmental Science: Water Research & Technology*, **4**(7), 1047–1056, https://doi.org/10.1039/C8EW00132D

Vu M. T., Price W. E., He T., Zhang X. and Nghiem L. D. (2019). Seawater-driven forward osmosis for pre-concentrating nutrients in digested sludge centrate. *Journal of Environmental Management*, **247**, 135–139.

Vu M. T., Vu H. P., Nguyen L. N., Semblante G. U., Johir M. A. H. and Nghiem L. D. (2020). A hybrid anaerobic and microalgal membrane reactor for energy and microalgal biomass production from wastewater. *Environmental Technology & Innovation*, **19**, 100834, https://doi.org/10.1016/j.eti.2020.100834

Vu M. T., Nguyen L. N., Hasan Johir M. A., Ngo H. H., Skidmore C., Fontana A., Galway B., Bustamante H. and Nghiem L. D. (2021a). Phosphorus removal from aqueous solution by steel making slag – mechanisms and performance optimisation. *Journal of Cleaner Production*, **284**, 124753, https://doi.org/10.1016/j.jclepro.2020.124753

Vu M. T., Nguyen L. N., Hasan Johir M. A., Zhang X., Nghiem L. D. and Elimelech M. (2021b). Biogas sparging to control fouling and enhance resource recovery from anaerobically digested sludge centrate by forward osmosis. *Journal of Membrane Science*, **625**, 119176.

Vu M. T., Nguyen L. N., Mofijur M., Johir M. A. H., Ngo H. H., Mahlia T. M. I. and Nghiem L. D. (2022). Simultaneous nutrient recovery and algal biomass production from anaerobically digested sludge centrate using a membrane photobioreactor. *Bioresource Technology*, **343**, 126069, https://doi.org/10.1016/j.biortech.2021.126069

Walsh J. J., Jones D. L., Chadwick D. R. and Williams A. P. (2018). Repeated application of anaerobic digestate, undigested cattle slurry and inorganic fertilizer N: impacts on pasture yield and quality. *Grass and Forage Science*, **73**(3), 758–763, https://doi.org/10.1111/gfs.12354

Wang H., Brown S. L., Magesan G. N., Slade A. H., Quintern M., Clinton P. W. and Payn T. W. (2008). Technological options for the management of biosolids. *Environmental Science and Pollution Research – International*, **15**(4), 308–317, https://doi.org/10.1007/s11356-008-0012-5

Ward A. J., Arola K., Thompson Brewster E., Mehta C. M. and Batstone D. J. (2018). Nutrient recovery from wastewater through pilot scale electrodialysis. *Water Research*, **135**, 57–65, https://doi.org/10.1016/j.watres.2018.02.021

Xie M., Nghiem L. D., Price W. E. and Elimelech M. (2014). Toward resource recovery from wastewater: extraction of phosphorus from digested sludge using a hybrid forward osmosis–membrane distillation process. *Environmental Science & Technology Letters*, **1**(2), 191–195, https://doi.org/10.1021/ez400189z

Xie S., Wickham R. and Nghiem L. D. (2017). Synergistic effect from anaerobic co-digestion of sewage sludge and organic wastes. *International Biodeterioration & Biodegradation*, **116**, 191–197, https://doi.org/10.1016/j.ibiod.2016.10.037

Yasar A., Rasheed R., Tabinda A. B., Tahir A. and Sarwar F. (2017). Life cycle assessment of a medium commercial scale biogas plant and nutritional assessment of effluent slurry. *Renewable and Sustainable Energy Reviews*, **67**, 364–371, https://doi.org/10.1016/j.rser.2016.09.026

Yeoman S., Stephenson T., Lester J. N. and Perry R. (1988). The removal of phosphorus during wastewater treatment: A review. *Environmental Pollution*, **49**(3), 183–233, https://doi.org/10.1016/0269-7491(88)90209-6

Yuan M.-H., Chen Y.-H., Tsai J.-Y. and Chang C.-Y. (2016). Removal of ammonia from wastewater by air stripping process in laboratory and pilot scales using a rotating packed bed at ambient temperature. *Journal of the Taiwan Institute of Chemical Engineers*, **60**, 488–495, https://doi.org/10.1016/j.jtice.2015.11.016

Zarebska A., Romero Nieto D., Christensen K. V., Fjerbæk Søtoft L. and Norddahl B. (2015). Ammonium fertilizers production from manure: a critical review. *Critical Reviews in Environmental Science and Technology*, **45**(14), 1469–1521, https://doi.org/10.1080/10643389.2014.955630

Zhang T., He X., Deng Y., Tsang D. C. W., Jiang R., Becker G. C. and Kruse A. (2020). Phosphorus recovered from digestate by hydrothermal processes with struvite crystallization and its potential as a fertilizer. *Science of the Total Environment*, **698**, 134240, https://doi.org/10.1016/j.scitotenv.2019.134240

Zhao Q. and Liu Y. (2019). Is anaerobic digestion a reliable barrier for deactivation of pathogens in biosludge? *Science of the Total Environment*, **668**, 893–902, https://doi.org/10.1016/j.scitotenv.2019.03.063

doi: 10.2166/9781789062755_0151

Chapter 7

Ammonia recovery from anaerobic digestate

*Vinay Pratap[1,2], Nitesh Machrika[2], Bholu Ram Yadav[2] and Sunil Kumar[2]**

[1]Academy of Science and Innovative Research (AcSIR), Ghazlabad 201002, India
[2]CSIR- National Environmental Engineering Research Institute (CSIR- NEERI), Nehru Marg, Nagpur 440020, India
**Corresponding author e-mail: s_kumar@neeri.res.in*

ABSTRACT

To meet the market demands and discharge regulations, anaerobic digestion (AD) plants have been improvised to recover nutrients such as ammonia and phosphate from the digestate. These recovered nutrients are vital for agriculture, and recovery of these nutrients will help in the reduction of synthetic fertilizers. In wastewater, ammonia is a significant inorganic pollutant; the recovery of ammonia from anaerobic digestate after treatment is an economical and environmental advancement. There are various techniques applied for ammonia recovery from waste streams, consisting of gas–liquid stripping, ion exchange, electrodialysis, membrane separation, and adsorption and struvite precipitation. Among these technologies, ammonia stripping is an emerging technology for removing ammonia from the digestate. This chapter discusses the various factors affecting ammonia stripping along with different configurations in AD. The chapter aims to educate readers about the existing and emerging technologies for ammonia extraction from the anaerobic digestate.

Keywords: Digestate, nurtients, ammonia stripping, recovery

7.1 INTRODUCTION

An increase in population and modernization has put pressure on natural resources. This growth has led to the generation of a large amount of organic waste such as food, agriculture, livestock, human waste, and so on. The direct disposal of these wastes into the environment causes soil, water, and air pollution. Various physicochemical, thermochemical, and biochemical methods are available for the treatment of these wastes. Generally, for these organic wastes, biochemical processes such as anaerobic digestion (AD) is a well-established technology for the valorization of organic waste. The benefits of AD include generating renewable energy in terms of power or heat. The effluent produced after AD also known as digestate constitutes organic material. The digestate can be used directly in agriculture but this practice has negative effects on the environment. The ammonia present in the digestate may contribute to air pollution by the formation of ammonium aerosols and particulate matter by reaction of volatile free ammonia with atmospheric compounds. Furthermore, the leaching of ammonia-rich streams to water bodies can cause eutrophication and harm to aquatic life by the decrease in the level of dissolved oxygen.

The AD process consists of four stages (hydrolysis, acidogenesis, acetogenesis and methanogenesis), where organic matter is transformed into different products in the absence of oxygen. During the first stage of hydrolysis, complex organic compounds are degraded to simpler compounds by extracellular enzymes. This stage is followed by the acidogenic stage where the soluble compounds are converted to volatile fatty acids (VFAs) and hydrogen. In the next step further transformation of VFAs to acetate, CO_2 and/or hydrogen. In the final stages of methanogenesis, methane is liberated by the conversion of acetate, CO_2 and hydrogen. During hydrolysis of nitrogenous organic matter consisting of amino acids, proteins and urea is transformed to ammoniacal nitrogen. The ammoniacal nitrogen consists of dissolved ammonia (NH_3) in gaseous form also known as free ammonia in equilibrium with ammonium (NH_4^+) ions. The ammonia released during this process is in maximum concentration, but it is harmful to the environment and also has potential for renewable sources of energy, thus instituting a necessity for ammonia recovery (Lauterböck *et al.*, 2014). In this context, valorization of the digestate produced after AD may provide different benefits such as the recovery of nutrients for benefits of agriculture sector and finding the end use of the digestate. The ammonia recovered can be valorized to find its energy potential. The recovered ammonia can be degraded for hydrogen production which is a clean and renewable fuel, with direct use as fuel or in devices like solar cells or refrigeration systems (Afif *et al.*, 2016).

7.2 AVAILABLE TECHNOLOGIES FOR AMMONIA REMOVAL FROM THE DIGESTATE

Various techniques have been considered for ammonia recovery from waste streams, comprising ammonia stripping, ion exchange, electrodialysis, membrane separation, adsorption, struvite precipitation, and biological processes. Recently, ammonia stripping method is the most efficient and economically viable technology to remove and recover ammonia from anaerobic digestate in the form of commercial bio-fertilizer.

7.2.1 Ammonia stripping

The ammonia stripping technology is applied for nitrogen gas recovery by desorption on a liquid fractionation column. The packed columns are ideal for high rate of mass transfer, however, column fouling and scaling are major concerns. Column fouling can be prevented by including an effective solid–liquid separation stage (De la Rubia *et al.*, 2010). Additionally, configurations such as gas–liquid contactors spray towers, air bubble producers, diffusers, aerators and aspirators are reported to help impede fouling. The scaling can be kept in check by removing calcium, magnesium and carbonates using CaO softener prior to the stripping step. The pH and temperature are most important factors governing the efficiency of ammonia stripping process (Guštin & Marinšek-Logar, 2011). Numerous factors including the NH_4-N content, column packing, hydraulic loading, rate of liquid and air flow, retention time, liquid to air ratios and mass transfer area also determine ammonia stripping efficacy. Optimal ranges of pH is above 9 and temperature is more than 20 °C. At 80 °C, the complete recovery of ammonia nitrogen has been reported. Studies have reported 70 °C and pH 10–11.5 as optimum for the stripping process (Lauterböck *et al.*, 2014). The stripping process in the presence of CO_2 could achieve high rates (40–50%) of nitrogen recovery along with generation of ammonium sulphate (35%) without the addition of alkalis or requiring stripping tower. Sulphuric acid scrubbers are responsible for generation of ammonium sulphate during the process. The ammonium sulphate solution is utilized for biofertilizer production, thereby having a fair market penetration and aligns well with the notions of circular economy. However, the market value of this end product are still low due to problems of inconsistent sulphur and nitrogen content (61–100 kg S/t; 30–100 kg N/t) and high reactivity (corrosive and acidic nature) (Siegrist *et al.*, 2005). Additionally, the use of H_2SO_4 causes decrease in pH (< 2) making ammonium sulphate solution unsuitable for use in fertilizers. This is also responsible for corrosive and acidic behaviour of the solution. Soil acidification, leaf burning and rust in the application instrument are some effects of ammonium sulphate-based fertilizer applications. Further, neutralization of ammonium sulphate solution is advised prior to its field application. To prevent acidity, nitric acid has been used instead of sulphuric acid. The high value of end product NH_4NO_3 is attributable to its high nitrogen content. The process of ammonia stripping has significant chemical (1.5 L H_2SO_4/kg NH_3) and energy (0.057 Wh/m^3 air) demand, therefore responsible for elevating the operational cost. The effluent of the stripping process contains nitrogen and other nutrients used for making bio-fertilizers (Jiang *et al.*, 2014). Thus, the net cost of recovering nitrogen by stripping process is 2–7 €/kg N removed. A 90–120 €/t can be profited after sufficient market penetration of the ammonium sulphate. Efficient process configurations are reported to achieve 98% N recovery and cut-down models can give up to 80% N recovery (Wang *et al.*, 2019). The United States and Europe have advanced to full-scale plants for recovering ammonia from the digestate. Currently, 10 full scale plants are operational in the United States.

7.2.1.1 Factors affecting ammonia stripping

The major factors affecting recovery of ammonia during stripping are temperature, pH, and gas flow rate. Ammonia stripping is also affected by factors such as CO_2 in the stripping gas and the digestate composition. The effect of these factors is being discussed in the upcoming sections.

7.2.1.1.1 Temperature

Increase in temperature is associated with an increase in amount of N recovered after the stripping process. Temperature is demonstrated to affect desorption of free ammonia (FA) by decreasing its solubility in the liquid phase. The enhancement in transfer rate of FA to gas phase is due to increase in vapour pressure upon temperature elevation. Further, rate of mass transfer and diffusion of molecules from liquid to gaseous state are increased at high temperatures (Provolo *et al.*, 2017). Several studies have demonstrated linear relationship of increase in ammonia recovery with increase in the temperatures of liquid phase from 15 to 100 °C (Zeng *et al.*, 2006; Zhao *et al.*, 2015).

7.2.1.1.2 pH

The pH is reported to affect FA concentration in the liquid phase. The pH ranges from 8 to 12 is demonstrated to give optimal ammonia recovery. Few studies have showed that pH above 10 did not proportionally increase the amount of ammonia recovered (Jiang *et al.*, 2014). Although, not all studies are definite for the relationship between pH and ammonia recovery. It was found at 57.5 °C and pH 9, 74.5% FA was present for a given amount of total ammonia. At pH 10 and 11, 96.5% and 99.6% of FA was present per given amount of TN, respectively. It is evident that, above pH 10 the amount of ammonia recovered does not increase significantly with an increase in pH. Therefore, pH 10 is regarded as a critical pH above the p*K*a of ammonium ions (pH 9.2). Above p*K*a, FA concentration proportionally increases, but the rise is not substantial after the critical pH (10) is achieved (Provolo *et al.*, 2017).

7.2.1.1.3 Gas flow rate

Escalation in rate of gas flow results in a rise in the amount of ammonia recovered. The stripping gas does not affect the FA concentration in liquid phase. It is regarded to drive mass transfer rates of ammonia from liquid to gaseous phase. The gas flow acts to maintain the concentration gradient of free ammonia across the two states. With increase in gas flow, the boundary layer depletes thereby decreasing the resistance for mass transfer (Lei *et al.*, 2007). Gas flow rate influences the interfacial layer between the liquid and gaseous phase, therefore causing more ammonia to diffuse to the gas phase. Moreover, gas flow rates are reported to have negative effects on stripping process, such as decrease in liquid phase temperature, evaporation and foaming (Provolo *et al.*, 2017). An equation (Equation (7.1)) is provided to relate gas flow rate Q_g, volume of liquid stripped V_l and the stripping time (t) in a dimensionless quantity. The quantity is the volume of gas used (G) per volume of liquid stripped (L).

$$G / L = Q_g \cdot t / V_l \tag{7.1}$$

It is suitable to compare diverse studies using the dimensionless quantity.

7.2.1.1.4 Carbon dioxide in the stripping gas

The digestate comprises of many ionic and non-ionic carbon species CO_2, CO_3^{2-} and HCO_3^-. Equations (7.2) and (7.3) show the equilibrium between the carbon species in the liquid phase.

$$CO_2 + H_2O \leftrightarrow HCO_3^- + H^+ \tag{7.2}$$

$$HCO_3^- \leftrightarrow CO_3^{2-} + H^+ \tag{7.3}$$

The CO_2 present in the stripping gas influences the concentration gradient of CO_2 in the liquid and gaseous phase. The rate of mass transfer of CO_2 from liquid to gaseous phase is increased with the decrease in the CO_2 present in the stripping gas. The CO_2 in the stripping gas drives the carbon species equilibrium to the left causing increase in liquid phase pH. Finally, the increase in pH proportionally increases the amount of ammonia recovered. Various studies have discussed the relation between CO_2 in the stripping gas and ammonia recovery; however, very few are performed on the digestate (Cohen & Kirchmann, 2004).

7.2.1.1.5 Composition of digestate

The source of waste affects the nature and composition of the digestate. The digestate may vary in its soluble, non-soluble, organic and inorganic components depending from where the waste is sourced (Table 7.1). There is lack in studies that investigated the influence of composition, type and source of digestate on the ammonia stripping process. Few studies are done to determine the effect of percentage of total solids (TS) on the ammonia recovery. The study by Georgiou *et al.* (2019) reported no significant increase in ammonia removal with variation in digestate composition (suspended solids, total solids and dissolved solids, etc.) compared to model solutions and a given digestate. On the contrary, the low organic matter (10 g COD/mL) content of the digestate gives 90% ammonia recovery compared to high organic matter (27 g COD/mL) which gives 50% of the ammonia recovery (Laureni *et al.*, 2013). This effect can be attributed to the ability of NH_4^+ ions to bind to organic matter, thereby lowering the availability of FA for high organic matter containing digestate. Theoretically, the ammonia removal rates can be influenced by TS content of the digestate. The TS content of the digestate substrate is in range of 1–5%. Therefore, decreasing the TS content will allow for trapping of FA by the stripping gas due to enhanced gas–liquid interchange and liquidity (Yao *et al.*, 2017). Additionally, because the ammonium ions are adsorbed onto the organic matter of the digestate,

Table 7.1 Composition of digestate's liquid and solid fractions (Source: Barampouti *et al.*, 2020).

Parameter	Liquid fraction	Solid fraction
pH	7.8–7.9	7.7–8.5
TS (%)	3.3–6.6	19.3–24.7
Total C (% DM)	2.64–3.15	9.0–10.1
Total N (% DM)	0.32–0.51	0.33–0.65
Total NH_4 (% DM)	0.17–0.3	0.13–0.3
NH_4 (% Total N)	40–80	26–49.4
Total P (% DM)	0.03–0.1	0.08–0.25
Total K (% DM)	0.29–0.52	0.25–0.48
Total Mg (% DM)	0.03–0.05	0.09–0.10
Total Ca (% DM)	0.04–0.06	0.16–0.19

separating liquid portion of the digestate is hypothesized to yield high amounts of ammonia. Conversely, studies by Huang *et al.* (2016) and Abouelenien *et al.* (2009) reported 95% and 85.5% ammonia recovery efficiency using high TS containing digestate (20%) and without segregation of liquid fraction. In another study, Zhao *et al.* (2015) demonstrated that TS content of the substrate does not have significant influence on ammonia volatilization. Thus, enough studies are lacking that give consensus reports on the influence of complex composition of digestate on ammonia removal efficiency. An exhaustive study is therefore required for providing information on the effects of digestates with varying composition of TS and organic content on ammonia recovery.

7.2.1.1.6 Stripping column design

The contact intensity of liquid and gas phases in a column is affected by the design of the column. High contact intensity between the phases gives increased rates of mass transfer at low energy utilization (Baldi *et al.*, 2018). Different stripping columns are available as shown in Figure 7.1.

The packed columns are largely available and have been extensively used for the stripping process by maximizing the surface area for mass transfer. The packed columns are hollow column filled with materials such as Raschig rings, plastic rings, and so on. The major disadvantage of packed columns is that it is prone to fouling by particulate matters of the digestate. Fouling reduces the effective surface area accessible for the transfer of mass (Siegrist *et al.*, 2005). The recovery of ammonia upto 80% was reported using the packed column for stripping from dairy manure (Jiang *et al.*, 2014). The decrease in TS solids in the effluent is also reported, therefore signifying the deposition of solid particles of the digestate in the packed column. The advancements in the column design is recommended for discouraging fouling. The problems of solid particles deposition can also be reduced applying liquid fractionate to the column for ammonia stripping (Laureni *et al.*, 2013). The studies by Guštin and Marinšek-Logar (2011) demonstrated ammonia stripping in a lab-scale packed column from the liquid phase of pig slurry. The ammonia recovery was reported to be 90%. In addition, the study by Zeng *et al.* (2006) showed 96% ammonia recovery from liquid fractions of cattle manure by stripping applied with steam in a packed column. About 90% ammonia was recovered from liquid fractions of waste from slaughter house using a plastic ring stacked column. The column design is only a small factor influencing the ammonia recovery process, as several stated studies have reported approximately 90% ammonia recovery. Major factor limiting the ammonia recovery rate and for its commercial applications is the packing material of the column, the fouling and clogging of the column material. The application of liquid fractions of the stripping columns is one way for mitigation of negative impacts of solid contents of the digestate. The liquid fractions are obtained by additional solid–liquid separation step, and the feasibility of this process determines its future implementations. The other way to reduce the negative impact of solids on column packing material is the consistent cleaning regime of the column. There is also a patented technique, wherein the N_2 or air is bubbled through the

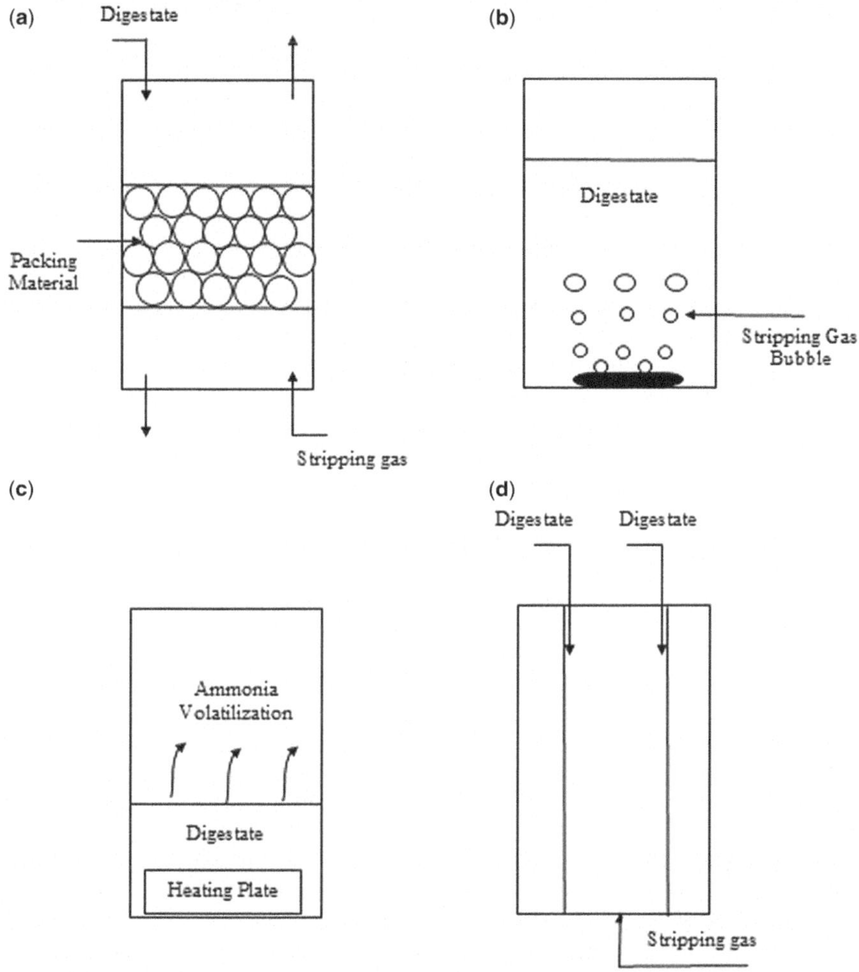

Figure 7.1 Design of stripping columns (a) packed-bed column, (b) bubble aeration column, (c) wet-walled column and (d) thermal stripping (Source: Palakodeti *et al.*, 2021).

clogged column to force removal of the debris. The specific gravity of the packed media must be less than that of the flooding gas to allow for the development of fluidized beds. It is recommended to clean the column 2 h per week but may vary depending on the column material, nature and level of clogging.

The bubble aeration is a widely adopted technology for ammonia stripping. The gas diffusers, air stone or submerged pipes are used for generating gas bubbles in the liquid phase of the digestate. The ammonia molecules are trapped in the gas bubbles that are accumulated at the column outlet. The bubble aeration is advantageous due to ease of its implementation. However, the drawbacks of this technology includes, foaming and cooling of digestate, water evaporation and volume loss. Numerous studies have been done for ammonia stripping by bubble aeration (De la Rubia *et al.*, 2010; Lei *et al.*, 2007). Some studies have reported application of air diffusers for upto 95% ammonia recovery from liquid supernatant of pig waste digestate. Another study by De la Rubia *et al.* (2010) demonstrated 44% ammonia recovery from synthetic biogas using air bubble aeration technology. High amount of ammonia (99%) was recovered from raw pig manure using air stone for bubbling air through the liquid fractionate. Maximum ammonia removal (95%) was reported from activated sludge digestate by using air stone and increase in pH was achieved by an electrolyser system (Park & Kim, 2015). The difference between packed column and bubble aeration column design is shown in Table 7.2.

Several novel stripping technologies have been tested for ammonia recovery. For instance, water-sparged aerocyclone configuration involves the sparging of ammonia containing liquid through pores inside the reactor vessel in addition with the tangent flow of stripping gas across the vessel. The mass transfer enhancement is regarded to the increase in contact area between the liquid and gas phases. The amount of ammonia removed was up to 91% from the liquid part of anaerobic swine wastewater. The turbulent mixing of the liquid digestate from poultry manure has been used to recover maximum amount of ammonia (97%). The vortex at the liquid surface was used to generate turbulence. Another configuration for ammonia stripping is the thermal stripping based on increase in the temperature (above 90°C) that helps in FA volatilization. In a lab-scale study, the liquid digestate of dairy manure was heated in a vessel at varying temperatures between 76°C and its boiling temperature. Maximum ammonia recovery (>95%) was achieved at 102°C and at pH 9–11. At the boiling temperature, the coefficients of mass transfer rate and ammonia recovery were 10–110 times more than at sub-boiling temperature range. This effect is observed due to turbulence created in the liquid fraction upon boiling that reduces the resistance for mass transfer. Another innovative configuration applied for ammonia stripping from swine wastewater is based on spraying of the droplets of liquid digestate over heated spiral water pipes (Cao *et al.*, 2019). The droplets formation is ascribed to increase the surface area and collision force for enhanced mass transfer. Further, the heat is liberated due to the temperature difference that allows FA release by breaking the equilibrium

Table 7.2 Difference between packed column and bubble aeration configuration for ammonia stripping.

Packed column	Bubble aeration
Liquid and gas phases circulated in co- or counter-current manner	Liquid phase is stationary, while stripping gas is bubbled through
Extensively researched and implemented	Easily implemented, and widely used
Suspended particles in digestate may cause fouling in the column	Cause foaming and cooling of digestate

between the liquid and gaseous phases. The ammonia removal achieved by the novel spraying technology was 95% at pH 8.06 and temperature within the hot spiral pipes was maintained at 55°C. In another study, a isothermal wet-walled column configuration was used wherein the column walls provided large surface area for transfer of masses between the gas and liquid phases. The study reported 98% ammonia removal at 80°C and pH 11.5 (Bonmatí & Flotats, 2003). The efficiency of ammonia recovery process is thus, affected by the process conditions (pH and temperature) and the stripping gas. The influence of column design is limited on the amount of ammonia recovered. The stripping column to be used is dictated by the nature and amount of suspended/total solids in the digestate, column material, manufacturing and maintaining the cost of the column. The constraints to the ammonia stripping process may also influence the choice of column. For instance, the aeration is ideal with packed columns as the solid–liquid separation is less practical.

7.2.1.2 Anaerobic digester: stripping column configurations

Ammonia inhibition in anaerobic digestor can be performed by different configuration of ammonia stripping inhibition. The different configurations of stripping columns are pre-AD stripping, in-situ stripping, side-stream stripping and post-AD stripping which are discussed elaborately in further sections (Figure 7.2).

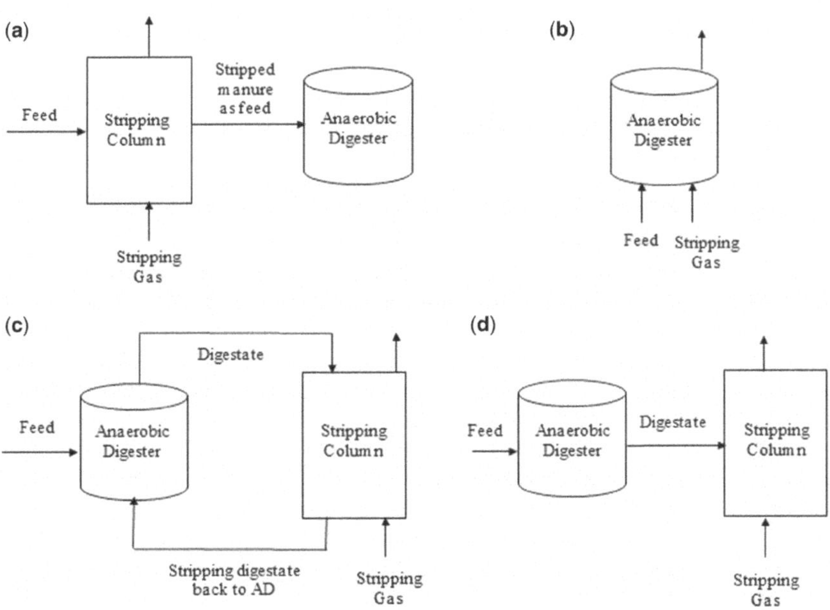

Figure 7.2 Configurations of stripping column in anaerobic digester (a) pre-AD stripping, (b) in-situ stripping, (c) side-stream stripping, (d) post-AD stripping (Source: Palakodeti *et al.*, 2021).

7.2.1.2.1 Pre-AD stripping

The ammonia is removed from the raw organic matter followed by its feeding to the anaerobically operated digester. The prior removal of ammonia from raw substrate promotes the AD process due to removal of excess ammonia. Numerous studies have reported high ammonia yield by employing pre-AD configuration. For instance, Zhang *et al.* (2012) found that the methane generation (170 mL methane/g COD) was thrice after the pre-AD stripping was applied compared to the conventional (49 mL methane/g COD) procedure for biomethane generation from piggery wastewater in a semicontinuous reactor vessel. Additionally, the study showed enhanced removal of organic carbon (total COD, soluble COD, and VFA) from the effluent of pre-AD system compared to the effluent from conventional system. The effluent from conventional system had higher values for TCOD (45%), soluble COD (50%) and VFA (57%) while pre-AD system effluent had lesser values for TCOD (20%), soluble COD (13%) and VFA (10%). Similarly, pre-AD treatment of piggery wastewater increased the methane yield by five times (250 mL methane/g COD) as compared to the non-stripped wastewater (50 mL methane/g COD) (Zhang & Jahng, 2010). The high methane yield (667 mL methane/g COD) and substrate degradation (1.8 times) were obtained after pre-AD ammonia stripping of activated sludge digestate. Only 234 mL methane/g COD was produced in a non-stripped system (Park & Kim, 2015). The increase in the amount of methane produced and organic matter removal have been attributed to the low ammonia concentrations of the digestate fed to the anaerobic reactor after ammonia recovery. The ammonia removed at the initial stages inhibits the accumulation of inhibitory concentration of ammonia responsible for disrupting methanogenesis in the anaerobic process. However, few studies were unable to find enhancement in methane yield with prior ammonia stripping opposed to no stripping process. The study by Bonmatí and Flotats, (2003) observed a 75% decline in methane generation (22 L methane/kg VS) post pre-AD treatment of swine waste slurry whereas 88.8 L methane/kg VS was produced with no ammonia stripping. In this study, the alkali treatment before ammonia stripping was responsible for addition of metal cations (Cu, Zn and Ca) and, thus, increases in pH. Therefore, adverse effects of high pH on both ammonia stripping and AD are the possible reason for decrease in ammonia production. Similarly, Li *et al.* (2018) found no significant increase in biomethane production after ammonia stripping from raw poultry manure. The addition of alkaline agent ($Ca(OH)_2$) decreased the amount of organic matter (COD) and increase the pH of the waste required for optimally carrying out its AD. Despite few studies claiming the opposite, it can be concluded that the pre-AD ammonia stripping is an effective solution for ammonia inhibition during AD. This configuration is particularly useful for substrates rich in nitrogen like poultry and swine manure for ammonia removal/recovery prior to AD without its inhibitory accumulation. Nonetheless, the alkaline treatment of raw feed before AD could limit the efficacy of pre-AD ammonia stripping configuration (Bonmatí & Flotats, 2003). The use of alkaline agents like NaOH and KOH affects methane yield due to cation inhibition of the microorganisms, whereas the $Ca(OH)_2$ causes precipitation of active microbial cells from the anaerobic reactor. Other alkaline agents may be used to overcome such problems. It is noted that the use of pre-AD ammonia stripping configuration for maximum ammonia recovery is not significant as the ammonia released by the AD process is difficult to trap. Therefore, this configuration is effective for improved performance of AD and methane production, but not for ammonia recovery. Finally, the studies are limited to batch or semi-continuous modes and no information is available for its efficacy in continuous mode.

7.2.1.2.2 In-situ stripping

The in-situ ammonia stripping system allows for stripping process and AD simultaneously in a reactor vessel. The advantage of this process is the inhibition of ammonia accumulation and therefore preventing reticence of microbial methanogenic activity. Additionally, as the process takes place in a single reaction vessel, no added equipment and post-treatment steps are required, making the process potentially economical. One drawback of the in-situ configuration is that the operational conditions of both AD and ammonia stripping are the same causing inhibition of one of the process. For instance, high pH and temperatures are suitable for ammonia stripping that might not be beneficial for AD because of inhibition of microbial degradation at high temperature ranges (Jiang *et al.*, 2014). The study by Abouelenien *et al.* (2010) determined that high efficacy of in-situ stripping for maximum ammonia recovery and anaerobic degradation of poultry manure. The authors however did not conduct a study with a control reactor; thus the actual increase in ammonia recovery and organic matter degradation upon in-situ stripping treatment cannot be quantified. In-situ ammonia stripping was employed for AD of chicken manure with high content of TS (15%). In the study by Bi *et al.* (2020), the ammonia recovery was not directly quantified but the TAN concentration in-situ stripping digester was 14.7% less than in the control digester (5.8 and 6.8 g N/L, respectively). In addition, the amount of methane produced in-situ stripping digester was 34% more compared to the control (250 mL/g VS feed and 187 mL/g VS feed, respectively). The VFA concentration in the effluent was 30% less (15.2 g/L) in the effluent from control digester (21.6 g/L). The decrease in TAN concentration in the in-situ stripping anaerobic digester allowed more rigorous microbial activity and thus, more methane yield as opposed to the control. This result is also substantiated by decrease in concentration of VFA in the in-situ stripping digester. The investigations on alteration in microbial community during ammonia stripping are being done. For example, in the prominent microbial metabolic pathway in the control digester was syntrophic acetate oxidation coupled with hydrogenotrophic methanogenesis (SAO-HM) was partially altered to acetoclastic methanogenesis in the in-situ mode. The partial shift in microbial pathways is attributed to decline in TAN concentration in-situ digester. The comparison of two in-situ stripping with different configurations was conducted by Sun *et al.* (2014) and Jiang *et al.* (2013). The first one was operated by circulation of biogas through liquid phase of the reactor and the second, involved biogas circulation only in the headspace of the reactor. The two reactor modes were compared on the basis of their performance rather than quantifying the amount of ammonia recovered. The methane yield proportionally increased with the increase in organic loading rate (OLR). However, a steady-state methane production (2900 mL/L/d) at 4 g volatile TS/L/d OLR was obtained in the reactor with circulation of biogas through the liquid phase. The accumulation of VFA (2820 mg/L) and low levels of TAN in the first digester in contrast to the second one (2050 mg/L VFA) is regarded as the possible reason for steady-state methane yield. The amount of ammonia extracted from the first digester is twice as much as in the second digester, thereby substantiating for a highly stable AD degradation process.

7.2.1.2.3 Side-stream stripping

In this mode, the fraction of the digestate from the AD reactor is separated in a stripping column and then fed back to the AD reactor. Evidently, the stripping column and AD reactor can work at independent operational conditions (pH and temperature). Multiple studies have claimed low level of ammonia in the liquid phase of AD digester when installed with side-stream stripping column. The study by Serna-Maza *et al.* (2014) found that the implementation of side stream stripping resulted in concentration reduction of TAN in the liquid fractions (initial; 5 g N/kg, final; 3 g N/kg). The studies by both Pedizzi *et al.* (2017) and Fernandez-Gonzalez *et al.* (2019) reported a decline in TAN concentration (2.5–1.1 g N/L) after recycling of the stripped digestate in the AD reactor. On the contrary, the changes in the yield of methane were not significant in the side stream mode compared to that in the in-situ and pre-AD stripping modes. This can be attributed to inhibition of microbial methanogenic activity due to high pH and temperature conditions of the stripping process. Nearly negligible methanogenic activity of microbial community of the stripping process is compensated by its recirculation to the AD reactor, thus showing no change in methane yield. The changes in the methogenic microbial community of the AD digester in the side stream stripping configuration are claimed by many studies. This can be explained by potential shift in microbial activity towards acetoclastic methanogenesis caused by a decrease in VFA and TAN concentrations and promotion of facultative anaerobic microorganisms due to exposure to air during digestate recirculation. The study by Li *et al.* (2018) reported the abundance of hydrogenotrophic methanogens in the three-side stream stripping columns operated at different conditions. The column operated at more rigorous conditions showed relatively high abundance of hydrogenotrophic methanogens than in the other two columns. In contrast, the study by Serna-Maza *et al.* (2014) found that the microbial community changed from syntrophic acetate oxidation to acetoclastic methanogenesis when no stripping conditions were altered. The high OLR concentration (9 g VS/L/d) and VFA concentration (12 g/L) compared to 2 g VS/L/d OLR and 250 mg/L in harsh conditions during the stripping process than in the less harsh conditions could account for the shift in microbial metabolism. Despite, the fact that appropriate comparisons are not possible with large differences in the operating conditions of side stream stripping columns, the alterations in microbial community metabolism and composition are significant.

7.2.1.2.4 Post AD stripping

The post-AD ammonia stripping is a post-treatment step applied on the AD digester effluent. The advantage of this configuration is the optimization of the stripping process without considering reaction conditions for carrying out stable AD. Other major pros is the use of post-treated digestate as a crop fertilizer and high temperatures during the stripping process acts a pasteurization step for sterilized end products. However, as the ammonia is not recovered from the system prior to AD, the ammonia inhibition during the AD process is still a matter of concern in the implementation of post-AD stripping configuration. The effluent from anaerobic digester fed with wheat straw substrate was subjected to ammonia stripping column and the subsequent digestate was applied as a fertilizer to the agricultural lands (Wang *et al.*, 2019).

7.2.1.2.5 Struvite precipitation

Struvite precipitation has been studied for the recovery of solubilized phosphate and ammonia nitrogen. Struvite is a slow-release fertilizer for agriculture and horticulture. The salts of magnesium are added to form magnesium ammonium phosphate (MAP) or struvite ($MgNH_4PO_4{\cdot}6H_2O$), which is an insoluble compound that precipitates from the solution. Struvite is formed by the aggregation of several suspended small particles in the digestate into the larger particles. Consequently, struvite is harvested as a complex of solids and precipitates. MAP can be recovered as green fertilizer with a stoichiometric ratio of Mg:N:P of 1:1:1, similar to guano, and it can be regarded as a potential commercial fertilizer. Struvite separation from water has been tested in several effluents from leather industries, swine farms, sludge liquor, and digester supernatant. The factors such as pH, temperature, liquid turbulence, suspended solids and ions concentration are limiting for struvite formation (Kinidi *et al.*, 2018). The precipitation of PO_4^{3-} ions ensues after pH increases from 5.0 to 7.5, while the optimal pH for struvite precipitation is in the range of 8.5–9.5. Continuous alkali additions are made to keep in check the rapid H^+ production during struvite precipitation. Ideally, higher ion concentration is more favourable for the struvite precipitation, along with the mole ratio of N/P and Mg/P maintained above 3/1 and 1/1, respectively. The concentrations of PO_4^{3-} ions are less compared to the NH_4^+ ions in the digestate, whereas Mg^{2+} is inadequate for struvite formation. Therefore, the supplementations of $MgCl_2$, $MgSO_4$ or MgOH are supplied to induce struvite formation. Additionally, a high concentration of Mg^{2+} abrogate the coprecipitation due to Ca^{2+}, that is addressed to cause negative impact on recovery yield of struvite. Recent studies have been focused on the maximization of the recovery yield and purification of struvite (Zhang *et al.*, 2015). Ammonia ions in the digestate are sufficient to support struvite formation. However, only 10% of P is available in the liquid fraction due to the high pH of the digestate (approximately 8). For instance, studies on the struvite recovery from pig manure digestate were operated by Song *et al.*, while the recovered P was only a small fraction of the total P. Besides struvite, increasing attention has been drawn to the other P precipitates in the solid fraction, such as $MgKPO_4$, $MgHPO_4$, $MgHPO_4{\cdot}3H_2O$ (newberyite), $Ca_3(PO_4)_2$, $K_2NH_4PO_4$, $Ca_5(PO_4)_3OH$, $CaHPO_4{\cdot}2H_2O$, and

$CaHPO_4^{3-}$. Increasing the concentration of PO_4^{3-} in the liquid phase of the digestate is needed for improving the yield of struvite. Various pretreatment methods have been verified for increasing the concentration of soluble P in the digestate, including microwave heating, acidification, orbital shaking and ultrasonic treatment. The acidification pretreatment is more powerful than the other approaches. Despite this the purification of struvite would be difficult due to the presence of Ca^{2+} and suspended particles in the digestate. Although Ca^{2+} can be chelated by EDTA or selectively reduced by oxalic acid, the high concentration of Ca^{2+} would lead to increase in chemical costing, and the EDTA toxicity has not been studied till now (Morales *et al.*, 2013).

7.2.2 Membrane-based technique

The separation of nutrients based on the membrane filtration is also an alternative for extraction of nutrients. Two types of membrane technology are applied for nutrient recovery. The two types of membrane systems are pressure based and non-pressure-based membrane technologies.

7.2.2.1 Pressure-based membrane filtration

The most prominent pressure-based filtration technologies are microfiltration (MF), ultrafiltration (UF), nanofiltration (NF) and reverse osmosis (RO). These technologies produce high-quality water along with concentration of nutrients from the digestate in the permeate suitable to be used as bio-based fertilizers. Porous membrane such as MF and UF removes suspended solids and macromolecules with dissolved components remaining in the permeate. For further purification of permeate NF and RO are applied. In NF 5–23% efficiency for separation of NH_4^+-N is achieved. RO produces a permeate free from pollutants with recovery of around 99–100% as dissolved solids are retained via a semi-permeable membrane at 10–100 m [99]. The RO systems require energy around 4–6 kWh/m^3 (Vaneeckhaute *et al.*, 2017). It has been found that in anaerobically digested manure about 95% of the total nitrogen can be recovered using a tubular RO membrane. Although the removal efficiency is affected by pH, at pH 8 removal efficiency ranges in 75–96% and at pH 4 efficiency upto 100% is reported. Improvement in removal efficiency may be achieved by adding surfactants and water-soluble polymers (chitosan, polyethylenimine and polyacrylamide) with ability to complex with metals. In MF and UF complete elimination of parasites (e.g., nematodes) and of pathogens (e.g., *Escherichia coli*) have been reported (Minardi *et al.*, 2015). In the treatment of anaerobically digested manure by UF complete removal of suspended solids and coliforms is achieved without any effect on COD. The permeate quality after pressure membrane filter is suitable to be used for irrigation, cleaning or direct disposal. For countering fouling issues generally NF and RO are preceded by MF and UF. Although fouling remains a major challenge in membrane processes. Fouling can be mitigated by increasing tangential velocities on cross flow stream or by intermittent backwashing by water or air. For UF membrane cleaning can be performed by ultrasonication. If physical methods are inadequate for recovering the membrane capacity, chemical methods of cleaning such as alkaline and acidic media, enzymes, surfactants, oxidants, chelants or mixture can be used (Fuchs & Drosg, 2013). Efficiency of membranes can be reduced by biofouling due to bacterial colonization or by scaling by low soluble salts precipitation. Scaling phenomena can be mitigated by pH regulation along with the use of antiscalants. In wastewater treatments these processes are well established but the potential of these processes is yet to be demonstrated at a large scale for the digestate treatment. Recently, ammonia stripping method has been applied in combination with RO for liquid digestate treatment. The critical issues in these processes are economic viability and energy requirement during extraction of nutrients from the digestate (Minardi *et al.*, 2015).

7.2.2.2 Non-pressure membrane technologies

For overcoming the drawbacks of conventional membrane filtration processes focus has been given to non-pressure membrane filtration. Electrodialysis (ED) uses electric potential along with ion-exchange membranes for concentrating ions (NH_4^+, K^+, HCO_3^-) of the liquid digestate. Concentration of ammonium ions has been found higher (upto 16–21 g/L) in ED compared to RO (Törnwall *et al.*, 2017). The digestate treatment using ED is generally reported at the lab scale only except for pilot-scale treatment reported in the Netherlands. In the Netherlands, recovery of potassium carbonates and ammonium has also been reported. Apart from ED there is another advanced technique known as bipolar membrane electrodialysis based on the principles of ED only. This technique utilizes a bipolar membrane promoting dissociation of water in H^+ and OH^-. The products (acid and base) derived can be utilized in various applications. In ED processes fouling is also of major concern. The production of chloride at the anode is considered as a reagent able to reduce fouling, but cracking and pitting phenomena on membranes are the key bottlenecks for its usage. For nutrient recovery bioelectrical systems are also an alternative membrane configuration. The application of microbial fuel cells in anaerobic digestate may achieve total recovery of ammonia. Despite the potential and low energy requirements of this method full-scale plant are yet to be established for this. In anaerobic digestate application of membrane distillation for nutrient recovery is an encouraging option as the concentration of ammonium in permeate can reach upto 18.3 g/L (pH $>$ 9.68 and 45°C) (Vaneeckhaute *et al.*, 2017). The fouling issues for this approach are not as important as for the other membrane-based processes. However, fouling issues and issues related to the customized design of membranes and draw solution still exist.

7.2.3 Ion exchange and adsorption

In ion exchange and adsorption a column bed filled with solid sorbents is used for the extraction of the targeted compounds from the liquid digestate. The sorbents generally used for these processes are resins, clays and zeolites. Sorbents modified either chemically or thermally can also be used. When after use the sorbents reach their capacity they have to be regenerated before starting a new cycle of the process. Depending upon the end products and sorbents, regeneration can be carried out with HNO_3, NaCl washing, or through biological means. Adsorption and the ion-exchange process are identical as adsorption is governed by intermolecular forces and in ion-exchange ionic forces govern the process. Both these processes can occur simultaneously in a single media such as resins and zeolites. In adsorption zeolites are the mostly evaluated material for extraction of nutrients such as ammonia and phosphate. Zeolite carries negative charge and consists of massive cavities which aid in efficient adsorption of cations. After application to land zeolite may potentially discharge adsorbed ammonium ions to the soil. The natural zeolite has an adsorption capacity of 19 g NH_4^+/kg and 21 g/kg for Na-zeolite with the removal efficiency lying in the range of 71 and 91% (Guo *et al.*, 2007). Various successful studies have been undertaken for the removal of ammonia from the digestate. Zeolite also adsorbs phosphorus but the removal efficiency is much lower in comparison to ammonia. Recently, biochar and hydrochar has also been studied for the removal of ammonia. The sorption capacities of ammonia using biochar have been found in the range of 100–160 g/kg [43]. The adsorption is affected by pH with 6.5–7 as optimum values as the pH value beyond this range affect the adsorption efficiency significantly. Adsorption and ion exchange are challenging technologies. The main bottlenecks of this technology are fouling due to the presence of high suspended solids in liquid digestate, reduction in ion-exchange capacity of resins after regeneration and foreign ion completion leading to reduction in efficiency. In addition to the above mentioned challenges, zeolite adsorption in comparison to ammonia stripping and struvite formation has an higher operating cost as more quantity of zeolite is utilized in comparison to Mg and lime. However, due to ease of operation and lower energy consumption and lower labour requirement ion exchange and adsorption may fare better (Wirthensohn *et al.*, 2009).

7.3 CONCLUSION

Based on the information compiled above, currently only ammonia stripping and absorption in a packed/unpacked tower, struvite crystallization and pressure-driven membrane filtration has been applied for ammonia recovery from the AD digestate. Among these technologies, only ammonia stripping and struvite are economically viable for implementation at full scale for digestate management. Membrane technologies due to their traditional drawbacks are yet to be established as viable options for treatment of the digestate. Till date depending upon adsorbent availability sorption of ammonia and recovery of nitrogen by zeolites are the most viable option for ammonia recovery from the anaerobic digestate. However for implementation of these technologies further research is required as these methods have low efficiency of ammonia recovery from the digestate. Although it may be beneficial to use these technologies in combination with stripping or struvite for reduction in cost and better removal efficiency. Among the full-scale technology acid air scrubbing is the most feasible method as odour-removing facilities often require air. The capital cost for adsorption and stripping method is usually low but the operational cost of these technologies are higher than the operational cost of struvite precipitation. In struvite recovery the dosage of chemicals contributes to operational expenditures whereas in ammonia stripping and scrubbing requirement of air is the key factor in increasing the operational cost. For these technologies to emerge as a viable option optimization of cost is a critical factor. Finally it can be concluded in view of the technical perspective all the existing technologies require fine tuning for minimizing cost especially due to chemical and energy demand, efficiency, fertilizer quality and recovery of ammonia and various nutrients. However, focus must be shifted on the recovery of ammonia along with other nutrients.

The most established and easily available technologies for ammonia recovery from the digestate are ammonia stripping adsorption and struvite precipitation. These technologies have potential to produce marketable fertilizers along with a better technical performance for ammonia recovery.

REFERENCES

Abouelenien F., Kitamura Y., Nishio N. and Nakashimada Y. (2009). Dry anaerobic ammonia-methane production from chicken manure. *Applied Microbiology and Biotechnology*, **82**, 757–764, https://doi.org/10.1007/s00253-009-1881-3

Abouelenien F., Fujiwara W., Namba Y., Kosseva M., Nishio N. and Nakashimada Y. (2010). Improved methane fermentation of chicken manure via ammonia removal by biogas recycle. *Bioresource Technology*, **101**, 6368–6373, https://doi.org/10.1016/j.biortech.2010.03.071

Afif A., Radenahmad N., Cheok Q., Shams S., Kim J. H. and Azad A. K. (2016). Ammonia-fed fuel cells: A comprehensive review. *Renewable and Sustainable Energy Reviews*, **60**, 822–835, https://doi.org/10.1016/j.rser.2016.01.120

Baldi M., Collivignarelli M. C., Abbà A. and Benigna I. (2018). The valorization of ammonia in manure digestate by means of alternative stripping reactors. *Sustainability*, **10**, 1–14, https://doi.org/10.3390/su10093073

Barampouti E. M., Mai S., Malamis D., Moustakas K. and Loizidou M. (2020). Exploring technological alternatives of nutrient recovery from digestate as a secondary resource. *Renewable and Sustainable Energy Reviews*, **134**, 110379, https://doi.org/10.1016/j.rser.2020.110379

Bi S., Qiao W., Xiong L., Mahdy A., Wandera S. M., Yin D. and Dong R. (2020). Improved high solid anaerobic digestion of chicken manure by moderate in situ ammonia stripping and its relation to metabolic pathway. *Renewable Energy*, **146**, 2380–2389, https://doi.org/10.1016/j.renene.2019.08.093

Bonmatí A. and Flotats X. (2003). Air stripping of ammonia from pig slurry: characterisation and feasibility as a pre- or post-treatment to mesophilic anaerobic digestion. *Waste Management*, **23**, 261–272, https://doi.org/10.1016/S0956-053X(02)00144-7

Cao L., Wang J., Zhou T., Li Z., Xiang S., Xu F., Ruan R. and Liu Y. (2019). Evaluation of ammonia recovery from swine wastewater via a innovative spraying technology. *Bioresource Technology*, **272**, 235–240, https://doi.org/10.1016/j.biortech.2018.10.021

Cohen Y. and Kirchmann H. (2004). Increasing the pH of wastewater to high levels with different gases – CO_2 stripping. *Water, Air, & Soil Pollution*, **159**, 265–275, https://doi.org/10.1023/B:WATE.0000049185.69759.b0

De la Rubia MÁ, Walker M., Heaven S., Banks C. J. and Borja R. (2010). Preliminary trials of in situ ammonia stripping from source segregated domestic food waste digestate using biogas: effect of temperature and flow rate. *Bioresource Technology*, **101**, 9486–9492, https://doi.org/10.1016/j.biortech.2010.07.096

Fernandez-Gonzalez N., Pedizzi C., Lema J. M. and Carballa M. (2019). Air-side ammonia stripping coupled to anaerobic digestion indirectly impacts anaerobic microbiome. *Microbial Biotechnology*, **12**, 1403–1416, https://doi.org/10.1111/1751-7915.13482

Fuchs W. and Drosg B. (2013). Assessment of the state of the art of technologies for the processing of digestate residue from anaerobic digesters. *Water Science and Technology*, **67**, 1984–1993, https://doi.org/10.2166/wst.2013.075

Georgiou D., Liliopoulos V. and Aivasidis A. (2019). Investigation of an integrated treatment technique for anaerobically digested animal manure: lime reaction and settling, ammonia stripping and neutralization by biogas scrubbing. *Bioresource Technology Reports*, **5**, 127–133, https://doi.org/10.1016/j.biteb.2019.01.001

Guo X., Zeng L., Li X. and Park H. S. (2007). Removal of ammonium from RO permeate of anaerobically digested wastewater by natural zeolite. *Separation Science and Technology*, **42**, 3169–3185, https://doi.org/10.1080/01496390701514949

Guštin S. and Marinšek-Logar R. (2011). Effect of pH, temperature and air flow rate on the continuous ammonia stripping of the anaerobic digestion effluent. *Process Safety and Environmental Protection*, **89**, 61–66, https://doi.org/10.1016/j.psep.2010.11.001

Huang W., Zhao Z., Yuan T., Lei Z., Cai W., Li H. and Zhang Z. (2016). Effective ammonia recovery from swine excreta through dry anaerobic digestion followed by ammonia stripping at high total solids content. *Biomass and Bioenergy*, **90**, 139–147, https://doi.org/10.1016/j.biombioe.2016.04.003

Jiang X., Hayashi J., Sun Z. Y., Yang L., Tang Y. Q., Oshibe H., Osaka N. and Kida K. (2013). Improving biogas production from protein-rich distillery wastewater by decreasing ammonia inhibition. *Process Biochemistry*, **48**, 1778–1784, https://doi.org/10.1016/j.procbio.2013.08.014

Jiang A., Zhang T., Zhao Q. B., Li X., Chen S. and Frear C. S. (2014). Evaluation of an integrated ammonia stripping, recovery, and biogas scrubbing system for use with anaerobically digested dairy manure. *Biosystems Engineering*, **119**, 117–126, https://doi.org/10.1016/j.biosystemseng.2013.10.008

Kinidi L., Tan I. A. W., Abdul Wahab N. B., Tamrin K. F. B., Hipolito C. N. and Salleh S. F. (2018). Recent development in ammonia stripping process for industrial wastewater treatment. *International Journal of Chemical Engineering*, **2018**, 1–14, https://doi.org/10.1155/2018/3181087

Laureni M., Palatsi J., Llovera M. and Bonmatí A. (2013). Influence of pig slurry characteristics on ammonia stripping efficiencies and quality of the recovered ammonium-sulfate solution. *Journal of Chemical Technology & Biotechnology*, **88**, 1654–1662, https://doi.org/10.1002/jctb.4016

Lauterböck B., Nikolausz M., Lv Z., Baumgartner M., Liebhard G. and Fuchs W. (2014). Improvement of anaerobic digestion performance by continuous nitrogen removal with a membrane contactor treating a substrate rich in ammonia and sulfide. *Bioresource Technology*, **158**, 209–216, https://doi.org/10.1016/j.biortech.2014.02.012

Lei X., Sugiura N., Feng C. and Maekawa T. (2007). Pretreatment of anaerobic digestion effluent with ammonia stripping and biogas purification. *Journal of Hazardous Materials*, **145**, 391–397, https://doi.org/10.1016/j.jhazmat.2006.11.027

Li K., Liu R., Yu Q. and Ma R. (2018). Removal of nitrogen from chicken manure anaerobic digestion for enhanced biomethanization. *Fuel*, **232**, 395–404, https://doi.org/10.1016/j.fuel.2018.05.142

Minardi E. R., Chakraborty S., Calabrò V., Curcio S. and Drioli E. (2015). Membrane applications for biogas production and purification processes: an overview on a smart alternative for process intensification. *RSC Advances*, **5**, 14156–14186, https://doi.org/10.1039/C4RA11819G

Morales N., Boehler M. A., Buettner S., Liebi C. and Siegrist H. (2013). Recovery of N and P from urine by struvite precipitation followed by combined stripping with digester sludge liquid at full scale. *Water*, **5**, 1262–1278, https://doi.org/10.3390/w5031262

Palakodeti A., Azman S., Rossi B., Dewil R. and Appels L. (2021). A critical review of ammonia recovery from anaerobic digestate of organic wastes via stripping. *Renewable and Sustainable Energy Reviews*, **143**, 110903, https://doi.org/10.1016/j.rser.2021.110903

Park S. and Kim M. (2015). Innovative ammonia stripping with an electrolyzed water system as pretreatment of thermally hydrolyzed wasted sludge for anaerobic digestion. *Water Research*, **68**, 580–588, https://doi.org/10.1016/j.watres.2014.10.033

Pedizzi C., Lema J. M. and Carballa M. (2017). Enhancing thermophilic co-digestion of nitrogen-rich substrates by air side-stream stripping. *Bioresource Technology*, **241**, 397–405, https://doi.org/10.1016/j.biortech.2017.05.113

Provolo G., Perazzolo F., Mattachini G., Finzi A., Naldi E. and Riva E. (2017). Nitrogen removal from digested slurries using a simplified ammonia stripping technique. *Waste Management*, **69**, 154–161, https://doi.org/10.1016/j.wasman.2017.07.047

Serna-Maza A., Heaven S. and Banks C. J. (2014). Ammonia removal in food waste anaerobic digestion using a side-stream stripping process. *Bioresource Technology*, **152**, 307–315, https://doi.org/10.1016/j.biortech.2013.10.093

Siegrist H., Hunziker W. and Hofer H. (2005). Anaerobic digestion of slaughterhouse waste with UF-membrane separation and recycling of permeate after free ammonia stripping. *Water Science and Technology*, **52**, 531–536, https://doi.org/10.2166/wst.2005.0563

Sun Z. Y., Yamaji S., Cheng Q. S., Yang L., Tang Y. Q. and Kida K. (2014). Simultaneous decrease in ammonia and hydrogen sulfide inhibition during the thermophilic anaerobic digestion of protein-rich stillage by biogas recirculation and air supply at 60°c. *Process Biochemistry*, **49**, 2214–2219, https://doi.org/10.1016/j.procbio.2014.09.006

Törnwall E., Pettersson H., Thorin E. and Schwede S. (2017). Post-treatment of biogas digestate – aAn evaluation of ammonium recovery, energy use and sanitation. *Energy Procedia*, **142**, 957–963, https://doi.org/10.1016/j.egypro.2017.12.153

Vaneeckhaute C., Lebuf V., Michels E., Belia E., Vanrolleghem P. A., Tack F. M. G. and Meers E. (2017). Nutrient recovery from digestate: systematic technology review and product classification. *Waste and Biomass Valorization*, **8**, 21–40, https://doi.org/10.1007/s12649-016-9642-x

Wang D., Xin Y., Shi H., Ai P., Yu L., Li X. and Chen S. (2019). Closing ammonia loop in efficient biogas production: recycling ammonia pretreatment of wheat straw. *Biosystems Engineering*, **180**, 182–190, https://doi.org/10.1016/j.biosystemseng.2019.02.010

Wirthensohn T., Waeger F., Jelinek L. and Fuchs W. (2009). Ammonium removal from anaerobic digester effluent by ion exchange. *Water Science and Technology*, **60**, 201–210, https://doi.org/10.2166/wst.2009.317

Yao Y., Yu L., Ghogare R., Dunsmoor A., Davaritouchaee M. and Chen S. (2017). Simultaneous ammonia stripping and anaerobic digestion for efficient thermophilic conversion of dairy manure at high solids concentration. *Energy*, **141**, 179–188, https://doi.org/10.1016/j.energy.2017.09.086

Zeng L., Mangan C. and Li X. (2006). Ammonia recovery from anaerobically digested cattle manure by steam stripping. *Water Science and Technology*, **54**, 137–145, https://doi.org/10.2166/wst.2006.852

Zhang L. and Jahng D. (2010). Enhanced anaerobic digestion of piggery wastewater by ammonia stripping: effects of alkali types. *Journal of Hazardous Materials*, **182**, 536–543, https://doi.org/10.1016/j.jhazmat.2010.06.065

Zhang L., Lee Y. W. and Jahng D. (2012). Ammonia stripping for enhanced biomethanization of piggery wastewater. *Journal of Hazardous Materials*, **199–200**, 36–42, https://doi.org/10.1016/j.jhazmat.2011.10.049

Zhang H., Lo V. K., Thompson J. R., Koch F. A., Liao P. H., Lobanov S., Mavinic D. S. and Atwater J. W. (2015). Recovery of phosphorus from dairy manure: a pilot-scale study. *Environmental Technology*, **36**, 1398–1404, https://doi.org/10.1080/09593330.2014.991354

Zhao Q. B., Ma J., Zeb I., Yu L., Chen S., Zheng Y. M. and Frear C. (2015). Ammonia recovery from anaerobic digester effluent through direct aeration. *Chemical Engineering Journal*, **279**, 31–37, https://doi.org/10.1016/j.cej.2015.04.113

doi: 10.2166/9781789062755_0171

Chapter 8

Renewable fuels recovery from anaerobic digestate

Nongmaithem Debeni Devi[1,#], Sutapa Das[2,#], Angana Chaudhuri[1] and Vaibhav V. Goud[1,2,]*

[1]School of Energy Science and Engineering, Indian Institute of Technology Guwahati, Guwahati, Assam 781039, India
[2]Department of Chemical Engineering, Indian Institute of Technology Guwahati, Guwahati, Assam 781039, India
**Corresponding author e-mail: vvgoud@iitg.ac.in*
Department of Chemical Engineering, Indian Institute of Technology Guwahati, Guwahati 781039, Assam, India.
[#]Both are co-first authors

ABSTRACT

Anaerobic digestion is a promising technology for the production of bioenergy by utilizing various feedstock and biowastes. During this process, a huge amount of digestate is generated as residue. The anaerobic digestate is rich in organic and inorganic compounds that will be harmful to the environment if not treated properly. However, innovative technologies are developed for digestate management. Appropriate treatment of the digestate provides soil fertility to agriculture, offers cost-effective nutrients to microalgae cultivation, feedstock for pyrolysis to generate biochar to adsorb pollutants, and renewable energy production. This study summarized the current trends and efforts on digestate valorization and aspects of resource recovery.

Keywords: Anaerobic digestate, microalgae, pyrolysis, biochar, feedstock, renewable energy

8.1 INTRODUCTION

Anaerobic digestion (AD) is an emerging technology for the production of renewable energy (biogas) from various feedstocks via microorganisms in the absence of oxygen. This technology provides an environmentally and economically feasible route for the conversion of wastes to value-added products, thus fulfilling one of the sustainable development goals of energy (Kaur *et al.*, 2019). In comparison to incineration and land-filling, AD technology provides a more effective treatment option (Obulisamy *et al.*, 2016). Nonetheless, one of the severe drawbacks of this technology relates to the management of post-digested residues or digestates (Tampio *et al.*, 2016). The organic-rich digestate has mostly been utilized for agricultural purposes such as soil conditioners (Monlau *et al.*, 2015) or as electron donors in denitrification employed for treating landfill leachate (Peng *et al.*, 2020). However, this method of utilization of digestate raises concern as it not only causes pollution of soil due to excessive-mineral-rich content, but also leads to the generation of greenhouse gases such as CO_2, CH_4, and N_2O that may seriously contribute toward environmental pollution (Cho *et al.*, 2013). The problem affects the environment as almost 80 MT (million tons) of waste is being generated from around 115 plants processing anaerobic digestate (Pérez-Camacho & Curry, 2018). With the restriction on land availability and ever-increasing generation of digestate, there is an urgent need to develop a sustainable approach for the management of anaerobic digestate so that it does not become another 'under-utilized waste product'. Conversion or valorization of the produced digestate to high-value products will not only improve the economic scenario of digestate plants and improve their sustainability but will also create newer opportunities for effectively utilizing additional waste and improving bioprocessing plants. Globally, this problem has inspired the creation of programmes for a circular-economy approach such as The Waste and Resources Action Programme along with EU Vision 2020, to deal with bioresource management of the digestate (Kaur *et al.*, 2019). This study summarizes the current scenario concerning the valorization of anaerobic digestate, recovery of potential renewable fuels and management of nutrient removal as biorefinery approach.

8.2 CHARACTERISTICS OF ANAEROBIC DIGESTATE

The digestate is the main by-product obtained from AD. The characteristics of the digestate depend on the feedstock utilized for AD and their valorized products have been illustrated in Figure 8.1. After filtering and centrifuging, the digestate is divided into parts that is 10–20% of solid fraction, and 80–90% of liquid fraction (Doble & Kumar, 2005). The pH of the digestate is in the range of 7.5–8.9. The alkalinity nature of this digestate will be beneficial to alleviate soil acidification. The solid fraction contains 55–65% phosphorus and 20–30% of the total ammonium ions (Chong *et al.*, 2022). This solid digestate can be reutilized and valorized as a feedstock for ethanol production (Cesaro & Belgiorno, 2015). The solid digestates can further be pyrolyzed or co-pyrolyzed to form other valuable products like biochar. This biochar can again be reutilized as its carbon content is high (40–50%) (Logan & Visvanathan, 2019). While, the main constituents of the liquid digestate are 70–80% of nitrogen and 35–45% of residual phosphorus (Peng & Pivato, 2019). The liquid fraction has a higher COD value due to the presence of volatile fatty acids like acetate in the range of 210–6900 mg/L (Xia & Murphy, 2016). The ammonia content in the liquid digestate was estimated to be about 60–80% of the total nitrogen content available (Makdi *et al.*, 2012). The liquid digestates have high carbon content and therefore, these could be mixed with feedstocks containing lower organic content (Chong *et al.*, 2022).

The characteristic features of anaerobic digestate originated from AD of different feedstock have been presented in Table 8.1. The variability in the

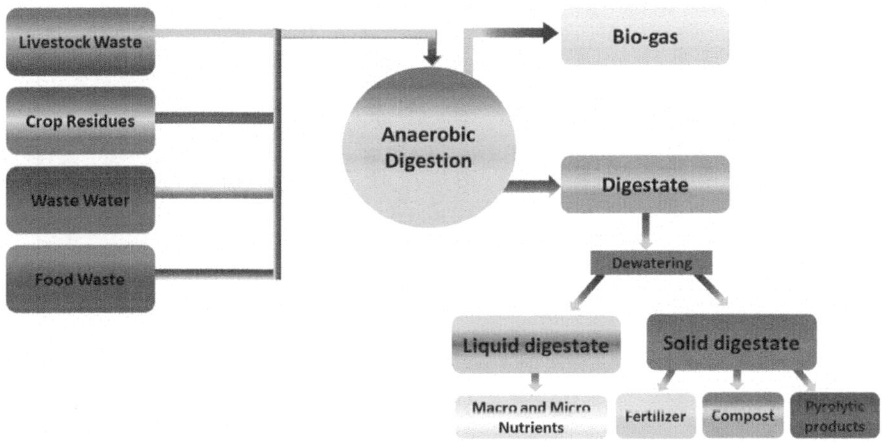

Figure 8.1 Types of anaerobic digestates based on the different feedstocks and their applications.

digestates lies mainly in the nutrient composition such as ammonia nitrogen (NH_3^+-N), total nitrogen (TN), COD, and so on. For instance, the digestate obtained from the AD of municipal sludge has considerable amounts of volatile solids (~63%), total solids content (0.72–51.2%), and ammonium (1.7–7.5 g/L). The C/N ratio of the solid fraction ranges between 1.3 and 29.8 where the COD value is 1.62 g/g VS. The liquid digestate part has a lower C/N ratio of about 2.7 (Zuliani *et al.*, 2016). While, swine manure digestate produced via mesophilic AD has pH ranging from 7.3 to 8.1, which contains total nitrogen (2.7–3.4 g/L), total ammonium (1.4–2.7 g/L), total phosphorus (0.7–1.3 g/L), and COD (13.4–38.0 g/L) (Loria *et al.*, 2007). Food waste anaerobic digestate from digestion has comparable levels of ammonia and nitrogen when compared with cattle and swine manure digestate. The pH of food waste digestate varies from about 6.7–8.4 owing to the processing of ammonia and degradation of volatile fatty acids (Tampio *et al.*, 2016). Total solids obtained were as low as 8.7%. This was in a lower range when compared with the animal manures and other waste digestates (Turovskiy & Mathai, 2005).

The digestate obtained from the co-digestion of multiple feedstocks such as manures of bulls and cows along with energy crops like maize and other lignocellulosic crops has high amounts of high total solids (TS) content (~70%), volatile fatty acids (VFA) such as acetic acid and valeric acid (Chong *et al.*, 2022). Another digestate obtained from the AD of primary and waste-activated sludge under mesophilic conditions shows a total solid content of about 53% and a high amount of nitrogen and phosphorus (Zuliani *et al.*, 2016). One of the studies had utilized calf manure along with the food waste. The digestate produced had a pH of 8.3 with ammonia 1.06 g/L. The total phosphorus level was obtained to be 0.46 which is lower as compared to other digestates (Siciliano & De Rosa, 2014).

Table 8.1 Characteristic features of different anaerobic digestates.

Feedstocks	Digestion type	pH of the digestate	Total solid (g/L)	Volatile solid (g/L)	Carbon content (%)	Total P	Total nitrogen (g/ L)	NH_4-N (g/ L)	C/N ratio	References
Municipal sludge	Thermophilic	8.05	–	–	–	0.23	0.16 g/g TS		–	(Botheju *et al.*, 2010)
Municipal activated sludge	Mesophilic	7.2	6.4	3.8	–	0.21	0.8	0.54	–	(Leite *et al.*, 2017)
Swine manure	Mesophilic	8.1	54 g/Kg	37.6 g/kg	–	0.029	5.3	3.9	–	(Timmerman *et al.*, 2015)
Swine manure	Mesophilic	8.1	–	–	–	0.93	2.93	2.23	–	(Loria *et al.*, 2007)
Cattle manure	Mesophilic	7.8	53.2 g/Kg	40 g/kg	–	–	3.5	2.0	–	(Timmerman *et al.*, 2015)
Food waste	Mesophilic	–	–	66.1%	0.24	–	1.92%	–	13.02	(Liu *et al.*, 2020)
Food waste	Mesophilic	7.9	11	–	–	–	–	0.3 g/Kg	–	(Li *et al.*, 2020)
Food waste	Thermophilic	–	18.5	73.1	–	–	2.43	–	17.6	(Svensson *et al.*, 2018)
Calf manure and food waste	–	8.3	–	–	–	0.46	–	1.06	–	(Siciliano and De Rosa, 2014)
Agricultural waste, crops, cow slurry, and municipal sludge	Thermophilic	8.3	36 g/Kg	684 g/kg	–	11.79 g/kg	110 g/kg	2.427	3.43	(Tambone *et al.*, 2009)

Furthermore, the digestate generated from the AD of bioethanol converts most of the residual organics into volatile fatty acids (Cheng & Brewer, 2021). While, the digestate from AD of lignocellulosic feedstocks contains microorganisms, polysaccharides, and holo-cellulose. The lignin degrades during the AD and hence the digestates have a very low amount of it present in them (Khan & Ahring, 2019). The composition of this digestate depends on the solid retention time during AD. Akyol *et al.* (2016) reported that the digestate from AD of barley was found to have 5.4% total solids, 4% volatile substance, C/N ratio of 16:1, TKN of 8327 mg/L, and pH of 8.1. The presence of a higher concentration of C/N ratio can be beneficial for algae cultivation in the waste.

8.3 POTENTIAL APPLICATION OF THE DIGESTATE

Owing to the varying composition of the digestate, the valorization of the digestate can be performed through various applications. This section summarizes the feasibility of the digestate in various aspects thereby increasing the economic viability and environmental management.

8.3.1 Digestate as feedstock

It may be noted that agricultural-derived waste such as animal dung, energy crops may be directly employed as bio-fertilizer (Nkoa, 2014). However, for other biomass, treatment of digestate usually takes place before its decomposition. The dewatering process is initially performed to separate the solid and liquid fractions. It is a capital-intensive process that employs technologies such as screw press separators and decanter centrifuges (Pérez-Camacho & Curry, 2018). However, this step not only reduces the overall weight of the actual feedstock but also improves the economic outlook of the transportation involved. Thereafter, the feedstock is sterilized or its constituents are concentrated via processes including membrane technology. As an example, micro- and ultrafiltration processes have been carried out to prepare a medium for the production of polyhydroxyalkanoate (PHA) bioplastic generation. The separation step involving microfiltration is carried out primarily to remove the solid residues and also, to modulate the nitrogen/phosphate ratio of the digestate, depending upon the suitability of the microorganism (Stiles *et al.*, 2018). This separation technique can be commercialized on a large scale as much research and development in the field of membrane technology in the last decade has demonstrated promising results in this regard (Veronesiv *et al.*, 2017). In other places, the usage of solid content within the digestate has been utilized as a substrate for fermentation. These processes often involve the hygienization of feedstock as it is maintained at 70°C for 1 h, followed by its valorization. Biosurfactants, biopesticides and enzymes are a few different valuable products generated via this step (Rodríguez *et al.*, 2019). Another technique for digestate treatment involves the usage of autoclaves for sterilization of nutrients belonging to synthetic medium. For instance, reports in the literature present autoclaving of digestate for fungal fermentation to generate chitin. Processes such as hygienization and autoclaving lead to maximum removal of pathogens. Although these processes are energy and cost-intensive, autoclaving may be used for learning different techniques for excessive solubilization of digestate solids as they contain maximum nutrient content. This step also increases the availability of nutrients to ferment micro-organisms when autoclaved digestate is utilized in the bioprocess (Kaur *et al.*, 2019).

8.3.2 Digestate as a substrate for microalgae cultivation

The integration of AD with microalgal cultivation is an emerging approach for biorefinery management. Microalgae are eukaryotic microorganisms that can convert light energy and carbon dioxide (CO_2) into biomass through photosynthesis and assimilation of nutrients (Molina Grima *et al.*, 2003). Microalgae can produce high lipids, carbohydrates, proteins, carotenoids, vitamins, and so on (Devi *et al.*, 2021). In addition, the biofuels derived from microalgae are renewable and sustainable with low emissions of greenhouse gases like CO_2, CO, SO_x, NO_x, and so on. The availability of macro and micronutrients is the key element for the commercial growth of microalgae on large scale (Brennan & Owende, 2010). Hence, anaerobic digestate may act as an alternate source of nutrients and overcome the high-cost issue for microalgae cultivation.

Several researchers reported the cultivation of microalgae by utilizing different types of anaerobic digestates (Ayre *et al.*, 2021; Chuka-ogwude *et al.*, 2021; Debowski *et al.*, 2020; Gu *et al.*, 2021; Sayedin *et al.*, 2020; Sekine *et al.*, 2020; Tan *et al.*, 2020). Various macronutrients (nitrogen, phosphorus, potassium, sulfur, magnesium, calcium, etc.), micronutrients (boron, chlorine, copper, iron, manganese molybdenum, nickel, zinc, etc.), and bioactive compounds (including phytohormones, vitamins, free amino acids, monosaccharides, fulvic acid, volatile fatty acid, etc.) are highly abundant in anaerobic digestate. These components influence the availability of nutrients, salinity, and pH (Chuka-ogwude *et al.*, 2020) of the medium. In addition, appropriate nutrient concentration in AD has been proven to correlate with the enhancement of biomass production and lipid content of microalgae (Geider & La Roche, 2002; Praveen & Loh, 2016). Nevertheless, there is a concern when using anaerobic digestate directly for microalgal growth due to inappropriate nutrient concentrations, high toxicity, ammonia toxicity, turbidity, light penetration inhibition, contamination of unwanted microbes, and so on (Chuka-ogwude *et al.*, 2020). Therefore, the prerequisite factors to utilize anaerobic digestate as the substrate for efficient growth of microalgae is discussed in the upcoming sections.

8.3.2.1 Balanced nutrient composition

The microalgal growth is influenced by the C/N and N/P ratio of the digestate. According to Silaban *et al.* (2014), the optimal C/N ratio for microalgal growth was between 4 and 15, while the P/N ratio ranged between 7 and 16. However, most of the digestates were reported to have relatively low C/N and N/P ratios that could not meet the optimal ratios (Khalid *et al.*, 2011). Hence, researchers suggested that the imbalanced C/N and P/N can be resolved by supplementing extra nutrients to the digestate (Silaban *et al.*, 2014). External carbon sources can be provided either with inorganics or organic carbon into the digestate (Tan *et al.*, 2020). However, due to the low atmospheric CO_2 concentration, the rate of CO_2 dissolution from the air into the medium is relatively low. Hence, diffusion of CO_2 into the medium can be performed through gas bubbling. Another possible source of CO_2 can be the flue gas liberated from a boiler or industry. The flue gas consists of ~4–15% of CO_2 which could be sequestered by the microalgae, eliminating the cost of pure CO_2 and reducing the carbon footprint (Mata *et al.*, 2010; Yen *et al.*, 2015). Supplementation of organic carbon sources can be achieved by adding commercial organic compounds such as glucose, glycerol, ethanol, fructose, and so on (Deng *et al.*, 2019; Koutra *et al.*, 2019; Tan *et al.*, 2020). However, the addition of artificial compounds will increase the cost of cultivation. Hence, some studies investigated the use of volatile fatty acid (VFA) (such as propionic, acetic, and n-butyric acids) and dissolved carboxylic acids from the sludge of anaerobic fermentation to reduce the production cost (Chalima *et al.*, 2017). Ramanan *et al.* (2013) reported that supplementation of acetate derived from AD of sludge could enhance the lipid content of *Chlamydomonas reinhardtii* up to 42%. Tan *et al.* (2020) investigated the growth of *Chlorella pyrenoidosa* in anaerobic digestate by supplementing acidified starch wastewater as a waste organic carbon source in the ratio of 1:1 (v/v). In their study, the biomass production and lipid yield increased by 0.5-fold and 4.2-fold, respectively. Nevertheless, the study on the wastewater-to-digestate ratio or optimal concentration of VFA for effective algal growth is less reported, therefore, further research is yet to be explored.

Besides, the compositions of the nutrients are entirely dependent on the types of digestate source, thus the optimal stoichiometric ratios of C/N or N/P can be achieved either by altering the feedstock for the AD or by utilizing the digestate derived from co-digestion (Malolan *et al.*, 2020; Zhang *et al.*, 2021). The advantages of utilizing co-digestion digestate are that the presence of various substrates including low C/N (e.g., sewage or animal manure) or high C/N ratio (organic solid waste) will produce a mutual effect to balance the appropriate propositions, reducing the effects of the individual substrate (Khalid *et al.*, 2011; Silaban *et al.*, 2014). Malolan *et al.* (2020) cultivated *Chlorella pyrenoidosa* by utilizing digestate derived from the co-digestion of cow dung and chicken manure. After effective growth, *Chlorella pyrenoidosa* could result in 2.4 g/L of biomass yield, reducing COD, nitrate and phosphate from 2850, 41.3 and 73.4 mg/L to 572, 16.5 and 23.1 mg/L, respectively.

8.3.2.2 Pretreatment of the digestate

Light is the limiting factor for the growth of microalgae. At the light saturation phase (50–400 μmol/m²/s), microalgae perform photosynthesis efficiently, resulting in maximum biomass productivity (Goldman, 1979; Liu *et al.*, 2012). The distribution of light throughout the culture medium or the digestate will be affected by high turbidity. The high turbidity of the digestate can be caused by high concentrations of ammonia, phosphorus, high dissolved suspended materials, dark color compounds, clay, microorganisms, and so on. (Chuka-ogwude *et al.*, 2020; Wang *et al.*, 2010; Zhang *et al.*, 2010). These scattered impurities can absorb the light, reducing the light transmittance through the water, thus inhibiting the photosynthesis of microalgae and reducing biomass yield. Akhiar *et al.* (2017) measured the turbidity of manure digestates collected from 11 digestion plants through a TN-100/T-100 turbidimeter. They observed that the values ranged between 2960 and 51 400 NTU. Hence, before direct use of the digestate for microalgae cultivation, a pre-treatment process is required. Generally, pretreatment processes such as precipitation, adsorption, centrifugation, filtration, are employed to manage the turbidity of the digestate (Cheng *et al.*, 2015; Sturm & Lamer, 2011). However, these techniques are expensive, energy-intensive, and require extra operation (Cheng *et al.*, 2015). Researchers introduced a new promising technique to diminish the turbidity by diluting the digestate with either freshwater, synthetic media, saline water, or various types of wastewaters (aquaculture, domestic, municipal wastewater, etc.). This technique not only alleviates the high nitrogen, ammonia, phosphorus concentration, photoinhibition, turbidity but also provides an appropriate proportion of nutrients in the culture medium (Torres Franco *et al.*, 2018; Uggetti *et al.*, 2014; Wang *et al.*, 2010). The digestate collected from a piggery anaerobic treatment plant was diluted by Xu *et al.* (2015) to grow *Scenedesmus obliquus* (FACHB-31) at several dilutions with distilled water to obtain different initial chemical oxygen demand (COD) concentration levels (3200, 2200, 1600, 1200, 800, and 400 mg/L COD). They found that the microalgal growth achieved its maximum biomass productivity (311.28 ± 12.36 mg/L/d) with the initial COD concentration of 1600 mg/ L. Their study also demonstrated the efficient removal rates of COD (61.58–75.29%), CO_2 (54.26–73.81%), total phosphorous (70.09–88.79), and nitrogen (58.39–74.63%). Cho *et al.* (2013) reported the cultivation of *Chlorella vulgaris* in the digestate diluted with wastewater released from sludge concentrate tanks (10:90, v/v). The microalgae exhibited a maximum biomass concentration of 3.01 g/L. There have been several reports on the dilution of the digestate. However, when considering the large microalgal cultivation, dilution using fresh water, synthetic media is not cost effective and water-intensive process. Thus, further research should be focused on utilizing different types of wastewaters to expand the economic and environmental security. Moreover, proper care should be taken while selecting a suitable dilution ratio that can alleviate the inhibitory effect of impurities and enhance the microalgal growth.

8.3.3 Selection of potential microalgal strains

The selection of robust microalgal strain is the key factor to compete for high salinity, high nutrient concentration, ammonia and metal toxicity, and so on when utilizing anaerobic digestate for microalgal growth. The genera *Scenedesmus*, *Chlorella*, and *Desmodesmus* species are famous to thrive well in extreme environment. Several studies reported the cultivation of these species in wastewater, digestate to maximize high lipid yield and biomass production (Duan, 2012; Pancha *et al.*, 2015; Xia *et al.*, 2014). Apart from this, Osundeko and Pittman (2014) confirmed the survival of *Chlorella luteoviridis* and *Parachlorella hussii* in highly toxic wastewater (centrate liquor) expelled from the activated sludge. Despite the resilient microalgae, trouble arises due to cross-contamination by other unwanted microbes while cultivating in an open or pond system (Rodolfi *et al.*, 2009). Hence, researchers introduced the application of mixed or consortium cultures to withstand harsh environments. The advantages of microalgal consortium or mixed cultures are as follows: (1) have higher diversity and stability than monocultures, (2) less susceptible to invasion, (3) easy to operate, (4) utilization wide absorption spectrum of light, and (5) production of various bioactive compounds by the microbes to facilitate the growth of microalgal partner (Chuka-ogwude *et al.*, 2020; Kang *et al.*, 2018; Ogbonna & Tanaka, 2000). These advantages will benefit the large-scale cultivation of microalgal consortia in anaerobic digestate. Wang *et al.* (2021a) showed an increase in biomass yield of microalgal consortia by 81% when grown in pretreated dairy manure digestate and municipal wastewater sludge digestate. Another study by Wang *et al.* (2021b) found that the mixed culture of *Chlorella sorokiniana* and indigenous bacteria from the activated sludge could efficiently remove 29% of nitrogen, and 34% of phosphorus when cultivated in municipal anaerobic digestate diluted with activated sludge. Qin *et al.* (2016) found high content of saturated fatty acids in the microalgal consortium and suggested the suitability for biodiesel production. In addition, coupling of nitrifying bacteria with microalgae cultivation showed a promising solution, avoiding microalgae from NH_4^+-N toxicity due to conversion of NH_4^+-N into non-toxic NO_3^- (Svehla *et al.*, 2017). Likewise, Praveen *et al.* (2018) demonstrated the nitrification of NH_4^+-N to NO_3^- by coupling nitrifying bacteria with *Chlorella vulgaris* ATCC 13482 cultivation in anaerobic digestate. Thus, the nitrification process will allow the direct use of anaerobic digestate as a substrate for microalgae cultivation.

8.3.4 Anaerobic digestate for the production of biofuels

The utilization of thermochemical methods for anaerobic digestate treatment has been carried out to improve the economic and environmental outlook of the conversion process. Processes such as pyrolysis (Feng & Lin, 2017; Monlau *et al.*, 2016; Salman *et al.*, 2017), gasification (Alfa *et al.*, 2014; Pecchi & Baratieri, 2019; Yao *et al.*, 2017), and carbonization (Reza *et al.*, 2014) have been analyzed with regards to overall efficiency and sustainability of the process.

8.3.4.1 Pyrolysis-based digestate treatment

The pyrolysis-based digestate by-products find utility across different fields such as for energy generation (bio-oil and syngas) or soil conditioner (bio-char) (Monlau *et al.*, 2015). The confluence of pyrolysis technology with anaerobic digestate treatment aids in enhancing the value addition of the final product. In a particular study, a pyrolytic environment at 500°C was employed to convert the digestate feedstock into bio-oil having a calorific value of 7.8 MJ/kg and yield of 52% of total mass (Opatokun *et al.*, 2017). Further improvement in the system was carried out by subjecting the solid digestate to liquefaction in the presence of polyethylene glycol, glycerol, and sulfuric acid (catalyst) (Barbanera *et al.*, 2018). The resultant treatment improved the calorific value (28.5 MJ/kg) and bio-oil yield (60%), respectively. In another study, agricultural waste digestate was subjected to pyrolysis at 500°C, which resulted in the generation of biochar having a heating value of 19 MJ/m^3 and yield of 20% (Neumann *et al.*, 2016). The utility of pyrolytic digestate products goes beyond fuels and fertilizers, as the liquid by-product of the pyrolytic digestate can be fed to the anaerobic digesters for the generation of biogas (Hübner & Mumme, 2015). However, care needs to be taken toward proper disposal of the wastewater/liquids generated from the pyrolysis of digestate source as reports have suggested contamination due to aromatic and refractory compounds (Jayakody *et al.*, 2018). The incorporation of digestate-pyrolyzed biochar in AD also improved the overall methane production and agronomic parameter of the digestate (Masebinu *et al.*, 2019). In principle, the pyrolytic-biochar facilitated the AD process by addressing the alkaline and acidic stresses (Luo *et al.*, 2015). Moreover, the solid digestate when pyrolyzed at 800°C produces porous biochar with high surface area, ideal for functioning as supports for microbial colonization in anaerobic digesters (Luo *et al.*, 2015).

Nonetheless, there are disadvantages associated with the working of this system as it leads to environmental concerns. Pre-treatment of the digestate in terms of drying is needed before utilizing it for processes such as pyrolysis or gasification. The drying process generates lots of ammonia-nitrogen into the air (Li *et al.*, 2011). Acid washing allows retention of the nitrogen from the ammonia-rich digestate. Also, once the digestate is pyrolyzed and products in the form of bio-oil, biochar, or syngas are obtained, their consumption will lead to further environmental detriment and affect sustainable biogas production. Moreover, acid washing also yields HCN and N-containing wastes that may lead to pollution (Tian *et al.*, 2014). In this regard, studies have reported employing thermochemical methods for alleviating the nitrogen emission by converting the digestate having large quantities of C and N.

8.3.4.2 Gasification-based digestate treatment

Gasification includes partial oxidation of digestate and subsequently, conversion of the solid residues to produce syngas. The produced syngas can be utilized as fuel as it primarily contains CO, CO_2, and H_2. The gasification process occurs at ~800°C, thereby preventing the formation of furans, aromatic hydrocarbons (Grycova *et al.*, 2013). At low temperatures 400°C–500°C, the gasification process may not enable phosphorous volatilization thus reducing its utility as a nutrient (Lacroix *et al.*, 2014). Additionally, the presence of the toxic metal in the digested sludge may lead to its concentration in the generated char or ash on completion of the gasification process (Marrero *et al.*, 2004). As a result, the solid residue from the gasification of the sewage sludge may not be utilized as a soil conditioner. Among the parameters affecting the efficiency of the gasification process, the air equivalence ratio (~0.25–0.3) plays an important role as it measures the amount of airflow compared to the theoretical amount necessary for full oxidation (Antoniou *et al.*, 2019). Among the mentioned feedstocks such as sewage sludge, agricultural waste, mostly are the solid digestates arising from the solid–liquid separation unit. For improving the process outlook, replacing the solid digestate with the high solid anaerobic digestate leads to a higher amount of syn-gas generation. In a particular study, ~5 MJ/kg of calorific value along with 1.5 N m^3/kg of syn-gas yield was generated (Li *et al.*, 2018). In brief, the overview, design, and operation of the integrated systems must consider the entire energy, material balance for a fixed period, for easy processing of the whole operation.

8.4 FUTURE PERSPECTIVES AND CONCLUSIONS

Over the past decades, direct use of the digestate toward agriculture or farming has been carried out, however, the consequence is hazardous to the environment. Now the disposal of digestate directly on land is governed by rules and regulations. Many technologies and processes have been discussed and their implementation in valorizing waste. This study highlighted the perspectives of digestate as biofertilizer production, nutrient extraction, and microalgae cultivation. Using digestate as a source of substrate for microalgae cultivation showed a promising solution to the commercialization of microalgal biomass and energy production. Moreover, this integrated strategy provides the precursor for biorefinery products. Currently, researchers are focusing on energy storage and conversion, bio-coal production (hydrochars), biochar. Future research should be focused on the production and conversion process of VFA, free fatty acids (FFAs) (including capric acid, lauric acid, myristic acid, palmitic acid, palmitoleic acid, stearic acid, oleic acid, linoleic acid, linolenic acid, arachidic acid, gadoleic acid, behenic acid, and erucic acid), PHA, and several enzyme syntheses. This will provide a wide angle on pollution remediation and its potential purposes as well. Furthermore, anaerobic digestate sludge is high in metal contents. This direct discharge will deteriorate the environment. Studies should be focused on using and converting these metals as a catalyst. The anaerobic digestates are abundant in NH_3^+-N as residue generated during bio-methane and bio-ethanol production. This shows the prospect of generating biohydrogen from NH_3^+-N thereby, solving the issues of ammonia toxicity. In conclusion, AD digestate is not a waste; rather it is the treasured bio-residue recovered from anaerobic digester plants which can aid as a precursor for many products. The anaerobic digestate has much economic potential and putting it to good practical use has to be our goal.

ACKNOWLEDGEMENTS

The authors are thankful to the Library and Open Access facilities of the Institute for providing free access to journals during the preparation of the manuscript. The authors are also thankful to the Ministry of Education (MoE), Government of India for the students' fellowship.

REFERENCES

Akhiar A., Battimelli A., Torrijos M. and Carrere H. (2017). Comprehensive characterization of the liquid fraction of digestates from full-scale anaerobic co-digestion. *Waste Management*, **59**, 118–128. https://doi.org/10.1016/j.wasman.2016.11.005

Akyol Ç., Ozbayram E. G., Ince O., Kleinsteuber S. and Ince B. (2016). Anaerobic co-digestion of cow manure and barley: effect of cow manure to barley ratio on methane production and digestion stability. *Environmental Progress & Sustainable Energy*, **35**, 589–595. https://doi.org/10.1002/ep.12250

Alfa M. I., Adie D. B., Igboro S. B., Oranusi U. S., Dahunsi S. O. and Akali D. M. (2014). Assessment of biofertilizer quality and health implications of anaerobic digestion effluent of cow dung and chicken droppings. *Renewable Energy*, **63**, 681–686. https://doi.org/10.1016/j.renene.2013.09.049

Antoniou N., Monlau F., Sambusiti C., Ficara E., Barakat A. and Zabaniotou A. (2019). Contribution to circular economy options of mixed agricultural wastes management: coupling anaerobic digestion with gasification for enhanced energy and material recovery. *Journal of Cleaner Production*, **209**, 505–514. https://doi.org/10.1016/j.jclepro.2018.10.055

Ayre J. M., Mickan B. S., Jenkins S. N. and Moheimani N. R. (2021). Batch cultivation of microalgae in anaerobic digestate exhibits functional changes in bacterial communities impacting nitrogen removal and wastewater treatment. *Algal Research*, **57**, 102338. https://doi.org/10.1016/j.algal.2021.102338

Barbanera M., Pelosi C., Taddei A. R. and Cotana F. (2018). Optimization of bio-oil production from solid digestate by microwave-assisted liquefaction. *Energy Conversion and Management*, **171**, 1263–1272. https://doi.org/10.1016/j.enconman.2018.06.066

Botheju D., Svalheim O. and Bakke R. (2010). Digestate nitrification for nutrient recovery. *The Open Waste Management Journal*, **3**, 1–12. https://doi.org/10.2174/1876400201003010001.

Brennan L. and Owende P. (2010). Biofuels from microalgae – a review of technologies for production, processing, and extractions of biofuels and co-products. *Renewable and Sustainable Energy Reviews*, **14**, 557–577. https://doi.org/10.1016/j.rser.2009.10.009

Cesaro A. and Belgiorno V. (2015). Combined biogas and bioethanol production: opportunities and challenges for industrial application. *Energies*, **8**, 8121–8144. https://doi.org/10.3390/en8088121

Chalima A., Oliver L., De Castro L. F., Karnaouri A., Dietrich T. and Topakas E. (2017). Utilization of volatile fatty acids from microalgae for the production of high added value compounds. *Fermentation*, **3**, 54. https://doi.org/10.3390/fermentation3040054

Cheng F. and Brewer C. E. (2021). Conversion of protein-rich lignocellulosic wastes to bio-energy: review and recommendations for hydrolysis + fermentation and anaerobic digestion. *Renewable and Sustainable Energy Reviews*, **146**, 111167. https://doi.org/10.1016/j.rser.2021.111167

Cheng J., Xu J., Huang Y., Li Y., Zhou J. and Cen K. (2015). Growth optimisation of microalga mutant at high CO_2 concentration to purify undiluted anaerobic digestion effluent of swine manure. *Bioresource Technology*, **177**, 240–246. https://doi.org/10.1016/j.biortech.2014.11.099

Cho S., Lee N., Park S., Yu J., Luong T. T., Oh Y. K. and Lee T. (2013). Microalgae cultivation for bioenergy production using wastewaters from a municipal WWTP as nutritional sources. *Bioresource Technology*, **131**, 515–520. https://doi.org/10.1016/j.biortech.2012.12.176

Chong C. C., Cheng Y. W., Ishak S., Lam M. K., Lim J. W., Tan I. S., Show P. L. and Lee K. T. (2022). Anaerobic digestate as a low-cost nutrient source for sustainable microalgae cultivation: a way forward through waste valorization approach. *Science of the Total Environment*, **803**, 150070. https://doi.org/10.1016/j.scitotenv.2021.150070

Chuka-ogwude D., Ogbonna J. and Moheimani N. R. (2020). A review on microalgal culture to treat anaerobic digestate food waste effluent. *Algal Research*, **47**, 101841. https:// doi.org/10.1016/j.algal.2020.101841

Chuka-ogwude D., Ogbonna J. C. and Moheimani N. R. (2021). Depth optimization of inclined thin layer photobioreactor for efficient microalgae cultivation in high turbidity digestate. *Algal Research*, **60**, 102509. https://doi.org/10.1016/j.algal.2021.102509

Debowski M., Zielinski M., Kisielewska M., Kazimierowicz J., Dudek M., Swica I. and Rudnicka A. (2020). The cultivation of lipid-rich microalgae biomass as anaerobic digestate valorization technology – a pilot-scale study. *Processes*, **8**, 517. https://doi.org/10.3390/PR8050517

Deng X. Y., Xue C. Y., Chen B., Amoah P. K., Li D., Hu X. L. and Gao K. (2019). Glucose addition-induced changes in the growth and chemical compositions of a freshwater microalga *Chlorella kessleri*. *Journal of Chemical Technology & Biotechnology*, **94**, 1202–1209. https://doi.org/10.1002/jctb.5870

Devi N., Goud V. V. and Hu B. (2021). Algal cultivation systems and photobioreactor designs. In: Algal Biorefinery: Developments, Challenges and Opportunities, A. K. Dalai, V. V. Goud, S. Nanda and V. B. Borugadda (eds.), Routledge,United Kingdom, p. 24.

Doble M. and Kumar A. (2005). Aerobic and anaerobic bioreactors. In: Biotreatment of Industrial Effluents, Doble M. and Kumar A. (eds), Butterworth-Heinemann, United Kingdom, pp. 19–38. https://doi.org/10.1016/b978-075067838-4/50004-x

Duan X. (2012). Salt-induced osmotic stress for lipid overproduction in batch culture of *Chlorella vulgaris* [WWW document]. *African Journal of Biotechnology*, **11**, 7072–7078. https://doi.org/10.5897/ajb11.3670

Feng Q. and Lin Y. (2017). Integrated processes of anaerobic digestion and pyrolysis for higher bioenergy recovery from lignocellulosic biomass: a brief review. *Renewable and Sustainable Energy Reviews*, **77**, 1272–1287. https://doi.org/10.1016/j.rser.2017.03.022

Geider R. J. and La Roche J. (2002). Redfield revisited: variability of C:N:P in marine microalgae and its biochemical basis. *European Journal of Phycology*, **37**, 1–17. https://doi.org/10.1017/S0967026201003456

Goldman J. C. (1979). Outdoor algal mass cultures-II. Photosynthetic yield limitations. *Water Research*, **13**, 119–136. https://doi.org/10.1016/0043-1354(79)90083-6

Grycova B., Stepkova K., Kubonova L., Kuca R., Veronesiová S. and Obroucka K. (2013). Evaluation of process products from pyrolysis and gasification of dried digestates. International Multidisciplinary Scientific GeoConference Surveying Geology and Mining Ecology Management, SGEM, Vol. **1**, pp. 437–444. https://doi.org/10.5593/SGEM2013/BE5.V1/S20.058

Gu Z., Liu Y., Zou G., Zhang Q., Lu R., Yan H., Cao L., Liu T. and Ruan R. (2021). Enhancement of nutrients removal and biomass accumulation of *Chlorella vulgaris* in pig manure anaerobic digestate effluent by the pretreatment of indigenous bacteria. *Bioresource Technology*, **328**, 124846. https://doi.org/10.1016/j.biortech.2021.124846

Hübner T. and Mumme J. (2015). Integration of pyrolysis and anaerobic digestion – use of aqueous liquor from digestate pyrolysis for biogas production. *Bioresource Technology*, **183**, 86–92. https://doi.org/10.1016/j.biortech.2015.02.037

Jayakody L. N., Johnson C. W., Whitham J. M., Giannone R. J., Black B. A., Cleveland N. S., Klingeman D. M., Michener W. E., Olstad J. L., Vardon D. R., Brown R. C., Brown S. D., Hettich R. L., Guss A. M. and Beckham G. T. (2018). Thermochemical wastewater valorization: via enhanced microbial toxicity tolerance. *Energy and Environmental Science*, **11**, 1625–1638. https://doi.org/10.1039/c8ee00460a

Kang D., Kim K., Jang Y., Moon H., Ju D. and Jahng D. (2018). Nutrient removal and community structure of wastewater-borne algal-bacterial consortia grown in raw wastewater with various wavelengths of light. *International Biodeterioration & Biodegradation*, **126**, 10–20. https://doi.org/10.1016/j.ibiod.2017.09.022

Kaur G., Luo L., Chen G. and Wong J. W. C. (2019). Integrated food waste and sewage treatment – a better approach than conventional food waste-sludge co-digestion for higher energy recovery via anaerobic digestion. *Bioresource Technology,* **289**, 121698. https://doi.org/10.1016/j.biortech.2019.121698

Khalid A., Arshad M., Anjum M., Mahmood T. and Dawson L. (2011). The anaerobic digestion of solid organic waste. *Waste Management,* **31**, 1737–1744. https://doi.org/10.1016/j.wasman.2011.03.021

Khan M. U. and Ahring B. K. (2019). Lignin degradation under anaerobic digestion: influence of lignin modifications – a review. *Biomass and Bioenergy,* **128**, 105325. https://doi.org/10.1016/j.biombioe.2019.105325

Koutra E., Kopsahelis A., Maltezou M., Grammatikopoulos G. and Kornaros M. (2019). Effect of organic carbon and nutrient supplementation on the digestate-grown microalga, *Parachlorella kessleri. Bioresource Technology,* **294**, 122232. https://doi.org/10.1016/j.biortech.2019.122232

Lacroix N., Rousse D. R. and Hausler R. (2014). Anaerobic digestion and gasification coupling for wastewater sludge treatment and recovery. *Waste Management & Research,* **32**, 608–613. https://doi.org/10.1177/0734242X14538308

Leite W., Magnus B. S., Guimarães L. B., Gottardo M. and Belli Filho P. (2017). Feasibility of thermophilic anaerobic processes for treating waste activated sludge under low HRT and intermittent mixing. *Journal of Environmental Management,* **201**, 335–344. https://doi.org/10.1016/j.jenvman.2017.06.069

Li Q., Zheng L., Qiu N., Cai H., Tomberlin J. K. and Yu Z. (2011). Bioconversion of dairy manure by black soldier fly (Diptera: Stratiomyidae) for biodiesel and sugar production. *Waste Management,* **31**, 1316–1320. https://doi.org/10.1016/j.wasman.2011.01.005

Li W., Lu C., An G., Zhang Y. and Tong Y. W. (2018). Integration of high-solid digestion and gasification to dispose horticultural waste and chicken manure. *Chinese Journal of Chemical Engineering,* **26**, 1145–1151. https://doi.org/10.1016/j.cjche.2017.09.020

Li C., Li J., Pan L., Zhu X., Xie S., Yu G., Wang Y., Pan X., Zhu G. and Angelidaki I. (2020). Treatment of digestate residues for energy recovery and biochar production: from lab to pilot-scale verification. *Journal of Cleaner Production,* **265**, 12. https://doi.org/10.1016/j.jclepro.2020.121852

Liu J., Yuan C., Hu G. and Li F. (2012). Effects of light intensity on the growth and lipid accumulation of microalga *Scenedesmus* sp. 11–1 under nitrogen limitation. *Applied Biochemistry and Biotechnology,* **166**, 2127–2137. https://doi.org/10.1007/s12010-012-9639-2

Liu J., Huang S., Chen K., Wang T., Mei M. and Li J. (2020). Preparation of biochar from food waste digestate: pyrolysis behavior and product properties. *Bioresource Technology,* **302**, 122841. https://doi.org/10.1016/j.biortech.2020.122841

Logan M. and Visvanathan C. (2019). Management strategies for anaerobic digestate of organic fraction of municipal solid waste: current status and future prospects. *Waste Management & Research,* **37**, 27–39. https://doi.org/10.1177/0734242X18816793

Loria E. R., Sawyer J. E., Barker D. W., Lundvall J. P. and Lorimor J. C. (2007). Use of anaerobically digested swine manure as a nitrogen source in corn production. *Agronomy Journal,* **99**, 1119–1129. https://doi.org/10.2134/agronj2006.0251

Luo C., Lü F., Shao L. and He P. (2015). Application of eco-compatible biochar in anaerobic digestion to relieve acid stress and promote the selective colonization of functional microbes. *Water Research,* **68**, 710–718. https://doi.org/10.1016/j.watres.2014.10.052

Makádi M., Tomcsik A. and Orosz V. (2012). Digestate: a new nutrient source – Review. In: Biogas, S. Kumar (ed.), *InTech*, London, pp. 295–310. https://doi.org/10.5772/31355

Malolan R., Jayaraman R. S., Adithya S., Arun J., Gopinath K. P., SundarRajan P. S., Nasif O., Kim W. and Govarthanan M. (2020). Anaerobic digestate water for *Chlorella pyrenoidosa* cultivation and employed as co-substrate with cow dung and chicken manure for methane and hydrogen production: a closed loop approach. Chemosphere, **266**, 128963. https://doi.org/10.1016/j.chemosphere.2020.128963

Marrero T. W., McAuley B. P., Sutterlin W. R., Morris J. S. and Manahan S. E. (2004). Fate of heavy metals and radioactive metals in gasification of sewage sludge. *Waste Management*, **24**, 193–198. https://doi.org/10.1016/S0956-053X(03)00127-2

Masebinu S. O., Akinlabi E. T., Muzenda E. and Aboyade A. O. (2019). A review of biochar properties and their roles in mitigating challenges with anaerobic digestion. *Renewable and Sustainable Energy Reviews*, **103**, 291–307. https://doi.org/10.1016/j.rser.2018.12.048

Mata T. M., Martins A. A. and Caetano N. S. (2010). Microalgae for biodiesel production and other applications: a review. *Renewable and Sustainable Energy Reviews*, **14**, 217–232. https://doi.org/10.1016/j.rser.2009.07.020

Molina Grima E., Belarbi E. H., Acién Fernández F. G., Robles Medina A. and Chisti Y. (2003). Recovery of microalgal biomass and metabolites: process options and economics. *Biotechnology Advances*, **20**, 491–515. https://doi.org/10.1016/S0734-9750(02)00050-2

Monlau F., Sambusiti C., Antoniou N., Barakat A. and Zabaniotou A. (2015). A new concept for enhancing energy recovery from agricultural residues by coupling anaerobic digestion and pyrolysis process. *Applied Energy*, **148**, 32–38. https://doi.org/10.1016/j.apenergy.2015.03.024

Monlau F., Francavilla M., Sambusiti C., Antoniou N., Solhy A., Libutti A., Zabaniotou A., Barakat A. and Monteleone M. (2016). Toward a functional integration of anaerobic digestion and pyrolysis for a sustainable resource management. Comparison between solid-digestate and its derived pyrochar as soil amendment. *Applied Energy*, **169**, 652–662. https://doi.org/10.1016/j.apenergy.2016.02.084

Neumann J., Meyer J., Ouadi M., Apfelbacher A., Binder S. and Hornung A. (2016). The conversion of anaerobic digestion waste into biofuels via a novel thermo-catalytic reforming process. *Waste Management*, **47**, 141–148. https://doi.org/10.1016/j.wasman.2015.07.001

Nkoa R. (2014). Agricultural benefits and environmental risks of soil fertilization with anaerobic digestates: a review. *Agronomy for Sustainable Development*, **34**, 473–492. https://doi.org/10.1007/s13593-013-0196-z

Obulisamy P. K., Chakraborty D., Selvam A. and Wong J. W. C. (2016). Anaerobic co-digestion of food waste and chemically enhanced primary-treated sludge under mesophilic and thermophilic conditions. *Environmental Technology (United Kingdom)*, **37**, 3200–3207. https://doi.org/10.1080/09593330.2016.1181112

Ogbonna J. C. and Tanaka H. (2000). Light requirement and photosynthetic cell cultivation – development of processes for efficient light utilization in photobioreactors. *Journal of Applied Phycology*, Springer, Netherlands, **12**, 207–218. https://doi.org/10.1023/a:1008194627239

Opatokun S. A., Lopez-Sabiron A. M., Ferreira G. and Strezov V. (2017). Life cycle analysis of energy production from food waste through anaerobic digestion, pyrolysis and integrated energy system. *Sustainability*, **9**, 1804. https://doi.org/10.3390/su9101804

Osundeko O. and Pittman J. K. (2014). Implications of sludge liquor addition for wastewater-based open pond cultivation of microalgae for biofuel generation and pollutant remediation. *Bioresource Technology*, **152**, 355–363. https://doi.org/10.1016/j.biortech.2013.11.035

Pancha I., Chokshi K., Maurya R., Trivedi K., Patidar S. K., Ghosh A. and Mishra S. (2015). Salinity induced oxidative stress enhanced biofuel production potential of microalgae *Scenedesmus* sp. CCNM 1077. *Bioresource Technology*, **189**, 341–348. https://doi.org/10.1016/j.biortech.2015.04.017

Pecchi M. and Baratieri M. (2019). Coupling anaerobic digestion with gasification, pyrolysis or hydrothermal carbonization: a review. *Renewable and Sustainable Energy Reviews*, **105**, 462–475. https://doi.org/10.1016/j.rser.2019.02.003

Peng W. and Pivato A. (2019). Sustainable management of digestate from the organic fraction of municipal solid waste and food waste under the concepts of back to earth alternatives and circular economy. *Waste and Biomass Valorization*, **10**, 465–481. https://doi.org/10.1007/s12649-017-0071-2

Peng W., Pivato A., Cerminara G., Garbo F. and Raga R. (2020). Denitrification of mature landfill leachate with high nitrite in simulated landfill columns packed with solid digestate from organic fraction of municipal solid waste. *Waste and Biomass Valorization*, **11**, 411–421. https://doi.org/10.1007/s12649-018-0422-7

Pérez-Camacho M. N. and Curry R. (2018). Regional assessment of bioeconomy options using the anaerobic biorefinery concept. *Proceedings of the Institution of Civil Engineers – Waste and Resource Management*, **171**, 104–113. https://doi.org/10.1680/jwarm.17.00015

Praveen P. and Loh K. C. (2016). Nitrogen and phosphorus removal from tertiary wastewater in an osmotic membrane photobioreactor. *Bioresource Technology*, **206**, 180–187. https://doi.org/10.1016/j.biortech.2016.01.102

Praveen P., Guo Y., Kang H., Lefebvre C. and Loh K. C. (2018). Enhancing microalgae cultivation in anaerobic digestate through nitrification. *Chemical Engineering Journal*, **354**, 905–912. https://doi.org/10.1016/j.cej.2018.08.099

Qin L., Wang Z., Sun Y., Shu Q., Feng P., Zhu L., Xu J. and Yuan Z. (2016). Microalgae consortia cultivation in dairy wastewater to improve the potential of nutrient removal and biodiesel feedstock production. *Environmental Science and Pollution Research*, **23**, 8379–8387. https://doi.org/10.1007/s11356-015-6004-3

Ramanan R., Kim B. H., Cho D. H., Ko S. R., Oh H. M. and Kim H. S. (2013). Lipid droplet synthesis is limited by acetate availability in starchless mutant of *Chlamydomonas reinhardtii*. *FEBS Letters*, **587**, 370–377. https://doi.org/10.1016/j.febslet.2012.12.020

Reza M. T., Werner M., Pohl M. and Mumme J. (2014). Evaluation of integrated anaerobic digestion and hydrothermal carbonization for bioenergy production. *Journal of Visualized Experiments*, **88**, e51734. https://doi.org/10.3791/51734

Rodolfi L., Zittelli G. C., Bassi N., Padovani G., Biondi N., Bonini G. and Tredici M. R. (2009). Microalgae for oil: strain selection, induction of lipid synthesis and outdoor mass cultivation in a low-cost photobioreactor. *Biotechnology and Bioengineering*, **102**, 100–112. https://doi.org/10.1002/bit.22033

Rodríguez P., Cerda A., Font X., Sánchez A. and Artola A. (2019). Valorisation of biowaste digestate through solid state fermentation to produce biopesticides from *Bacillus thuringiensis*. *Waste Management*, **93**, 63–71. https://doi.org/10.1016/j.wasman.2019.05.026

Salman C. A., Schwede S., Thorin E. and Yan J. (2017). Enhancing biomethane production by integrating pyrolysis and anaerobic digestion processes. *Applied Energy*, **204**, 1074–1083. https://doi.org/10.1016/j.apenergy.2017.05.006

Sayedin F., Kermanshahi-pour A., He Q. S., Tibbetts S. M., Lalonde C. G. E. and Brar S. K. (2020). Microalgae cultivation in thin stillage anaerobic digestate for nutrient recovery and bioproduct production. *Algal Research*, **47**, 101867. https://doi.org/10.1016/j.algal.2020.101867

Sekine M., Yoshida A., Akizuki S., Kishi M. and Toda T. (2020). Microalgae cultivation using undiluted anaerobic digestate by introducing aerobic nitrification-desulfurization treatment. *Water Science & Technology*, **82**, 1070–1080. https://doi.org/10.2166/wst.2020.153

Siciliano A. and De Rosa S. (2014). Recovery of ammonia in digestates of calf manure through a struvite precipitation process using unconventional reagents. *Environmental Technology (United Kingdom)*, **35**, 841–850. https://doi.org/10.1080/09593330.2013.853088

Silaban A., Bai R., Gutierrez-Wing M. T., Negulescu I. I. and Rusch K. A. (2014). Effect of organic carbon, C: N ratio and light on the growth and lipid productivity of microalgae/cyanobacteria coculture. *Engineering in Life Science*, **14**, 47–56. https://doi.org/10.1002/elsc.201200219

Stiles W. A. V., Styles D., Chapman S. P., Esteves S., Bywater A., Melville L., Silkina A., Lupatsch I., Fuentes Grünewald C., Lovitt R., Chaloner T., Bull A., Morris C. and Llewellyn C. A. (2018). Using microalgae in the circular economy to valorise anaerobic digestate: challenges and opportunities. *Bioresource Technology*, **267**, 732–742. https://doi.org/10.1016/j.biortech.2018.07.100

Sturm B. S. M. and Lamer S. L. (2011). An energy evaluation of coupling nutrient removal from wastewater with algal biomass production. *Applied Energy*, **88**, 3499–3506. https://doi.org/10.1016/j.apenergy.2010.12.056

Svehla P., Radechovska H., Pacek L., Michal P., Hanc A. and Tlustos P. (2017). Nitrification in a completely stirred tank reactor treating the liquid phase of digestate: the way towards rational use of nitrogen. *Waste Management*, **64**, 96–106. https://doi.org/10.1016/j.wasman.2017.03.041

Svensson K., Kjørlaug O., Higgins M. J., Linjordet R. and Horn S. J. (2018). Post-anaerobic digestion thermal hydrolysis of sewage sludge and food waste: effect on methane yields, dewaterability and solids reduction. *Water Research*, **132**, 158–166. https://doi.org/10.1016/j.watres.2018.01.008

Tambone F., Genevini P., D'Imporzano G. and Adani F. (2009). Assessing amendment properties of digestate by studying the organic matter composition and the degree of biological stability during the anaerobic digestion of the organic fraction of MSW. *Bioresource Technology*, **100**, 3140–3142. https://doi.org/10.1016/j.biortech.2009.02.012

Tampio E., Marttinen S. and Rintala J. (2016). Liquid fertilizer products from anaerobic digestion of food waste: mass, nutrient and energy balance of four digestate liquid treatment systems. *Journal of Cleaner Production*, **125**, 22–32. https://doi.org/10.1016/j.jclepro.2016.03.127

Tan X. B., Yang L. B., Zhang W. W. and Zhao X. C. (2020). Lipids production and nutrients recycling by microalgae mixotrophic culture in anaerobic digestate of sludge using wasted organics as carbon source. *Bioresource Technology*, **297**, 122379. https://doi.org/10.1016/j.biortech.2019.122379

Tian K., Liu W. J., Qian T. T., Jiang H. and Yu H. Q. (2014). Investigation on the evolution of N-containing organic compounds during pyrolysis of sewage sludge. *Environmental Science and Technology*, **48**, 10888–10896. https://doi.org/10.1021/es5022137

Timmerman M., Schuman E., Van Eekert M. and Van Riel J. (2015). Optimizing the performance of a reactor by reducing the retention time and addition of glycerin for anaerobically digesting manure. *Environmental Technology (United Kingdom)*, **36**, 1223–1236. https://doi.org/10.1080/09593330.2014.983989

Torres Franco A. F., Da Encarnação Araújo S., Passos F., De Lemos Chernicharo C. A., Filho C. R. M. and Figueredo C. C. (2018). Treatment of food waste digestate using microalgae based systems with low-intensity light-emitting diodes. *Water Science & Technology*, **78**, 225–234. https://doi.org/10.2166/wst.2018.198

Turovskiy I. S. and Mathai P. K. (2005). Wastewater Sludge Processing. John Wiley & Sons, Inc, Germany. https://doi.org/10.1002/047179161X

Uggetti E., Sialve B., Trably E. and Steyer J. P. (2014). Integrating microalgae production with anaerobic digestion: a biorefinery approach. *Biofuels, Bioproducts and Biorefining*, **8**, 516–529. https://doi.org/10.1002/bbb.1469

Veronesiv D., D'Imporzano G., Salati S. and Adani F. (2017). Pre-treated digestate as culture media for producing algal biomass. *Ecological Engineering*, **105**, 335–340. https://doi.org/10.1016/j.ecoleng.2017.05.007

Wang L., Li Y., Chen P., Min M., Chen Y., Zhu J. and Ruan R. R. (2010). Anaerobic digested dairy manure as a nutrient supplement for cultivation of oil-rich green microalgae *Chlorella* sp. *Bioresource Technology*, **101**, 2623–2628. https://doi.org/10.1016/j.biortech.2009.10.062

Wang Q., Cherones J. and Higgins B. (2021a). Acclimation of an algal consortium to sequester nutrients from anaerobic digestate. *Bioresource Technology* **342**, 125921. https://doi.org/10.1016/j.biortech.2021.125921

Wang Q., Hyman M. and Higgins B. T. (2021b). Factors impacting the effectiveness of biological pretreatment for the alleviation of algal growth inhibition on anaerobic digestate. *Algal Research*, **53**, 102129. https://doi.org/10.1016/j.algal.2020.102129

Xia A. and Murphy J. D. (2016). Microalgal cultivation in treating liquid digestate from biogas systems. *Trends in Biotechnology*, **34**, 264–275. https://doi.org/10.1016/j.tibtech.2015.12.010

Xia L., Rong J., Yang H., He Q., Zhang D. and Hu C. (2014). NaCl as an effective inducer for lipid accumulation in freshwater microalgae *Desmodesmus abundans*. *Bioresource Technology*, **161**, 402–409. https://doi.org/10.1016/j.biortech.2014.03.063

Xu J., Zhao Y., Zhao G. and Zhang H. (2015). Nutrient removal and biogas upgrading by integrating freshwater algae cultivation with piggery anaerobic digestate liquid treatment. *Applied Microbiology and Biotechnology*, **99**, 6493–6501. https://doi.org/10.1007/s00253-015-6537-x

Yao Z., Li W., Kan X., Dai Y., Tong Y. W. and Wang C. H. (2017). Anaerobic digestion and gasification hybrid system for potential energy recovery from yard waste and woody biomass. *Energy*, **124**, 133–145. https://doi.org/10.1016/j.energy.2017.02.035

Yen H. W., Ho S. H., Chen C. Y. and Chang J. S. (2015). CO_2, NO_x and SO_x removal from flue gas via microalgae cultivation: A critical review. *Biotechnology Journal*, **10**, 829–839. https://doi.org/10.1002/biot.201400707

Zhang B., Ren J., Liu X., Guo Y., Guo Y., Lu G. and Wang Y. (2010). Novel sulfonated carbonaceous materials from p-toluenesulfonic acid/glucose as a high-performance solid-acid catalyst. *Catalysis Communications*, **11**, 629–632. https://doi.org/10.1016/j.catcom.2010.01.010

Zhang Y., Jiang Y., Wang S., Wang Z., Liu Y., Hu Z. and Zhan X. (2021). Environmental sustainability assessment of pig manure mono- and co-digestion and dynamic land application of the digestate. *Renewable and Sustainable Energy Reviews*, **137**, 110476. https://doi.org/10.1016/j.rser.2020.110476

Zuliani L., Frison N., Jelic A., Fatone F., Bolzonella D. and Ballottari M. (2016). Microalgae cultivation on anaerobic digestate of municipal wastewater, sewage sludge and agro-waste. *International Journal of Molecular Sciences*, **17**, 1692. https://doi.org/10.3390/ijms17101692

doi: 10.2166/9781789062755_0189

Chapter 9

Value-added hydrochar recovery from anaerobic digestate for environmental applications

*Parmila Devi, Huan Liu, Ibrahim Alper Basar and Cigdem Eskicioglu**

UBC Bioreactor Technology Group, School of Engineering, University of British Columbia, Okanagan Campus, 3333 University Way, Kelowna, BC V1V 1V7, Canada
**Corresponding author e-mail: cigdem.eskicioglu@ubc.ca, cigdem.eskicioglu@gmail.com*

ABSTRACT

Anaerobic digestion (AD) is commonly used for waste treatment and energy recovery via biogas production by valorization of lignocellulosic biomass, food waste, municipal and industrial sludges. A huge quantity of anaerobic digestate is generated during AD process as a residual by-product. The proper management and utilization of anaerobic digestate are mandatory to achieve the goal of circular economy and to reduce environmental and economic effects of digestate management and disposal. This chapter summarizes the composition of anaerobic digestate derived from different feedstocks and explored the literature related to hydrochar recovery from anaerobic digestate through hydrothermal processing (HTP) including, the effect of the key parameters (initial solid content, reaction temperature, reaction time, and catalyst) on hydrochar recovery. Also, the characteristics of digestate-derived hydrochar are discussed in detail as they are vital in determining their environmental applications. Furthermore, the usage of hydrochar in various environmental applications (e.g. microbial carrier in biological treatment, adsorption of pollutants, fuel, nutrient recovery, and soil amendment) and risk assessment of hydrochar application (in terms of bioavailability, leaching potential and phytotoxicity of heavy metals and other inhibitory compounds) are discussed in detail.

Keywords: Hydrochar, anaerobic digestate, sludge, hydrothermal treatment, adsorbent

9.1 INTRODUCTION

Anaerobic digestion (AD) for the production of biogas as renewable energy has attracted much attention in most countries due to its applicability in processing a broad range of high-strength organic waste including lignocellulosic biomass, food waste, municipal and industrial sludges, manures, and bio-waste. This

Editor(s): Vinay Kumar Tyagi, Kaoutar Aboudi and Cigdem Eskicioglu

has resulted in the installment of a large number of AD plants all over the world particularly in Europe (17 240 plants in 2014), the United States (1497 plants in 2013), and China (26.5 million plants in 2007) (Akhiar *et al.*, 2017; Magdziarz *et al.*, 2021; Mao *et al.*, 2015). However, even with pretreatment AD process can only convert about 50–75% of the energy in the raw feedstock into biogas; a significant portion of the input energy is retained in the residual digestate (Inyang *et al.*, 2010). Depending on the waste product processed in AD, the digestate can consist of a varied amount of organic fraction including cellulose (25–50%), hemicellulose (20–25%), lignin (10–15%), ash (10–20%) and nutrients (such as nitrogen (N), potassium (K), phosphorus (P), and so on) (Ma *et al.*, 2018). The considerable amount of organic matter in the digestate provides additional opportunities for thermochemical conversion into useful products and applications in energy sector. The estimated energy potential of the digestate ranges from 6 to 15 V/ton (Magdziarz *et al.*, 2021). Traditionally, the digestate is used as a fertilizer on arable lands but the prolonged application of the digestate can cause harmful effects on the environment due to nitrate leaching into the groundwater, heavy metals and other macro- and micro-contaminants (pharmaceuticals, hormones, dioxins, furans, etc.) transferred to soil, and ammonia volatilization into the atmosphere (Akhiar *et al.*, 2017). Land application is not an ideal or sustainable option because of the current interest in the application of the principles of circular economy as well as the environmental and economic costs of land application. Further, the use of digestate in energy applications is limited due to high water content in the digestate and the cost associated with its transportation to a processing facility.

Hydrothermal processing (HTP) of the digestate can overcome the aforementioned problems and is considered as a promising technology to process the wet digestate into value-added products like hydrochar and biocrude. Literature studies have reported numerous applications of hydrochar, including biological processes enhancement, as an adsorbent, as a fuel and soil amendment (Ni *et al.*, 2019; Takaya *et al.*, 2016; Yuan *et al.*, 2018). The physicochemical properties and molecular structure of hydrochar such as organic functional groups, chemical composition, and electron exchange capacities are remarkably varied in digestates derived from different feedstocks, thus influencing their practical applications (Devi & Saroha, 2015). For instance, manure-derived hydrochar is considered a low-quality fuel due to the presence of high ash and mineral content but is deemed most suitable for agricultural and bioreactors applications (Belete *et al.*, 2021). The anaerobic digesters utilizing ethanol as substrate supplemented with manure-derived hydrochar were reported to have 211 times higher methane production efficiency than woodchip-derived hydrochar supplemented AD, although both hydrochars had similar surface areas (Yuan *et al.*, 2018).

This book chapter provides a comprehensive review on the composition of anaerobic digestate derived from different feedstocks and discusses the production/recovery of hydrochar from anaerobic digestate through HTP, including the effects of the key process parameters on hydrochar recovery. Moreover, the characterization of hydrochar, its activation, usages in different environmental applications (microbial carrier for biological treatment, adsorption of pollutants, carbon dioxide (CO_2) capture, and agricultural applications), and risk assessment are discussed in detail.

9.2 COMPOSITION OF ANAEROBIC DIGESTATE

The composition analysis of the digestate is crucial for deciding the correct downstream digestate valorization or disposal process. For instance, a digestate with high macro- and trace-element concentrations can be used as fertilizer in agriculture, whereas a digestate with high carbon content can be a good feedstock for hydrothermal conversion processes, such as hydrothermal carbonization (HTC) or hydrothermal liquefaction (HTL) (Liu *et al.*, 2021a). Further, the high concentrations of leachable heavy metals in the digestate can limit the usage of the digestate in agriculture applications and cause leaching concerns in digestate disposal by landfilling. Therefore, it is essential to characterize the digestate carefully to choose the proper processing method for the valorization of the digestate.

The characteristics of the digestate are highly dependent on the type of substrate, digester efficiency, digester operation temperature, and retention time. Table 9.1 shows the proximate analysis of digestates obtained from processing different types of feedstocks. Generally, AD process is aimed to recover carbon in the form of methane and a well-treated digestate in AD with sufficient sludge retention time (SRT) is expected to have very low carbon content (around 25%). However, if the biodegradability of feedstock is low, such as lignocellulosic substrates (Eskicioglu *et al.*, 2017), the carbon concentration in the residual digestate can become higher than expected. It can be seen from Table 9.1 that some substrates like agricultural residues, energy crops, straws, silages, and garden wastes have high carbon concentrations. Digestates with high carbon concentrations can be processed further to recover more carbon as energy or value-added products such as hydrochar or biocrude. According to literature reports, the total solid contents of digestates typically ranged from 8.8 to 38.4%. However, the solid contents of dewatered digestate from AD utilizing mixed sludge ranged from 20 to 40% (Kor-Bicakci & Eskicioglu, 2019). Given the presence of high moisture content in the digestate, it would be more energy efficient to valorize digestates using hydrothermal processes that does not require drying of feedstock, instead of pyrolysis for further carbon recovery.

Further, the pH of digestates varied in the range of 7.1–9.8 in the literature studies. The volatile fatty acids (VFA) and ammonia concentrations are the key parameters affecting the pH of anaerobic digestate. In a properly operated digester, where anaerobic consortia activity and SRTs are sufficient, VFA concentrations in the effluent digestate should be lower than 100 mg/L. Therefore, ammonia stays as the most significant parameter that affects digestate pH and causes pH to be alkaline. The source of ammonia in AD is the protein deamination reaction. Hence, AD feedstocks with higher protein contents, such as waste-activated sludge and cattle manure (Belete *et al.*, 2021; Celletti *et al.*, 2021), have higher pH levels in the digestate.

Table 9.1 Proximate analysis results of digestates obtained from AD of various feedstocks.

Anaerobic digester feedstock	C (%db)	Fixed C (%db)	H (%db)	N (%db)	O (%db)	S (%db)	HHV (MJ/kg)	TS (%)	VS/TS (%)	Ash (%db)	References
Agricultural residue	47.6	–	7.1	–	32.6	–	21.8	–	–	10.5	Mumme *et al.* (2015)
	44.1	13.8	5.1	3.2	31.3	0.3	17.8	-	70.2	16.0	Parmar and Ross (2019)
	50.6	16.4	6.1	4.2	38.4	0.52	–	-	48.3	24.9	Correa *et al.* (2017)
Agro-industrial residue	46.7	21.8	5.5	1.2	16.9	0.5	19.2	23.8	68.9	9.3	Miliotti *et al.* (2020)
Wheat straw	48.4	-	6.5	1.0	43.9	0.2	20.1	-	-	16.0	Reza *et al.* (2015)
Energy crops	40.0	14	4.6	1.9	24.0	0.31	16.3	-	56.8	29.2	Cao *et al.* (2021)
	40.3	21.6	4.6	2.1	24.0	0.3	16.4	-	78.4	28.7	Cao *et al.* (2019)
Food – garden waste	29.5	9.0	3.0	2.0	21.3	0.3	14.9	-	47.2	43.8	Parmar and Ross (2019)
Cattle manure	26.8	-	3.4	1.7	22.8	0.4	-	20.0	-	44.8	Belete *et al.* (2021)
	36.6	-	4.9	1.9	25.5	0.7	-	8.8	-	26.8	Celletti *et al.* (2021)
	42.6	20.2	5.0	2.0	34.3	0.4	17.0	-	79.8	15.7	Cao *et al.* (2019)
	35.8	18.8	9.5	3.2	25.5	0.3	14.7	-	55.5	6.8	Hoffmann *et al.* (2013)
Swine manure	37.2	-	5.5	4.6	31.9	1.49	16.6	-	50.8	23.0	Hung *et al.* (2017)
	34.1	1.1	5.4	4.3	26.5	1.89	15.3	-	58.1	40.8	Tsai *et al.* (2018)
Maize–grass silage and cattle manure	50.5	25.9	6.4	3.6	39	0.5	-	-	74.1	27.7	Zhao *et al.* (2018)
Grass silage	54.4	-	6.9	5.4	33.3	-	15.2	-	68.3	31.7	Deng *et al.* (2020)
Maize silage	46.6	17.2	5.8	1.1	40.7	0.08	18.8	-	77.1	5.6	Cao *et al.* (2020)
	45.8	-	6.0	2.9	45.1	0.24	22.3	15.3	88.5	11.5	Mumme *et al.* (2011)

Municipal sludge	33.3	9.4	4.6	4.0	20.3	1.2	14.4		51.8	36.7	Aragón-Briceño *et al.* (2020)
	41.7	6.8	5.8	3.2	47.5	1.8	16.6	12.4	60.2	39.8	Zhai *et al.* (2013)
	29.6	–	4.2	3.7	21.2	0.1	–	–	–	–	Becker *et al.* (2019)
	28.6	2.1	3.1	3.4	16.4	1.5	14.9		51.0	46.9	Parmar and Ross (2019)
	30.3	–	4.4	3.5	59.5	2.3	11.5	25.6	57.0	43.0	Hämäläinen *et al.* (2021)
	34.8	7.8	3.9	5.0	20.3	1.15	16.6	–	55.8	31.8	Aragón-Briceño *et al.* (2017)
	27.6	10	4.2	3.8	23.6	0.8	11.4	–	–	40.0	Ovsyannikova *et al.* (2019)
	38.6	7.1	6.5	5.8	46.7	0.44	16.5	9.3	66.9	26.1	Park *et al.* (2018)
	38.7	–	5.7	4.5	27.9	1.63	–	16.0	–	28.1	Marrone *et al.* (2018)
	32.7	4.4	4.9	5.1	15.4	1	14.9	16.5	54.7	40.9	Marin-Batista *et al.* (2020)
	20.9	–	2.5	1.9	13.5	0.8	–	27.2	–	60.3	Wang *et al.* (2020a)
	18.3	–	2.9	2.7	14.6	0.3	7.8	–	–	–	Ekpo *et al.* (2016)
	25.6	4.8	3.9	3.6	21.2	–	10.7	29.0	50.3	44.9	Merzari *et al.* (2020)
	35.9	5.9	5.4	5.8	24.4	–	–	22.0	65.6	28.4	Ferrentino *et al.* (2020)
	38.6	7.1	6.5	8.05	46.5	0.44	16.5		66.9	26.1	Kim *et al.* (2014)
	32.5	–	4.7	5	15.2	1.7	14.3	24.3	–	40.9	Lühmann and Wirth (2020)
Municipal solid waste	52.8	3.7	6.7	5	34.4	1.1	–	24.0	–	33.8	Bernardo *et al.* (2020)
	34.4	14.9	4.0	1.9	23.9	0.2	13.4	–	85.1	35.8	Cao *et al.* (2019)
	24.1	8.3	1.7	1.5	16.9	0.2	15.6	–	36.2	55.5	Parmar and Ross (2019)
	–	–	–	–	–	–	15.7	23.0	63.3	36.7	Reza *et al.* (2016)
Poultry litter, maize silage and food waste	39	–	–	1.51	–	1.2	–	38.4	45.9	–	Gurmessa *et al.* (2021)

db: dry basis, TS: total solids, VS: volatile solids.

Table 9.2 shows the macro and trace nutrient concentrations in different digestates, such as plant biomass-derived digestate, manure-derived digestate, municipal sludge-derived digestate. It can be seen from Table 9.2 that plant-based digestates have varying amounts of macro and trace-element contents. This variation in macro and trace elements in plant biomass is caused by the difference in soil type used for growing the plant biomass and concentration of these elements in the soil (Kabata-Pendias, 2004). Likewise, trace-element concentrations in manure-based digestates depend on the animal feed being used (Sheppard & Sanipelli, 2012). High N, P, and K concentrations in manure-based digestates indicate that they can be a suitable feedstock for composting. Further, higher Al and Fe concentrations were reported in municipal sludge-based digestates due to chemical dewatering agents being used in treatment plants, such as polymers, ferric chloride ($FeCl_3$), and aluminum sulfate ($Al_2(SO_4)_3$).

Based on the literature results in Tables 9.1 and 9.2, lignocellulosic substrate-based digestates are suitable for direct combustion and HTP due to their higher heating values (HHV). However, valorization of lignocellulosic substrate digestates into biofuels produces low-quality fuel with high oxygen concentrations as these digestates were reported to have high oxygen concentrations compared to other digestates (biosolids, manure, food waste etc.) (Basar *et al.*, 2021). Additionally, digestates from highly biodegradable feedstocks (e.g., manure, food waste, etc.) are enriched in nutrients and considered biologically more stable than other digestate types. Therefore, these types of digestates are suitable to be used in soil amendment, before and after HTP (Makdi *et al.*, 2012)

9.3 RECOVERY OF HYDROCHAR FROM ANAEROBIC DIGESTATE THROUGH HTP

HTP represents a promising technology for digestate treatment, which can convert organics into aqueous, biocrude, gas, and a carbonaceous char referred to as hydrochar. Thus, value-added products can be separated and recovered from different liquid and solid product streams. HTP can treat wet biomass efficiently by utilizing hot pressurized water as a reaction medium at elevated temperature and autogenous pressure (Mathimani & Mallick, 2019). This ability allows HTP to be more economical and sustainable than other techniques that require dry feedstocks (e.g., incineration and pyrolysis). Depending on the reaction temperature (150–280°C, 280–375°C, and >375°C), HTP can be classified as HTC, HTL, and hydrothermal gasification (HTG), respectively (Liu *et al.*, 2021a). In each process, the yield and energy recovery of each target product (i.e., hydrochar, biocrude, and syngas) can be maximized.

Considering complex compositions of the digestate, the reaction pathways during HTP can be vast. Figure 9.1 summarizes some of the major potential pathways. Generally, large organic molecules (e.g., carbohydrates, protein, lipids, and lignin) are depolymerized and form reaction intermediates (e.g., furfurals, amines, alcohols, fatty acids, and organic acids) mainly through hydrolysis, decomposition, dehydration, decarboxylation, and deamination

Table 9.2 Macro and trace-element contents of digestates obtained from AD of various feedstock

Anaerobic digester feedstock	Al (%db)	Ba (%db)	Na (%db)	Ca (%db)	Fe (%db)	K (%db)	Mg (%db)	Mn (%db)	P (%db)	Si (%db)	Ti (%db)	Zn (%db)	Sr (%db)	References
Agricultural residue	3.7	-	0.1	3.4	2.1	0.4	0.6		3.2	-	-	-	-	Correa *et al.* (2017)
	-	-	0.3	1.3	2.3	1.6	0.7		1.2	1.8	-	-	-	Parmar and Ross (2019)
Energy crop	1.2	0.03	0.4	6.5	0.8	13.2	2.2	0.08	3.1	19.6	0.09	0.05	0.013	Cao *et al.* (2021)
	0.3	0.01	0.08	2.5	0.2	3.7	0.7	0.02	0.9		-	0.01	-	Cao *et al.* (2019)
Food – garden waste	-	-	0.5	4.3	3	0.7	0.8	-	2.6	7.1	-	-	-	Parmar and Ross (2019)
Cattle manure	-	-	0.4	0.04	0.01	-	0.01	0.0003	1	0.01	-	0.001	-	Belete *et al.* (2021)
	-	-	-	-	-	-	-	-	-	-		0.02	-	Celletti *et al.* (2021)
	0.1	0.001	0.2	1.7	0.1	2.8	0.5	0.02	0.7	-	-	0.01	-	Cao *et al.* (2019)
Swine manure	0.2	-	0.1	26.3	0.5	0.1	0.9	0.31	11.5	0.69	0.01	0.15	0.08	Hung *et al.* (2017)
	0.3	-	0.2	9.9	0.8	0.8	1.3	0.12	4.09	1.06	0.02	1.48	0.03	Tsai *et al.* (2018)
Maize–grass silage and cattle manure	0.5	-	-	2.01	0.3	5.2	0.9	-	1.4	6.71	-	-	-	Zhao *et al.* (2018)
Municipal sludge	4.0	-	-	3.5	1.1	0.5	0.6	-	3.1	-	-	-	-	Becker *et al.* (2019)
	4.2	-	0.2	3.0	1.6	0.3	0.3	-	4.6	1.94	-	-	-	Marin-Batista *et al.* (2020)
	-	-	0.5	4.6	3.3	0.7	0.9	-	2.7	7.6	-	-	-	Parmar and Ross (2019)
	1.4	-	0.8	1.5	17	0.3	0.4	0.04	2.4	7.4	-	0.13	-	Wang *et al.* (2020a)
	1.9	-	0.6	2.4	3.1	0.4		0.51	1.6	-	-	-	-	Ekpo *et al.* (2016)
Municipal solid waste	-	-	0.9	10.4	3.2	1.6	1.4	-	0.7	10.2	-	-	-	Parmar and Ross (2019)
	0.9	0.02	0.7	4.3	0.7	1.8	0.5	0.05	0.4	-	-	0.01	-	Cao *et al.* (2019)
Poultry litter, maize silage and food waste	0.01	0.00001	-	0.3	0.2	14.1	0.29	0.003	1	-		0.01		Gurmessa *et al.* (2021)

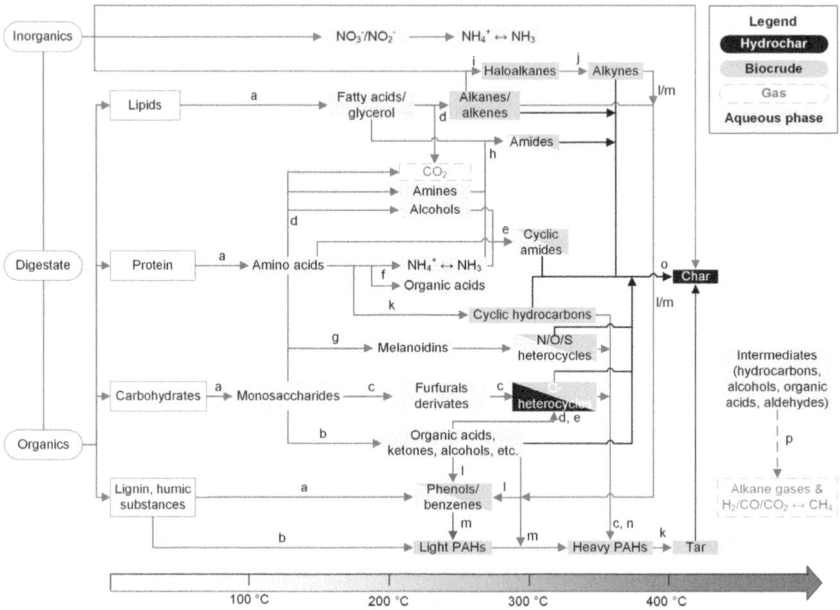

Figure 9.1 Potential reaction pathways for HTP of digestate: (a) hydrolysis; (b) decomposition; (c) dehydration; (d) decarboxylation; (e) cyclization; (f) deamination; (g) Maillard reaction; (h) aminolysis; (i) halogenation; (j) dehydrohalogenation; (k) condensation; (l) Diels–Alder reaction; (m) hydrogen abstraction acetylene addition reaction; (n) aromatization; (o) polymerization; (p) free radical reactions or metal-catalyzed C-C/C-O cleavage. Modified from Liu *et al.* (2021a).

(Basar *et al.*, 2021). The intermediates can eventually recombine and distribute into hydrochar by cyclization, aromatization, polymerization, and condensation. However, HTP process parameters, such as feedstock nature, initial solids/moisture content, initial pH, reaction severity (temperature, residence time, and pressure), ramping time, catalyst, and solvent, can largely shift reaction pathways, which can essentially affect the formation and properties of hydrochar.

9.3.1 Initial solids content

Water acts as the reaction medium of HTP and its content primarily impacts the cleavage of hydrogen bonds through ionic reactions (particularly hydrolysis) at subcritical conditions or free-radical reactions at supercritical conditions (Wang *et al.*, 2019). Typically, lower initial solid content or higher moisture content in digestate generates a lower yield of hydrochar. Aragón-Briceño *et al.* (2020) increased the initial solids content of sludge digestate from 2.5 to 30 wt% for HTC treatment (250°C for 30 min) and found that hydrochar yield gradually increased from 68 to 76%, dry basis (db), with improved HHV

(15.4–16.5 MJ/kg) and energy recovery (65.9–76.7%). For the process efficiency and economic benefit, it is recommended to dewater digestate prior to HTP treatment. The current continuous-flow HTP systems can process biomass with a solids content of up to 25% (Li *et al.*, 2021). For the pumpability of digestate, pumps with a maximum solid content of 45% (biosolids) are available from commercial vendors (Berglin *et al.*, 2012).

9.3.2 Temperature, pressure, and residence time (severity factor)

Reaction temperature and residence time are considered the most critical parameters for HTP, which govern the reaction pathways. Reaction pressure is typically autogenous in a closed system, which is directly related to reaction temperature. Increasing temperature and pressure above the critical point of water (374.3°C and 22.1 MPa) can shift the reaction mechanism from ionic to free-radical, which significantly decreases the hydrochar yield and enhances gas yield (Watson *et al.*, 2020). Reaction temperature limits the energy required for degradation and recombination of chemical bonds during HTP. For example, the optimal temperature range for hydrolysis is 160–180°C, but to hydrothermally decompose carbohydrates, proteins, lignin, and lipids, a temperature of >180°C, >180°C, >280°C, and >640°C, respectively, is necessary (Liu *et al.*, 2021a). Polymerization also occurs with increasing temperature and residence time under subcritical conditions. Residence time is the duration of the process held at the reaction temperature before cooling, which gives a fine adjustment of the reaction severity. A severity factor has been introduced to show the collective effects of temperature and residence time on hydrochar recovery. The severity factor (*f*, dimensionless) can be calculated by Eq. (10.1) (Ruyter, 1982):

$$f = 50t^{0.2}e^{-3500/T} \tag{10.1}$$

where T is the reaction temperature (K), and t is the residence time at target temperature (s).

Figure 9.2 illustrates hydrochar recovery from the digestate in terms of severity factor. The hydrochar yield decreases drastically with increasing severity factor (Figure 9.2a). However, hydrochar yield from different types of digestate can vary significantly even under the same reaction severity due to diverse compositions (e.g., high content) of the digestate. Therefore, hydrochar yield can be normalized on a dry ash-free basis (daf) to indicate the recovery of organics in hydrochar as shown in Figure 9.2b. Similarly, energy recovery in hydrochar from most digestates also decreases significantly with increasing severity factor (Figure 9.2c). Generally, a severity factor of <0.5 could lead to a good hydrochar yield and energy recovery. Therefore, processes aiming for hydrochar recovery mostly were conducted under HTC conditions (150–280°C). It should be noted that higher reaction severity could promote the condensation of a carbonaceous matrix with the formation of a highly condensed aromatic structure (Tasca *et al.*, 2019).

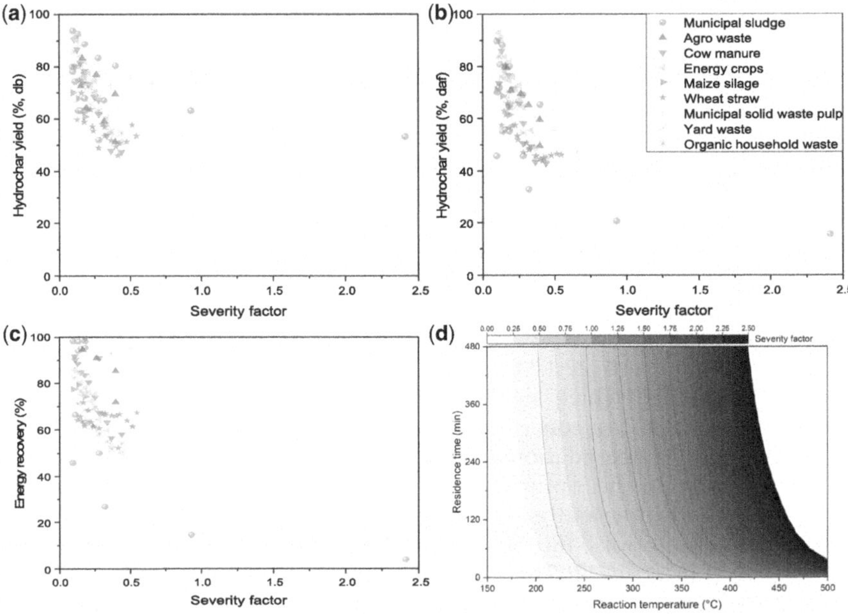

Figure 9.2 Effect of reaction severity on the (a) hydrochar yield (%, db), (b) hydrochar yield (%, daf), and (c) energy recovery of hydrochar from various types of digestate and (d) counter plot of severity factor in terms of reaction temperature and residence time. Data from Table 9.3. db = dry basis; daf = dry ash free.

9.3.3 Ramping time

Ramping time represents the heating duration before reaching the target reaction temperature and after reaction (cooling period), which also affects the hydrochar recovery. Generally, a longer ramping time or slow heating/cooling rate can lead to a longer reaction period at lower temperatures, and thus promote hydrochar formation due to insufficient decomposition. Another mechanism could be the enhancement of re-polymerization, condensation, and oligomerization that occurred during the longer ramping period, which forms more hydrochar (Liu *et al.*, 2021a).

9.3.4 Catalyst

Various types of catalysts have been applied to improve the yield and quality of hydrochar. In most cases, acids and bases/alkalis are used as catalysts by changing the reaction pH. An acidic condition could stimulate the hydrolysis, deamination, and dehydration reactions, but an alkaline environment could promote water–gas shift reaction and condensation (higher aromaticity). Acids could promote the protonation effect and improve the structural stability of hydrochar for adsorption applications (Chen *et al.*, 2020). Acids could decrease ash content in hydrochar, while alkalis, such as $Ca(OH)_2$, may increase ash content in hydrochar due to enhanced mineral precipitation (Becker *et al.*, 2019). The formation of fatty substances could be enhanced under an acidic environment, while more N-containing organic compounds and ketone organics could be generated in hydrochar at an alkaline condition (Liu *et al.*, 2020). Raising initial pH could also affect the species, mobility, and distribution of nutrients (e.g., P) and heavy metals in hydrochar, thus allowing its land application by reducing associated environmental risks.

9.4 CHARACTERIZATION OF HYDROCHAR

Numerous studies have shown that the characteristics of hydrochar are highly related to the initial digestate (e.g., C and ash contents) and the hydrothermal conditions (e.g., reaction temperature and residence time). A characterization summary of various types of digestate and their derived hydrochar is presented in Table 9.3. Figure 9.3 is showing the violin distribution of hydrochar properties with the average and 10/90 percentiles. Most studies (10–90%) found that hydrochar from digestate had ash, volatile matter (VM), fixed carbon (FC), and HHV contents of 18–49.5%, 36.2–61.8%, 6.4–24.4%, and 12.2–20.8 MJ/kg on a dry basis, respectively. Besides ash, digestate-derived hydrochar consists of 29.1–53% C, 2.9–5.6% H, 11.8–31.1% O, 0.9–3.8% N, and 0.1–0.5% S on a dry basis. Among the reported feedstocks, digestates from cow manure, agro waste (mixture of manure and crop residues), and lignocellulosic biomass (e.g., energy crops, maize silage, wheat straw, and yard waste) tend to generate hydrochar with above-mean levels of C, VM, FC, and HHV, with low ash contents. However, digestates (e.g., municipal sludge, municipal solid waste pulp, and organic household waste digestates) with high contents of ash could lead to the hydrochar with below-mean levels of C, VM, FC, and HHV. Digestate-derived hydrochar also contains up to 6.3% (db) of total phosphorus (TP), indicating a great potential for phosphorus recovery to mitigate future P demands and environmental challenges (Liu *et al.*, 2021b). Therefore, the application of hydrochar will be dependent on its feedstock nature and chemical properties.

The van Krevelen diagram for digestate and hydrochar showing H/C and O/C atomic ratios is presented in Figure 9.4. Due to the primary hydrothermal reactions (dehydration and decarboxylation), both H/C and O/C ratios decreased in hydrochar compared to its initial digestate feedstock. Regardless of the feedstock type, such reduction generally continues with increasing reaction severity (temperature and residence time). However, hydrochar from various types of digestates showed a significant difference in terms of H/C and O/C ratios. The performance of hydrochar could also be improved due to the removal of oxygen, with similar quality to peat, lignite, and coal. The O/C ratio could be used to indicate the hydrochar stability for carbon sequestration. Spokas (2010) suggested that biochar with an O/C of <0.2, 0.2–0.6, and >0.6 could have a half-life of >1000, 100–1000, and <100 years, respectively. Therefore, hydrochar could be a much better source for carbon sequestration compared to raw digestate.

Table 9.3 Chemical properties (in dry basis) of raw AD digestates and their derived hydrochar (TS = total solids; VM = volatile matter; FC = fixed carbon; HHV = higher heating value; TP = total phosphorus).

Digestate feedstock	TS (%)	Reaction temperature (°C)	Residence time (min)	Hydrochar yield (%)	C (%)	H (%)	O (%)	N (%)	S (%)	Ash (%)	VM (%)	FC (%)	HHV (MJ/kg)	TP (%)	O/C	H/C	References
Municipal sludge	20	Raw	-	-	34.2	6.4	30.3	4.2	0.7	24.2	68.9	6.9	13.3	1.4	0.66	2.25	Bhatt *et al.* (2018)
		180	30	78	42.3	5.3	15.2	4.1	1.1	32.1	54.2	13.7	19.9	1.6	0.27	1.50	
		200	15	91	43.1	5.5	13.8	3.7	1.0	32.8	54.5	12.7	19.9	1.7	0.24	1.52	
		200	30	63	42.2	5.3	13.7	4.0	1.1	33.9	53.6	12.6	19.9	1.7	0.24	1.51	
		200	60	63	44.0	5.6	12.0	4.2	1.1	33.1	53.4	13.5	20.1	1.8	0.21	1.51	
		220	30	63	43.9	5.4	11.8	3.8	1.1	33.9	52.7	13.4	20.2	1.9	0.20	1.48	
	3.8	Raw	-	-	28.5	4.8	34.5	5.3	0.2	26.1	66.9	7.1	16.5	-	0.91	2.01	Kim *et al.* (2014)
		180	30	93.9	28.2	4.2	33.0	5.1	0.2	29.4	62.3	8.4	17.3	-	0.88	1.78	
		200	30	92.6	28.4	4.1	32.8	5.0	0.1	29.6	61.1	9.3	17.5	-	0.87	1.72	
		220	30	88.7	27.7	3.6	31.1	4.1	0.1	33.3	57.2	9.5	18.3	-	0.84	1.56	
		250	30	83.4	27.6	2.9	26.7	3.8	0.1	38.9	50.4	10.7	20.2	-	0.72	1.27	
		280	30	80.4	29.1	2.5	25.5	3.0	0.0	40.0	47.3	12.7	22.4	-	0.66	1.02	
	16.5	Raw	-	-	32.7	4.9	15.4	5.1	1.0	40.9	54.7	4.4	14.8	4.6	0.35	1.80	Marin-Batista *et al.* (2020)
		180	60	74.2	30.8	4.3	16.8	4.2	1.0	42.9	50.9	6.2	13.2	5.28	0.41	1.68	
		210	60	67.8	31.8	4.3	14.9	4.1	1.0	43.9	48.3	7.8	13.7	5.32	0.35	1.62	
		240	60	51.8	32.6	4.1	10.0	4.1	1.1	48.1	44.0	7.9	14.2	6.3	0.23	1.51	
	-	Raw	-	-	18.3	2.9	14.6	2.7	0.3	61.9	34.8	3.3	7.0	1.6	0.60	1.90	Ekpo *et al.* (2016)
	10	170	60	80	12.0	1.8	6.6	1.0	0.2	78.3	20.6	0.7	4.0	1.7	0.41	1.80	
	10	250	60	67	10.0	1.4	6.4	0.6	0.2	81.4	17.4	1.1	2.8	1.9	0.48	1.68	
	10	350	60	63	7.4	1.0	3.2	0.5	0.1	87.7	12.0	0.3	1.6	1.9	0.32	1.62	
	6.25	500	30	53	5.6	0.7	4.2	0.5	0.1	89.0	10.5	0.5	0.5	2.0	0.56	1.50	
Cow manure	25	Raw	-	-	26.8	3.4	22.8	1.7	0.4	44.8	-	-	10.1	0.8	0.64	1.52	Belete *et al.* (2021)
		180	60	80	28.9	3.4	16.2	1.8	0.2	49.5	-	-	11.4	1.3	0.42	1.41	
		210	60	77	31.4	3.5	9.1	1.9	0.4	53.7	-	-	13.0	1.6	0.22	1.34	
		240	60	67	32.5	3.3	3.2	1.9	0.5	58.6	-	-	13.7	1.8	0.07	1.22	

Organic household waste	15	Raw	-	-	34.3	4.0	23.9	1.9	0.2	35.8	54.6	9.6	13.5	0.4	0.52	1.40	Cao *et al.* (2019)
		170	120	85.9	33.9	3.6	23.3	1.7	0.2	37.3	49.7	13.0	12.9	-	0.52	1.27	
		170	300	80.5	32.3	3.3	23.2	1.5	0.1	39.6	47.8	12.6	11.9	-	0.54	1.23	
		190	120	80.4	34.3	3.4	22.7	1.5	0.2	37.9	48.4	13.7	12.8	-	0.50	1.19	
		190	300	75.2	33.1	3.2	21.6	1.7	0.1	40.3	45.3	14.4	12.2	-	0.49	1.16	
		210	120	72.2	32.7	3.0	21.1	1.6	0.1	41.5	44.4	14.1	11.9	-	0.48	1.10	
		210	300	70.4	35.6	3.2	19.2	1.7	0.2	40.1	43.4	16.5	13.4	-	0.40	1.08	
		230	120	62.9	33.6	2.9	17.1	1.7	0.2	44.5	39.5	16.0	12.4	-	0.38	1.04	
		230	300	57.8	34.6	2.7	14.4	2.0	0.2	46.1	36.2	17.7	12.8	-	0.31	0.94	
		250	120	55.0	34.3	2.6	13.1	1.8	0.2	48.0	34.7	17.3	12.7	-	0.29	0.91	
		250	300	54.5	35.7	2.7	12.6	2.0	0.2	46.8	35.1	18.1	13.3	-	0.26	0.91	
Cow manure	15	Raw	-	-	42.6	5.0	34.3	2.0	0.4	15.7	67.3	17.0	16.9	0.7	0.60	1.41	
		170	120	90.6	43.6	4.8	33.2	2.1	0.4	16.0	64.8	19.2	17.1	-	0.57	1.32	
		170	300	86.7	44.1	4.7	32.1	2.2	0.4	16.5	63.1	20.4	17.3	-	0.55	1.28	
		190	120	81.4	45.2	4.5	30.2	2.5	0.4	17.2	61.8	21.0	17.6	-	0.50	1.19	
		190	300	78.3	45.5	4.7	29.2	2.4	0.5	17.7	59.7	22.6	18.0	-	0.48	1.24	
		210	120	73.2	46.9	4.6	27.5	2.6	0.4	18.0	58.6	23.4	18.6	-	0.44	1.18	
		210	300	68.3	47.2	4.2	26.3	2.8	0.4	19.1	56.5	24.4	18.3	-	0.42	1.07	
		230	120	61.7	47.7	4.0	24.1	2.8	0.5	20.8	51.6	27.6	18.4	-	0.38	1.01	
		230	300	53.5	50.7	4.0	19.1	3.1	0.5	22.6	47.5	29.9	20.0	-	0.28	0.95	
		250	120	47.7	51.0	3.8	17.9	3.2	0.5	23.7	45.9	30.4	19.9	-	0.26	0.89	
		250	300	47.3	52.4	3.9	16.1	3.3	0.4	23.9	44.6	31.5	20.7	-	0.23	0.89	
Energy crops	15	Raw	-	-	40.3	4.6	24.0	2.1	0.3	28.7	55.9	15.4	16.4	0.9	0.45	1.37	
		170	120	91.0	39.7	4.2	26.5	2.0	0.3	27.3	55.3	17.4	15.5	-	0.50	1.27	
		170	300	90.0	39.7	4.1	25.9	2.0	0.3	28.0	54.1	17.9	15.4	-	0.49	1.24	
		190	120	83.9	40.5	4.0	24.1	2.1	0.4	28.9	51.8	19.3	15.8	-	0.45	1.19	
		190	300	82.1	40.4	3.9	24.1	2.2	0.3	29.0	50.3	20.7	15.6	-	0.45	1.16	
		210	120	77.6	41.2	3.9	21.9	2.2	0.3	30.4	49.6	20.0	16.1	-	0.40	1.14	
		210	300	74.7	42.4	3.8	20.2	2.3	0.3	30.9	47.8	21.3	16.5	-	0.36	1.08	

(Continued)

Table 9.3 Chemical properties (in dry basis) of raw AD digestates and their derived hydrochar (TS = total solids; VM = volatile matter; FC = fixed carbon; HHV = higher heating value; TP = total phosphorus) (*Continued*).

Digestate feedstock	TS (%)	Reaction temperature (°C)	Residence time (min)	Hydrochar yield (%)	C (%)	H (%)	O (%)	N (%)	S (%)	Ash (%)	VM (%)	FC (%)	HHV (MJ/kg)	TP (%)	O/C	H/C	References
		230	120	67.9	40.8	3.5	20.4	2.4	0.3	32.6	45.5	21.9	15.6	-	0.38	1.03	
		230	300	60.4	42.0	3.4	16.8	2.6	0.4	34.8	41.9	23.3	16.2	-	0.30	0.97	
		250	120	54.6	41.6	3.2	15.8	2.5	0.3	36.5	39.8	23.7	15.9	-	0.28	0.92	
		250	300	51.1	42.0	3.1	14.0	2.8	0.3	37.8	37.9	24.3	16.1	-	0.25	0.89	
Energy crops	15	Raw	-	-	40.0	4.6	24.0	1.9	0.3	29.2	56.8	14.0	16.3	-	0.45	1.38	Cao *et al.* (2021)
		210	30	75.4	41.9	3.8	22.0	2.3	0.3	29.6	49.9	20.5	16.2	-	0.39	1.09	
		210	60	75.6	42.4	3.7	21.1	2.3	0.3	30.2	49.3	20.5	16.3	-	0.37	1.05	
		210	180	73.0	42.3	3.5	20.4	2.4	0.3	31.2	47.8	21.0	16.1	-	0.36	0.99	
		210	300	70.0	42.2	3.4	19.8	2.4	0.3	32.0	45.6	22.4	16.0	-	0.35	0.97	
Maize silage	16.64	Raw	-	-	46.6	5.8	40.7	1.1	0.1	5.6	77.1	17.2	18.8	-	0.66	1.49	Cao *et al.* (2020)
		180	30	70.0	51.2	5.8	36.8	1.7	0.1	4.5	76.0	19.4	20.8	-	0.54	1.36	
		200	30	67.7	54.4	5.6	33.5	1.8	0.1	4.6	71.7	23.6	22.0	-	0.46	1.24	
		220	30	60.7	56.4	5.6	31.4	1.9	0.1	4.7	70.3	25.0	22.9	-	0.42	1.19	
Municipal solid waste pulp	23	Raw	-	-	27.2	3.5	23.1	2.6	-	36.7	48.8	14.5	10.4	-	0.64	1.54	Reza *et al.* (2016)
		200	30	88.1	29.6	4.1	22.6	2.8	-	43.4	48.1	8.5	11.9	-	0.57	1.66	
		200	120	78.1	37.6	4.4	22.0	2.9	-	45.0	47.9	7.1	15.0	-	0.44	1.40	
		250	30	78.1	30.3	3.9	19.4	2.1	-	44.6	42.9	12.5	12.2	-	0.48	1.54	
		250	120	72.6	35.4	3.3	18.1	2.2	-	47.4	37.0	15.6	13.3	-	0.38	1.12	
		300	30	75.3	33.5	3.6	15.4	2.5	-	53.6	40.0	6.4	13.2	-	0.34	1.29	
		300	120	73.4	35.4	3.2	14.3	2.3	-	59.5	39.1	1.4	13.4	-	0.30	1.08	

Wheat straw	6.25	Raw	-	-	40.7	5.5	36.9	0.8	0.2	16.0	-	-	16.5	-	0.68	1.61	Reza *et al.* (2015)
		180	120	59.6	45.9	4.4	29.4	0.9	0.1	19.3	-	-	17.7	-	0.48	1.14	
		180	240	74.6	40.0	3.3	26.6	0.6	0.1	29.4	-	-	14.5	-	0.50	1.00	
		180	360	69.8	41.1	3.5	28.3	0.7	0.1	26.4	-	-	15.0	-	0.52	1.03	
		180	480	73.6	42.9	3.7	27.4	0.7	0.1	25.1	-	-	16.0	-	0.48	1.05	
		200	120	58.7	45.4	4.2	29.7	0.8	0.1	19.8	-	-	17.3	-	0.49	1.10	
		200	240	63.6	44.7	4.4	25.3	0.8	0.1	24.7	-	-	17.7	-	0.42	1.19	
		200	360	63.5	43.5	3.9	27.5	0.8	0.2	24.2	-	-	16.4	-	0.47	1.07	
		200	480	57.0	46.0	4.2	27.1	0.9	0.2	21.6	-	-	17.8	-	0.44	1.10	
		220	120	55.3	47.1	4.2	26.3	0.9	0.2	21.4	-	-	18.2	-	0.42	1.06	
		220	240	48.8	51.6	4.4	23.2	1.2	0.2	19.4	-	-	20.4	-	0.34	1.03	
		220	360	57.9	49.2	3.7	19.1	1.0	0.2	26.8	-	-	19.0	-	0.29	0.91	
		220	480	53.3	53.0	3.8	17.3	1.1	0.2	24.6	-	-	20.7	-	0.24	0.87	
		240	120	56.8	50.4	3.4	16.4	1.1	0.2	28.6	-	-	19.3	-	0.24	0.80	
		240	240	51.6	54.3	3.6	14.9	1.3	0.2	25.8	-	-	21.1	-	0.21	0.79	
		240	360	49.3	55.8	3.8	14.2	1.4	0.2	24.7	-	-	22.0	-	0.19	0.83	
		240	480	50.5	55.4	3.7	13.7	1.3	0.2	25.7	-	-	21.7	-	0.19	0.81	
		260	120	45.9	51.2	3.2	23.9	1.3	0.1	20.3	-	-	18.7	-	0.35	0.75	
		260	240	51.1	53.6	3.5	16.7	1.2	0.2	25.0	-	-	20.6	-	0.23	0.79	
		260	360	57.6	47.6	2.8	15.9	1.0	0.2	32.6	-	-	17.5	-	0.25	0.70	
		260	480	53.3	53.4	3.5	13.9	1.2	0.2	27.8	-	-	20.8	-	0.19	0.79	
Yard waste	13	Raw	-	-	35.9	5.0	36.8	3.8	-	18.3	65.1	16.6	14.2	-	0.77	1.68	Sharma *et al.* (2020)
		180	360	68.2	41.6	5.7	32.1	1.4	-	19.3	58.7	22.0	17.5	-	0.58	1.65	
		200	360	65.6	42.6	5.7	29.7	1.3	-	20.2	53.6	26.2	18.0	-	0.52	1.60	
Agro waste	12	Raw	-	-	36.5	4.6	28.2	2.6	0.4	27.7	53.6	18.7	14.7	1.4	0.58	1.52	Zhao *et al.* (2018)
		190	180	83.2	41.5	4.2	19.8	3.0	0.4	31.1	49.2	19.7	16.7	1.56	0.36	1.20	
		220	180	76.7	42.6	4.1	15.2	3.0	0.4	34.7	45.4	19.9	17.4	1.66	0.27	1.16	
		250	180	69.4	43.8	4.0	10.6	3.1	0.3	38.2	41.2	20.6	18.0	1.80	0.18	1.09	

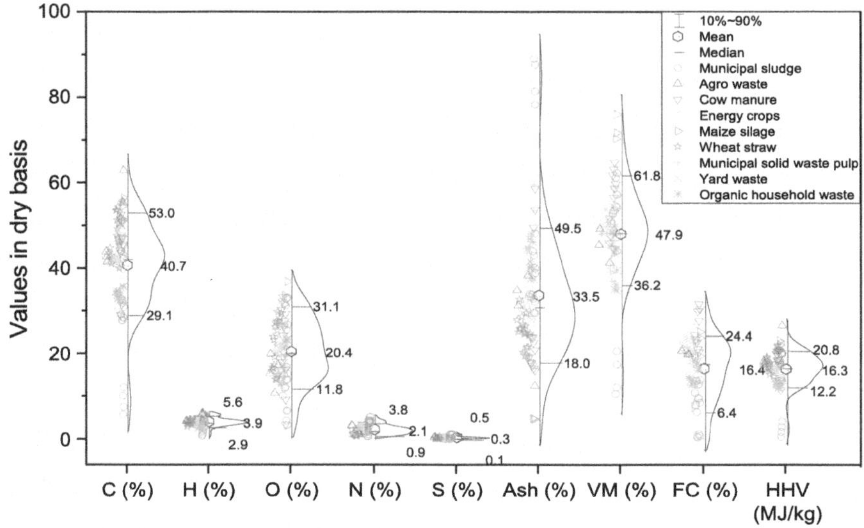

Figure 9.3 Violin distribution plot of characteristics of hydrochar from various types of digestate. VM = volatile matter; FC = fixed carbon; HHV = higher heating value.

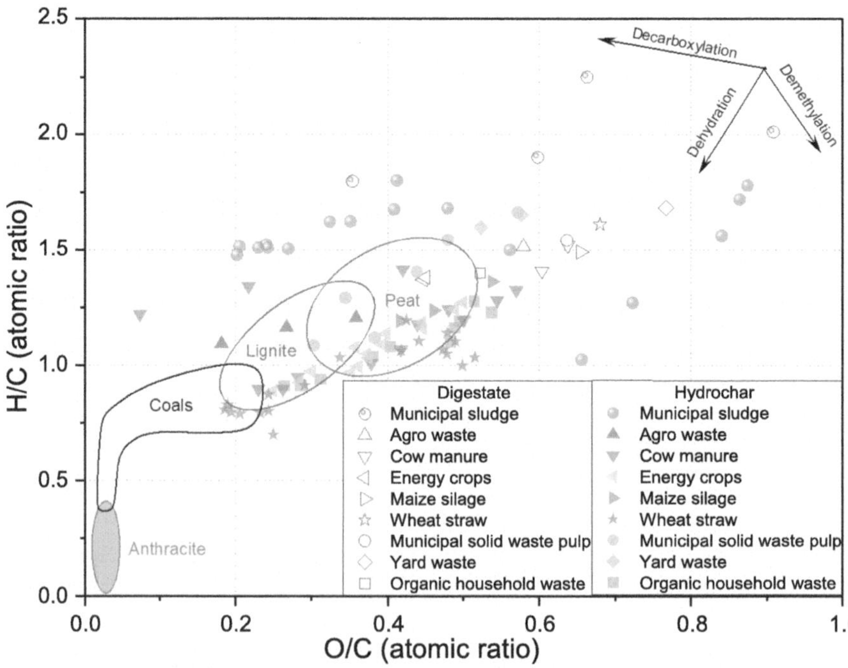

Figure 9.4 van Krevelen diagram of various types of digestate (with open interior) and the corresponding hydrochar (with solid interior).

9.5 ACTIVATION OF HYDROCHAR

Hydrochar from HTP of digestate generally has a small surface area ($\leq$31 m^2/g) and porosity, which could limit its applications in adsorption processes. Hydrochar's physicochemical properties can be enhanced by various physical and chemical methods to improve the surface area (porosity) and enhance adsorption ability (Table 9.4).

9.5.1 Physical activation

Physical activation (or thermal activation) uses heated gas (700–1200°C), such as CO_2, air or O_2, and steam, to reduce the surface heterogeneity and enhance the porous properties of hydrochar (Azzaz *et al.*, 2020). The application of an oxidant atmosphere (e.g., CO_2, O_2, and air) could increase specific surface area, pore volume, and micropores by partial gasification of carbon precursor, while applying water steam could enlarge pore diameters. Khoshbouy *et al.* (2019) found that physically activated (CO_2 at 900°C) sludge hydrochar significantly increased the surface area (from 6 to 262 m^2/g), pore volume (from 0.015 to 0.195 cm^3/g), and the maximum adsorption capacity of methylene blue (from 63 to 122 mg/g). However, physical activation may have limited enhancement on ash-rich hydrochar since it mostly affects the organic fraction. Although physical activation is less popular than chemical activation, it has the outstanding advantages of enhanced micropores, economic, no chemicals input, and no secondary process (e.g., washing) and pollution.

Hydrochar can also be upgraded by a thermochemical method. Under high-temperature pyrolysis conditions, volatile compounds on the hydrochar surface oxidize, leaving an increased surface area. This oxidation can be assisted by various gasses such as CO_2, steam, and O_2 (Xiong *et al.*, 2017). After the application of these treatments, hydrochar can be metal impregnated to convert it into a heterogeneous catalyst for chemical and biochemical reactions (Xu *et al.*, 2020).

9.5.2 Chemical activation

As shown in Table 9.4, chemical activation has been widely applied to AD digestate hydrochar activation for its effective results of high porosity and large specific surface area. The activation process is usually conducted as follows: (i) impregnate hydrochar with acid (e.g., HCl and H_3PO_4), alkali (e.g., KOH, NaOH, $NaHCO_3$, and Na_2CO_3) or salt (e.g., $ZnCl_2$ or NaCl); (ii) heat-impregnated hydrochar under an inert atmosphere (e.g., N_2) at 400–900°C for 1–2 h; and (iii) wash the carbonized material with water to remove chemicals and obtain remaining activated hydrochar.

Acid treatment is a promising hydrochar upgrading method performed with strong acids, such as HNO_3, H_2SO_4, HF, HPO_4, and HCl. Depending on the feedstock and thermochemical conversion conditions, hydrochar can have up to 80% inorganics content (Liu *et al.*, 2021a). Acid treatment of hydrochar dissolves the inorganic fraction (Song *et al.*, 2019) and leaves a porous, carbon-rich structure. Acid activation could enhance oxygenated functional groups, increase the point of zero charge (pH_{PZC}), and improve micropore structure

Table 9.4 Characteristics of activated hydrochar derived from AD digestate

Digestate feedstock	Hydrothermal conditions		Activation	BET surface area (m^2/g)	Micropore volume (cm^3/g)	Average pore diameter (nm)	Reference
	Temperature (°C)	Residence time (min)					
Municipal sludge	190	180	No	31	–	–	Ferrentino *et al.* (2020)
	220	180		8.8	–	–	
	250	180		11.9	–	–	
	190	180	Chemical washing (2 M KOH for 1 h)	0.3	–	–	
	220	180		2.7	–	–	
	250	180		13.4	–	–	
Organic fraction of municipal solids waste	250	320	Chemical (KOH:hydrochar mass ratio = 3:1 at 600°C for 2 h)	1106	0.28 (0.54)[a]	–	Bernardo *et al.* (2020)
Agro waste	190	360	No	14	–	–	Correa *et al.* (2017)
	220	360		8	–	–	
	250	180		8	–	–	
	250	360		9	–	–	
	190	360	Chemical (KOH:hydrochar mass ratio = 4:1 at 600°C for 2 h)	930	0.35	–	
	220	360		1038	0.4	–	
	250	180		1265	0.5	–	
	250	360		1351	0.42 (0.63)[a]	–	
Agro waste	190	180	Chemical (KOH:hydrochar mass ratio = 4:1 at 600°C for 2 h)	854	0.32	–	Zhao *et al.* (2018)
	220	180		1178	0.44	–	
	250	180		1609	0.59	–	
	190	180	Acid leaching (H_2SO_4 at pH 2) + chemical (KOH:hydrochar mass ratio = 4:1 at 600°C for 2 h)	1344	0.50	–	
	220	180		1516	0.56	–	
	250	180		1945	0.67	–	
Cattail	Raw	–	No	0.45	–	10.7	Zhang *et al.* (2020a)
	250	240	No	10.09	–	13.2	
			Chemical (Na_2CO_3:hydrochar mass ratio = 3:1 at 900°C for 1 h)	628.49	0.19	3.22	
			Chemical ($NaHCO_3$:hydrochar mass ratio = 3:1 at 900°C for 1 h)	441.09	0.13	3.76	
			Chemical (NaCl:hydrochar = 3:1 w/w at 900°C for 1 h)	251.73	0.09	2.5	

[a]Numbers in parentheses indicate total pore volume.

and surface area of the hydrochar (Azzaz *et al.*, 2020). Molten salt is mainly used to condense and scaffold C precursors and develop mesopore structures in hydrochar (Wang *et al.*, 2020b). Zhang *et al.* (2020a) reported that NaCl activated hydrochar could have a surface area (251.73 m^2/g) 25 times higher than that of the original hydrochar. Salts are relatively cheap, but they are generally less effective than alkalis for hydrochar activation. NaOH, Na_2CO_3, KOH, and K_2CO_3 are the chemicals commonly used for the alkaline treatment of hydrochar. Although acid treatment provides larger pores, it was reported that alkaline treatment provides higher total pore volumes (Liu *et al.*, 2020). Alkaline activation mainly contributes to pore formation through: (i) altering carbon framework by redox reactions with C; (ii) forming water and CO_2 and promoting C gasification; and (iii) intercalating metallic element (e.g., K) into C layers for further reactions (Zhao *et al.*, 2018). Among alkaline reagents, KOH is the most popular as it gives the best performance on hydrochar activation, leading to a surface area of up to 1609 m^2/g. Combining acid washing before alkaline activation could even further increase specific surface area and pore volume. It is also noted that chemical activation has a greater impact on hydrochar generated from a higher HTP temperature. After the same activation process, the specific surface area of hydrochar produced at 250°C nearly doubled compared to that of hydrochar generated at 190°C, which could be resulted from the more FC at higher HTP temperature (Zhao *et al.* 2018). Activated hydrochar has been found to have the equivalent capacity to commercial activated carbon in water remediation (Liu *et al.*, 2021a). Further, $ZnCl_2$ treatment was also found effective on hydrochar activation. Liu *et al.* (2020) reported a 52% adsorption performance increase by the application of $ZnCl_2$ treatment.

9.6 HYDROCHAR IN ENVIRONMENTAL APPLICATIONS

An increasing number of studies are published in literature on the usage of hydrochar for various applications including microbial carrier in biological treatment, adsorbent, fuel, nutrient recovery, and soil amendment. This section discusses the environmental application of AD digestate-derived hydrochar in detail.

9.6.1 Hydrochar as a microbial carrier for biological treatment

Hydrochar can be used in various environments as a microbial carrier or support medium for prolonged retention. Depending on its characteristics, hydrochar can be utilized as microbial carrier in AD and water/wastewater treatment processes. Factors that determine the effect of hydrochar as a microbial carrier can be listed as porosity and particle size (total surface area), adsorption ability, nutrient, and inhibitory compound contents. Due to its porous structure, hydrochar is a suitable attached growth surface for microorganisms. Hydrochar's conductive structure allows the attached microorganisms to exchange electrons. This electron transfer between species is an essential process in AD, and it is known as direct interspecies electron transfer (DIET). Anaerobic microorganisms can use protons as electron acceptors while utilizing organic compounds and forming hydrogen (Stams *et al.*, 2006). In syntrophic communities, the hydrogen produced by bacteria is utilized by archaea and converted into methane. Decreasing hydrogen concentration as a result of this utilization reduces the inhibition effect on the hydrogen-producing bacteria and therefore increases the overall reaction rate. The same mechanism between species works for most of the intermediate molecules in the AD reactions, such as organic acids and alcohols (Chen *et al.*, 2014). When an attached growth surface is provided, such as hydrochar, syntrophic microorganisms share their products easily and decrease the inhibition effects. This allows for a significant enhancement of the AD process (Kumar *et al.*, 2021). For instance, Xu *et al.* (2020) used swine manure digestate-based hydrochar in the co-digestion of swine manure and sewage sludge and observed a 49% increase in methane production.

The hydrochar can be added to digester inflow (lose medium) or used as fixed-bed medium. The optimum dose and the effect of hydrochar vary depending on the hydrochar's source. The desired type of microbial carrier media hydrochar should have high carbon content, high durability (low biodegradability), high surface area, sufficient nutrients, and low heavy metal leaching and other potential toxic compounds. Hydrochars have large surface area and are also rich in phosphorus (Liu *et al.*, 2021b), which is deficient in some AD substrates, such as sugar industry wastewater (Fito *et al.*, 2018). Hydrochar from digestate can also be used in biofilters for water and wastewater treatment. Biofilters are efficient, low energy consuming, and simple to construct treatment units. They use the biofilm growing on the packing media to treat organic pollutants. Owing to its ability to treat the contaminants of emerging concern (CEC), natural organic matter (NOM) (Basu *et al.*, 2016; Kirisits *et al.*, 2019; Liu *et al.*, 2001), disinfection by-products (DBP) (Devi & Dalai, 2021; Fu *et al.*, 2017; Liu *et al.*, 2017), and pharmaceuticals (Binelli *et al.*, 2014; Fu *et al.*, 2019; McKie *et al.*, 2016; Reungoat *et al.*, 2011), biofilters became popular in the last decade.

9.6.2 Adsorption of pollutants

Hydrochar has typical physico-chemical and structural properties that impart them superior adsorption capacity for the removal of various nutrients and heavy metals from water and wastewater systems. Especially, hydrochar processed at higher temperatures are highly desirable for adsorption applications due to the higher surface areas and well-developed pore structure available for adsorption of pollutants (Liu *et al.*, 2021a).

9.6.2.1 Nutrients removal

Ammonium-nitrogen (NH_4-N) and phosphate (PO_4-P) are essential nutrients for biomass growth, however, excessive concentration of these nutrients can cause eutrophication problems in water bodies. Previous literature studied the influence of HTP temperature and physicochemical properties of hydrochar (surface functionality, surface area, porosity, elemental composition, and cation exchange capacity) on NH_4-N and PO_4-P removal capacity of the digestate based hydrochars (Takaya *et al.*, 2016; Zeng *et al.*, 2013; Zhang *et al.*, 2020b). The hydrochars prepared at lower temperatures (<250°C) were reported to have higher NH_4-N and PO_4-P removal capacity compared to hydrochars derived at higher temperatures (>250°C). Reportedly, polar functional groups enriched in lower temperature hydrochar interact with NH_4-N and improve the NH_4-N adsorption capacity. It is increasingly evident from the previous literature that surface area and porosity are not the key factors influencing NH_4-N adsorption capacity rather surface functional groups play an important role and are directly correlated with adsorption capacity of hydrochar. Some reports found that calcium (Ca^{2+}) and magnesium (Mg^{2+}) contents may influence PO_4-P adsorption (Takaya *et al.*, 2016). For instance, hydrochar derived from two different feedstocks (digestate filter press cake with surface area-2.5 m^2/g and oak wood with surface area – 280 m^2/g) had similar NH_4-N adsorption capacities although they had different surface areas indicating that physisorption (physical adsorption) or ion exchange was not the dominant adsorption mechanism (Takaya *et al.*, 2016). Contrary to this, Zeng *et al.* (2013) found that some of the functional groups disappeared in FTIR spectra of exhausted hydrochar after NH_4-N adsorption indicating the involvement of aromatic functional groups (C=O and C=C, -CH_2 and -CO) in NH_4-N adsorption. Similarly, Zhang *et al.* (2020b) found that NH_4-N adsorption is mainly associated with the interaction between oxygen-containing functional groups and NH_4-N. The adsorption of NH_4-N was found endothermic and spontaneous process, and adsorption data was well described by Langmuir isotherm signifying the single layer adsorption of NH_4-N on hydrochar surface. Despite the fact that physicochemical properties and processing conditions of hydrochars were different, the maximum adsorption capacity of hydrochar ranged from around 0 to 30 mg/g for PO_4-P and 105.8–146.4 mg/g for NH_4-N (Takaya *et al.*, 2016; Zhang *et al.*, 2020b).

9.6.2.2 Heavy metals

Hydrochar and biochar look alike, however, they have different inherent structures, chemical characteristics, and pollution removal capacities. Interestingly, very few studies are available in literature on the usage of digestate-derived hydrochar for heavy metal removal, however, several studies are available on the usage of digestate-derived biochar for heavy metal removal. For instance, Inyang *et al.* (2012) evaluated the dairy waste digestate and sugar beet digestate-based biochars for lead (Pb^{2+}), copper (Cu^{2+}), nickel (Ni^{2+}), and cadmium (Cd^{2+}) removal and found that both biochars were efficient in Pb^{2+} and Cu^{2+} removal, however, sugar beet-derived biochar was found more effective for Ni^{2+} and Cd^{2+} removal. Surface precipitation on biochar surface was depicted as the main mechanism for Pb^{2+} removal for both biochars. Further, Ni *et al.* (2019) investigated the competitive adsorption of Pb^{2+} and Cd^{2+} on anaerobically digested sludge-derived biochar in single and binary systems. Biochar showed a substantially better adsorption capacity for Pb^{2+} compared to Cd^{2+} in a single-metal system. Binary metal study revealed that both Pb^{2+} and Cd^{2+} had an affinity for the same adsorption sites and adsorption of Cd^{2+} was substantially inhibited in the presence of Pb^{2+}. Equilibrium adsorption data revealed the monolayer adsorption of metal ions on biochar surface. Pb^{2+} and Cd^{2+} adsorption was depicted as spontaneous and exothermic phenomena based on the thermodynamic parameters.

Additionally, several researchers investigated the effects of chemical activation of biochar on their heavy metal removal potential. Jin *et al.* (2016) reported that adsorption capacity of anaerobic digestate-derived biochar significantly improved (increased 3×) after KOH activation due to enhancement of biochar's surface area and porous structure, and improvement in overall surface functionality of biochar. Chemisorption (chemical adsorption) was found as the main mechanism for Cu^{2+} removal owing to their strong interaction with functional groups present on biochar surface. Similarly, Wongrod *et al.* (2018) compared the efficiency of KOH-treated and H_2O_2-treated biochars for Pb^{2+} removal. KOH treatment was more efficient than H_2O_2-treated biochar and adsorption capacity improved by 10× and 2×, respectively, compared to untreated biochar. It was observed that insufficient washing after KOH treatment can obstruct Pb^{2+} sorption due to the release of organic materials from the activated biochar matrix. The released organic matter further interacts with Pb^{2+} present in solution and inhibits its adsorption on biochar surface. Proper washing of biochar post-activation seemed to resolve the problem. As mentioned, the structurally dependent pollutant removal performances on digestate-derived hydrochars and related mechanisms have received significantly less attention, which confines comparative analysis of the hydrochar and biochar performances in heavy metal removal.

9.6.2.3 Hydrochar for carbon capture

An elevated concentration of CO_2 in atmosphere is responsible for climate change phenomena such as global warming and unusual weather patterns. In literature, hydrochars derived from anaerobic digestate were tested as adsorbents for CO_2 capture in gas phase applications. Apparently, HTP temperature and surface functional groups play an important role in determining the CO_2 uptake capacity of hydrochar. The presence of N-containing functional groups on biochar surface helps to boost CO_2 uptake to a major extent. The concentration of N-containing functional groups was found to reduce with an increase in HTP temperature resulting in decrease in CO_2 uptake capacity of hydrochar prepared at elevated temperatures (Qiao *et al.*, 2020). Correa *et al.* (2017) reported that hydrochar generated at 250°C was able to achieve a maximum CO_2 capture capacity of 8.80 mol CO_2/kg. Further, chemical activation agents like KOH and urea were also reported to improve the CO_2 uptake capacity of hydrochar. KOH activation enhances the uptake capacity by improving the surface area and porous structure of the biochar. On the other hand, urea treatment enriches the hydrochar with N-containing functional groups which enables the stronger chemical interactions between biochar and CO_2 molecules. Reportedly, urea-treated biochar enriched in N-containing functional groups and have a stronger ability to capture CO_2 compared to un-treated hydrochars (Qiao *et al.*, 2020).

9.6.3 Hydrochar as a fuel source

Generally, hydrochar is considered a good fuel source due to the high carbon content, low oxygen content, hydrophobicity, and heating value. Previous literature indicated that the digestate feedstock type and HTP conditions play a major role in determining the future applications of digestate-derived hydrochar. For example, the digestate-derived hydrochar from an AD utilizing municipal sludge is considered as low-quality solid fuel due to the low carbon content, high ash content, poorly developed aromatic structure, and low HHV. The ash content in municipal sludge-derived hydrochar from AD was found to vary between 52.3 and 67.8% making it a low-quality fuel source as coal must have less than 50% ash concentration in order to be utilized as a fuel (Wang *et al.*, 2019). Similarly, digestate-derived hydrochar from AD utilizing corn starch as feedstock showed poor combustion behavior than the corn starch hydrochar. The cornstarch contained a high amount of cellulose and hemicellulose which decreased significantly after the AD. AD appeared to weaken the pyrolysis-like reactions of cellulose and enriched the insoluble ash content in corn starch digestate-derived hydrochar resulting in a poor combustion behavior compared to corn starch hydrochar. To understand the effect of HTP temperature on HHV of hydrochar, data collected on HHV of digestate-derived hydrochar was plotted against temperature, and the plot is shown in Figure 9.5. In most cases, the HHV of hydrochar increased compared to the raw anaerobic digestate resulting in improved ignition and burnout temperatures. However, HHV seemed to decrease at higher temperatures (>250°C) mainly due to increased ash content in hydrochar indicating that elevated temperature was not suitable for converting digestate to fuel.

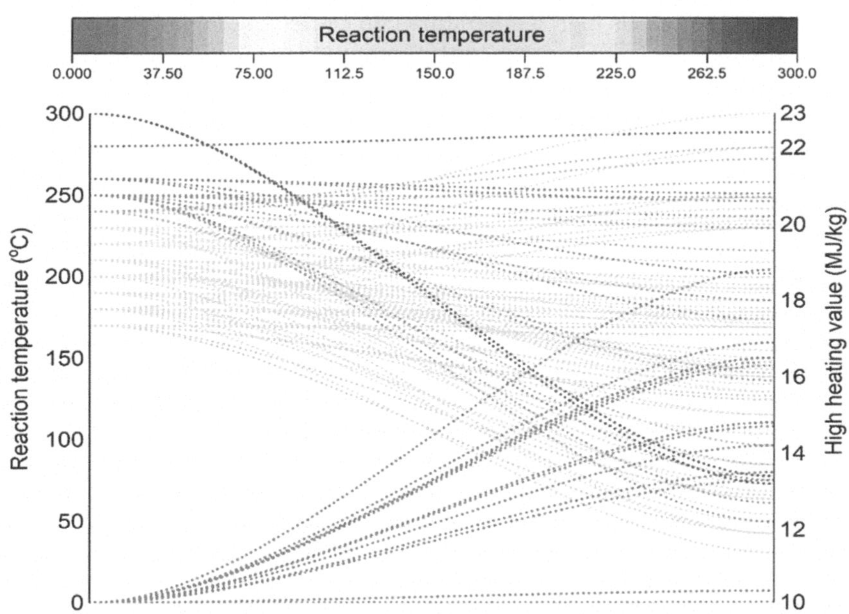

Figure 9.5 High heating value (HHV) of hydrochar prepared from HTP of anaerobic digestate.

Hydrochar is generally converted into fuel pellets to increase its durability, compressive strength, and easy transportation. The mechanical strength (durability and compressive strength) of the fuel pellets is very important to transport them without any breakage. Usually, mechanical pressure or chemical binders are applied during the pellet-making process to develop strong chemical interactions (H-bonding, van-der Waals forces) and mechanical interlocking between the particles (Mittapalli *et al.*, 2021; Zhang *et al.*, 2018). Zhang *et al.* (2018) reported that mechanically pressured digestate-derived hydrochar pellets had higher compressive strength (3.10 MPa) compared to raw anaerobic sludge pellets (2.44 MPa). However, compressive strength of digestate-derived hydrochar was found to decrease when hydrochar was co-palletized with yard waste. Hypothetically, compressive strength should increase as lignin content may promote bonding mechanisms during palletization (Mittapalli *et al.*, 2021). Further research on co-palletization of hydrochar and lignocellulosic waste is required to validate the hypothesis.

9.6.4 Hydrochar for soil amendment

The presence of high concentrations of N, P, K, and other elements in the digestate-derived hydrochars makes them suitable to use as soil amendment. According to literature reports, the majority of P and K fractions (89%–96%) was retained in the hydrochar during HTP of the digestate and only a small fraction was solubilized in the aqueous phase (Huezo *et al.*, 2021). Hydrochars processed at lower temperatures were found to retain more N compared to higher temperatures, as high temperature can cause a loss N, likely volatilized as ammonia (Table 9.5). Contrary to this, retention capacity of P in hydrochar was found to increase with an increase of temperature. Hydrochar is considered a better P source than raw digestate as it releases P slowly into soil. In many cases, plants cannot directly utilize all of the P released from hydrochar. Some reports mentioned that HTP of anaerobic digestate further reduced the bioavailable P for plants (Martin-Batista *et al.*, 2020; Zhang *et al.*, 2020c). Therefore, P-rich hydrochar is considered an ineffective P fertilizer in some cases. Some researchers suggested combusting the hydrochar to make the P bioavailable for plants. Combustion of hydrochar converts non-bioavailable P to more crystalline and soluble hydroxyl apatite that can be easily absorbed by the plants (Wu *et al.*, 2021). Similarly, ammonium was found to release continually from saturated activated hydrochar during the nutrient extraction procedure. Zhang *et al.* (2020b) reported the recovery of 13.1 mg/g of ammonium from saturated hydrochar due to nitrogen solubilization, capture, and release. Only 10% of ammonium was recovered after five extractions at 260°C. But it is expected to work best in an alkaline environment. Further, hydrochar provides essential trace and macro elements for microorganisms and plants (Fang *et al.*, 2018). Particularly, hydrochar application is reported to lower N leaching by rainfall (Yao *et al.*, 2021). An increase in biological activity was reported in soil by the application of hydrochar (Islam *et al.*, 2021), which is critical for optimal agricultural production. Likewise, hydrochar is reported to enhance biological activities in paddy fields while decreasing CH_4 and N_2O emissions (Zhou *et al.*, 2018). The benefits and downsides of incorporating hydrochar into soil have

Table 9.5 Recovery of nutrients from hydrochar derived from anaerobic digestate.

Anaerobic digester feedstock	Recovery	Nutrient	Recovery (%)	Remarks	Reference
Food and agricultural waste	HTC*	N	60%	pH and T dominated N recovery efficiency	Stutzenstein *et al.* (2018)
Food and agricultural waste	HTC	P	57%	pH played a key role in P recovery	Zhao *et al.* (2018)
Sewage sludge	HTC and acid leaching	P	25%	Acid leaching reduced ash content and improved HHV (23.1 MJ/kg)	Martin-Batista *et al.* (2020)
Swine manure	HTP** combined with struvite precipitation	P	87%	Higher struvite crystallization (99.3%)	Zhang *et al.* (2020c)
Swine manure	HTP with hydrochar adsorption	N	46%	Adsorption capacity improved after hydrochar activation	Zhang *et al.* (2020b)

*HTC, hydrothermal carbonization.
**HTP, hydrothermal processing.

been extensively investigated. Digestate-derived hydrochar's impact on soil microbial communities and the molecular structure of soil organic carbon has yet to be fully understood.

9.7 RISK ASSESSMENT OF DIGESTATE-DERIVED HYDROCHARS

It is well known that all hydrochars are not created equal and physicochemical properties of the hydrochars are highly dependent on operating conditions like reaction time and temperature. Interestingly, the reaction temperature has a great impact on the mobility and bioavailability of heavy metals, mineral composition, and concentrations of inhibitory compounds (e.g. hydroxymethylfurfural and furfural, polycyclic aromatic hydrocarbons (PAHs)) in the hydrochar than reaction time. According to literature reports, three methods are most commonly used to assess the risk due to hydrochar application (i) sequential extraction; (ii) leaching potential, and (iii) phytotoxicity (Figure 9.6). Sequential extraction is a five-step extraction process used to determine the bioavailability and speciation behavior of heavy metals (Devi & Saroha, 2014). Based on this, heavy metals can be divided into five fractions (i) water-soluble fraction (F1); (ii) exchangeable fraction (F2); (iii) reducible fraction (F3); (iv) oxidizable fraction

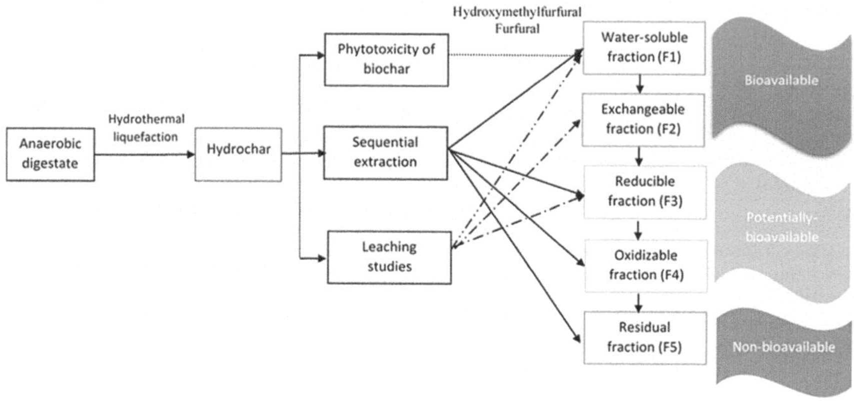

Figure 9.6 Risk assessment based on bioavailability, speciation behavior, and leaching potential of heavy metals and other inhibitory compounds.

(F4) and (v) residual fraction. Fractions F1 and F2 are classified as bioavailable fractions, fractions F3 and F4 are classified as potentially bioavailable fractions and fraction F5 is classified as the non-bioavailable fraction. The concentration of bioavailable heavy metals in hydrochar decreases with an increase in reaction temperature due to immobilization of heavy metals in the hydrochar matrix. The bioavailability and speciation behavior of the heavy metals in hydrochar determines the leaching potential and the environmental risk of usage of hydrochar in various applications.

9.7.1 Leaching of heavy metals

Basically, two methods are used to determine the leaching potential of heavy metals and other toxic compounds from hydrochar matrix (i) pH-based method and (ii) toxicity characteristics leaching procedure (TCLP method). Bioavailable and potentially bio-available fractions are more prone to leaching during leaching experiments. Jin *et al.* (2016) reported that most of the heavy metals are stabilized in the hydrochar matrix after HTL process except Zn and Cd. It is also reported that acid-soluble, exchangeable, and reducible heavy metals are transformed into stable fractions after HTL processing that leads to a lower leaching risk of these heavy metals from hydrochar matrix. On the other hand, acidic pH (2) was found to cause maximum leaching of heavy metals compared to neural and alkaline pH ranges.

9.7.2 Phytotoxicity of hydrochar

The presence of potentially phytotoxic compounds such as hydroxymethylfurfural and furfural could increase the phytotoxicity potential of hydrochars. Chromatographic analysis of hydrochar aqueous extracts is a proven method to determine the phytotoxicity of hydrochars. Previous literature reported that germination tests using hydrochars and hydrochar water extracts could be used to investigate the phytotoxic effects of the hydrochar (Belda *et al.*, 2016;

Celletti *et al.*, 2021). Reportedly, the phytotoxicity of hydrochars was higher than the hydrochar water extracts. The phytotoxicity caused due to application of hydrochar can be reduced by adjusting the amount and ratio of mineral content of hydrochar based on the physiological and metabolic requirements of the plants (Busch *et al.*, 2013; Celletti *et al.*, 2021). This can help to reduce the nutritional imbalances and resulting phytotoxicity. Further, the phytotoxic effect of the extracts could be decreased with an increasing extraction ratio and decreasing pH. Lowering the pH would enhance the bioavailability and solubility of nutrients, such as Mg^{2+}, PO_4^{2-}, iron (Fe^{2+}), Cu^{2+}, zinc (Zn^{2+}), and boron (B^{3+}) (Celletti *et al.*, 2021). Furthermore, the potential phytotoxicity of hydrochar caused from the combination of alkalinity and high electrical conductivity can be reduced by proper acidification and saline leaching treatments of hydrochars.

9.8 CONCLUSIONS

HTP has a great potential to reduce the volume of anaerobic digestate by converting them into value-added products (i.e. hydrochar), thus offering an attractive solution for anaerobic digestate management. It was observed that different feedstock types and process conditions affect hydrochar yield and quality. Process temperature has the highest effect on hydrochar yield, apart from the initial solids content. The residence time profile, and catalyst are the other process parameters that influence the hydrochar yield and quality. For instance, a temperature range of 180–220°C was found as optimum to achieve the maximum hydrochar yield. Further, the hydrochar derived from digestate of food and municipal solid waste seemed to have high HHV.

Several researchers successfully used hydrochar for the removal of heavy metals, nutrients, and CO_2 from wastewater and gas streams through adsorption of these contaminants to hydrochar's carbon matrix through weaker physical interactions or stronger chemical bonding. Chemical activation has proven to improve the contaminant removal efficiency of hydrochar by developing the surface area and improving the porosity of hydrochar. The addition of hydrochar in AD was found to enhance the process by promoting direct interspecies electron transfer (DIET), increasing tolerance to ammonia, adsorbing CO_2, reducing acid stress, providing attach growth surface, increasing alkalinity, and releasing micro-nutrients. Hydrochar was also used as a source of nutrients to aid plant growth. Reportedly, hydrochar obtained at lower HTP temperature tends to be rich in N-contents (<200°C), while the hydrochar obtained at high HTP temperature (>200°C) is usually rich in P-content.

Currently, very limited literature is available on the application of hydrochar as catalyst and adsorbent, and more in-depth research on surface properties and deactivation mechanisms is necessary to extend the usage of hydrochar in these applications. Furthermore, research should focus to optimize the nutrient availability of hydrochar and the feasibility of employing it in agricultural applications. Additionally, digestate is enriched in many organic (proteins, sugars, carbohydrates, etc.) and inorganic substances, and understanding the conversion processes of proteins and polyphenols during the HTP process requires extensive research.

ACKNOWLEDGMENTS

The authors are thankful for the financial support provided by The Natural Sciences and Engineering Research Council of Canada (NSERC) and our municipal partner (Metro Vancouver) through the Industrial Research Chair Program in Advanced Resource Recovery from Wastewater (IRCPJ 548816-18).

REFERENCES

Akhiar A., Battimelli A., Torrijos M. and Carrere H. (2017). Comprehensive characterization of the liquid fraction of digestates from full-scale anaerobic co-digestion. *Waste Management*, **59**, 118–128, https://doi.org/10.1016/j.wasman.2016.11.005

Aragón-Briceño C., Ross A. B. and Camargo-Valero M. A. (2017). Evaluation and comparison of product yields and bio-methane potential in sewage digestate following hydrothermal treatment. *Applied Energy*, **208**, 1357–1369, https://doi.org/10.1016/j.apenergy.2017.09.019

Aragón-Briceño C. I., Grasham O., Ross A. B., Dupont V. and Camargo-Valero M. A. (2020). Hydrothermal carbonization of sewage digestate at wastewater treatment works: influence of solid loading on characteristics of hydrochar, process water and plant energetics. *Renewable Energy*, **157**, 959–973, https://doi.org/10.1016/j.renene.2020.05.021

Azzaz A. A., Khiari B., Jellali S., Ghimbeu C. M. and Jeguirim M. (2020). Hydrochars production, characterization and application for wastewater treatment: A review. *Renewable and Sustainable Energy Reviews*, **127**, 109882, https://doi.org/10.1016/j.rser.2020.109882

Basar I. A., Liu H., Carrere H., Trably E. and Eskicioglu C. (2021). A review on key design and operational parameters to optimize and develop hydrothermal liquefaction of biomass for biorefinery applications. *Green Chemistry*, **23**, 1404–1446, https://doi.org/10.1039/D0GC04092D

Basu O. D., Dhawan S. and Black K. (2016). Applications of biofiltration in drinking water treatment – a review. *Journal of Chemical Technology & Biotechnology*, **91**, 585–595, https://doi.org/10.1002/jctb.4860

Becker G. C., Wüst D., Köhler H., Lautenbach A. and Kruse A. (2019). Novel approach of phosphate-reclamation as struvite from sewage sludge by utilising hydrothermal carbonization. *Journal of Environmental Management*, **238**, 119–125, https://doi.org/10.1016/j.jenvman.2019.02.121

Belda R. M., Lidón A. and Fornes F. (2016). Biochars and hydrochars as substrate constituents for soilless growth of myrtle and mastic. *Industrial Crops and Products*, **94**, 132–142, https://doi.org/10.1016/j.indcrop.2016.08.024

Belete Y. Z., Mau V., Yahav Spitzer R., Posmanik R., Jassby D., Iddya A., Kassem N., Tester J. W. and Gross A. (2021). Hydrothermal carbonization of anaerobic digestate and manure from a dairy farm on energy recovery and the fate of nutrients. *Bioresource Technology*, **333**, 125164, https://doi.org/10.1016/j.biortech.2021.125164

Berglin E., Enderlin C. and Schmidt A. (2012). Review and Assessment of Commercial Vendors / Options for Feeding and Pumping Biomass Slurries for Hydrothermal Liquefaction, Pacific Northwest National Laboratory.

Bernardo M., Correa C. R., Ringelspacher Y., Becker G. C., Lapa N., Fonseca I., Esteves I. A. A. C. and Kruse A. (2020). Porous carbons derived from hydrothermally treated biogas digestate. *Waste Management*, **105**, 170–179, https://doi.org/10.1016/j.wasman.2020.02.011

Bhatt D., Shrestha A., Dahal R. K., Acharya B., Basu P. and MacEwen R. (2018). Hydrothermal carbonization of biosolids from waste water treatment plant. *Energies*, **11**, 2286, https://doi.org/10.3390/en11092286

Binelli A., Magni S., Soave C., Marazzi F., Zuccato E., Castiglioni S., Parolini M. and Mezzanotte V. (2014). The biofiltration process by the bivalve *D. polymorpha* for the removal of some pharmaceuticals and drugs of abuse from civil wastewaters. *Ecological Engineering*, **71**, 710–721, https://doi.org/10.1016/j.ecoleng.2014.08.004

Busch D., Stark A., Kammann C. I. and Glaser B. (2013). Genotoxic and phytotoxic risk assessment of fresh and treated hydrochar from hydrothermal carbonization compared to biochar from pyrolysis. *Ecotoxicology and Environmental Safety*, **97**, 59–66, https://doi.org/10.1016/j.ecoenv.2013.07.003

Cao Z., Jung D., Olszewski M. P., Arauzo P. J. and Kruse A. (2019). Hydrothermal carbonization of biogas digestate: effect of digestate origin and process conditions. *Waste Management*, **100**, 138–150, https://doi.org/10.1016/j.wasman.2019.09.009

Cao Z., Hülsemann B., Wüst D., Illi L., Oechsner H. and Kruse A. (2020). Valorization of maize silage digestate from two-stage anaerobic digestion by hydrothermal carbonization. *Energy Conversion and Management*, **222**, 113218, https://doi.org/10.1016/j.enconman.2020.113218

Cao Z., Hülsemann B., Wüst D., Oechsner H., Lautenbach A. and Kruse A. (2021). Effect of residence time during hydrothermal carbonization of biogas digestate on the combustion characteristics of hydrochar and the biogas production of process water. *Bioresource Technology*, **333**, 125110, https://doi.org/10.1016/j.biortech.2021.125110

Celletti S., Bergamo A., Benedetti V., Pecchi M., Patuzzi F., Basso D., Baratieri M., Cesco S. and Mimmo T. (2021). Phytotoxicity of hydrochars obtained by hydrothermal carbonization of manure-based digestate. *Journal of Environmental Management*, **280**, 111635, https://doi.org/10.1016/j.jenvman.2020.111635

Chen S., Rotaru A. E., Liu F., Philips J., Woodard T. L., Nevin K. P. and Lovley D. R. (2014). Carbon cloth stimulates direct interspecies electron transfer in syntrophic co-cultures. *Bioresource Technology*, **173**, 82–86, https://doi.org/10.1016/j.biortech.2014.09.009

Chen C., Liu G., An Q., Lin L., Shang Y. and Wan C. (2020). From wasted sludge to valuable biochar by low temperature hydrothermal carbonization treatment: insight into the surface characteristics. *Journal of Cleaner Production*, **263**, 121600, https://doi.org/10.1016/j.jclepro.2020.121600

Correa C. R., Bernardo M., Ribeiro R. P., Esteves I. A. and Kruse A. (2017). Evaluation of hydrothermal carbonization as a preliminary step for the production of functional materials from biogas digestate. *Journal of Analytical and Applied Pyrolysis*, **124**, 461–474, https://doi.org/10.1016/j.jaap.2017.02.014

Deng C., Kang X., Lin R. and Murphy J. D. (2020). Microwave assisted low-temperature hydrothermal treatment of solid anaerobic digestate for optimising hydrochar and energy recovery. *Chemical Engineering Journal*, **395**, 124999, https://doi.org/10.1016/j.cej.2020.124999

Devi P. and Dalai A. K. (2021). Implications of breakpoint chlorination on chloramines decay and disinfection by-products formation in brine solution. *Desalination*, **504**, 114961, https://doi.org/10.1016/j.desal.2021.114961

Devi P. and Saroha A. K. (2014). Risk analysis of pyrolyzed biochar made from paper mill effluent treatment plant sludge for bioavailability and eco-toxicity of heavy metals. *Bioresource Technology*, **162**, 308–315, https://doi.org/10.1016/j.biortech.2014.03.093

Devi P. and Saroha A. K. (2015). Effect of pyrolysis temperature on polycyclic aromatic hydrocarbons toxicity and sorption behaviour of biochars prepared by pyrolysis of paper mill effluent treatment plant sludge. *Bioresource Technology*, **192**, 312–320, https://doi.org/10.1016/j.biortech.2015.05.084

Ekpo U., Ross A. B., Camargo-Valero M. A. and Williams P. T. (2016). A comparison of product yields and inorganic content in process streams following thermal hydrolysis and hydrothermal processing of microalgae, manure and digestate. *Bioresource Technology*, **200**, 951–960, https://doi.org/10.1016/j.biortech.2015.11.018

Eskicioglu C., Monlau F., Barakat A., Ferrer I., Kaparaju P., Trably E. and Carrère H. (2017). Assessment of hydrothermal pretreatment of various lignocellulosic biomass with CO_2 catalyst for enhanced methane and hydrogen production. *Water Research*, **120**, 32–42, https://doi.org/10.1016/j.watres.2017.04.068

Fang J., Zhan L., Ok Y. S. and Gao B. (2018). Mini review of potential applications of hydrochar derived from hydrothermal carbonization of biomass. *Journal of Industrial and Engineering Chemistry*, **57**, 15–21, https://doi.org/10.1016/j.jiec.2017.08.026

Ferrentino R., Ceccato R., Marchetti V., Andreottola G. and Fiori L. (2020). Sewage sludge hydrochar: An option for removal of methylene blue from wastewater. *Applied Sciences*, **10**, 3445, https://doi.org/10.3390/app10103445

Fito J., Tefera N., Kloos H. and Van Hulle S. W. H. (2018). Anaerobic treatment of blended sugar industry and ethanol distillery wastewater through biphasic high rate reactor. *Journal of Environmental Science and Health, Part A*, **53**, 676–685.

Fu D., Chen Z., Xia D., Shen L., Wang Y. and Li Q. (2017). A novel solid digestate-derived biochar-Cu NP composite activating H_2O_2 system for simultaneous adsorption and degradation of tetracycline. *Environmental Pollution*, **221**, 301–310, https://doi.org/10.1016/j.envpol.2016.11.078

Fu J., Lee W. N., Coleman C., Nowack K., Carter J. and Huang C. H. (2019). Removal of pharmaceuticals and personal care products by two-stage biofiltration for drinking water treatment. *Science of the Total Environment*, **664**, 240–248, https://doi.org/10.1016/j.scitotenv.2019.02.026

Gurmessa B., Cocco S., Ashworth A. J., Foppa Pedretti E., Ilari A., Cardelli V., Fornasier F., Ruello M. L. and Corti G. (2021). Post-digestate composting benefits and the role of enzyme activity to predict trace element immobilization and compost maturity. *Bioresource Technology*, **338**, 125550, https://doi.org/10.1016/j.biortech.2021.125550

Hämäläinen A., Kokko M., Kinnunen V., Hilli T. and Rintala J. (2021). Hydrothermal carbonisation of mechanically dewatered digested sewage sludge – energy and nutrient recovery in centralised biogas plant. *Water Research*, **201**, 117284, https://doi.org/10.1016/j.watres.2021.117284

Hoffmann J., Rudra S., Toor S. S., Holm-Nielsen J. B. and Rosendahl L. A. (2013). Conceptual design of an integrated hydrothermal liquefaction and biogas plant for sustainable bioenergy production. *Bioresource Technology*, **129**, 402–410, https://doi.org/10.1016/j.biortech.2012.11.051

Huezo L., Vasco-Correa J. and Shah A. (2021). Hydrothermal carbonization of anaerobically digested sewage sludge for hydrochar production. *Bioresource Technology Reports*, **15**, 100795, https://doi.org/10.1016/j.biteb.2021.100795

Hung C. Y., Tsai W. T., Chen J. W., Lin Y. Q. and Chang Y. M. (2017). Characterization of biochar prepared from biogas digestate. *Waste Management*, **66**, 53–60, https://doi.org/10.1016/j.wasman.2017.04.034

Inyang M., Gao B., Pullammanappallil P., Ding W. and Zimmerman A. R. (2010). Biochar from anaerobically digested sugarcane bagasse. *Bioresource Technology*, **101**, 8868–8872, https://doi.org/10.1016/j.biortech.2010.06.088

Inyang M., Gao B., Yao Y., Xue Y., Zimmerman A. R., Pullammanappallil P. and Cao X. (2012). Removal of heavy metals from aqueous solution by biochars derived from anaerobically digested biomass. *Bioresource Technology*, **110**, 50–56, https://doi.org/10.1016/j.biortech.2012.01.072

Islam M. A., Limon M. S. H., Romić M. and Islam M. A. (2021). Hydrochar-based soil amendments for agriculture: a review of recent progress. *Arabian Journal of Geosciences*, **14**, 102, https://doi.org/10.1007/s12517-020-06358-8

Jin H., Hanif M. U., Capareda S., Chang Z., Huang H. and Ai Y. (2016). Copper(II) removal potential from aqueous solution by pyrolysis biochar derived from anaerobically digested algae-dairy-manure and effect of KOH activation. *Journal of Environmental Chemical Engineering*, **4**, 365–372, https://doi.org/10.1016/j.jece.2015.11.022

Kabata-Pendias A. (2004). Soil-plant transfer of trace elements – An environmental issue. *Geoderma*, **122**, 143–149, https://doi.org/10.1016/j.geoderma.2004.01.004

Khoshbouy R., Takahashi F. and Yoshikawa K. (2019). Preparation of high surface area sludge-based activated hydrochar via hydrothermal carbonization and application in the removal of basic dye. *Environmental Research*, **175**, 457–467, https://doi.org/10.1016/j.envres.2019.04.002

Kim D., Lee K. and Park K. Y. (2014). Hydrothermal carbonization of anaerobically digested sludge for solid fuel production and energy recovery. *Fuel*, **130**, 120–125, https://doi.org/10.1016/j.fuel.2014.04.030

Kirisits M. J., Emelko M. B. and Pinto A. J. (2019). Applying biotechnology for drinking water biofiltration: advancing science and practice. *Current Opinion in Biotechnology*, **57**, 197–204, https://doi.org/10.1016/j.copbio.2019.05.009

Kor-Bicakci G. and Eskicioglu C. (2019). Recent developments on thermal municipal sludge pretreatment technologies for enhanced anaerobic digestion. *Renewable and Sustainable Energy Reviews*, **110**, 423–443, https://doi.org/10.1016/j.rser.2019.05.002

Kumar V., Nabaterega R., Khoei S. and Eskicioglu C. (2021). Insight into interactions between syntrophic bacteria and archaea in anaerobic digestion amended with conductive materials. *Renewable and Sustainable Energy Reviews*, **144**, 110965, https://doi.org/10.1016/j.rser.2021.110965

Li S., Jiang Y., Snowden-Swan L. J., Askander J. A., Schmidt A. J. and Billing J. M. (2021). Techno-economic uncertainty analysis of wet waste-to-biocrude via hydrothermal liquefaction. *Applied Energy*, **283**, 116340, https://doi.org/10.1016/j.apenergy.2020.116340

Liu X., Huck P. M. and Slawson R. M. (2001). Factors affecting drinking water biofiltration. *Journal – American Water Works Association*, **93**, 90–101, https://doi.org/10.1002/j.1551-8833.2001.tb09358.x

Liu C., Olivares C. I., Pinto A. J., Lauderdale C. V., Brown J., Selbes M. and Karanfil T. (2017). The control of disinfection byproducts and their precursors in biologically active filtration processes. *Water Research*, **124**, 630–653, https://doi.org/10.1016/j.watres.2017.07.080

Liu X., Zhai Y., Li S., Wang B., Wang T., Liu Y., Qiu Z. and Li C. (2020). Hydrothermal carbonization of sewage sludge: effect of feed-water pH on hydrochar's physicochemical properties, organic component and thermal behavior. *Journal of Hazardous Materials*, **388**, 122084, https://doi.org/10.1016/j.jhazmat.2020.122084

Liu H., Basar I. A., Nzihou A. and Eskicioglu C. (2021a). Hydrochar derived from municipal sludge through hydrothermal processing: A critical review on its formation, characterization, and valorization. *Water Research*, **199**, 117186, https://doi.org/10.1016/j.watres.2021.117186

Liu H., Hu G., Basar I. A., Li J., Lyczko N., Nzihou A. and Eskicioglu C. (2021b). Phosphorus recovery from municipal sludge-derived ash and hydrochar through wet-chemical technology: a review towards sustainable waste management. *Chemical Engineering Journal*, **417**, 129300, https://doi.org/10.1016/j.cej.2021.129300

Lühmann T. and Wirth B. (2020). Sewage sludge valorization via hydrothermal carbonization: optimizing dewaterability and phosphorus release. *Energies*, **13**, 1–16, https://doi.org/10.3390/en13174417

Ma Q., Song W., Wang R., Zou J., Yang R. and Zhang S. (2018). Physicochemical properties of biochar derived from anaerobically digested dairy manure. *Waste Management*, **79**, 729–734, https://doi.org/10.1016/j.wasman.2018.08.023

Magdziarz A., Mlonka-Mędrala A., Sieradzka M., Aragon-Briceño C., Pożarlik A., Bramer E. A., Brem G., Niedzwiecki Ł and Pawlak-Kruczek H. (2021). Multiphase analysis of hydrochars obtained by anaerobic digestion of municipal solid waste organic fraction. *Renewable Energy*, **175**, 108–118, https://doi.org/10.1016/j.renene.2021.05.018

Makádi M., Tomócsik A. and Orosz V. (2012). Digestate: a new nutrient source-review. In: Biogas, S. Kumar (ed.), InTech, Rijeka, Croatia, pp. 295–312.

Mao C., Feng Y., Wang X. and Ren G. (2015). Review on research achievements of biogas from anaerobic digestion. *Renewable and Sustainable Energy Reviews*, **45**, 540–555, https://doi.org/10.1016/j.rser.2015.02.032

Marin-Batista J. D., Mohedano A. F., Rodríguez J. J. and de la Rubia M. A. (2020). Energy and phosphorous recovery through hydrothermal carbonization of digested sewage sludge. *Waste Management*, **105**, 566–574, https://doi.org/10.1016/j.wasman.2020.03.004

Marrone P. A., Elliott D. C., Billing J. M., Hallen R. T., Hart T. R., Kadota P., Moeller J. C., Randel M. A. and Schmidt A. J. (2018). Bench-scale evaluation of hydrothermal processing technology for conversion of wastewater solids to fuels. *Water Environment Research*, **90**, 329–342, https://doi.org/10.2175/106143017X15131012152861

Mathimani T. and Mallick N. (2019). A review on the hydrothermal processing of microalgal biomass to bio-oil – knowledge gaps and recent advances. *Journal of Cleaner Production*, **217**, 69–84, https://doi.org/10.1016/j.jclepro.2019.01.129

McKie M. J., Andrews S. A. and Andrews R. C. (2016). Conventional drinking water treatment and direct biofiltration for the removal of pharmaceuticals and artificial sweeteners: A pilot-scale approach. *Science of the Total Environment*, **544**, 10–17, https://doi.org/10.1016/j.scitotenv.2015.11.145

Merzari F., Goldfarb J., Andreottola G., Mimmo T., Volpe M. and Fiori L. (2020). Hydrothermal carbonization as a strategy for sewage sludge management: influence of process withdrawal point on hydrochar properties. *Energies*, **13**, 2890, https://doi.org/10.3390/en13112890

Miliotti E., Casini D., Rosi L., Lotti G., Rizzo A. M. and Chiaramonti D. (2020). Lab-scale pyrolysis and hydrothermal carbonization of biomass digestate: characterization of solid products and compliance with biochar standards. *Biomass and Bioenergy*, **139**, 105593, https://doi.org/10.1016/j.biombioe.2020.105593

Mittapalli S., Sharma H. B. and Dubey B. K. (2021). Hydrothermal carbonization of anaerobic granular sludge and co-pelletization of hydrochar with yard waste. *Bioresource Technology Reports*, **14**, 100691, https://doi.org/10.1016/j.biteb.2021.100691

Mumme J., Eckervogt L., Pielert J., Diakité M., Rupp F. and Kern J. (2011). Hydrothermal carbonization of anaerobically digested maize silage. *Bioresource Technology*, **102**, 9255–9260, https://doi.org/10.1016/j.biortech.2011.06.099

Mumme J., Titirici M. M., Pfeiffer A., Lüder U., Reza M. T. and Mašek O. (2015). Hydrothermal carbonization of digestate in the presence of zeolite: process efficiency and composite properties. *ACS Sustainable Chemistry & Engineering*, **3**, 2967–2974, https://doi.org/10.1021/acssuschemeng.5b00943

Ni B. J., Huang Q. S., Wang C., Ni T. Y., Sun J. and Wei W. (2019). Competitive adsorption of heavy metals in aqueous solution onto biochar derived from anaerobically digested sludge. *Chemosphere*, **219**, 351–357, https://doi.org/10.1016/j.chemosphere.2018.12.053

Ovsyannikova E., Arauzo P. J., Becker G. and Kruse A. (2019). Experimental and thermodynamic studies of phosphate behavior during the hydrothermal carbonization of sewage sludge. *Science of the Total Environment*, **692**, 147–156, https://doi.org/10.1016/j.scitotenv.2019.07.217

Park K. Y., Lee K. and Kim D. (2018). Characterized hydrochar of algal biomass for producing solid fuel through hydrothermal carbonization. *Bioresource Technology*, **258**, 119–124, https://doi.org/10.1016/j.biortech.2018.03.003

Parmar K. R. and Ross A. B. (2019). Integration of hydrothermal carbonisation with anaerobic digestion; opportunities for valorisation of digestate. *Energies*, **12**, 1586, https://doi.org/10.3390/en12091586

Qiao Y., Zhang S., Quan C., Gao N., Johnston C. and Wu C. (2020). One-pot synthesis of digestate-derived biochar for carbon dioxide capture. *Fuel*, **279**, 118525, https://doi.org/10.1016/j.fuel.2020.118525

Reungoat J., Escher B. I., Macova M. and Keller J. (2011). Biofiltration of wastewater treatment plant effluent: effective removal of pharmaceuticals and personal care products and reduction of toxicity. *Water Research*, **45**, 2751–2762, https://doi.org/10.1016/j.watres.2011.02.013

Reza M. T., Mumme J. and Ebert A. (2015). Characterization of hydrochar obtained from hydrothermal carbonization of wheat straw digestate. *Biomass Conversion and Biorefinery*, **5**, 425–435, https://doi.org/10.1007/s13399-015-0163-9

Reza M. T., Coronella C., Holtman K. M., Franqui-Villanueva D. and Poulson S. R. (2016). Hydrothermal carbonization of autoclaved municipal solid waste pulp and anaerobically treated pulp digestate. *ACS Sustainable Chemistry & Engineering*, **4**, 3649–3658, https://doi.org/10.1021/acssuschemeng.6b00160

Ruyter H. P. (1982). Coalification model. *Fuel*, **61**, 1182–1187, https://doi.org/10.1016/0016-2361(82)90017-5

Sharma H. B., Panigrahi S., Sarmah A. K. and Dubey B. K. (2020). Downstream augmentation of hydrothermal carbonization with anaerobic digestion for integrated biogas and hydrochar production from the organic fraction of municipal solid waste: a circular economy concept. *Science of the Total Environment*, **706**, 135907, https://doi.org/10.1016/j.scitotenv.2019.135907

Sheppard S. C. and Sanipelli B. (2012). Trace elements in feed, manure, and manured soils. *Journal of Environmental Quality*, **41**, 1846–1856, https://doi.org/10.2134/jeq2012.0133

Song E., Park S. and Kim H. (2019). Upgrading hydrothermal carbonization (HTC) hydrochar from sewage sludge. *Energies*, **12**, 2383, https://doi.org/10.3390/en12122383

Spokas K. A. (2010). Review of the stability of biochar in soils: predictability of O:C molar ratios. *Carbon Management*, **1**, 289–303, https://doi.org/10.4155/cmt.10.32

Stams A. J. M., de Bok F. A. M., Plugge C. M., van Eekert M. H. A., Dolfing J. and Schraa G. (2006). Exocellular electron transfer in anaerobic microbial communities. *Environmental Microbiology*, **8**, 371–382, https://doi.org/10.1111/j.1462-2920.2006.00989.x

Stutzenstein P., Bacher M., Rosenau T. and Pfeifer C. (2018). Optimization of nutrient and carbon recovery from anaerobic digestate via hydrothermal carbonization and investigation of the influence of the process parameters. *Waste and Biomass Valorization*, **9**, 1303–1318, https://doi.org/10.1007/s12649-017-9902-4

Takaya C. A., Fletcher L. A., Singh S., Anyikude K. U. and Ross A. B. (2016). Phosphate and ammonium sorption capacity of biochar and hydrochar from different wastes. *Chemosphere*, **145**, 518–527, https://doi.org/10.1016/j.chemosphere.2015.11.052

Tasca A. L., Puccini M., Gori R., Corsi I., Galletti A. M. R. and Vitolo S. (2019). Hydrothermal carbonization of sewage sludge: a critical analysis of process severity, hydrochar properties and environmental implications. *Waste Management*, **93**, 1–13, https://doi.org/10.1016/j.wasman.2019.05.027

Tsai W. T., Fang Y. Y., Cheng P. H. and Lin Y. Q. (2018). Characterization of mesoporous biochar produced from biogas digestate implemented in an anaerobic process of large-scale hog farm. *Biomass Conversion and Biorefinery*, **8**, 945–951, https://doi.org/10.1007/s13399-018-0344-4

Wang C., Zhu W., Chen C., Zhang H., Fan Y., Mu B. and Zhong J. (2019). Behavior of phosphorus in catalytic supercritical water gasification of dewatered sewage sludge: The conversion pathway and effect of alkaline additive. *Energy and Fuels*, **33**, 1290–1295, https://doi.org/10.1021/acs.energyfuels.8b04054

Wang S., Persson H., Yang W. and Jönsson P. G. (2020a). Pyrolysis study of hydrothermal carbonization-treated digested sewage sludge using a Py-GC/MS and a bench-scale pyrolyzer. *Fuel*, **262**, 116335, https://doi.org/10.1016/j.fuel.2019.116335

Wang P., Zhang G., Chen W., Chen Q., Jiao H., Liu L., Wang X. and Deng X. (2020b). Molten salt template synthesis of hierarchical porous nitrogen-containing activated carbon derived from chitosan for CO_2 capture. *ACS Omega*, **5**, 23460–23467, https://doi.org/10.1021/acsomega.0c03497

Watson J., Wang T., Si B., Chen W. T., Aierzhati A. and Zhang Y. (2020). Valorization of hydrothermal liquefaction aqueous phase: pathways towards commercial viability. *Progress in Energy and Combustion Science*, **77**, 100819, https://doi.org/10.1016/j.pecs.2019.100819

Wongrod S., Simon S., Guibaud G., Lens P. N. L., Pechaud Y., Huguenot D. and van Hullebusch E. D. (2018). Lead sorption by biochar produced from digestates: consequences of chemical modification and washing. *Journal of Environmental Management*, **219**, 277–284, https://doi.org/10.1016/j.jenvman.2018.04.108

Wu S. L., Wei W., Xu Q., Huang X., Sun J., Dai X. and Ni B. J. (2021). Revealing the mechanism of biochar enhancing the production of medium chain fatty acids from waste activated sludge alkaline fermentation liquor. *ACS ES&T Water*, **1**, 1014–1024, https://doi.org/10.1021/acsestwater.0c00278

Xiong X., Yu I. K. M., Cao L., Tsang D. C. W., Zhang S. and Ok Y. S. (2017). A review of biochar-based catalysts for chemical synthesis, biofuel production, and pollution control. *Bioresource Technology*, **246**, 254–270, https://doi.org/10.1016/j.biortech.2017.06.163

Xu S., Wang C., Duan Y. and Wong J. W. C. (2020). Impact of pyrochar and hydrochar derived from digestate on the co-digestion of sewage sludge and swine manure. *Bioresource Technology*, **314**, 123730, https://doi.org/10.1016/j.biortech.2020.123730

Yao Y., Dai Q., Gao R., Gan Y. and Yi X. (2021). Effects of rainfall intensity on runoff and nutrient loss of gently sloping farmland in a karst area of SW China. *PLoS ONE*, **16**, e0246505.

Yuan H. Y., Ding L. J., Zama E. F., Liu P. P., Hozzein W. N. and Zhu Y. G. (2018). Biochar modulates methanogenesis through electron syntrophy of microorganisms with ethanol as a substrate. *Environmental Science & Technology*, **52**, 12198–12207, https://doi.org/10.1021/acs.est.8b04121

Zeng Z., Zhang S., Li T., Zhao F., He Z., Zhao H., Yang X., Wang H., Zhao J. and Rafiq M. T. (2013). Sorption of ammonium and phosphate from aqueous solution by biochar derived from phytoremediation plants. *Journal of Zhejiang University Science B*, **14**, 1152–1161, https://doi.org/10.1631/jzus.B1300102

Zhai Y., Wang C., Chen H., Li C., Zeng G., Pang D. and Lu P. (2013). Digested sewage sludge gasification in supercritical water. *Waste Management and Research*, **31**, 393–400, https://doi.org/10.1177/0734242X12471097

Zhang D., Wang F., Shen X., Yi W., Li Z., Li Y. and Tian C. (2018). Comparison study on fuel properties of hydrochars produced from corn stalk and corn stalk digestate. *Energy*, **165**, 527–536, https://doi.org/10.1016/j.energy.2018.09.174

Zhang B., Joseph G., Wang L., Li X. and Shahbazi A. (2020a). Thermophilic anaerobic digestion of cattail and hydrothermal carbonization of the digestate for co-production of biomethane and hydrochar. *Journal of Environmental Science and Health Part A Toxic/Hazardous Substances & Environmental Engineering*, **55**, 230–238.

Zhang T., Wu X., Shaheen S. M., Zhao Q., Liu X., Rinklebe J. and Ren H. (2020b). Ammonium nitrogen recovery from digestate by hydrothermal pretreatment followed by activated hydrochar sorption. *Chemical Engineering Journal*, **379**, 122254, https://doi.org/10.1016/j.cej.2019.122254

Zhang T., He X. Y., Deng Y. X., Tsang D. C. W., Jiang R. F., Becker G. C. and Kruse A. (2020c). Phosphorus recovered from digestate by hydrothermal processes with struvite crystallization and its potential as a fertilizer. *Science of the Total Environment*, **698**, 134240, https://doi.org/10.1016/j.scitotenv.2019.134240

Zhao X., Becker G. C., Faweya N., Rodriguez C. C., Yang S., Xie X. and Kruse A. (2018). Fertilizer and activated carbon production by hydrothermal carbonization of digestate. *Biomass Conversion and Biorefinery*, **8**, 423–436, https://doi.org/10.1007/s13399-017-0291-5

Zhou B., Feng Y., Wang Y., Yang L., Xue L. and Xing B. (2018). Impact of hydrochar on rice paddy CH_4 and N_2O emissions: A comparative study with pyrochar. *Chemosphere*, **204**, 474–482, https://doi.org/10.1016/j.chemosphere.2018.04.056

doi: 10.2166/9781789062755_0225

Chapter 10

Anaerobic digestate pre-treatments for enhanced energy and resources recovery

Barbara Ruffino, Giuseppe Campo and Alberto Cerutti*

DIATI, Politecnico di Torino, Corso Duca degli Abruzzi 24, 10129 Torino, Italy
**Corresponding author email: barbara.ruffino@polito.it*

ABSTRACT

Digestate, one of the products of anaerobic digestion (AD), is a substrate rich in carbon, macronutrients and micronutrients. Notwithstanding its interesting characteristics, the digestate remains a bottleneck in the biogas industry. European countries have an urgent need of processing plants, capable of treating the digestate with appropriate methods in order to recover material and/or energy. This chapter presents a series of processes intended to an initial solid–liquid separation of the digestate and a subsequent treatment of the separated fractions, in the view of a material or energy valorization. The solid–liquid separation can be obtained in settling tanks or by using centrifuges. Furthermore, the solid fraction can be mechanically separated by forced filtration, using screw presses, or by drainage, through fabric belts or screens, eventually with the aid of flocculation or precipitation agents. The solid and liquid fractions of the digestate, after separation, have different destinations. The solid fraction has a total solids concentration that ranges from 20 to 30%, up to 35–40%. This fraction is already partially stabilized, so that appropriate storage, or direct application on agricultural land as biofertilizer or soil improver, are facilitated. Nevertheless, the solid fraction still contains biodegradable material, consequently further processing is necessary, namely composting, drying or other forms of stabilization, if a stable and marketable biofertilized product is desired. Drying is often used as a pre-treatment before thermal conversion processes, namely combustion, hydrothermal carbonization and pyrolysis, used to produce energy from the digestate so as to improve the overall energy balance of an AD process. Finally, the principal technologies, available or under development, for the treatment of the liquid fraction of the digestate in the view of nutrient recovery are evaporation, gas stripping and absorption, chemical crystallization and membrane separation.

Keywords: Solid–liquid separation, mechanical devices, thermal processes, struvite, biofertilizer, material and energy recovery

Editor(s): Vinay Kumar Tyagi, Kaoutar Aboudi and Cigdem Eskicioglu

10.1 INTRODUCTION

Together with biogas, digestate is one of the products of anaerobic digestion (AD). The digestate is a substrate rich in carbon, macronutrients, namely nitrogen and phosphorous, and other macro and micronutrients. Möller and Müller (2012) reported a content of macronutrients in the raw digestate in the order of 1.2–9.1 kg/t for nitrogen, 0.4–2.6 kg/t for phosphorous and of 1.2–11.5 kg/t for potassium. Furthermore, the digestate is characterized by high biological stability and appropriate fertilizing quality. Because of these characteristics, it can be applied directly onto farmland, as an alternative to mineral fertilizers. According to Riva *et al.* (2016), the digestate did not negatively affect the characteristics of soil and air, provided that correct spreading procedures are followed, with subsurface injection to limit ammonia emissions. Furthermore, fertilizers based on the digestate achieve comparable or even higher biomass production rates compared to synthetic mineral fertilizer, for example, the production of wheat (Abubaker *et al.*, 2012) or silage corn (Riva *et al.*, 2016). However, an uncontrolled and rapid release of nutrients from the digestate to soils can increase the risk of nutrient runoff and penetration into groundwater resources, leading to final soil and water eutrophication (Gienau *et al.*, 2018). Furthermore, some physical–chemical features of digestates generated from substrates of animal origin (pig slurries), namely the high content of ammonium, salinity, chemical oxygen demand (COD), phosphate and color, seem to have toxic effects on some organisms (algae, seeds) (Tigini *et al.*, 2016).

Up to now, the agricultural use of digestates has not been specifically regulated by the European Community and only local guidelines have been proposed, with the exception of the application of Good Agricultural Practices (FAO, 2005), as well as the European 'Nitrate Directive' (91/676/EEC), that strongly limits the amount of zootechnical nitrogen that can be applied per hectare of agricultural soil (Gardoni & Guarino, 2016). With the aim of guaranteeing good fertilizing procedures, without local over-fertilization, and preserving a high quality of groundwater, recent revisions of the fertilizer ordinance in Germany have limited field applications of organic residues to 170 kgN/ha y (DüV, 2017). Regarding phosphorus, no surplus application on fields with high or very high phosphate content will be allowed (saldo = 0 kgP/ha y). Consequently, proper fertilizing practices that make use of the digestate require the transportation of the substrate from the area of production to the area of utilization. Because the digestate contains more than 90% by weight of water, depending on the process used for AD and subsequent digestate pre-treatment, transportation is neither economically nor ecologically sustainable (Gienau *et al.*, 2018). In Italy, the limits for the application of the digestate in agriculture are still controversial, especially when the digestate comes from a co-digestion process that includes also organic waste such as sewage sludge from wastewater treatment plants (Campo *et al.*, 2021; Gardoni & Guarino, 2016). For these reasons, many industries that co-digest organic waste have been required to dispose in landfills or municipal solid waste incinerators, or post-treat their digestate. These practices have a significant economic impact, therefore, a number of different digestate treatment processes have been developed.

Recent works have identified digestate as the bottleneck of the biogas industry (Logan & Visvanathan, 2019; Vondra *et al.*, 2019). In European countries, such as Austria, Germany, Italy and Switzerland, there still is an urgent need of digestate processing plants, which are capable of treating the digestate with appropriate methods in order to recover material and/or energy.

The digestate processing can be partial, primarily for the purpose of volume reduction, or it can be complete, refining the digestate to for example pure water, a solid biofertilizer fraction, and fertilizer concentrates (Drosg *et al.*, 2015). While partial processing uses relatively simple and cheap technologies, complete processing requires methods and technologies with various degrees of technical maturity, higher energy input, and higher investment and operating costs.

A recent review (Monfet *et al.*, 2018) identified 14 possible categories of processes for the treatment of the digestate in the view of an energy or material valorization. Among those, the most diffused treatment processes, which will be discussed in this chapter, are: thickening, dewatering, membrane filtration, struvite precipitation, ammonia stripping, thermal drying, combustion, pyrolysis, hydrothermal carbonization (see Figure 10.1).

The first step in the processing of the digestate for energy or resource recovery is the separation of the solid phase from the liquid phase, that can be obtained through a thickening process most often followed by dewatering. Most of the technologies used for solid–liquid separation originates from manure processing, sewage sludge treatment and wastewater treatment. Solid–liquid separation can be performed in settling tanks, where solids are removed from the bottom of the tank. Settling can be forced by using centrifuges. Solids can also be removed mechanically by forced filtration, using screw presses, or by

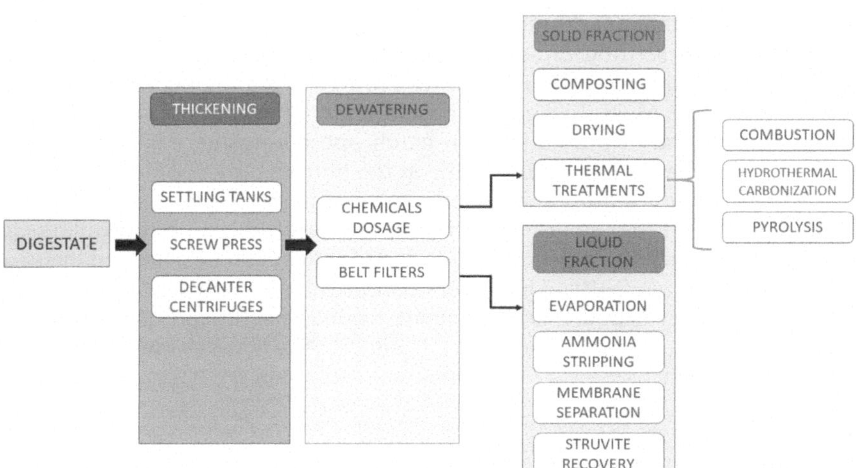

Figure 10.1 Categories of processes for the treatment of the digestate in the view of energy or resource recovery.

drainage, through fabric belts or screens (Hjorth *et al.*, 2010). Flocculation or precipitation agents can be added to improve solid–liquid separation.

To identify the best solid–liquid separation process, the focus has to be set on finding the right technology, or combination of technologies, for an efficient but cost-effective separation. Especially for a subsequent membrane treatment or evaporation, the right degree of separation of the solid fraction is essential.

When comparing the results obtained with different processes, it is an advantage to use one separation parameter capable to express the efficiency of separation. The removal efficiency (R) expresses the efficiency of removal of a specific compound (x) from a slurry to the solid fraction. It is defined as in Equation (10.1):

$$R(x) = 1 - \frac{c(x)_{\text{liquid}}}{c(x)_{\text{slurry}}} \tag{10.1}$$

where $c(x)$slurry and $c(x)$liquid are the concentrations (g/L) of the species in consideration (dry matter, P, N) in the slurry being treated in the separator and the liquid fraction that is produced, respectively, Hjorth *et al.*, 2010. The greater the removal efficiency, the lower will be the amount of compound x remaining in the liquid fraction.

Equation (10.1) characterizes the efficiency of the separator with respect to the liquid fraction; however, the removal efficiency parameter does not give any indication of the production of the solid fraction.

The following section presents some of the processes used for digestate thickening and dewatering.

10.2 PROCESSES FOR SOLID–LIQUID SEPARATION

10.2.1 Settling tanks: sedimentation thickeners

Sedimentation in settling tanks or sedimentation thickeners is an attractive solid–liquid separation solution, because of low capital and operating costs and simple technology. Most thickeners consist of a container that is cylindrical at the top and conical at the bottom. In batch operation, slurry is added to the top of the thickener and the solids settle at the bottom of the conical part, from which the solids are removed. To encourage settling and increase the transfer of settled solids on the upper part of the conical section, small thickeners can be vibrated; conversely, in larger thickeners this effect can be achieved by using a rake (Hjorth *et al.*, 2010). Thickeners can also be operated in a continuous mode, where slurry is added continuously, while the solid and liquid products are removed at the same rate as slurry is added. Alternatively, lagoons can be used as settling basins.

10.2.2 Screw press

Screw press separators are often used if the digestate has a high fiber content (Drosg *et al.*, 2015). In this kind of apparatus, a screw presses the fibers against a cylindrical screen. The liquid fraction of the digestate leaves the separator through the screen. Because of the increasing diameter of the screw, the pressure is increased as the fibers advance along the separator. Finally, the solid fraction, containing the fibers, exits at the end of the separator, where the resistance can be adjusted mechanically. The degree of separation is affected by the mesh size of the screen (Drosg *et al.*, 2015). Compared to other devices, such as decanter centrifuges, screw press separators have only a limited capacity in the separation of small particles (0.5–1 mm). Interest on screw presses relies on the low investment costs, approximately 20 000 € for a 500 kWel biogas plant, and low energy consumption, in the order of 0.4–0.5 kWh/m^3 (Drosg *et al.*, 2015).

10.2.3 Decanter centrifuges

Decanter centrifuges are frequently used in digestate processing because of their capacity of separating also small particles and colloids from the liquid fraction. In such a device the separation is obtained thanks to a centrifugal force that reduces the settling time necessary to obtain a given separation efficiency. Both vertical and horizontal types of decanter centrifuges are available.

Horizontal decanter centrifuges use a closed cylinder with a continuous turning motion. The digestate enters the centrifuge via a central inlet and it is fed into the center of the centrifuge. The centrifugal force separates solids and liquids at the wall into an inner layer, with a high dry matter concentration, and an outer layer, consisting of a liquid containing a suspension of colloids, organic components and salts (Hjorth *et al.*, 2010). The solid and liquid phases are transported to either end of the centrifuge by rotating the entire centrifuge at high speed and by simultaneously rotating the conveyor at a speed that differs slightly from the speed of the bowl (outer conical shell). The solid particles are conveyed towards the conical end and let out through the solid discharge openings, whereas the supernatant flows toward the larger end of the cylinder formed by the bowl and the flights of the conveyor. During the transport of the slurry, the particles are separated from the liquid and the liquid phase is discharged through liquid-discharge openings at the wide end of the decanter centrifuge (Hjorth *et al.*, 2010).

The separation performance depends on particle size and shape, difference in density between particles and fluid, and fluid viscosity. According to Drosg *et al.* (2015), the energy consumption of a decanter centrifuge, in the order of 3–5 kWh/m^3, is rather high compared with other solid–liquid separation technologies.

10.2.4 Belt filters

Belt filter presses and vacuum belt filters are commercially available separation equipment for draining liquids from solid digestate. A belt filter press consists of a closed loop of textile belt wound around cylinders. Digestate is applied continuously at the start of the belt filter. The first pre-dewatering occurs by gravitation, subsequently the material is pressed between two filter belts. Varying mechanical forces are applied so that the filter cake is dewatered further. Finally, the dewatered cake is removed from the filter belt by a scraper. The filter belt is then cleaned by spray-washing and it is used again for filtration; often the filtrate is used for the washing operation (Drosg *et al.*, 2015).

With a belt separator, the filter cake is continuously removed as the belt rotates, so that the raw-slurry loading area and the solid-fraction unloading area change over and are continuously cleaned. In vacuum belt filters, the digestate is applied onto a filter belt and vacuum is applied to its underside, whereby water is sucked through the filter and the filter cake remains on the belt (Drosg *et al.*, 2015).

Digestate processing onto belt filters requires a prior dosage of a combination of chemical products (mostly polyelectrolytes and organic polymers) in order to improve the separation efficiency. The most efficient coagulant and flocculant products used for sludge dewatering are polyaluminum chloride (PAC), polyacrylamide (PAM) and epichlorohydrine-dimethylamine with ethylendiamine (DEED) (Beggio *et al.*, 2022; Bu *et al.*, 2016). Chemical conditioning improves the efficiency of the solid–liquid separation by achieving particles coagulation, that is destabilization/neutralization of particles' charged surface, and further aggregation of destabilized particles, that is patch flocculation and polymers bridging (Beggio *et al.*, 2022).

The separation efficiency of a belt filter is affected by the characteristics of the digestate, amount and type of added precipitating and flocculating agents and mesh size of the filter. A belt filter has a higher separation efficiency compared to a screw press, with an energy demand of 1.5–2 kWh/m^3, which is lower than that of a decanter centrifuge (Drosg *et al.*, 2015).

10.3 TREATMENT PROCESSES FOR THE SOLID FRACTION OF THE DIGESTATE

The solid and liquid fractions of the digestate, after separation, have different destinations and are subjected to different processes for the final recovery of material or heat. The solid fraction, coming from solid–liquid separation processes, has a total solids concentration that ranges from 20 to 30%, up to 35–40% (Drosg *et al.*, 2015). This fraction is already partially stabilized, so that appropriate storage, or direct application on agricultural land as biofertilizer or soil improver are facilitated. Nevertheless, this fraction still contains biodegradable material, consequently, it is subjected to microbial activity and generates odorous emissions. Further processing is necessary, namely composting, drying or other forms of stabilization, if a stable and marketable biofertilized product is desired.

10.3.1 Composting

Composting is a good choice, among different processes developed to post-treat the solid fraction of the digestate, in order to improve its quality in the view to be used as a fertilizer. In fact, composting increases the maturity and stability of the digestate. In the composting process a consortium of microorganisms degrades and transforms the organic matter of the digestate, under aerobic conditions, into mature compost, which is stable and safe and contains humic substances and slow-releasing nutrients.

Compost is an ideal biofertilizer and shows good performances in improving the soil structure. Mixing digestate with fresh organic feedstock can increase the composting rate because of higher nutrient contents (Arab & McCartney, 2017). A special application of composting is vermiculture, using earthworms.

However, composting of the digestate can show some shortcomings:

- The solid fraction of the digestate has a high moisture content, which is a factor that impacts the aerobic microbial activity of the composting. Consequently, the addition of bulking material, such as woodchips, is necessary for a stable process (Drosg *et al.*, 2015).
- In general, composting increases the concentration of nutrients, but may also result in nitrogen loss. In fact, digestate is usually alkaline: under composting conditions part of the NH_4^+/NH_3 can be volatilized in the form of NH_3.

Zeng and coauthors (2016) demonstrated that the use of woodchips as bulking material can help to simultaneously stabilize digestate and reduce nitrogenous emissions. Dry woodchips improved the characteristics of the composting environment, by serving as free-water and nitrogen sinks, helped air to enter the compost heap and maintain the aerobic conditions more easily. Depending on the local availability of bulking material, it may be advantageous to do the composting at a centralized composting facility.

Pecorini and coauthors (2020) investigated the stability, the status of nutrient cycles (N, P, S) and the biological activity of three digestate-based mixtures, containing fresh or mature compost. They demonstrated that the mixtures responded to the main threshold values of biological stability fixed in the EC fertilizer Decree (Fertilizer Regulation 1009/2019) for EC marked fertilizers, with the exception of self-heating factor. The residual heavy metal content did not pose environmental risk and the physical and nutritional properties of the digestate-based mixtures were comparable with those usually found in high-quality composts. Pecorini and coauthors (2020) concluded that mixing digestates with composts helped in mitigating the environmental risk posed by each starting material and allowed to valorize the nutrient content of the digestate.

Recently, Gurmessa *et al.* (2021a) demonstrated that digestate composting or co-composting with biogas feedstocks (such as maize silage, food processing waste or poultry litter) was a promising strategy to improve quality of the digestate for fertilizer use. They observed that the temperature dropped earlier in the digestate and maize silage co-composting pile, reducing the time to maturity by 20 days. Furthermore, composting and co-composting reduced phytotoxicity and C/N ratio, but increased immobilization of Al, Ba, Fe, Zn, and Mn at least by 40% in all piles. The same authors, in another study, found that some selected enzymes could be used as indicators of compost maturity and immobilization of trace elements (Gurmessa *et al.*, 2021b).

10.3.2 Drying

The drying process is carried out in order to reduce the total mass of the digestate, make the product stable and concentrate nutrients. Because of the high energy demand, the drying process often involves only the solid fraction of the digestate. Sometimes, where there is large availability of heat, the whole digestate, that is before the solid–liquid separation, is subjected to the drying process (Drosg *et al.*, 2015). The increase of the digestate total solids from an average content of 30–90% needs the evaporation of 2.2 kg water/kg TS. Since the latent heat of evaporation of water is 2.5 MJ/kg water, the theoretical amount of energy necessary to dry the digestate is in the order of 5.5 MJ/kg TS. However, available drying technologies, that is belt dryer, drum dryer, feed-and-turn dryer and fluidized-bed dryer, generally have higher energy requirements, between 2.9 and 3.6 MJ/kg of evaporated water (Đurđević *et al.*, 2019). For digestate applications, belt dryer is the more commonly applied technology (Drosg *et al.*, 2015).

Heat to be used for the drying process can be recovered from a combined heat and power (CHP) unit, if electrical power is produced at the biogas plant. For this purpose, Monlau *et al.* (2015b) have demonstrated, through an energy balance of a real anaerobic digester, that the heat surplus (in the order of 13 300 kWh_{th}/d) produced from the CHP after the self-consumption of the digester, was sufficient to ensure the complete drying of the solid fraction of the digestate. Where the excess heat from the CHP unit is not sufficient, the support of solar drying seems to be a suitable option. Solar energy can integrate the existing available energy, in the form of waste heat from a cogeneration unit or micro turbine, thus avoiding additional energy costs. The results of a recent study (Maurer & Müller, 2019) showed that the combination of solar energy and waste heat from a CHP unit was a suitable way to reduce the moisture content of the digestate to less than 15% after 13 days of drying. That residual water content was a safe condition for further handling and storage.

According to the German Biogas Association, in 2018 a number of 500–700 dryers were in use in the Germany's biogas industry (Salamat *et al.*, 2020). Thermal drying can meet the requirements for hygienization and produce a transportable and storable fertilizer. Dried digestate can be marketed as it is or can be pelletized for better marketability. Such products are already available as biofertilizers in horticulture or gardening markets. The dry product can also be used for special cultivation systems, such as mushroom production (Drosg *et al.*, 2015). Furthermore, the greenhouse gas emissions from dried digestate are considerably lower than emissions from the moist digestate (Salamat *et al.*, 2020).

Drying of the digestate has some drawbacks, in fact, in addition to the high energy demand, the emissions that are generated during drying result in a decrease in the fertilizing value of the digestate and have some negative effects on the atmosphere environment. As the exhaust gases of digestate dryers contain dust, ammonia, and other volatile substances (e.g., volatile acids), cleaning systems have to be applied in order to reduce emissions. Such systems contain a dust filter as well as washer/scrubber units.

Drying is often a pre-treatment before thermal conversion processes, namely combustion, hydrothermal carbonization, HTC, and pyrolysis, used to produce energy from the digestate and to improve the overall energy efficiency of an AD process.

10.3.3 Thermal conversion processes

10.3.3.1 Combustion

Combustion is a thermochemical process based on the complete oxidation of organic substrates to produce energy in the form of heat. Contrarily to hydrothermal processes, combustion (and also pyrolysis) can be efficiently carried out only on solid fractions of the digestate with a low moisture content. Consequently, highly efficient drying-up pre-treatments are necessary (Fiori *et al.*, 2014). After the drying phase, solid digestate can be converted into pellets for combustion. The digestate pellets have a heating value of the same order of that of wood, that is 16.5–17.3 MJ/kg. Notwithstanding the apparent benefit produced by digestate combustion, only very few case studies are reported in literature. Kratzeisen *et al.* (2010) characterized the heating value, ash properties and emission behavior of two kinds of digestate pellets and proved the feasibility of using digestate in the combustion unit of a conventional solid biomass.

Pedrazzi and coauthors (2015) used the digestate produced in an Italian 1 MW biogas plant as a biofuel in an ordinary domestic air furnace. The digestate was dried out and pelletized before the combustion in a wood pellet furnace with 29 kW_{th} of nominal power. The digestate proved not to be a suitable fuel because of the low temperature melting point of ashes and high emissions (three times higher than those produced by the combustion of a certified 'ENplus A1' wood pellet). Conversely, Gardoni and Guarino (2016) reported the results of a two-year successful test carried out in a demonstrative-scale drying-and-combustion plant that treated the digestate produced in a cattle slaughterhouse. The composition of raw and dried digestate, ash and flue gases was monitored, and energy and chemicals consumption were accounted in order to obtain representative scenarios and to assess the practicability of that in-situ approach. Among the technologies considered, the drying-combustion solution resulted to be of interest under some conditions: (a) the lower heating value (LHV) of the digestate was higher than 17–18 MJ/kg of dry matter, (b) the anaerobic plant had a production of digestate of more than 10 t/d, (c) the cost of conventional disposal routes of the digestate was higher than 50 €/t. Finally, the presented technology appeared to be adequate to respect the regulatory limits regarding emissions.

More recently, Wei and coauthors (2021) tested a co-combustion process with digestate and municipal solid waste. They demonstrated that a co-combustion process could overcome the environmental impacts, caused by the only digestate used as a feedstock, thus significantly reducing the hazardous substances in the emissions. Furthermore, the facility costs for waste-to-energy plants can be saved by retrofitting existing equipment rather than constructing new processing equipment. In fact, among the existing equipment for thermochemical processes, waste-to-energy plants have the widest adaption for fuel, the most advanced pollutants purification systems, and the strictest emission standards (Chen *et al.*, 2020). However, with regard to the high nitrogen content and associated increased NO_x emissions, the suitability of the digestate pellets for thermal recovery is an issue that has still to be better clarified.

10.3.3.2 Hydrothermal carbonization

Hydrothermal carbonization is a thermochemical process carried out at temperatures lower than 250°C, where carbonization reactions convert wet organic feedstocks into carbon-rich products called 'hydrochars'. An alternative to HTC is vapothermal carbonization (VTC), where the organic matter is saturated in vapor instead of being submerged in water as for HTC. VTC allows higher initial solid content (25% vs. 5%) (Monlau *et al.*, 2015a).

The physicochemical properties of hydrochar, similar to those of fossil coal, would allow it to be used as a solid fuel. In their review, Monlau and coauthors (2015a) reported hydrochar yields from 66 to 74 g/100 g dry matter, for processes carried out at temperature values of 180–190°C. The study of Funke and coauthors (2013) highlighted that a rich-lignin biomass is beneficial in hydrochar production, conversely an abundance of carbohydrates determined lower yields. Hydrochars produced at moderate temperatures contain primarily crystalline cellulose and lignin. Higher temperatures (>250°C) cause the degradation of crystalline cellulose to more aliphatic carbon and an increase in the content of lignin (Toufiq-Reza *et al.*, 2015). As a general rule, elemental carbon and oxygen contents are related to the quality of an hydrochar; high carbon to oxygen ratios are desirable because hydrochar can be used as a fuel.

Hydrochar produced at various HTC operating parameters from agricultural digestates had a heating value ranging from 25 to 35 MJ/kg. The production of hydrochar from digestate can improve the overall energy efficiency of an AD process. Toufiq-Reza and coauthors (2014) observed that the combination of the AD and HTC processes yielded 13.2 MJ/kg of the original feedstock, representing at least 20 and 60% more than for HTC and AD alone, respectively.

10.3.3.3 Pyrolysis

Thermal valorization of the solid fraction of the digestate can be obtained by using pyrolysis (Monlau *et al.*, 2015a). Pyrolysis process allows to recover energy from organics, regardless of the biodegradable or non-biodegradable nature of the substrate. Pyrolysis is an endothermic thermo-chemical process that occurs in an inert, oxygen-free atmosphere, where the organic matter within the feedstock is converted into a solid phase, namely biochar, and vapor. After vapor cooling, polar and high-molecular-weight compounds condense out as a liquid, namely bio-oil, while low-molecular-weight volatile compounds remain as gas phase, namely syngas (Giwa *et al.*, 2019; Monlau *et al.*, 2015a).

Until now, most attention has been drawn on the solid product of pyrolysis, namely biochar, which shows multiple applications such as soil amendment, carbon sequestration, and adsorbent precursor. However, accumulation of ash in the digestate can worsen the quality of the biochar product. The production of bio-oil opens a new horizon toward digestate utilization, because the processes not only minimize the limitation caused by ash, but also prevent ash appearing in the desired product (Wei *et al.*, 2018). Bio-oil generally contains an aqueous phase and an organic phase, the latter can be of interest for energy production or engine applications. The organic phase is generally composed of a broad range of compounds, including mainly sugars, acids, ketones, phenols and

Table 10.1 Pyrolysis methods and conditions influencing the product yields.

Pyrolysis process	Process conditions			Product yield		
	Temperature (°C)	Heating rate	Retention time	Char (%)	Bio-oil (%)	Gas (%)
Slow pyrolysis	400–600	5–7°C/min	5–30 min	<35	<30	<40
Fast pyrolysis	400–600	300°C/min	<5 s	<25	<75	<20
Flash pyrolysis	400–950	1000°C/s	30 ms–1.5 s	<25	<70	<16

Source: Giwa *et al.* (2019).

furans. Specifically, phenolic compounds come from the three phenylpropenyl units, namely coumaryl alcohol, coniferyl alcohol and sinapyl alcohol, of which lignin is found in the solid digestates (Wei *et al.*, 2018) Bio-oil obtained from AD residues contained more phenol compounds (i.e., mainly 4-vinyl phenol) than the bio-oil obtained by the original feedstock, due to the higher content of lignin in digestate (Ghysels *et al.*, 2020; Liang *et al.*, 2015).

Syngas is essentially a mixture of H_2 and CO, but also typically contains CH_4, CO_2, H_2O, and several low-molecular-weight volatile organic compounds. The distribution of products depends on many factors such as heating rate, temperature, operating pressure, residence times of the vapors, and the converting biomass/residues and their states of mixing (Giwa *et al.*, 2019). Pyrolysis processes can be classified into 'slow', 'fast' or 'flash', depending on the operating conditions. As shown in Table 10.1, slow pyrolysis occurs at low heating rates (<10°C/min) and residence times of several minutes, whereas fast pyrolysis occurs at high heating rates (≈300°C/min) during seconds. Desired products obtained by slow pyrolysis are generally char (≈35%) and syngas (≈35%), whereas bio-oil is mainly produced through fast or flash pyrolysis (≈70%) (Giwa *et al.*, 2019).

Pyrolysis processes have been recently applied on solid digestate for energy production and to improve the energy balances of AD plants. It was found that a pyrolysis process could transfer over 91% of the digestate energy into solid, liquid or gaseous products (Li *et al.*, 2017). The amount of potentially valuable products as energy sources (i.e., syngas, bio-oil) ranged from 54.5 to 82 g 100 g dried matter (Monlau *et al.*, 2015a).

Conversion of bio-oil and syngas through recirculation into an AD process for biogas production has been proposed in the view of obtaining an integrated process. Pyrolysis syngas contains hydrogen as one of the major components that can biologically be converted into methane via enhanced hydrogenotrophic methanogenesis. Syngas has also been investigated for production of biofuels (i.e., bioethanol, butanol, biohydrogen) and chemicals through syngas fermenting microorganisms (Giwa *et al.*, 2019; Monlau *et al.*, 2015a).

Some authors investigated the conversion of the aqueous phase of pyrolysis bio-oil to methane in AD process. That liquid phase was rich in both high water-soluble and low water-soluble organic substances. The first contains rapidly degradable COD that can be used in an AD process, the latter are made of recalcitrant compounds, like phenols and furans. Hubner and Mumme (2015) observed a methane production of 200 mL/gCOD from the aqueous phase of a pyrolysis bio-oil obtained with processes carried out at 330 and 430°C. They found that pyrolysis processes carried out at higher temperature values (530°C) produced an aqueous phase richer in recalcitrant compounds that inhibited the generation of methane. A similar inhibition phenomenon was observed by Torri and Fabbri (2014). They suggested that the inhibition could be overcome by adsorbing the problematic substances with pyrochar.

Preliminary energy balances using anaerobic digestate as potential feedstock for energy recovery through pyrolysis have been drawn. Li and coauthors (2014), on the basis of lab-scale data, concluded that the pyrolysis of a digestate coming from the digestion of chicken manure and corn stover provided an additional energy yield of 6.1 MJ/kg VS added compared with the AD process alone. Nonetheless, in that study only syngas was valorized as energy without considering the bio-oil upgrade. Recently, González-Arias and coauthors (2020) combined the co-digestion of sewage sludge and grease from veal meat with the pyrolysis of the digestate. They demonstrated that the grease content in the substrate mixture was the key parameter to guarantee the energy self-sustainability of the drying unit preceding the pyrolysis reactor and to increase the electricity production.

In their energy balance, Monlau *et al.* (2015b) showed that coupling AD with a pyrolysis process at 500°C increased the electricity yield by 59% compared with the AD process alone. In the energy balance the authors distinguished between electrical and heat inputs/outputs; in fact, electrical energy could be dispatched to the public grid, providing extra income for the farmer, whereas thermal energy, not used to support AD and drying, was lost.

A recent study evaluated the environmental sustainability of a multifunctional process that combined anaerobic co-digestion and pyrolysis for the valorization of sewage sludge and quinoa residues. The results demonstrated that the main contributor to environmental impacts was the treatment of the bio-oil, while other impacts could be counterbalanced by the energy production (Caiardi *et al.*, 2021).

10.4 TREATMENT PROCESSES FOR NUTRIENT RECOVERY FROM THE LIQUID FRACTION OF THE DIGESTATE

The principal technologies, available or under development, for the treatment of the liquid fraction of the digestate in the view of nutrient recovery are evaporation, gas stripping and absorption, chemical crystallization and membrane separation.

10.4.1 Evaporation

Vacuum evaporation can be used for the treatment of the liquid fraction or whole digestate. However, typically, the solid fraction of the digestate is removed before evaporation, because fibers in the digestate can cause clogging of the evaporators. The solid fraction is removed by using one or a combination of solid–liquid separation processes, for example, a screw press and a vibrating screen.

Evaporation is carried out with the aim of concentrating nitrogen for a subsequent recovery, for that reason it is necessary that all nitrogen is in the form of NH_4-N. Consequently, the pH is lowered to 4.5 with the addition of H_2SO_4 and CO_2 is stripped out. After acidification, digestate is concentrated via a three-step low-pressure evaporation system and the generated vapor is condensed (Vondra *et al.*, 2018). As low pressure is applied, excess low-grade heat at 90°C can be used for evaporation. The residual concentrations of ammonia nitrogen NH_4-N (30–50 mg/L) and volatile acids (<1000 mg/L as COD) in the condensed vapor do not allow its discharge in receiving water bodies, unless it is treated with advanced processes, namely reverse osmosis or ion exchange. Condensed vapor can be used as a process water in the digestion plant or discharged to a wastewater treatment plant (Vondra *et al.*, 2019).

For digestate evaporation either forced circulation evaporators or natural recirculation evaporators are used (Drosg *et al.*, 2015). In these processes, the digestate is heated above the water evaporation temperature in a heat exchanger and then held in the evaporation vessel. In forced circulation evaporators, the circulation of the digestate is obtained by driving a pump, conversely, in natural circulation evaporators, circulation occurs spontaneously thanks to the rise of the digestate mixture's vapor into the evaporation vessel.

Evaporators decrease the amount of water in the liquid digestate, thus reducing the initial digestate volume by 50–85% depending on the used heat source (Heviánková *et al.*, 2014). Both forced and natural recirculating evaporators are of simple construction and robust operation. However, vacuum evaporators are suitable only for those biogas plants which produce excess of heat energy or which have waste heat available from other processes. In fact, the process requires 300–350 kWh_{th} per ton of evaporated water (Drosg *et al.*, 2015). Furthermore, it must be taken into account that there may be a shortage of the available heat especially during the winter season. Other than the availability of heat, other decisive economic factors for the integration of an evaporator in the line of treatment of the digestate are transport distances, incentives for customers purchasing the digestate (Auburger *et al.*, 2015), costs related to purchase of necessary chemical components (Chiumenti *et al.*, 2013a) or electric power demands (Vondra *et al.*, 2018). On the basis of the results of a mathematical model applied to vacuum evaporation, Vondra and coauthors (2018) concluded that the transportation distance had a big impact on the profitability of the investment and that, for a feed-in tariff of 0.12 €/kWh, a minimum of 20 km in the transport distance would be required to make the economic balance positive. Under the examined operational conditions, the maximum reasonable project price could vary between 30 and 240 k€. The results of the study also showed that incentives for electricity production discouraged plant owners from adopting energy efficient and environmentally friendly solutions for digestate management.

10.4.2 Ammonia stripping

The combination of the processes of ammonia stripping and absorption are used to extract nitrogen from digestate and recover it in the form of a concentrate solution of ammonia salts, usually ammonium sulfate ($(NH_4)_2SO_4$, AmS) or ammonium nitrate (NH_4NO_3). Ammonia salts end-products can be used as marketable fertilizers for land application. The implementation of the stripping-absorption technology for the treatment of nitrogen-rich manures and digestate is on the rise (Baldi *et al.*, 2018; Vaneeckhaute *et al.*, 2017).

Ammonia stripping is a process where nitrogen, in the form of ammonia (NH_3), is removed from a liquid (waste stream) by a physical transfer to a gas flow. The ammonia stripping process includes the four operations listed as follows (Limoli *et al.*, 2016):

(i) conversion of ammonium ions (NH_4^+) to ammonia gas (NH_3) (ammonia dissociation equilibrium),
(ii) diffusion of NH_3 to the air–water interface (water-side mass transfer),
(iii) release of NH_3 to the air at the interface (volatilization), and
(iv) diffusion of NH_3 from the air–water interface into the air above (air-side mass transfer).

The volatility of ammonia in an aqueous solution can be enhanced by increasing temperature and pH. Specifically, in digestate processing, the excess heat from the AD process can be used for heating up the digestate and the pH can be risen by degassing to remove carbon dioxide, CO_2, or by the addition of alkali. The gas is subsequently transferred to an air scrubber, where NH_3 is absorbed to an acidic solution, often sulfuric acid, in order to obtain and recover a concentrated solution of ammonium sulfate $(NH_4)_2SO_4$, AmS. AmS is an inorganic salt that has an undeniable commercial value and can be used as a fertilizer for agricultural uses. AmS provides a valuable substitute for chemical fertilizers based on fossil resources, being rich in direct available macronutrients, N and S, and having low organic contamination. Depending on local legislation or options for land application, ammonia salts different from AmS, such as ammonium nitrate from nitric acid, can be produced. Stripping and subsequent absorption of ammonia in an acidic solution are carried out in packed-bed towers and the clean gas can be reused for further absorption cycles.

According to Vaneeckhaute and coauthors (2017), the commercially available stripping technologies for raw or digested sludge and manure treatment are (1) AMFER, Colsen (NL), (2) ANAStrip, GNS (DE), and the (untitled) stripping processes developed by the manufacturers: (3) Anaergia (Canada, CA), (4) Branch Environmental Corp (USA), (5) Europe Environnement (France, FR), and (6) RVT Process Equipment (DE). Theoretically, these systems may achieve NH_3-removal efficiencies up to 98%, but they are generally operated to reach 80–90% removal in order to reduce the operating costs. At present, most stripping processes implemented at full-scale aims at the production of AmS-fertilizer. The AmS-content in the recovered solution ranges from approximately 25% AmS (ANAStrip, GNS) and 30% AmS (Branch Environmental Corp) to 38% AmS (Anaergia; RVT Process Equipment) and 40% AmS (AMFER, Colsen; Europe Environnement) (Vaneeckhaute *et al.*, 2017).

In the stripping and absorption reactors the extent of contact between the liquid and the gas phase is maximized by using columns filled with packing materials, that can offer a high surface area for the ammonia mass transfer.

However, scaling and fouling phenomena occurring in the packing material are to date recognized as the major technical bottlenecks observed in NH_3 stripping and absorption (Baldi *et al.*, 2018). Scaling can be avoided through the introduction of a lime softening step before stripping, which removes a large part of calcium, magnesium, carbonic acids and carbonates, and increases the pH. In case of high buffering capacity, a preceding CO_2-stripper might be envisaged. Fouling can be limited by using high-efficiency systems for solid–liquid separation so as to retain many suspended solids as possible in the solid fraction. Nonetheless, high maintenance and cleaning efforts for the regeneration of the packing material are unavoidable. Because of these constraints, some of the above manufacturers have developed a stripping process without internal packing (Anaergia; Colsen). As such, the process developed by Anaergia would be capable of handling waste flows containing up to 8–9% total suspended solids (Vaneeckhaute *et al.*, 2017).

Capital costs of stripping are relatively low compared to biological activated sludge systems and depend on the method used for pH increment. That can occur (1) chemically, by use of an alkali agent, mostly sodium hydroxide (NaOH) (Branch Environmental Corp; Europe Environnement; RVT Process Equipment), or (2) physically, by simultaneous stripping of CO_2 (Anaergia; Colsen), optionally in combination with the addition of low-quality gypsum ($CaSO_4$) for parallel recovery of $CaCO_3$ (GNS) (Vaneeckhaute *et al.*, 2017).

A comparison of budget proposals provided by the above suppliers for NH^3-stripping and absorption systems treating a digestate flow of 800 m^3/d at 2400 mg NH_4^+-N/L (90% recovery) resulted in a capital cost ranging from 500 k€ to 1.58 M€, if the pH-increase is conducted chemically, and from 3.5 M€ to 11–15 M€, if the pH-increase is obtained physically. Electricity consumption for this case was estimated by the suppliers at 127–400 kW_{el}, heat consumption at 2115–2333 kW_{th}, and H_2SO_4-consumption (concentrated at 95–97%) at 5.5–6.8 ton/d or 7.0–10 kg/m^3 digestate. If NaOH is used for pH-increase, its consumption would amount to 6.0–6.5 kg/m^3. Consequently, operational costs range from 1.4 to 2.5 M€/year depending on the system, equivalent to 4.5–8.6 €/m^3 of the digestate (Vaneeckhaute *et al.*, 2017).

10.4.3 Struvite recovery

Nutrient recovery through phosphorus precipitation and crystallization is a mature technology, known as struvite ($MgNH_4PO_4 \cdot 6H_2O$) or MAP (magnesium ammonia phosphate) precipitation. In this process magnesium is added in excess to a solution containing soluble orthophosphate (P-PO_4) and ammonia, so that nutrient concentrations are approximately in a ratio of 1.3:1.0:0.9 for Mg:N:P. Caustic soda (NaOH) is used to increase the pH to values of 8.3–9.5 and induce the precipitation of struvite (Le Corre *et al.*, 2009). The chemicals, that is magnesium in the form of $MgO/MgCl_2$ and NaOH, can be added either in a first step, with a subsequent separation by centrifugation, or in the same vessel where also the sedimentation of struvite crystals takes place. The design involves fluidized-bed reactors and continuously stirred tank reactors (Drosg *et al.*, 2015).

The process has been implemented at full-scale installations for the liquid phase of digestate and manure treatment, as well as at pilot-scale for the treatment of raw digestate. According to Vaneeckhaute and coauthors (2017), to date the most established processes commercially available are: (1) AIRPREX, Berliner Wasserbetriebe (Germany, DE), (2) ANPHOS, Colsen (the Netherlands, NL), (3) CAFR, NALVA (DE), (4) Ceres, Ceres Milieutechniek (Belgium, BE), (5) NuReSys, Akwadok (BE), (6) Nutritec, Sustec (NL), (7) Pearl, Ostara (Canada, CA), (8) Phosnix, Unitika (Japan, JP), (9) PHOSPAQ, Paques (NL), and (10) PRISA, Aachen University (DE). Moreover, in Gelderland (NL) four installations are available for the production of K-struvite ($KMgPO_4{\cdot}6H_2O$) from calf manure. These processes have the capacity to remove and recover over 80–90% of the soluble phosphorous in the waste(water) flow, but only 10–40% of the NH_4-N. Crystal/pellet sizes range from 0.5 to 5 mm and above, depending on the final end-use (Vaneeckhaute *et al.*, 2017).

The main disadvantage of struvite precipitation is the large amount of chemicals required, and this translates into high operational costs. Struvite recovery proved to be economical only on streams with a P-load of more than 20% by weight. Operational costs and payback times are highly dependent on the input composition (e.g., available P, Mg, and pH), as it determines the chemical (NaOH, Mg) and energy costs. Operating costs were estimated at 0.19–0.28 €/m^3 digestate (Battistoni *et al.*, 2005a, 2005b). Dockhorn (2009) reported a unit cost of struvite production of 520 €/ton for an initial PO_4-P concentration of 800 mg/L. Important technical challenges in the process of struvite precipitation remain the reduction of chemical requirements, the control of the production process and the guarantee of a pure product (Drosg *et al.*, 2015).

10.4.4 Membrane technologies

Pressure-driven membranes perform a physical separation on the liquid fraction of the digestate, thus generating a nutrient concentrate and a purified process water. Conventional solid–liquid separation methods, such as gravity sedimentation, centrifugation, plate filtration and screw extrusion, can only remove coarse particles, and the liquid fraction still contains large amount of colloids (size of 0.001–1.0 μm) and fine particles (Fuchs & Drosg, 2013). Depending on the pore sizes, membranes can be categorized into microfiltration (MF), ultrafiltration (UF), nanofiltration (NF) and reverse osmosis (RO).

MF membranes can retain particles with diameters of more than 0.1 μm, UF membranes are able to separate colloids even down to diameters of 0.01 μm. NF and RO can separate dissolved salts from pure water (Drosg *et al.*, 2015).

Membrane filtration is a complex process that consists of several steps. First, a decanter centrifuge is used for preliminary solid–liquid separation. Subsequently, the liquid fraction of the digestate is subjected to enhanced solids removal with the aid of coagulation and flocculating agents. This is a crucial stage in membrane purification processes in order to limit membrane fouling. The next step is ultrafiltration and finally reverse osmosis is used for removal of ammonia and COD (Drosg *et al.*, 2015).

Even membrane filtration is an established technology in wastewater treatment, especially for wastewater reuse purposes, it has not yet proven to be a valuable option for the treatment of the digestate. Only a few commercial pilots have been installed at full-scale manure and digestate processing facilities, mostly on a short-term basis because of excessive operational costs (Vaneeckhaute *et al.*, 2017). Nevertheless, the produced membrane filtration concentrates are an interesting source of nutrients, which could potentially be reused as chemical fertilizer substitutes.

Only few authors published the results of pilot or full-scale trials (Gienau *et al.*, 2018). Chiumenti *et al.* (2013b) reported nutrient balances of a digestate treatment chain at an agricultural biogas plant fed with swine and poultry manure and corn silage, Ledda *et al.* (2013) reported the results obtained from the separation of a digestate originated from cattle and swine manure. Apart from the feedstock characteristics, the experimental set-ups consisted of different membranes used for the ultrafiltration step. Chiumenti *et al.* (2013b) used ceramic modules with 0.1 μm pore size and Ledda *et al.* (2013) used 40 kDa grafted polyacrylonitrile membranes. Although the set-up was quite similar, the results of the mass and nutrient balances varied considerably. Ledda *et al.* (2013) separated the digestates into 38–51% water, 12–32% solid, % fertilizer, and 30–37% liquid concentrate, while Chiumenti *et al.* (2013b) reported 48% water, 35% solid % fertilizer, and 17% liquid concentrate.

More recently, Gienau and coauthors (2018) analyzed the performances of a membrane pilot plant that was operated at a 2.5 Mwe agricultural biogas plant over a period of 7 months. It consisted of a screw press separator, a decanter centrifuge, an ultrafiltration unit ($\alpha Al_2O_3/TiO_2$, pore size = 50 nm) and a three-stage reverse osmosis unit. Solid/liquid separation resulted in separation efficiencies of 70% for total solids and 80% for phosphorus. The solid fraction contained high concentrations of organics and particle-ligated nutrients (20% TS, 8 kg/t Ntotal, 5.5 kg/t P_2O_5). The retentate of the reverse osmosis had high concentrations of dissolved ammonia and potassium (4 kg/t NH_4–N and 10 kg/t K_2O). An amount equal to 38% of the sludge volume was recovered as clean water. The study of Gienau and coauthors (2018) demonstrated that the membrane pilot plant successfully produced a solid N/P-fertilizer, a liquid N/K-fertilizer and clean water. However, the biggest technical problem found in membrane filtration was clogging and fouling of the membrane, resulting in significant chemical and energy requirements. The equivalent energy requirement for an RO-system was estimated at 4–6 kWh/m^3 digestate (Drosg *et al.*, 2015) and the treatment costs at approximately 7 €/m^3 digestate (Bolzonella *et al.*, 2018).

10.5 CONCLUSIONS

The solid–liquid separation for digestate can be obtained in settling tanks or by using centrifuges. The solid fraction can be mechanically separated by forced filtration, using screw presses, or by drainage, through fabric belts or screens, eventually with the aid of flocculation or precipitation agents. The solid and liquid fractions of the digestate, after separation, have different destinations. The solid fraction has a total solids concentration that ranges from 20 to 30%, up to 35–40%. This fraction is already partially stabilized, so that appropriate storage, or direct application on agricultural land as biofertilizer or soil improver, are facilitated. Nevertheless, the solid fraction still contains biodegradable material, consequently further processing is necessary, namely composting, drying or other forms of stabilization, if a stable and marketable biofertilized product is desired. Drying is often used as a pre-treatment before thermal conversion processes, namely combustion, hydrothermal carbonization and pyrolysis, used to produce energy from the digestate so as to improve the overall energy balance of an AD process. The principal technologies, available or under development, for the treatment of the liquid fraction of the digestate in view of nutrient recovery are evaporation, gas stripping and absorption, chemical crystallization and membrane separation.

REFERENCES

Abubaker J., Risberg K. and Pell M. (2012). Biogas residues as fertilisers – effects on wheat growth and soil microbial activities. *Applied Energy*, **99**, 126–134, https://doi.org/10.1016/j.apenergy.2012.04.050

Arab G. and McCartney D. (2017). Benefits to decomposition rates when using digestate as compost co-feedstock: part i – focus on physicochemical parameters. *Waste Management*, **68**, 74–84, https://doi.org/10.1016/j.wasman.2017.07.018

Auburger S., Wustholz R., Petig E. and Bahrs E. (2015). Biogas digestate and its economic impact on farms and biogas plants according to the upper limit for nitrogen spreading – the case of nutrient-burdened areas in north-west Germany. *AIMS Energy*, **3**, 740–759, https://doi.org/10.3934/energy.2015.4.740

Baldi M., Collivignarelli M. C., Abbà A. and Benigna I. (2018). The valorization of ammonia in manure digestate by means of alternative stripping reactors. *Sustainability*, **10**, 3073, https://doi.org/10.3390/su10093073

Battistoni P., Bocadoro R., Fatone F. and Pavan P. (2005a). Auto-nucleation and crystal growth of struvite in a demonstrative fluidized bed reactor (FBR). *Environmental Technology*, **26**, 975–982, https://doi.org/10.1080/09593332608618486

Battistoni P., Paci B., Fatone F. and Pavan P. (2005b). Phosphorus removal from supernatants at low concentration using packed and fluidised bed reactors. *Industrial & Engineering Chemistry Research*, **44**, 6701–6707, https://doi.org/10.1021/ie050186g

Beggio G., Peng W., Lü F., Cerasaro A., Bonato T. and Pivato A. (2022). Chemically enhanced solid–liquid separation of digestate: suspended solids removal and effects on environmental quality of separated fractions. *Waste Biomass Valororization*, **13**, 1029–1041, https://doi.org/10.1007/s12649-021-01591-y

Bolzonella D., Fatone F., Gottardo M. and Frison N. (2018). Nutrients recovery from anaerobic digestate of agro-waste: techno-economic assessment of full scale applications. *Journal of Environmental Management*, **216**, 111–119, https://doi.org/10.1016/j.jenvman.2017.08.026

Bu F., Gao B., Li R., Sun S. and Yue Q. (2016). Impacts of epichlorohydrin–dimethylamine on coagulation performance and membrane fouling in coagulation/ultrafiltration combined process with different Al-based coagulants. *Chemosphere*, **159**, 228–234, https://doi.org/10.1016/j.chemosphere.2016.06.001

Caiardi F., Belaud J. P., Vialle C., Monlau F., Tayibi S., Barakat A., Oukarroum A., Zeroual Y. and Sablayrolles C. (2021). Environmental assessment of a multifunctional process coupling anaerobic digestion and pyrolysis. *Chemical Engineering Transactions*, **86**, 709–714.

Campo G., Cerutti A., Lastella C., Leo A., Panepinto D., Zanetti M. and Ruffino B. (2021). Production and destination of sewage sludge in the Piemonte region (Italy): the results of a survey for a future sustainable management. *International Journal of Environmental Research and Public Health*, **18**, 3556, https://doi.org/10.3390/ijerph18073556

Chen L., Liao Y., Xia Y. and Ma X. (2020). Combustion characteristics of co-combusted municipal solid wastes and sewage sludge. *Energy Sources, Part A: Recovery, Utilization and Environmental Effects*, 1–13, https://doi.org/10.1080/15567036.2020.1739175

Chiumenti A., da Borso F., Chiumenti R., Teri F. and Segantin P. (2013a). Treatment of digestate from a codigestion biogas plant by means of vacuum evaporation: tests for process optimization and environmental sustainability. *Waste Management*, **33**, 1339–1344, https://doi.org/10.1016/j.wasman.2013.02.023

Chiumenti A., Da Borso F., Teri F., Chiumente R. and Piaia B. (2013b). Fullscale membrane filtration system for the treatment of digestate from a co-digestion plant. *Applied Engineering in Agriculture*, **29**, 985–990, https://doi.org/10.13031/aea.29.10117

Dockhorn T. (2009). About the economy of phosphorus recovery. In: International Conference on Nutrient Recovery From Wastewater Streams, D. Mavinic, F. Koch and, S. Ashley (eds), IWA Publishing, London, pp. 145–158.

Drosg B., Fuchs W., Al Seadi T., Madsen M. and Linke B. (2015). Nutrient recovery by biogas digestate processing. In: IEA Bioenergy. http://www.iea-biogas.net/files / daten -redak tion/downl oad/Techn icalB rochu res/NUTRI ENT_RECOV ERY_RZ_web1.pdf (accessed October 30, 2021)

Đurđević Đ., Blecich P. and Jurić Ž. (2019). Energy recovery from sewage sludge: the case study of Croatia. *Energies*, **12**, 1927, https://doi.org/10.3390/en12101927

DüV. (2017). Fertiliser Ordinance: (in German) Verordnung über die Anwendung von Düngemitteln, Bodenhilfsstoffen, Kultursubstraten und Pflanzenhilfsmitteln nach den Grundsätzen der guten fachlichen Praxis beim Düngen (Düngeverordnung–DüV).

FAO. (2005). Sustainable Agriculture and Rural Development (SARD) and Good Agricultural Practices (GAPs). (Rome, 13-16/04/2005). https://www.fao.org/3/J4236e/j4236e00.htm

Fiori L., Basso D., Castello D. and Baratieri M. (2014). Hydrothermal carbonization of biomass: design of a batch reactor and preliminary experimental results. *Chemical Engineering Transactions*, **37**, 55–60.

Fuchs W. and Drosg B. (2013). Assessment of the state of the art of technologies for the processing of digestate residue from anaerobic digesters. *Water Science and Technology*, **67**(9), 1984–1993, https://doi.org/10.2166/wst.2013.075

Funke A., Mumme J., Koon M. and Diakité M. (2013). Cascaded production of biogas and hydrochar from wheat straw: energetic potential and recovery of carbon and plant nutrients. *Biomass & Bioenergy*, **58**, 229–237, https://doi.org/10.1016/j.biombioe.2013.08.018

Gardoni D. and Guarino M. (2016). Drying and combustion of an anaerobic digestate: results and economical evaluation of a demonstrative-scale plant. *International Journal of Engineering Research & Science (IJOER)*, **2**(8), 1–9.

Ghysels S., Acosta N., Estrada A., Pala M., De Vrieze J., Ronsse F. and Rabaey K. (2020). Integrating anaerobic digestion and slow pyrolysis improves the product portfolio of a cocoa waste biorefinery. *Sustainable Energy & Fuels*, **4**, 3712–3725, https://doi.org/10.1039/D0SE00689K.

Gienau T., Brüß U., Kraume M. and Rosenberger S. (2018). Nutrient recovery from anaerobic sludge by membrane filtration: pilot tests at a 2.5 MWe biogas plant. *International Journal of Recycling of Organic Waste in Agriculture*, **7**, 325–334, https://doi.org/10.1007/s40093-018-0218-6

Giwa S. A., Xu H., Chang F., Zhang X., Ali N., Yuan J. and Wang K. (2019). Pyrolysis coupled anaerobic digestion process for food waste and recalcitrant residues: fundamentals, challenges, and considerations. *Energy Science & Engineering*, **7**, 2250–2264, https://doi.org/10.1002/ese3.503

González-Arias J., Gil M. V., Fernández R. A., Martínez E. J., Fernández C., Papaharalabos G. and Gómez X. (2020). Integrating anaerobic digestion and pyrolysis for treating digestates derived from sewage sludge and fat wastes. *Environmental Science and Pollution Research International*, **27**(26), 32603–32614, https://doi.org/10.1007/s11356-020-09461-1

Gurmessa B., Cocco S., Ashworth A. J., Foppa Pedretti E., Ilari A., Cardelli V., Fornasier F., Ruello M. L. and Corti G. (2021a). Post-digestate composting benefits and the role of enzyme activity to predict trace element immobilization and compost maturity. *Bioresource Technology*, **338**, 125550, https://doi.org/10.1016/j.biortech.2021.125550

Gurmessa B., Milanovic V., Foppa Pedretti E., Corti G., Ashworth A. J., Aquilanti L., Ferrocino I., Corvaglia M. R. and Cocco S. (2021b). Post-digestate composting shifts microbial composition and degrades antimicrobial resistance genes. *Bioresource Technology*, **340**, 125662, https://doi.org/10.1016/j.biortech.2021.125662

Heviánková S., Kyncl M. and Kodymová J. (2014). Study and research on cleaning procedures of anaerobic digestion products. *GeoScience Engineering*, **60**, 39–50, https://doi.org/10.1515/gse-2015-0005

Hjorth M., Christensen K. V., Christensen M. L. and Sommer S. G. (2010). Solid-liquid separation of animal slurry in theory and practice. A review. *Agronomy for Sustainable Development*, **30**, 153–180, https://doi.org/10.1051/agro/2009010

Hubner T. and Mumme J. (2015). Integration of pyrolysis and anaerobic digestion – use of aqueous liquor from digestate pyrolysis for biogas production. *Bioresource Technology*, **183**, 86–92, https://doi.org/10.1016/j.biortech.2015.02.037

Kratzeisen M., Starcevic N., Martinov M., Maurer C. and Mãller J. (2010). Applicability of biogas digestate as solid fuel. *Fuel*, **89**(9), 2544–2548, https://doi.org/10.1016/j.fuel.2010.02.008

Le Corre K. S., Valsami-Jones E., Hobbs P. and Parsons S. A. (2009). Phosphorus recovery from wastewater by struvite crystallization: a review. *Critical Reviews in Environmental Science and Technology*, **39**(6), 433–477, https://doi.org/10.1080/10643380701640573

Ledda C., Schievano A., Salati S. and Adani F. (2013). Nitrogen and water recovery from animal slurries by a new integrated ultrafiltration reverse osmosis and cold stripping process: a case study. *Water Research*, **47**, 6157–6166, https://doi.org/10.1016/j.watres.2013.07.037

Li Y., Zhang R., He Y., Zhang C., Liu X., Chen C. and Liu G. (2014). Anaerobic co-digestion of chicken manure and corn stover in batch and continuously stirred tank reactor (CSTR). *Bioresource Technology*, **156**, 342–347, https://doi.org/10.1016/j.biortech.2014.01.054

Li Y., Su D., Luo S., Jiang H., Qian M., Zhou H., Street J., Luo Y. and Xu Q. (2017). Pyrolysis gas as a carbon source for biogas production via anaerobic digestion. *RSC Advances*, **7**, 41889–41895, https://doi.org/10.1039/C7RA08559A

Liang J., Lin Y., Wu S., Liu C., Lei M. and Zeng C. (2015). Enhancing the quality of bio-oil and selectivity of phenols compounds from pyrolysis of anaerobic digested rice straw. *Bioresource Technology*, **181**, 220–223, https://doi.org/10.1016/j.biortech.2015.01.056

Limoli A., Langone M. and Andreottola G. (2016). Ammonia removal from raw manure digestate by means of a turbulent mixing stripping process. *Journal of Environmental Management*, **176**, 1–10, https://doi.org/10.1016/j.jenvman.2016.03.007

Logan M. and Visvanathan C. (2019). Management strategies for anaerobic digestate of organic fraction of municipal solid waste. Current status and future prospects. *Waste Management & Research*, **37**(1), 27–39, https://doi.org/10.1177/0734242X18816793

Maurer C. and Müller J. (2019). Drying characteristics of biogas digestate in a hybrid waste-heat/solar dryer. *Energies*, **12**, 1294, https://doi.org/10.3390/en12071294

Möller K. and Müller T. (2012). Effects of anaerobic digestion on digestate nutrient availability and crop growth: a review. *Engineering in Life Sciences*, **12**, 242–257, https://doi.org/10.1002/elsc.201100085

Monfet E., Aubry G. and Avalos Ramirez A. (2018). Nutrient removal and recovery from digestate: a review of the technology. *Biofuels*, **9**(2), 247–262, https://doi.org/10.1080/17597269.2017.1336348

Monlau F., Sambusiti C., Ficara E., Aboulkas A., Barakat A. and Carrère H. (2015a). New opportunities for agricultural digestate valorization: current situation and perspectives. *Energy and Environmental Sciences*, **8**, 2600–2621, https://doi.org/10.1039/c5ee01633a

Monlau F., Sambusiti C., Antoniou N., Barakat A. and Zabaniotou A. (2015b). A new concept for enhancing energy recovery from agricultural residues by coupling anaerobic digestion and pyrolysis process. *Applied Energy*, **148**, 32–38, https://doi.org/10.1016/j.apenergy.2015.03.024

Pecorini I., Peruzzi E., Albini E., Doni S., Macci C., Masciandaro G. and Iannelli R. (2020). Evaluation of MSW compost and digestate mixtures for a circular economy application. *Sustainability*, **12**, 3042, https://doi.org/10.3390/su12073042

Pedrazzi S., Allesina G., Belló T., Rinaldini C. A. and Tartarini P. (2015). Digestate as bio-fuel in domestic furnaces. *Fuel Processing Technology*, **130**, 172–178, https://doi.org/10.1016/j.fuproc.2014.10.006

Riva C., Orzi V., Carozzi M., Acutis M., Boccasile G., Lonati S., Tambone F., D'Imporzano G. and Adani F. (2016). Short-term experiments in using digestate products as substitutes for mineral (N) fertilizer: agronomic performance, odours, and ammonia emission impacts. *Science of the Total Environment*, **547**, 206–214, https://doi.org/10.1016/j.scitotenv.2015.12.156

Reza M. T., Werner M., Pohl M. and Mumme J. (2014). Evaluation of integrated anaerobic digestion and hydrothermal carbonization for bioenergy production. *Journal of Visualized Experiments*, **88**, 1–9, https://doi.org/10.3791/51734

Reza M. T., Mumme J. and Ebert A. (2015). Characterization of hydrochar obtained from hydrothermal carbonization of wheat straw digestate. *Biomass Conversion and Biorefinery*, **1**, 1–11.

Salamat R., Scaar H., Weigler F., Berg W. and Mellmann J. (2020). Drying of biogas digestate: a review with a focus on available drying techniques, drying kinetics, and gaseous emission behavior. *Drying Technology*, **40**(1), 5–29, https://doi.org/10.1080/07373937.2020.1781879

Tigini V., Franchino M., Bona F. and Varese G. C. (2016). Is digestate safe? A study on its ecotoxicity and environmental risk on a pig manure. *Science of the Total Environment*, **551–552**, 127–132, https://doi.org/10.1016/j.scitotenv.2016.02.004

Torri C. and Fabbri D. (2014). Biochar enables anaerobic digestion of aqueous phase from intermediate pyrolysis of biomass. *Bioresource Technology*, **172**, 335–341, https://doi.org/10.1016/j.biortech.2014.09.021

Vaneeckhaute C., Lebuf V., Michels E., Belia E., Vanrolleghem P. A., Tack F. M. G. and Meers E. (2017). Nutrient recovery from digestate: systematic technology review and product classification. *Waste and Biomass Valorization*, **8**, 21–40, https://doi.org/10.1007/s12649-016-9642-x

Vondra M., Masa V., Tous M. and Konecna E. (2018). Vacuum evaporation of a liquid digestate from anaerobic digestion: a techno-economic assessment. *Chemical Engineering Transactions*, **70**, 769–774, https://doi.org/10.3303/CET1870129

Vondra M., Tous M. and Teng S. Y. (2019). Digestate evaporation treatment in biogas plants: a techno-economic assessment by Monte Carlo, neural networks and decision trees. *Journal of Cleaner Production*, **238**, 117870, https://doi.org/10.1016/j.jclepro.2019.117870

Wei Y., Hong J. and Ji W. (2018). Thermal characterization and pyrolysis of digestate for phenol production. *Fuel*, **232**, 141–146, https://doi.org/10.1016/j.fuel.2018.05.134

Wei C., Yu Z., Zhang X. and Ma X. (2021). Co-combustion behavior of municipal solid waste and food waste anaerobic digestates: combustion performance, kinetics, optimization, and gaseous products. *Journal of Environmental Chemical Engineering*, **9**, 106028, https://doi.org/10.1016/j.jece.2021.106028

Zeng Y., De Guardia A. and Dabert P. (2016). Improving composting as a post-treatment of anaerobic digestate. *Bioresource Technology*, **201**, 293–303, https://doi.org/10.1016/j.biortech.2015.11.013

doi: 10.2166/9781789062755_0247

Chapter 11

Effect of digestate recirculation on anaerobic digestion performance

Pallavi Gahlot[1], *Kaoutar Aboudi*[2] *and Vinay Kumar Tyagi*[3,*]

[1]Department of Civil Engineering, Environmental Biotechnology Group, Indian Institute of Technology Roorkee, Roorkee 247667, India
[2]Department of Chemical Engineering and Food Technology, University of Cadiz, Puerto Real, Cadiz 11510, Spain
[3]Environmental Hydrology Division, National Institute of Hydrology, Roorkee, 247667, India
**Corresponding author: Scientist D, Environmental Hydrology Division, National Institute of Hydrology, Roorkee 247667, Uttarakhand, India, Email: vinayiitrp@gmail.com; vtyagi.nihr@gov.in*

ABSTRACT

Anaerobic digestion (AD) is the most worldwide spread technology for waste management and valorization practices. The conversion of organic wastes could be limited due to recalcitrant materials such as lignocellulosic biomass. Therefore, only 40–50% of the feedstock is efficiently converted into biogas and the residual effluent, named digestate, remained unused, leading to the overall efficiency of the AD process. Digestate recirculation has been shown to increase AD overall potential. Thus, digestate post-treatment allows reaching high biomass degradability and increasing biogas production. The digestate post-treatment and recirculation contribute to the sustainability of large-scale biogas plants in the circular economy framework.

Keywords: Anaerobic digestate, recirculation, post-treatment, sustainability, biogas

11.1 INTRODUCTION

Wastes and wastewater generation has markedly increased in recent decades due to overpopulation, economic growth, and humans' overconsumption habits. In this context, organic waste management and treatment are a major concern worldwide. Anaerobic digestion (AD) has proved to be a promising strategy for organic waste valorization to simultaneously restrain greenhouse gas emissions (GHG), tackling organic waste issues and supplying clean energy for humans necessary uses (Tiwari *et al.*, 2021). AD has successfully proved its application to a wide variety of organic wastes such as organic fraction of municipal solid wastes (OFMSW), agro-wastes, food waste, livestock manure, and wastewater sludge (Aboudi *et al.*, 2016; Paul & Dutta, 2018; Stanchev *et al.*, 2020; Tiwari *et al.*, 2021). In the AD process, the breakdown of organic

Editor(s): Vinay Kumar Tyagi, Kaoutar Aboudi and Cigdem Eskicioglu

material by specific microbial consortia and enzymatic activity produces, alongside the biogas, a semi-solid residual effluent termed digestate, which is a nutrient-rich product (e.g. nitrogen, phosphorous, potassium, etc.), and a carbon-stabilized material (Chong *et al.*, 2022; Dahlin *et al.*, 2017). The amount of digestate generated in AD and its specific properties depend on the feedstock characteristics and the anaerobic process conditions (Möller, 2015). Digestate management and treatment are also a matter of concern due to the possible presence of undesired components (e.g. pollutants) or elevated mineral content in digestates (Dahlin *et al.*, 2017; Herbes *et al.*, 2020). Therefore, its direct application for improving agricultural soils is conditioned by certain criteria and standards (e.g. EU Nitrate directives to avoid excessive nitrogen oxide and ammonia gas emissions and nitrogen percolating into soils and groundwater) (Brémond *et al.*, 2021). Several life-cycle assessment studies of AD have shown that the economic feasibility of the AD process involves the suitable strategy of digestate management and uses (Li *et al.*, 2018, 2021). Some strategies for digestate management and valorization are its use as a fertilizer, its recirculation into the digester to improve nutrient availability and humidity, nutrients and other valuable product recovery, and its use as a feedstock for microalgae biomass cultivation (Chong *et al.*, 2022; Dahlin *et al.*, 2017).

The digestate separation into solid and liquid fractions allows a fruitful valorization of this material (Herbes *et al.*, 2020; Tambone *et al.*, 2017). Thus, nutrient recovery can be performed from both fractions depending on the desired product. The solid fraction could be composted, while the liquid fraction could be recirculated upon the anaerobic digester to improve system performance and biogas production, mainly in dry AD (Chen *et al.*, 2020; Diamantis *et al.*, 2021; Zheng *et al.*, 2020). Scientists have also tested the effect of recirculating the solid fraction of digestate on AD (Brémond *et al.*, 2021). Coupling AD with other technologies for digestate valorization also revealed a great potential (Antoniou *et al.*, 2019). This chapter will discuss the potential role of digestate recirculation strategy in optimizing the AD process.

11.2 AN OVERVIEW OF ANAEROBIC DIGESTION AND PROCESS LIMITATION

Among other sustainable approaches (e.g. aerobic digestion, composting, pyrolysis, gasification), AD has been established as a reliable process for renewable energy production and waste valorization. Therefore, it has been largely justified that AD exhibits several advantages such as valorizing low-cost organic waste streams largely and worldwide available, being a naturally occurring biological process, producing clean energy, and allowing recovery of valuable biochemicals (Diamantis *et al.*, 2021; Wu *et al.*, 2019).

AD process is carried out in four main steps: (1) Hydrolysis, in which complex organic compounds are degraded by hydrolytic bacteria and enzymatic activity into simple monomers. (2) Acidogenesis, where simple monomers are converted by acidogenic bacteria into intermediates, mainly to volatile fatty acids (VFAs), hydrogen, and carbon monoxide. (3) Acetogenesis, in this phase the products of the acidogenesis step are converted into acetic acid, hydrogen, and carbon dioxide. (4) Methanogenesis, where the products of acetogenesis, hydrogen and acetate, are transformed by archaea into methane. Nevertheless, there are also aspects to challenge for a successful AD process. This involves studying and optimizing process design and various parameters, including temperature, total solids content, the application of pre-treatments, the co-digestion of different waste types, and other methodologies (Cucina *et al.*, 2021; Fdez.-Güelfo *et al.*, 2011a; Fernández-Rodríguez *et al.*, 2016; Tiwari *et al.*, 2021). Similarly, ensuring an economically viable and cost-effective process is also a challenge to be reckoned with. AD limitations lie in both waste stream characteristics and the process conditions. Key drawbacks related to the intrinsic characteristics of organic wastes include the complex structure of lignocellulosic biomass, the lack of some essential nutrients (e.g. nitrogen deficiency in carbonaceous materials or lack of carbon in manure), and the fast hydrolysis of some wastes which release inhibitory compounds such as ammonia or VFAs (Cirne *et al.*, 2007; Cucina *et al.*, 2021; Paul & Dutta, 2018; Xu *et al.*, 2019; Yenigün & Demirel, 2013). The main difficulties in AD due to operation parameters include mainly the fast pH drop and acidification leading to system inhibition and methanogens blockage, unsuitable C/N ratio, unsuitable organic loading rate (OLR) or hydraulic retention time (HRT). Acidification results from high VFAs production rates and the lack of sufficient buffer capacity. Moreover, degradation of lignocellulosic biomass could release intermediate inhibitors such as phenolic compounds (furfural), heterocyclic compounds (hydroxymethylfurfural 'HMF'), or vanillin, resulting from the degradation of xylose, glucose, and lignin, respectively (Ghasimi *et al.*, 2016; Paul & Dutta, 2018). The decomposition of nitrogen-rich substrates or even OFMSW may lead to higher ammonia nitrogen release, which subsequently inhibits sensitive methanogenic archaea (Yenigün & Demirel, 2013). AD of lipid-rich wastes may generate long-chain fatty acids (LCFA), which, at elevated concentrations, may inhibit microbial communities, mostly those of methanogens (Cirne *et al.*, 2007). Limitation or inhibition of AD could also be produced due to the co-effect of two or more of the above-mentioned factors.

These roadblocks to an efficient AD had conducted scientists and engineers to optimize the process by the application of pre-treatments (Ahmed *et al.*, 2021; Suthar *et al.*, 2022), the co-digestion of different wastes with complementary characteristics (Aboudi *et al.*, 2016; Tyagi *et al.*, 2018), the supplementation of digesters by organic and inorganic additives (Suthar *et al.*, 2022; Tiwari *et al.*, 2021), the separation of AD phases (Fernández-Rodríguez *et al.*, 2016; Li *et al.*, 2020), and many other strategies. Some of these approaches are further detailed in the following. Moreover, it should be recalled that to avoid the above-mentioned obstacles, a proper design of each process is needed taking into account the properties of each feedstock. Hydrolysis is the first step in AD and has been recognized as a limiting step due to the low degradation of some complex organic material. Lignocellulosic biomass (LB) comprises mainly cellulose, hemicellulose and lignin. LB has shown to be recalcitrant to biological and enzymatic degradation due to the low accessible surface area of cellulose and its high polymerization and crystallinity, together with crosslinkages of hemicellulose and lignin (Ghasimi *et al.*, 2016; Paul & Dutta,

2018). Similarly, OFMSW is a complex waste stream and may be recalcitrant to the anaerobic degradation (Tyagi *et al.*, 2018). In this sense, pre-treatments have been extensively studied for the improvement of AD of LB and OFMSW. Pre-treatment methods comprise physical (e.g. mechanical, sonication), chemical (e.g. acidic/alkaline, ozone), biological (e.g. fungi, enzymes), or combined pre-treatments. Lee and Park (2020) applied hydrothermal and mechanical (milling) pre-treatments to sunflower residue (mixture of stem, leaf, and flower). The authors found that thermally pre-treated substrate at 180°C showed a significant increase in methane yield (289.5 mL/gVS) and biodegradability (63%) in comparison with the non-thermally pre-treated biomass. Moreover, authors highlighted that temperature condition of thermal pre-treatment is a critical variable and that temperature increase above 200°C led to the formation of HMF and furfural, which therefore significantly decreased the methane yield and biodegradability. For the mechanically pre-treated sunflower residue, methane yields were 129, 112, and 106 mL/gVS for particle sizes of <3, 3–6, and 6–10 mm, respectively.

Fdez.-Güelfo *et al.* (2011a) studied the effect of biological pre-treatment on OFMSW using mature compost as a biological agent at a temperature of 55°C. They observed that the cumulative methane production highly increased by 190% in comparison to untreated OFMSW. In another study of the same authors (Fdez.-Güelfo *et al.*, 2011b), they compared the effect of thermochemical (different pressures, temperatures and alkaline agent dosage) and biological (mature compost and the fungus *Aspergillus awamori*) pre-treatments on the kinetics of AD of OFMSW. They deduced that the best pre-treatment method was pre-composting, increasing the microorganism's maximum specific growth rate by 160–205%, in comparison to that of OFMSW without pre-treatment. Ahmed *et al.* (2021) reviewed the application of thermal pre-treatment for the enhancement of OFMSW solubilization previously to AD. The authors highlighted that the temperature and reaction time of thermal pre-treatment plays a crucial role on substrate solubilization and overall process performance. They deduced that a thermal pre-treatment at temperatures below 100°C has no effect on substrate solubilization and methane production and at a temperature exceeding 150°C this pre-treatment led to the formation on inhibitory recalcitrant intermediates, inhibiting methane yield and the overall process performance. Therefore, they observed that temperature conditions around 120°C have a significant positive effect on the organic matter solubilization and methane yield increase.

In a recent study by Usman Khan and Kiaer Ahring (2021), who compared the effectiveness of different pre-treatment methods to manure fibers (the recalcitrant fraction separated out after AD of cow manure) in a semi-continuous anaerobic reactor. They found that all pre-treatments had a positive effect on AD. The highest increase in methane yield to 127% and VS conversion to 42.4% were obtained from substrates pre-treated with a combined chemical (3% NaOH) and thermal (180°C) pre-treatments. The authors reported a reduction in cellulose, hemicellulose and lignin in digested manure fibers composition by 25, 29 and 9%, respectively. They observed that around 76 and

84.9% of cellulose and hemicellulose was successfully converted into methane due to the pre-treatment application. Recently, the addition of carbon-based materials to improve AD has been widely emphasized as a suitable strategy to promote microbial activity in AD. Suthar *et al.* (2022) studied the effect of cattle dung biochar addition at different dosages together with dilute acid-thermal pre-treatment aiming to improve anaerobic co-digestion of water hyacinth wastes with waste-activated sludge. The authors found that the applied pre-treatment increased the soluble fraction of chemical oxygen demand (COD) by 4.9 folds in comparison to control tests. Thus, this pre-treatment improved the lignocellulose substrate solubilization (water hyacinth) by hydrolyzing 57, 87, and 38% of cellulose, hemicellulose, and lignin, respectively. Moreover, biochar addition to the pre-treated substrate increased cumulative biogas production and methane content by 73.4–98.7 and 9.3–19.8%, respectively. The highest methane yield was found as 235 ml/gVS from pre-treated LB and biochar addition at 1%. It should be realized that choosing a suitable pre-treatment for the studied biomass is of utmost importance to avoid excessive hydrolysis in which inhibitory intermediates could be released. In a recent review study of microbial community dynamics in AD of OFMSW carried out by Tyagi *et al.* (2021), authors deduced that both operational conditions and substrate properties highly affect the microbial community composition in AD. Moreover, they found that a suitable nutrient balance leads to more healthy and diverse conditions among microorganisms, which allows a high tolerance of the system to process disturbances.

Anaerobic co-digestion of complementary substrates has been applied to a wide range of organic wastes in order to counterbalance nutrient deficiencies of feedstock and improve methane yields and AD stability. Aboudi *et al.* (2016) compared AD performance and biogas production from the single digestion of sugar beet dried pulp and its co-digestion with cow manure in long-term semi-continuous assays. The authors found that single digestion of beet pulp requires higher HRT and low OLR in comparison with co-digestion. Thus, AD of beet pulp alone was only feasible at HRT of 20 days and lower HRT as 18 and 15 days led to VFA accumulation. Nevertheless, co-digestion allowed decreasing HRTs to 15 days (OLR of 4.97 gVS/L d) increasing methane yields by 32%. Cucina *et al.* (2021) studied co-digestion of the lignocellulosic residue buckwheat hull (BH) with different co-substrates (sewage sludge – SS, dairy manure – DM, fruit wastes – FW, brewery trub – BT, and slaughterhouse wastes – SW). Firstly, they compared the digestion of each substrate individually and they observed that cumulative methane yields increased in the order of DM > FW > SS > BH > SW > BT. Moreover, they compared different co-digestion assays with the AD of BH individually. In this case, they found that all co-digestion assays have improved methane yields in comparison to individual digestion of BH. They reported that the highest methane yield was reached in co-digestion with slaughterhouse wastes with 254% higher yields than BH individually, while co-digestion with FW and BT has significantly shortened the lag-phase periods. Nevertheless, the easy solubilization of FW and fast conversion of sugars into VFA led to excessive acidification, in which pH correction was necessary. Surprisingly to what it reported in other studies, dairy manure has shown the lowest synergistic effect on co-digestion. While SS and BT showed similar synergy in co-digestion with BH and both FW and SW showed the highest synergistic effect. Overall, authors deduced that chemical characteristics of co-substrates in anaerobic co-digestion should be meticulously considered, specifically taking into account the soluble organic matter content.

Two-phase AD configuration has been applied as an improved strategy for AD of complex organic wastes enhancing the process stability of each stage as well as the enrichment of the microorganisms involved in each phase, and thus, allowing the operation at shorter HRT and higher OLR (Demirer & Chen, 2005; Malinowsky *et al.*, 2021). In this framework, Fezzani and Cheikh (2010) studied two-phase anaerobic co-digestion of olive mill wastes in semi-continuous digesters at mesophilic temperature (37°C in the first acidification stage and +37°C in the second methanation stage). They found that two-phase AD has shown the best performances of methane yields, soluble chemical oxygen demand and phenol removal efficiencies, together with giving a high effluent quality, in comparison to those obtained by conventional one-phase AD. Li *et al.* (2020) studied two-phase AD of corn straw hydrolysate and found that the use of this configuration was more suitable for bioconversion of lignocellulosic hydrolysate into biogas in comparison with single-stage AD. They reported that VFAs produced in the first acidogenic stage were efficiently converted into methane in the second digester, reaching maximum methane yield of 290 ml/gVS at the highest OLR of 8 g/L d with 85% of COD removal. Demirer and Chen (2005) found that two-phase AD of unscreened dairy manure at a total HRT of 10 days (2 days-HRT in the first phase and 8 days-HRT in the second methanogenic phase) showed an increase in biogas production by 50 and 67% at OLRs of 5 and 6 gVS/L day, respectively. Moreover, they reported that two-phase systems could support operation at a higher OLR of 12.6 gVS/L d, which was not feasible in single-phase systems. Similar to two-phase AD, temperature-phased (TPAD) strategy has been also suggested for the same purpose, implying the use of high temperature conditions in the first stage and a lower temperature, usually mesophilic, in the second digester (Amodeo *et al.*, 2021; Fernández-Rodríguez *et al.*, 2016). This approach used the first step as a pre-treatment to enhance hydrolysis and solubilization of the substrate due to elevated temperature, together with producing H_2, and therefore, obtaining a higher methane yield in the subsequent step. Amodeo *et al.* (2021) studied high solids temperature-phased anaerobic co-digestion of OFMSW with digested sludge at three different temperatures in the first hydrolytic stage (37, 55 and 65°C) and mesophilic temperature in the second stage. They found that the first hydrolytic stage slightly increased the methane yield (0–15%, 15% at 55°C). Nevertheless, they observed that temperature-phased assays led to a significant increase in methane production kinetics, reaching 40% for the tests 55–37°C. Similarly, Fernández-Rodríguez *et al.* (2016) reported that TPAD (55–37°C) of OFMSW reached 60 and 30% higher methane yields and VS removal in comparison to single stage, respectively. Wang *et al.* (2020) studied anaerobic co-digestion of food waste and cow dung in two-phased systems together with recirculation of the digestate. They found that co-digestion and recirculation allowed maintaining pH values within favorable ranges, which led to reduce the amount of external alkali addition. Moreover, they reported that co-digestion and digestate recirculation could promote both the hydrogen production rate (HPR) and the methane production rate (MPR), reaching maximum values of 3.3 LH_2/L d and 3.1 LCH_4/L d, respectively. In this line, digestate recirculation has also shown to be an interesting strategy for AD improvement. For example, Diamantis *et al.* (2021) reported that digestate recirculation is one of the most suitable strategies for improving AD of lipid-rich wastes.

In the following section, the effect of digestate recirculation on AD will be deeply addressed.

11.3 EFFECT OF DIGESTATE RECIRCULATION ON ANAEROBIC DIGESTION PERFORMANCE

Globally, solid waste production is estimated to exceed 27 billion tons by 2050 due to population explosion, especially in Asian and African countries (Laurent *et al.*, 2014). Circular economy for waste recycling supports the synthesis of eco-friendly biofuels as well as the recovery of valuable by-products (Aghbashlo *et al.*, 2019; De Corato *et al.*, 2018). Currently, AD is the most widely used technology for waste management practices. Biogas rich in methane produced via AD is utilised for heat and electricity cogeneration, as a fuel in the automotive sector, and the digestate is utilized for land reclamation (Gontard *et al.*, 2018). However, AD does not completely transform recalcitrant molecules such as lignocellulosic compounds limiting their conversion into biogas. As a result, only 40–50% of the feedstock gets converted into biogas while the residual leftover biomass exits the reactor unused, thereby reducing the overall efficiency of the AD process (Ahring *et al.*, 2015). Therefore, recirculating the residual hydrolysate back to the anaerobic reactor increases biogas production. Post-treatment of the digestate is an upcoming methodology for maximizing biogas production, allowing sustainable plant operation as well as the circular economy. The main aim of post-treatment is to ameliorate biodegradation of refractory biomasses which otherwise are less often degraded during AD. It could help in overpowering conventional pre-treatments that demand expensive chemicals and high-energy input for biomasses degradation. Bolzonella *et al.* (2020) investigated the efficiency of hydro-thermal post-treatment of agricultural digestate under mesophilic conditions and in semi-continuous mode. They reported that at 65°C (HRT of 3 days), agro-waste rich in recalcitrant compounds, showed an increase in volatile fatty acids (VFAs) concentration and specific methane production by 65 and 30% (60–65 L CH_4/kg volatile organic matter), respectively. The dominant microbial communities during thermophilic post-treatment conditions were identified to be Clostridiales, Bacteroidales and Thermotogaceae eubacterial families, *Methanosarcina* and Methanothermobacter archeal genera. This approach is expected to have the potential to increase the energy cogeneration from biogas up to 3000 PJ from 2500 PJ.

Recirculation of post-treated solid digestate within digesters has been identified as a promising practice for anaerobically digesting recalcitrant feedstocks and for improving the energy efficiency of continuously stirred tank reactor (CSTR) agricultural biogas plants. Moreover, fermentation of recirculated digestate tends to show higher stability than the control reactor (Xue *et al.*, 2020). Finding a robust alternative method for mitigating GHGs emissions is crucial for a sustainable ecosystem. To accelerate the biodegradation, Xue *et al.* (2020) recirculated 50% liquid corn straw digestate and investigated the effects of recirculation on enzymatic catalysis (cellulase, xylanase, dehydrogenase and coenzyme F_{420}) and biogas synthesis with respect to the control reactor. They reported average biogas and methane production for recirculated digestate to be 7891 mL/d and 347 mL CH_4/gVS added/d, respectively. Crop waste is the most copious resource for bioenergy production by AD (Talebnia *et al.*, 2010), however, billion tons of lignocellulosic waste is globally produced every year but very less percent is utilized. The complex lignocellulosic structure is redundant to biodegradation using conventional systems. Also, discharging the digestate directly into water streams could pollute water, air and soil (Li *et al.*, 2018). Liquid digestate recirculation (LDR) into anaerobic digesters could reduce the direct discharge of liquid digestate into the environment, enhance the hydrolysis of organic matter and improve buffering capacity and stability of the AD process (Li *et al.*, 2018; Peng *et al.*, 2016). On the contrary, very high LDR often leads to the accumulation of $NH4^+$-N and/or VFAs, which might hamper microbial activities, thereby inhibiting the biogas production (Zamanzadeh *et al.*, 2016). Besides, limiting factors such as the formation of recalcitrant compounds, unoptimized semi-continuous operation conditions or economic barriers make the biomass degradation process difficult.

Thermal and chemical post-treatments are the most extensively used methods that have shown to be efficient in enhancing the degradability of lignocellulosic- and sludge-based digestates. Ultrasonication is a potential mechanical treatment method used for promoting digestate biodegradability, yet, the energy recovery (biogas) is often very low as compared to the energy input. Biological post-treatments, such as simple aeration, fungal solid-state fermentation (SSF), and exogenous action of enzymes, are under study and still needs to be explored. Physico-chemical properties of digestate also influence the treatment efficacy which in turn is dependent upon the type of feedstock given to the digester, inoculum source, AD operational conditions, and its configuration (Monlau *et al.*, 2015). Brémond *et al.* (2021) considered different strategies within recirculation of solid fraction of digestate (SFD) by: (1) addition of SFD to the existing feedstock, or (2) partially replacing feedstock by SFD. The former case resulted in the enhancement of biogas between 0.6% and 5.7%, while the latter allowed saving of 64–1431 tons/year of feedstock while maintaining constant biogas production. Romio *et al.* (2021) reviewed various post-treatment methods to reveal additional biogas recovery via digestate recirculation. Moderate thermal post-treatments does not significantly affect the methane yields, however, an inter-stage treatment (applied between two batch AD stages) in comparison to a single-step AD process, at the same temperature, could improve overall methane yield (Nielsen *et al.*, 2011). High temperatures ranging from 120 to 230°C (15–60 min) have shown to enhance methane yields of both agricultural and sludge-based digestates (whole or solid fractions) in both batch as well as continuous AD processes. Mechanical process, such as chipping, grinding and milling (comminution) and ultrasonication reduce the particle size of biomass and cellulose crystallinity, increase the surface area and decreases the polymerization degree. However, the application of mechanical post-treatment demands high-energy inputs, often higher than biogas energy gain. Also, ultrasonication of liquids results in cavitation, leading to particulate disruption (Ashokkumar, 2011). Similarly, chemical processes like ozonation, acid, alkali post-treatment and biological post-treatments like enzymatic, aerobics, fungal solid-state fermentation (SSF) have also been evaluated for enhancement in biogas synthesis using digestate recirculation methodology. Various studies conducted using different post-treatments are represented in Table 11.1.

11.4 CONCLUSIONS

AD is a well-established technology for organic waste treatment. In addition to biogas, a residual effluent named digestate in released at the end of the process. A proper digestate treatment and valorization strategy is necessary for a successful overall AD process. Digestate recirculation has proven to enhance AD performance as the recirculated digestate contains many microorganisms that serve as an inoculum that consistently instils the raw material, in addition to enrich the medium with the macro and micronutrient required for microorganisms' growth. The development of continuously evolving methodology is the need of the hour to maintain sustainable energy resources and promote recycling of renewable energy. Utilizing the AD process to convert waste to renewable energy (biogas) and organic fertilizer is the most widely used methodology for converting waste to energy. However, incomplete biomass degradation renders the digestate, still potential for some biogas production, to be further exploited/recirculated. In the AD process, the simpler fractions of biomass are degraded first, followed by the complex ones, but leading to the accumulation of refractory compounds. Post-treatment, such as digestate recirculation, assists in promoting the hydrolysis of the recalcitrant fractions and further utilization of the residual potential digestate for biogas/methane synthesis. Although, the use of this strategy remains limited due to knowledge gaps with respect to the effects of post-treatment on the final digestate composition as well as on digestate handling and utilization. Digestate post-treatment might lead to further degradation of amino compounds, increasing ammonium concentration and, thus, enhancing the quality of manure. Biological post-treatments have not been evaluated to a great extent, and are still not used commercially at a large scale. Mechanical, although more widely applied have higher efficiency of biodegradation, but the energy recovery is often low with respect to energy input. The most widely studied methodology includes thermal

Table 11.1 Various studies on post-treatment of digestate for additional biogas recovery and impact of digestate recirculation.

	Post-treatment	Digestate type	AD conditions	Post-AD conditions	Methane yield after AD (mL CH_4/gVS)	Methane yield after recirculation	Reference
Thermal	Moderate thermal	Agro-waste (whole digestate)	80°C, 1 h	Batch, 65 days, 35°C	70	−19%	Sambusiti *et al.* (2015)
	High thermal	Agro-waste (solid fraction)	100°C, 6 h+135°C, 1 h	Batch, 50 days, 37°C	62	48%	Khan and Ahring (2021)
	High thermal	Sludge (whole digestate)	120, 170, 190°C, 1 h	Batch, 60 days, 35°C	52	120°C = +246%; 170°C = +327%; 190°C = +304%	Bjerg-Nielsen *et al.* (2018)
	High thermal	Sludge (whole digestate)	130–210°C, 30 min	Batch, 30 days, 3546.47°C	46.47	+178 to +255%	Yuan *et al.* (2019)
Mechanical	Comminution (chipping/grinding/milling)	Agro-waste (solid fraction)	Frequency 20/s, 30 min	Batch, 25 days, 35°C	101.5	+31%	Monlau *et al.* (2019)
	Comminution (chipping/grinding/milling)	Agro-waste (solid fraction)	Manual shearing	Continuous, 53°C, 15 days, 2–2.5 gVS L d, co-digestion manure (0.7): SFD (0.3) VS basis	211 (manure) 143 (manure+raw SFD)	7%	Tsapekos *et al.* (2016)
	Ultrasonication	Agro-waste (whole digestate)	1500–3000 kJ/kg TS	Continuous, 37°C, 20–30 days, 2.4–3.6 g VS/L day, Co-digestion manure (0.7): SFD (0.3) VS basis	0.22 (20 days) 0.14 (30 days)	−17%, −18% and +18% (at 20, 20, 30 days and 1500, 3000, 1500 kJ/kg TS, respectively)	Azman *et al.* (2020)
	Ultrasonication	Agro-waste (whole digestate)	3000	−50,000 kJ/kg TS	Batch, 56 days, 37°C, treated:untreated=3	−4% to +21%	Boni *et al.* (2021)
Biological	Enzymatic (xylanase + glucanase)	Agro-waste (whole digestate)	0.15 mL/gTS, pH=5, 40°C, 24 h	Batch, 65 days, 35°C	70	+51%	Sambusiti *et al.* (2015)

Enzymatic (xylanase + glucanase) aeration	Agro-waste (solid fraction)	0.15 mL/gTS, pH=5, 40°C, 24 h1.5–33 L air/h/kgTS.1.75–6 days	Batch, 65 days, 35°CBatch, 35°C	90	13%	Sambusiti *et al.* (2015)Brémond *et al.* (2021)
	Agro-waste (solid fraction)			153–171	–21% and 0%	
Thermo-acidic (H_3PO_4) + enzymatic (laccases)	Agro-waste (solid fraction)	4% H_3PO_4, 160°C, 15 min+48–59 U/gTS, pH=5.5, 37°C, 20 h	Batch, 52°C	+18%		Bruni *et al.* (2010)
Thermo-alkaline (NaOH) + enzymatic (laccases)	Agro-waste (solid fraction)	4% NaOH, 160°C, 15 min+48–59 U/gTS, pH=5.5, 37°C, 20 h	Batch, 52°C		+34%	Bruni *et al.* (2010)
Fungal SSF (*Pleurotus ostreatus*)	Agro-waste (solid fraction)	Autoclave with 2% CaO, 20–25°C, 75%moisture, 5.5–21 days	Batch, 35°C	232	–15 to –51	Brémond *et al.* (2020)
Fungal SSF (*Stropharia rugosoannulata*)	Agro-waste (solid fraction)	Autoclave with 2% CaO, 20–25°C, 75% moisture, 5.5–21 days	Batch, 35°C	167	–17 to +1%	Brémond *et al.* (2020)
Fungal SSF (*Cephalotrichum stemonitis*)	Agro-waste (solid fraction)	No sterilization, 25°C, 70–75% moisture, 10–20 days	Batch, 75 days, 40°C	44	+166% and +214% (10 and 20 days)	Zanellati *et al.* (2020)

(*Continued*)

Table 11.1 Various studies on post-treatment of digestate for additional biogas recovery and impact of digestate recirculation (*Continued*).

	Post-treatment	Digestate type	AD conditions	Post-AD conditions	Methane yield after AD (mL CH_4/gVS)	Methane yield after recirculation	Reference
Chemical	Oxidative (H_2O_2)	Agro-waste (whole digestate)	5,10,30 g H_2O_2/kg/TS, 2 h	Batch, 13 days, 37°C, treated: untreated = 1	427	+8 to +12%	Somers *et al.* (2018)
	Acidic (HNO_2)	Sludge (whole digestate)	0, 0.77, 1.54, 2.31, 3.08, 3.85 mgN/L; pH=5.5; 22°C; 24 h	Batch, 4 days, 37°C	37(22°C; 24 h, absence of HNO_2)	38%–0.77, 22%–1.54, 3%–2.31, –3%–3.08, –19%–3.85	Zhang *et al.* (2016)
	Thermo-acidic (HCl+H_2SO_4)	Sludge (solid fraction)	170°C, 1 h, pH=5–6	Continuous, 15–20 days, recycling rate 30%(SFD/raw)	0.92–1.15 gCOD-CH_4/COD-substrate	+14 to +21%	Takashima and Tanaka (2014)
	Alkaline (NaOH)	Agro-waste (solid fraction)	20 g NaOH/kgVS, 25°C, 30 min	Continuous, 20 days, 53°C	80	+11%	Khan and Ahring (2021)
	Thermo-alkaline (NaOH)	Agro-waste (solid fraction)	20 g NaOH/kgTS, 55°C, 3 days	Batch, 35°C	120–150	+30 to +46%	Brémond *et al.* (2021)
	Thermo-alkaline (NaOH)	Agro-waste (solid fraction)	10–30 g NaOH/kgTS, 100°C for 6 h+135°C for 1 h	Batch, 50 days, 53°C	71	+89 to +180%	Khan and Ahring (2021)

Source: Modified from Romio *et al.* (2021).

and chemical post-that have been demonstrated to be efficient in promoting degradability of biomass (agro-waste and sludge-based digestates). Moreover, the efficacy of the AD process is dependent upon treatment conditions, digester configurations as well as digestate characteristics, when recirculated. Many of the studies conducted on digestate recirculation gave promising results regarding additional biogas generation, but the techno-economic feasibility of the process must also be considered. For the evaluation and comparison of the effects of post-treatments (digestate) and pre-treatments (raw feedstock) in real systems, further extended studies need to be done.

REFERENCES

Aboudi K., Álvarez-Gallego C. J. and Romero-García L. I. (2016). Biomethanization of sugar beet byproduct by semi-continuous single digestion and co-digestion with cow manure. *Bioresource Technology*, **200**, 311–319. https://doi.org/10.1016/J.BIORTECH.2015.10.051

Aghbashlo M., Tabatabaei M., Soltanian S. and Dadak A. (2019). Comprehensive exergoeconomic analysis of a municipal solid waste digestion plant equipped with a biogas genset. *Waste Management*, **87**, 485–498. https://doi.org/10.1016/j.wasman.2019.02.029

Ahmed B., Tyagi V. K., Aboudi K., Naseem A., Álvarez-Gallego C. J., Fernández-Güelfo L. A., Kazmi A. A. and Romero-García L. I. (2021). Thermally enhanced solubilization and anaerobic digestion of organic fraction of municipal solid waste. *Chemosphere*, **282**, 131136. https://doi.org/10.1016/J.CHEMOSPHERE.2021.131136

Ahring B. K., Biswas R., Ahamed A., Teller P. J. and Uellendahl H. (2015). Making lignin accessible for anaerobic digestion by wet-explosion pretreatment. *Bioresource Technology*, **175**, 182–188. https://doi.org/10.1016/j.biortech.2014.10.082

Amodeo C., Hattou S., Buffiere P. and Benbelkacem H. (2021). Temperature phased anaerobic digestion (TPAD) of organic fraction of municipal solid waste (OFMSW) and digested sludge (DS): effect of different hydrolysis conditions. *Waste Management*, **126**, 21–29. https://doi.org/10.1016/J.WASMAN.2021.02.049

Antoniou N., Monlau F., Sambusiti C., Ficara E., Barakat A. and Zabaniotou A. (2019). Contribution to circular economy options of mixed agricultural wastes management: coupling anaerobic digestion with gasification for enhanced energy and material recovery. *Journal of Cleaner Production*, **209**, 505–514. https://doi.org/10.1016/J.JCLEPRO.2018.10.055

Ashokkumar M. (2011). The characterization of acoustic cavitation bubbles – an overview. *Ultrasonics Sonochemistry*, **18**, 864–872. https://doi.org/10.1016/j.ultsonch.2010.11.016

Azman S., Milh H., Somers M. H., Zhang H., Huybrechts I., Meers E., Meesschaert B., Dewil R. and Appels L. (2020). Ultrasound assisted digestate treatment of manure digestate for increased biogas production in small pilot scale anaerobic digesters. *Renewable Energy*, **152**, 664–673. https://doi.org/10.1016/j.renene.2020.01.096

Bjerg-Nielsen M., Ward A. J., Møller H. B. and Ottosen L. D. M. (2018). Influence on anaerobic digestion by intermediate thermal hydrolysis of waste activated sludge and co-digested wheat straw. *Waste Management*, **72**, 186–192. https://doi.org/10.1016/j.wasman.2017.11.021

Bolzonella D., Battista F., Mattioli A., Nicolato C., Frison N. and Lampis S. (2020). Biological thermophilic post hydrolysis of digestate enhances the biogas production in the anaerobic digestion of agro-waste. *Renewable and Sustainable Energy Reviews*, **134**, 110174.

Boni M. R., Polettini A., Pomi R. and Rossi A. (2021). Effect of ultrasonic post-treatment on anaerobic digestion of lignocellulosic waste. *Waste Management and Research*, **39**, 221–232. https://doi.org/10.1177/0734242X20931940

Brémond U., Bertrandias A., Loisel D., Jimenez J., Steyer J. P., Bernet N. and Carrere H. (2020). Assessment of fungal and thermo-alkaline post-treatments of solid digestate in a recirculation scheme to increase flexibility in feedstocks supply management of biogas plants. *Renewable Energy*, **149**, 641–651. https://doi.org/10.1016/j.renene.2019.12.062

Brémond U., Bertrandias A., de Buyer R., Latrille E., Jimenez J., Escudié R., Steyer J. P., Bernet N. and Carrere H. (2021). Recirculation of solid digestate to enhance energy efficiency of biogas plants: strategies, conditions and impacts. *Energy Conversion and Management*, **231**, 113759. https://doi.org/10.1016/J.ENCONMAN.2020.113759

Bruni E., Jensen A. P. and Angelidaki I. (2010). Comparative study of mechanical, hydrothermal, chemical and enzymatic treatments of digested biofibers to improve biogas production. *Bioresource Technology*, **101**, 8713–8717. https://doi.org/10.1016/j.biortech.2010.06.108

Chen R., Li Z., Feng J., Zhao L. and Yu J. (2020). Effects of digestate recirculation ratios on biogas production and methane yield of continuous dry anaerobic digestion. *Bioresource Technology*, **316**, 123963. https://doi.org/10.1016/J.BIORTECH.2020.123963

Chong C. C., Cheng Y. W., Ishak S., Lam M. K., Lim J. W., Tan I. S., Show P. L. and Lee K. T. (2022). Anaerobic digestate as a low-cost nutrient source for sustainable microalgae cultivation: a way forward through waste valorization approach. *Science of the Total Environment*, **803**, 150070. https://doi.org/10.1016/J.SCITOTENV.2021.150070

Cirne D. G., Paloumet X., Björnsson L., Alves M. M. and Mattiasson B. (2007). Anaerobic digestion of lipid-rich waste – effects of lipid concentration. *Renewable Energy*, **32**, 965–975. https://doi.org/10.1016/J.RENENE.2006.04.003

Cucina M., Pezzolla D., Tacconi C. and Gigliotti G. (2021). Anaerobic co-digestion of a lignocellulosic residue with different organic wastes: relationship between biomethane yield, soluble organic matter and process stability. *Biomass and Bioenergy*, **153**, 106209. https://doi.org/10.1016/J.BIOMBIOE.2021.106209

Dahlin J., Nelles M. and Herbes C. (2017). Biogas digestate management: evaluating the attitudes and perceptions of German gardeners towards digestate-based soil amendments. *Resources, Conservation and Recycling*, **118**, 27–38. https://doi.org/10.1016/J.RESCONREC.2016.11.020

De Corato U., De Bari I., Viola E. and Pugliese M. (2018). Assessing the main opportunities of integrated biorefining from agro-bioenergy co/by-products and agroindustrial residues into high-value added products associated to some emerging markets: a review. *Renewable and Sustainable Energy Reviews*, **88**, 326–346. https://doi.org/10.1016/j.rser.2018.02.041

Demirer G. N. and Chen S. (2005). Two-phase anaerobic digestion of unscreened dairy manure. *Process Biochemistry*, **40**, 3542–3549. https://doi.org/10.1016/J.PROCBIO.2005.03.062

Diamantis V., Eftaxias A., Stamatelatou K., Noutsopoulos C., Vlachokostas C. and Aivasidis A. (2021). Bioenergy in the era of circular economy: anaerobic digestion technological solutions to produce biogas from lipid-rich wastes. *Renewable Energy*, **168**, 438–447. https://doi.org/10.1016/J.RENENE.2020.12.034

Fdez.-Güelfo L. A., Álvarez-Gallego C., Sales Márquez D. and Romero García L. I. (2011a). Biological pretreatment applied to industrial organic fraction of municipal solid wastes (OFMSW): effect on anaerobic digestion. *Chemical Engineering Journal*, **172**, 321–325. https://doi.org/10.1016/J.CEJ.2011.06.010

Fdez.-Güelfo L. A., Álvarez-Gallego C., Sales Márquez D. and Romero García L. I. (2011b). The effect of different pretreatments on biomethanation kinetics of industrial organic fraction of municipal solid wastes (OFMSW). *Chemical Engineering Journal*, **171**, 411–417. https://doi.org/10.1016/J.CEJ.2011.03.095

Fernández-Rodríguez J., Pérez M. and Romero L. I. (2016). Semicontinuous temperature-phased anaerobic digestion (TPAD) of organic fraction of municipal solid waste (OFMSW). Comparison with single-stage processes. *Chemical Engineering Journal*, **285**, 409–416. https://doi.org/10.1016/J.CEJ.2015.10.027

Fezzani B. and Cheikh R. B. (2010). Two-phase anaerobic co-digestion of olive mill wastes in semi-continuous digesters at mesophilic temperature. *Bioresource Technology*, **101**, 1628–1634. https://doi.org/10.1016/J.BIORTECH.2009.09.067

Ghasimi D. S. M., Aboudi K., de Kreuk M., Zandvoort M. H. and van Lier J. B. (2016). Impact of lignocellulosic-waste intermediates on hydrolysis and methanogenesis under thermophilic and mesophilic conditions. *Chemical Engineering Journal*, **295**, 181–191. https://doi.org/10.1016/J.CEJ.2016.03.045

Gontard N., Sonesson U., Birkved M., Majone M., Bolzonella D., Celli A., Angellier-Coussy H., Jang G. W., Verniquet A., Broeze J., Schaer B., Batista A. P. and Sebok A. (2018). A research challenge vision regarding management of agricultural waste in a circular bio-based economy. *Critical Reviews in Environmental Science and Technology*, **48(6)**, 614–654. https://doi.org/10.1080/10643389.2018.1471957

Herbes C., Roth U., Wulf S. and Dahlin J. (2020). Economic assessment of different biogas digestate processing technologies: a scenario-based analysis. *Journal of Cleaner Production*, **255**, 120282. https://doi.org/10.1016/J.JCLEPRO.2020.120282

Khan M. U. and Ahring B. K. (2021). Anaerobic digestion of digested manure fibers: influence of thermal and alkaline thermal pretreatment on the biogas yield. *BioEnergy Research*, **14**(3), 891–900.

Laurent A., Bakas I., Clavreul J., Bernstad A., Niero M., Gentil E., Hauschild M. Z. and Christensen T. H. (2014). Review of LCA studies of solid waste management systems – part I: lessons learned and perspectives. *Waste Management*, **34**(3), 573–588. https://doi.org/10.1016/j.wasman.2013.10.045

Lee J. and Park K. Y. (2020). Impact of hydrothermal pretreatment on anaerobic digestion efficiency for lignocellulosic biomass: influence of pretreatment temperature on the formation of biomass-degrading byproducts. *Chemosphere*, **256**, 127116. https://doi.org/10.1016/J.CHEMOSPHERE.2020.127116

Li Y., Manandhar A., Li G. and Shah A. (2018). Life cycle assessment of integrated solid state anaerobic digestion and composting for on-farm organic residues treatment. *Waste Management*, **76**, 294–305. https://doi.org/10.1016/J.WASMAN.2018.03.025

Li Y., Xu H., Hua D., Zhao B., Mu H., Jin F., Meng G. and Fang X. (2020). Two-phase anaerobic digestion of lignocellulosic hydrolysate: focusing on the acidification with different inoculum to substrate ratios and inoculum sources. *Science of the Total Environment*, **699**, 134226. https://doi.org/10.1016/J.SCITOTENV.2019.134226

Li Y., Qi C., Zhang Y., Li Y., Wang Y., Li G. and Luo W. (2021). Anaerobic digestion of agricultural wastes from liquid to solid state: performance and environ-economic comparison. *Bioresource Technology*, **332**, 125080. https://doi.org/10.1016/J.BIORTECH.2021.125080

Malinowsky C., Nadaleti W., Debiasi L. R., Gonçalves Moreira A. J., Bayard R. and Borges de Castilhos Junior A. (2021). Start-up phase optimization of two-phase anaerobic digestion of food waste: effects of organic loading rate and hydraulic retention time. *Journal of Environmental Management*, **296**, 113064. https://doi.org/10.1016/J.JENVMAN.2021.113064

Möller K. (2015). Effects of anaerobic digestion on soil carbon and nitrogen turnover, N emissions, and soil biological activity. A review. *Agronomy for Sustainable Development*, **35**, 1021–1041. https://doi.org/10.1007/s13593-015-0284-3

Monlau F., Sambusiti C., Ficara E., Aboulkas A., Barakat A. and Carrère H. (2015). New opportunities for agricultural digestate valorization: current situation and perspectives. *Energy and Environmental Science*, **8**, 2600–2621. https://doi.org/10.1039/C5EE01633A

Monlau F., Sambusiti C. and Barakat A. (2019). Comparison of dry versus wet milling to improve bioethanol or methane recovery from solid anaerobic digestate. *Bioengineering*, **6**, 80. https://doi.org/10.3390/bioengineering6030080

Nielsen H. B., Thygesen A., Thomsen A. B. and Schmidt J. E. (2011). Anaerobic digestion of waste activated sludge – comparison of thermal pretreatments with thermal inter-stage treatments. *Journal of Chemical Technology and Biotechnology*, **86**, 238–245. https://doi.org/10.1002/jctb.2509

Paul S. and Dutta A. (2018). Challenges and opportunities of lignocellulosic biomass for anaerobic digestion. *Resources, Conservation and Recycling*, **130**, 164–174. https://doi.org/10.1016/J.RESCONREC.2017.12.005

Peng X., Nges I. A. and Liu J. (2016). Improving methane production from wheat straw by digestate liquor recirculation in continuous stirred tank processes. *Renewable Energy*, **85**, 12–18.

Romio C., Kofoed M. V. W. and Møller H. B. (2021). Digestate post-treatment strategies for additional biogas recovery: a review. *Sustainability*, **13**(16), 9295. https://doi.org/10.3390/su13169295

Sambusiti C., Monlau F., Ficara E., Musatti A., Rollini M., Barakat A. and Malpei F. (2015). Comparison of various post-treatments for recovering methane from agricultural digestate. *Fuel Processing Technology*, **137**, 359–365. https://doi.org/10.1016/j.fuproc.2015.04.028

Somers M. H., Azman S., Sigurnjak I., Ghyselbrecht K., Meers E., Meesschaert B. and Appels L. (2018). Effect of digestate disintegration on anaerobic digestion of organic waste. *Bioresource Technology*, **268**, 568–576. https://doi.org/10.1016/j.biortech.2018.08.036

Stanchev P., Vasilaki V., Egas D., Colon J., Ponsá S. and Katsou E. (2020). Multilevel environmental assessment of the anaerobic treatment of dairy processing effluents in the context of circular economy. *Journal of Cleaner Production*, **261**, 121139. https://doi.org/10.1016/J.JCLEPRO.2020.121139

Suthar S., Sharma B., Kumar K., Rajesh Banu J. and Tyagi V. K. (2022). Enhanced biogas production in dilute acid-thermal pretreatment and cattle dung biochar mediated biomethanation of water hyacinth. *Fuel*, **307**, 121897. https://doi.org/10.1016/J.FUEL.2021.121897

Takashima M. and Tanaka Y. (2014). Acidic thermal post-treatment for enhancing anaerobic digestion of sewage sludge. *Journal of Environmental Chemical Engineering*, **2**, 773–779. https://doi.org/10.1016/j.jece.2014.02.018

Talebnia F., Karakashev D. and Angelidaki I. (2010). Production of bioethanol from wheat straw: an overview on pretreatment, hydrolysis and fermentation. *Bioresource technology*, **101**(13), 4744–4753.

Tambone F., Orzi V., D'Imporzano G. and Adani F. (2017). Solid and liquid fractionation of digestate: mass balance, chemical characterization, and agronomic and environmental value. *Bioresource Technology*, **243**, 1251–1256. https://doi.org/10.1016/J.BIORTECH.2017.07.130

Tiwari S. B., Dubey M., Ahmed B., Gahlot P., Khan A. A., Rajpal A., Kazmi A. A. and Tyagi V. K. (2021). Carbon-based conductive materials facilitated anaerobic co-digestion of agro waste under thermophilic conditions. *Waste Management*, **124**, 17–25. https://doi.org/10.1016/J.WASMAN.2021.01.032

Tsapekos P., Kougias P. G., Frison A., Raga R. and Angelidaki I. (2016). Improving methane production from digested manure biofibers by mechanical and thermal alkaline pretreatment. *Bioresource Technology*, **216**, 545–552. https://doi.org/10.1016/j.biortech.2016.05.117

Tyagi V. K., Fdez-Güelfo L. A., Zhou Y., Álvarez-Gallego C. J., Garcia L. I. R. and Ng W. J. (2018). Anaerobic co-digestion of organic fraction of municipal solid waste (OFMSW): progress and challenges. *Renewable and Sustainable Energy Reviews*, **93**, 380–399. https://doi.org/10.1016/J.RSER.2018.05.051

Tyagi V. K., Bhatia A., Kubota K., Rajpal A., Ahmed B., Khan A. A., Kazmi A. A. and Kumar M. (2021). Microbial community dynamics in anaerobic digesters treating organic fraction of municipal solid waste. *Environmental Technology & Innovation*, **21**, 101303. https://doi.org/10.1016/J.ETI.2020.101303

Usman Khan M. and Kiaer Ahring B. (2021). Improving the biogas yield of manure: effect of pretreatment on anaerobic digestion of the recalcitrant fraction of manure. *Bioresource Technology*, **321**, 124427. https://doi.org/10.1016/J.BIORTECH.2020.124427

Wang Y., Wang Z., Zhang Q., Li G. and Xia C. (2020). Comparison of bio-hydrogen and bio-methane production performance in continuous two-phase anaerobic fermentation system between co-digestion and digestate recirculation. *Bioresource Technology*, **318**, 124269. https://doi.org/10.1016/J.BIORTECH.2020.124269

Wu D., Li L., Zhao X., Peng Y., Yang P. and Peng X. (2019). Anaerobic digestion: a review on process monitoring. *Renewable and Sustainable Energy Reviews*, **103**, 1–12. https://doi.org/10.1016/J.RSER.2018.12.039

Xu N., Liu S., Xin F., Zhou J., Jia H., Xu J., Jiang M. and Dong W. (2019). Biomethane production from lignocellulose: biomass recalcitrance and its impacts on anaerobic digestion. *Frontiers in Bioengineering and Biotechnology*, **7**(191), 1–12.

Xue S., Qiu L., Guo X. and Yao Y. (2020). Effect of liquid digestate recirculation on biogas production and enzyme activities for anaerobic digestion of corn straw. *Water Science and Technology*, **82**(1), 144–156.

Yenigün O. and Demirel B. (2013). Ammonia inhibition in anaerobic digestion: a review. *Process Biochemistry*, **48**, 901–911. https://doi.org/10.1016/J.PROCBIO.2013.04.012

Yuan T., Cheng Y., Zhang Z., Lei Z. and Shimizu K. (2019). Comparative study on hydrothermal treatment as pre- and post-treatment of anaerobic digestion of primary sludge: focus on energy balance, resources transformation and sludge dewaterability. *Applied Energy*, **239**, 171–180. https://doi.org/10.1016/j.apenergy.2019.01.206

Zamanzadeh M., Hagen L. H., Svensson K., Linjordet R. and Horn S. J. (2016). Anaerobic digestion of food waste–effect of recirculation and temperature on performance and microbiology. *Water Research*, **96**, 246–254.

Zanellati A., Spina F., Rollé L., Varese G. C. and Dinuccio E. (2020). Fungal pretreatments on non-sterile solid digestate to enhance methane yield and the sustainability of anaerobic digestion. *Sustainability*, **12**, 8549. https://doi.org/10.3390/su12208549

Zhang T., Wang Q., Ye L. and Yuan Z. (2016). Enhancing post anaerobic digestion of full-scale anaerobically digested sludge using free nitrous acid treatment. *Journal of Industrial Microbiology and Biotechnology*, **43**, 713–717. https://doi.org/10.1007/s10295-016-1742-x

Zheng Z., Cai Y., Zhao Y., Meng X., Zhang Y., Lu C., Hu Y., Cui Z. and Wang X. (2020). Achieve clean and efficient biomethane production by matching between digestate recirculation and straw-to-manure feeding ratios. *Journal of Cleaner Production*, **263**, 121414. https://doi.org/10.1016/J.JCLEPRO.2020.121414

doi: 10.2166/9781789062755_0265

Chapter 12

Biochar-augmented anaerobic digestate treatment

Pallavi Gahlot[1], *Kaoutar Aboudi*[2], *Ahmed Tawfik*[3] *Vinay Kumar Tyagi*[4,*]

[1]*Department of Civil Engineering, Environmental Biotechnology Group, Indian Institute of Technology Roorkee, Roorkee 247667, India*
[2]*Department of Chemical Engineering and Food Technology, University of Cadiz, Puerto Real, Cadiz 11510, Spain*
[3]*National Research Centre, Water Pollution Research Department, 12622 Dokki, Cairo, Egypt*
[4]*Environmental Hydrology Division, National Institute of Hydrology, Roorkee 247667, India*
**Corresponding author email: Scientist D, Environmental Hydrology Division, National Institute of Hydrology, Roorkee 247667, Uttarakhand, India, vinayiitrp@gmail.com; vtyagi.nihr@gov.in*

ABSTRACT

Biochar (BC) is a waste-derived conductive carbon material, used in anaerobic digestion to enhance the process robustness against inhibitory compounds such as antibiotics, furans, microplastics, etc. BC aids in adsorbing these refractory substances, maintaining pH and improving the transfer of electrons between microbial communities via the direct interspecies electron transfer process. Furthermore, these refractory compounds must be removed/treated prior to the usage of anaerobic digestate to avoid soils contamination. Among the used methods, the augmentation of carbon-based conductive material such as BC is a pragmatic process ensuring an unobjectionable, stable, and nutrient-rich digestate production.

Keywords: Anaerobic digestate, biochar, augmentation.

12.1 INTRODUCTION

Anaerobic digestion (AD) is a mature technology and it is recognized as one of the most widespread sustainable and clean technologies for the treatment of organic wastes. AD is a simple process, which could be applied to various organic waste and wastewater types, giving clean energy as biogas, as well as other valuable products. In addition to biogas (biomethane, biohydrogen, or both in two-phased AD), a residual semi-solid effluent remains after degradation of the organic material by the consortium of specific microorganisms involved in AD. This anaerobic effluent is named 'digestate' and it is a nutrient-rich material, usually used as a fertilizer (Chong *et al.*, 2022). However, the digestate should be post-treated because its direct use in agriculture is regulated by specific

Editor(s): Vinay Kumar Tyagi, Kaoutar Aboudi and Cigdem Eskicioglu

standards to avoid the possible excessive content of nutrients or the presence of heavy metals and antibiotics, which may contaminate soils and groundwater (Johansen *et al.*, 2013; Riau *et al.*, 2021). AD of some heterogeneous or recalcitrant organic material could hinder the effectiveness of the process by affecting one or more steps of the process (hydrolysis, acidogenesis, acetogenesis, and methanogenesis). The use of specific additives to the digester (i.e., organic additives such as carbon-based conductive materials: biochar (BC), graphene, etc.) has been shown to enhance AD performance (Fagbohungbe *et al.*, 2017; Ma *et al.*, 2020; Tiwari *et al.*, 2021). In this sense, it has been found that carbon-based additives used as abiotic conductive material through the direct interspecies electron transfer (DIET) process, significantly improved the process stability and biogas generation (Gahlot *et al.*, 2020; Tiwari *et al.*, 2021). Thus, the DIET process using carbon-based components consists of using conductive materials as electron carriers, which promotes electron transfer between syntrophic microorganisms. Moreover, carbon-based materials are characterized by high porosity and surface area, buffering ability, together with high conductivity (Ambaye *et al.*, 2021; Qiu *et al.*, 2019; Zhao *et al.*, 2021a). Among the successful carbon-based additives in AD, BC has demonstrated a substantial increase in the process performance, improving both biogas production and digestate treatment and quality. Furthermore, compared to other carbon-based additives (e.g., graphene, single-walled carbon nanotube), BC has proven to be the most economically efficient due to its production from waste biomass (Wang *et al.*, 2018b, 2021a; Xu *et al.*, 2020). BC is a solid material obtained from pyrolysis of biomass and could be applied for several purposes such as heat and power production, CO_2 sequestration, soil conditioners, intoxicating harmful organic molecules in water, etc. (Rasapoor *et al.*, 2020; Weber & Quicker, 2018).

This chapter discusses the use of BC in the AD process, emphasizing its role in the treatment of anaerobic digestate.

12.2 PHYSICOCHEMICAL PROPERTIES OF BIOCHAR

BC is obtained through various thermochemical processes pyrolysis being the most established methodology (Qian *et al.*, 2015; Tripathi *et al.*, 2016). The temperature used for this purpose is usually around 300–650°C, which is considered as a slow pyrolysis heating process (heating rates ranged from 1 to 30°C/min) and it is performed under low oxygen or oxygen-free environment (Khalid *et al.*, 2021; Qian *et al.*, 2015; Zhao *et al.*, 2013). Slow pyrolysis has been shown to be effective to produce high BC yields at suitable costs, retaining the highest biomass carbon content in the material (Ghysels *et al.*, 2019; Kant Bhatia *et al.*, 2021; Tripathi *et al.*, 2016). Nevertheless, temperature conditions and residence time may vary according to biomass feedstock. Zhao *et al.* (2021b) compared BC properties after pyrolysis at three different temperatures, 400, 600, and 800°C. They found that the BC conductivity increased with temperature from 0.001 to 0.03, 0.8 and 6.3 S/m at 400, 600, 800°C, and 1000°C, respectively. Nevertheless, they observed that BC from low temperatures (400 and 600°C) contained magnetite (Fe_3O_4), which has been reported to promote the extracellular electron exchange. In BC obtained at 800 and 1000°C, they found that magnetite was replaced by Fe_3O_4 and Fe_3C, which are non-available iron minerals for microbial growth. Regarding the effect of biomass type on BC (Shen *et al.*, 2017), compared BC obtained from two different biomass feedstock (agro-plants and foresty) and they found that both BC types have a similar porosity and high surface area. However, they have different chemical compositions. BC derived from agro-plant had high nutrient elements in comparison to that obtained from forest biomass. Nutrients such as nitrogen, phosphorous, potassium, and sulfur, were 90, 900, 2600, 400% higher in BC derived from agro-plant, respectively. Along similar lines Fazeli Sangani *et al.* (2020) found that biomass used as feedstock mainly affects the physicochemical characteristics of BC and that physical properties of BC were influenced by particle size. Thus, BC obtained from plant-based material or lignocellulosic biomass has shown to be carbon-rich BC with high porosity, adsorption, and immobilization abilities, while BC derived from manure has higher pH values and a high content of ashes (Codignole Luz *et al.*, 2018; Qian *et al.*, 2013; Tripathi *et al.*, 2016). Moreover, BC derived from composted dairy manure has shown a higher ash content in comparison with BC from raw dairy manure (Kumar *et al.*, 2021a). Depending on the expected use of BC, these properties could have different effects (Qian *et al.*, 2015). From the above-mentioned studies, it could be deduced that biomass used as feedstock and the pyrolysis operation conditions are key factors on BC characteristics and quality for a successful AD process (Qian *et al.*, 2013; Zhao *et al.*, 2013). In this section, BC properties influencing biological processes such as AD are discussed.

The specific surface area (SSA) is one of the main physical properties that influence BC quality because it allows the adherence of microbes when BC is used as an additive in biological processes (Lim *et al.*, 2020; Paritosh & Vivekanand, 2019; Romero-Güiza *et al.*, 2016). As previously reported, the production of BC from pyrolysis of lignocellulosic biomass and plant-based material has shown to give BC a high SSA in comparison with BC produced from manure (Zhao *et al.*, 2021c). Similarly, porosity or pore-size distribution in BC has a significant effect on BC behavior when used as an additive for contaminants and inhibitory compounds adsorption (e.g., ammonia, toxins). In the same way, it has been found that BC from lignocellulosic biomass and plant-based feedstock is characterized by a high porosity (Fabbri & Torri, 2016; Romero-Güiza *et al.*, 2016; Zhao *et al.*, 2021a). Both higher SSA and porosity of BC lead to enhance the BC sorption and immobilization abilities. The immobilization ability of BC allows the survival of microorganisms such as methanogens in anaerobic digesters (Paritosh & Vivekanand, 2019; Qi *et al.*, 2021).

The elementary composition (EC) of BC contains mainly carbon, hydrogen and oxygen, in addition to other compounds (e.g., nitrogen, sulfur, and phosphorous) (Kumar *et al.*, 2021b). EC depends on both feedstock type and pyrolysis conditions. Enders *et al.* (2012) found that increasing the pyrolysis temperature leads to an increase in the total carbon content and a decrease in total hydrogen and oxygen. The same authors reported that agro-wastes from corn, pine, and oak, have a similar carbon content (43–49%), however, variation in the carbon content of BC after the pyrolysis of these biomass materials was more noticeable (60–91%). They found that BC produced from high-ash feedstock such as poultry manure has low carbon content due to the correction for inorganic carbon bound in carbonates, indicating that the ash content of this biomass is carbonates composite. Furthermore, the alkaline (Na^+, K^+) and earth (Ca^{2+}, Mg^{2+}) metals of BC have been shown to play important roles in BC buffering capacity (Giwa *et al.*, 2019; Zhao *et al.*, 2021b).

Functional groups abundance is another interesting property of BC in improving biological processes. Thus, after pyrolysis, the resulted BC acquires several functional groups such as hydroxyl (-OH), carbonyl (C=O), carboxylic (-COOH), phenolic (X-OH), amine (-NH_2), which confer on the BC the buffering ability, inhibitory compounds and contaminants removal, and electron transfer ability (Kumar *et al.*, 2021a; Wang *et al.*, 2020a; Zhao *et al.*, 2021c). The electron transfer ability of BC, resulting from electrochemical functional group distribution, has triggered an interest in using BC for DIET mechanism promotion (Wang *et al.*, 2018a; Zhao *et al.*, 2021a). DIET is the process in which electrons are transferred between syntrophic microorganisms without the intervening of any electron carriers or redox mediators (Gahlot *et al.*, 2020; Tiwari *et al.*, 2021).

12.3 BIOCHAR APPLICATION IN ANAEROBIC DIGESTION

BC has shown a great potential in enhancing the AD process. The beneficial addition of BC in AD includes lag-phase reduction and process rate acceleration, buffering potency, microbial functionalities enrichment, inhibitors adsorption, electron transfer mechanisms stimulation, and therefore, improving the overall process performance and biogas production. Luo *et al.* (2015) investigated the use of BC (10 g/L; pores size of 0.5–1 mm) in three mesophilic anaerobic digesters fed with 2, 4, 6, and 8 g/L of glucose. The authors reported a lag-phase shortening of 11.4, 30.3, and 21.6%, respectively in each digester. Moreover, an increase in methane production rates (MPR) was observed reaching 86.6, 21.4, and 5.2% higher values compared to control digesters, respectively. Furthermore, they reported that the pore size of BC plays a role in the obtained results. Thus, BC with 75 μm pores size used in a digester fed with 6 g/L of glucose decreased the lag-phase period by 38% and improved MPR by 70.6%. Microbial sequencing analysis revealed a higher abundance of methanogenic archaea in BC amended reactors. Thus, *Methanosarcina* genus was most abundant in the BC surface area in the tightly bound fractions of larger particles (2–5 mm), while *Methanosaeta* was abundant in all size particles of BC in the loosely bound fractions and within the tightly bound fractions in small size particles of BC. Qin *et al.* (2020) have searched for how physico-chemical properties of BC produced from woody and plant-based biomass affect AD. They reported that despite the high effectiveness of both BC types on reducing lag-phase, woody BC outperformed herbal BC, giving higher methane yields due to higher SSA and electron-donating ability (EDA) in comparison to plant-based BC. Thus, they found that the high SSA and EDC abilities allowed the enrichment and growth of bacterial communities promoting the daily methane production in the early stage of AD and facilitating the immobilization of functional microorganisms for a higher methane yield.

Regarding the effect of BC derived from manure, Suthar *et al.* (2022) reported that cattle dung BC when added to anaerobic digesters fed with mixtures of acid-thermally pre-treated water hyacinth and waste-activated sludge enhanced the cumulative biogas production and methane content by 73.4–98.7% and 9.33–19.8%, respectively in comparison to the control tests. Similarly, Sun *et al.* (2019) studied the effect of BC derived from cow manure on dry AD of beer lees at thermophilic and mesophilic temperatures. They found that addition of cow-dung BC of 10 g/L significantly improved methane production yields by 82.6 and 46.8% in mesophilic and thermophilic conditions, respectively. Wang *et al.* (2020b) performed a study on the effect of BC addition in promoting electron transfer between syntrophic bacteria during anaerobic co-digestion of sewage sludge and food waste at high organic loading rate (OLRs). In the first step of the study, they compared BC produced at different temperatures (300, 500, and 700°C) on AD in a batch assay. Subsequently, they applied the optimal conditions of batch assays in semi-continuous digesters. They found that, in comparison to control tests, all assays with BC addition have shown to shorten lag phases from 14 days to 10.5, 6.5, and 7.2 days for BC produced at 300, 500, and 700°C, respectively. They reported that the use of BC produced at 500°C gives the highest methane production rate, which was 1.8 fold higher than the control test. Furthermore, they observed that BC produced at 500°C was enriched with specific redox-active functional groups (e.g., quinone, phenazine, etc.), and therefore have promoted DIET mechanisms, in comparison to the low abundance of aromatic groups in BC produced at 700°C. At the steady-state operation conditions of the semi-continuous digester, the average methane yield obtained with BC addition was 60.4% higher than that of the control with a high tolerance to increase OLR and suitable pH values for the process. Furthermore, propionic acid was accumulated in control reactors while in amended BC reactor this inhibitory organic acid, at certain concentrations, was completely degraded. Besides, they found that despise increasing OLR, the relative abundance of the two typical syntrophic bacteria *Petrimonas* and *Syntrophomonas* were 8.5 and 11.2 folds higher than for the control test, respectively for each genus. Recently, Zhao *et al.* (2021b) studied the effect of BC supplemented by iron (Fe^{3+}) with different dosages (0, 1, 5, and 10 g/L) on AD performance of the azo dye acid orange 7. They found that the BC supplemented with iron (1 gFe^{3+}/L) showed the highest COD removal of 88.5%, which was 19% higher than the control digester. They deduced that supplementation with iron enhanced functional groups and electron transfer capability of BC and enriched exoelectrogens related to syntrophic metabolism, which improved DIET mechanism for acid orange 7 degradation. Furthermore, the bacterial analysis showed a higher relative abundance of fermentative bacteria in BC amended reactors, including *Bellilinea* genus, which is known for its high ability in removing phenolic compounds and decolouring azo dyes. Besides, the archaea community was also enriched due to BC addition with the predominance of *Methanobacterium* and the acetoclastic methanogen *Methanothrix*.

About the effect of BC on AD of sludge, Wei *et al.* (2020) tested BC derived from pyrolysis of corn stover on primary sludge AD and reported that the

BC addition increased methane content from 67% to 81–87% and methane production by 8.6–17.8%, in comparison to control tests. They found that BC alleviated ammonia inhibition and provided a strong buffering capacity to the digester. Besides, they deduced that BC addition to long-term continuous digesters of primary sludge allowed increasing VS removal by 14.9%, which allowed reducing the volume of the resulted digestate by 14%. In the same way, Shen *et al.* (2017) studied the thermophilic AD of sewage sludge in two-stage systems and long-term semi-continuous assays, they reported that BC produced from corn stover pyrolysis improved methane content and MPR by 25 and 37%, respectively, in comparison to control tests. Moreover, they observed that BC addition enriched the macro- and micro-nutrient availability in the produced digestate by up to 33%. Likewise, the microbial analysis showed a higher relative abundance of *Clostridia* and *Methanosarcina*. In another study carried out by Tiwari *et al.* (2021) on thermophilic anaerobic co-digestion of sewage sludge with wheat husk, they evaluated the effect of granular BC on the process performance and they reported that the highest biogas yield was observed at 20 g/L of BC addition with 273 ml/gVS$_{added}$, which was 27% higher than the control test.

Concerning the mitigation of inhibitory compounds released during AD, Ma *et al.* (2019) investigated the effect of BC on semi-continuous AD of dry chicken manure. They obtained around 12% higher methane yield in comparison to control assays at OLR of 3.2 and 6.3 gVS/L d. Furthermore, they noticed that BC alleviated propionic acid accumulation released from the AD of dry chicken manure due to providing a high buffer capacity to the system. The microbial analysis showed that the archaea *Methanosarcina* was enriched and dominated in BC amended digesters, while the strictly acetoclastic *Methanosaeta* dominated in control tests. Thus, they deduced that BC improved the methanogenesis phase by diversifying its metabolic pathways. Similarly, Lim *et al.* (2020) studied the effect of BC addition on the start-up and stabilization of thermophilic AD of food waste. They reported that BC has a significant role in promoting the growth of electroactive *Clostridia* and other electroactive bacteria, in comparison to the predominance of homoacetogenic *Clostridia* and syntrophic acetate oxidizing bacteria. Thus, they found that BC promoted DIET mechanisms and was responsible for the faster degradation of VFAs and process stability. Moreover, they reported that BC improved the thermodynamically favorable acetoclastic methanogenic pathway with a high abundance of *Methanosarcina*.

Wang *et al.* (2021b) revealed how BC can help in highly organic loading anaerobic reactors with up to 80 kg COD/m^3 d daily feedstock. In comparison to control tests, which showed an irreversible acidification and abrupt pH drop, amended reactors with BC could be restored in a short time returning to suitable pH values around 7 despite the organic loading shock. They demonstrated that addition of BC favored DIET mechanisms, including VFAs-oxidizing bacteria and methanogenic archaea, and allowing balanced rates of VFAs production and conversion into biogas. Besides, BC promoted the retention of *Methanosaeta* and *Methanosarcina* genus even under VFAs stress. Giwa *et al.* (2019) studied AD of food waste in mesophilic semi-continuous long-term operation assays with the BC amendment. They found that the use of BC only increased methane yield by 5% in comparison with control assays; however, they highlighted the role of BC in alleviating ammonia inhibition and fast degradation of VFAs. Moreover, they observed that *Methanothrix* genera were the most abundant methanogens with a high growth rate and inhibition tolerance. In addition, they found that the reactors amended with BC have shown high and stable values of alkalinity ranged between 6 and 8 g$CaCO_3$/L, while for control reactors, their alkalinity decreased from 5 to 1.8 g$CaCO_3$/L with OLR increase. In the presence of BC, anaerobic reactors operated properly at higher total ammonium nitrogen (TAN) above 2.5 g/L due to BC adsorption ability, while earlier TAN accumulation and concentrations of 1.5–2.4 g/L led to inhibit the process in control tests. Moreover, they deduced that the BC ability in promoting syntrophic VFAs utilizing bacteria and methanogens has benefitted the AD process. Cheng *et al.* (2020) investigated the effect of BC (derived from rice straw) on the mitigation of ammonia inhibition during AD of piggery wastewater. They found that biogas yield and COD removal proportionally increased with BC dosage in the digester, reaching a biogas yield and COD removal 1.8 and 1.2 folds higher in comparison to control tests at 15 g of BC addition. Furthermore, at a high ammonium stress level (3.5 g/L), biogas yield and COD removals were 4.5 and 1.8 folds higher in comparison with control tests also submitted to ammonium stress. Moreover and despite ammonium stress, VFAs concentration decreased from 3.3 g/L in control assays to 1.7, 1.3, 1.1, and 0.9 g/L at BC amendment of 2, 5, 10, and 15 g, respectively. Similarly, the final concentration of ammonium nitrogen decreased from 2.4 g/L in controls to 1.9, 1.7, 1.4, and 1.2 g/L for digester with BC at 2, 5, 10, and 15 g, respectively. They deduced that the high adsorption ability of BC allowed alleviating ammonia stress in AD. In the same way, Xu *et al.* (2021) recently studied the effect of BC in alleviating nitrite stress on the anaerobic ammonium oxidation (anammox) process. They found that BC addition allowed mitigating nitrite inhibition at concentrations as high as 600 mgN/L. At the higher nitrite concentration, BC addition allowed increasing total nitrogen removal in a range of 51–231%, depending on the type of BC used and in comparison with the control. This allowed a specific anammox activity to increase by 1.1–3.1-folds compared to the activity observed in the control test. Moreover, they observed that BC addition enriched the relative abundance of *Candidatus Kuenenia* and increased core functional genes, as well as alleviating the suppression of enzyme complexes for anammox electron transport chain, likely due to the contribution of redox-active moieties of BC to energy conversion metabolism for mitigating nitrite inhibition.

Lü *et al.* (2016) studied the effect of BC on the methanogenesis phase under double stresses of high ammonium and VFAs content. They found that BC addition allowed reducing lag-phase periods by 23.9, 23.8, 5.9% for BC with particles size of 2–5 mm, 0.5–1 mm, and 75–150 μm, respectively in comparison to the control. Moreover, BC addition improved the MPR by 23–47%, according to the particle sizes of BC. Despite the high concentration of ammonium up to 7 gN/L, the hydrogenotrophic genus of *Methanobacterium* was actively resistant to ammonium in suspended fractions. Moreover, the authors observed that the acetoclastic methanogens genus *Methanosaeta* can withstand to VFAs concentrations up to 60–80 mmol-C/L. They observed that once methanogenesis has been improved due to BC, the tightly bound material has been colonized by genus *Methanosarcina*. In a recent study carried out by Cui *et al.* (2021), they focused on the BC role in high-solid AD of food waste concerning cell viability, microbial community, and methanogenic pathways. They found that BC with a dosage of 25 g/L has promoted cell viability by increasing viable cells based on cell membrane integrity from 2.9 to 6.4%. Moreover, they observed that BC led to the digester enrichment by syntrophic bacteria such as *Syntrophomonas* and methanogens such as *Methanosarcina* and *Methanocelleus*, resulting in the diversification of methanogenic pathways and shifting the methanogens route from acetoclastic/hydrogenotrophic to more diverse pathways. Thus, *Methanosarcina* and *Methanocelleus* abundance increased by 49 and 186%, respectively in comparison to control tests. The authors deduced that *Methanosarcina* enrichment was the key factor in promoting DIET mechanisms and that *Methanocelleus* enrichment has promoted hydrogen-utilization methanogenic pathway. Besides, they also observed the abundance of the denitrifying bacteria *Pseudomonas* in BC amended digesters due to the selective immobilization ability of BC.

As can be seen from the previously reported works, the use of BC in AD has been extended to a wide variety of organic wastes. In a study by Lü *et al.* (2019) they applied powdered BC even in AD of oil, which is a substrate with high potential but its low solubility and release of long-chain fatty acids could hinder the process efficiency. Thus, they found that powdered BC with particle sizes under 5 μm allowed a higher microorganisms growth compared to granular BC (0.5–1 mm), enhancing quantification and distribution of both bacteria and archaea in the whole digester, specifically, at mesophilic temperature. Nevertheless, both BCs (powdered and granular) have been shown to alleviate acid stress due to propionic acid accumulation in thermophilic digesters, likely due to DIET mechanisms established between propionate oxidizing acetogens and carbon dioxide reducing methanogens. They observed that powdered BC mitigated the propionic acid accumulation and enhanced methane yield, while granular BC could only fasten propionate degradation. They reported that methane production increased by 13.3 and 32.5% in thermophilic digesters with powdered BC and mesophilic digesters with granulated BC, compared to controls, respectively. In a work approached to study the effect on each waste category, Xu *et al.* (2020) investigated the effect of BC on AD of carbohydrate-rich, protein-rich, and lipid-rich substrates. They showed that using BC enriched methanogens *archaea* growth in digesters fed with carbohydrate-rich and lipid-rich substrates, but not the protein-rich substrate. Specifically, BC added to carbohydrate-rich substrate reduced the lag-phase period by 48.6% and improved methane production by 4.7% in comparison to controls. Moreover, addition of BC improved methane production rates from lipid-rich substrate. They found that the predominant methanogenic genus varied according to the substrates. Thus, they observed that in the start-up of AD of carbohydrate-rich substrate, the microbial community was dominated by

Firmicutes, Bacteroidetes, and Fusobacteria. The most abundant methanogens, in this case, were *Methanosaeta* and *Methanobacterium*, which are known to convert acetate into methane and hydrogen and carbon dioxide exclusively as electron donors for methane production, respectively. Furthermore, they found that at the end of the AD process, *Methanosarcina* was dominant, which is known to be a metabolically and physiologically versatile methanogen allowing the conversion of different substrates into methane. On the other hand, for digesters fed with protein-rich substrate and amended with BC, the authors reported that there was an abundance of Bacteroidetes due to BC addition. Nevertheless, at the end of the process and due to substrate exhaustion, species in the phyla Proteobacteria and Spirochaetae became abundant. They reported that Proteobacteria are essential for AD, utilizing molecules such as glucose, propionate, butyrate to produce acetate, while Spirochaetae is known to convert VFAs into acetate. They deduced that despite enriching AD of protein-rich substrates by BC addition, this has not increased methane production due to a lower abundance of methanogens. For lipid-rich substrate, they showed that bacterial communities in digestate amended with BC were slightly higher than the controls. Besides, they observed that the higher methane production in BC amended digesters in the early stage of AD was related to the activities of *Methanosaeta* and *Methanobacterium* and that *Methanosaeta* genus predominated throughout the whole process.

Using BC to enhance AD performance in two-phase systems has also been reported. Sunyoto *et al.* (2016) studied the effect of BC in addition to both hydrogen and methane production in two-phase AD of aqueous carbohydrates food waste. They found that BC increased the maximum production rates of hydrogen and methane by 32.5 and 41.6%, respectively. Furthermore, hydrogen and methane yields were increased by 31 and 10% in comparison to controls, respectively. In both AD stages, the lag-phase periods were reduced by 36 and 41% in the acidogenesis and methanogenesis steps, respectively. They observed that BC improved VFAs production in the first hydrolytic–acidogenic system and their degradation and conversion into methane in the methanogenic system. In a later work by the same authors, they investigated the effect of BC dosage (0, 1.2, 5, 10, 15 and 18.6 g/L) on hydrogen production from the first phase of the two-phase AD of carbohydrates food waste. They found that the BC addition considerably enhanced hydrogen production rate (HPR) and hydrogen yield (HY) by up to 26 and 41%, respectively. Furthermore, they reported that the BC dosage at 10 g/L led to the highest HPR and HY with values of 820.6 mlH_2/L d and 1475 ml/L, respectively (Sunyoto *et al.*, 2017). In the same line and in circular approach, Zhao *et al.* (2021c) performed an assay about BC effect on hydrogen generation from the fermentation of cornstalk using BC derived from the digestate obtained after AD of the same substrate (cornstalk). They found that increasing BC dosage from 2.5 to 15 g/L enhanced HY and HPR by 83.2 and 62.8%, respectively. Furthermore, they deduced that BC provided more SSA for the immobilization of hydrogen-producing bacteria and promoted the cellulolytic enzyme activity by increasing cellulose and hemicellulose degradation. Wang *et al.* (2018a) studied the effect of 12 types of BC derived from four different biomass at three pyrolysis temperatures (300,

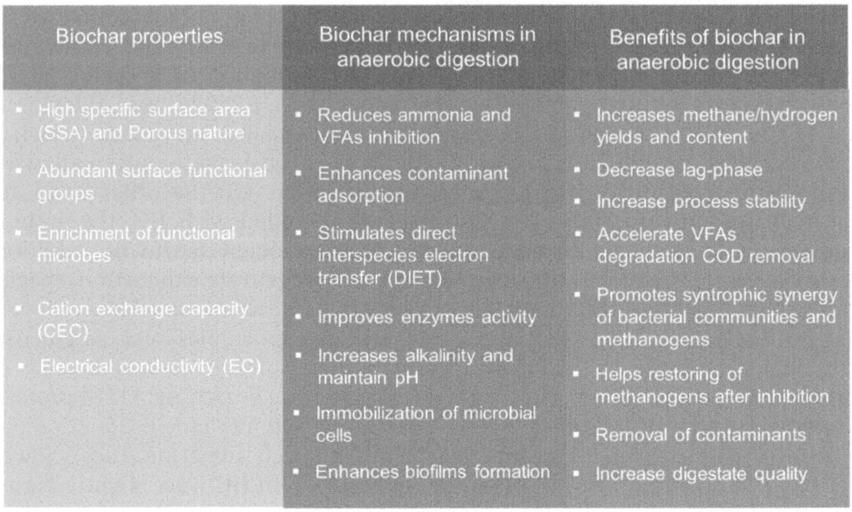

Figure 12.1 BC properties and role in improving AD.

500, and 700°C) on hydrogen production from co-digestion of food wastes and dewatered sewage sludge. They showed that most BC types has instigated the VFAs production and shifted the fermentative pathway from that of acetate type to butyrate type, leading therefore to an efficient hydrogen production process. Furthermore, they observed that BC addition could reduce the lag-phase periods by 18–62%, and increase HPR by 18–110%, according to the BC type. From the principal components analysis, they found that buffering ability of BC and its role in maintaining a suitable pH has a sustainable role in the acidogenic process. Furthermore, they observed that the BC obtained from pyrolysis at higher temperatures (700°C) was more efficient in promoting hydrogen production in comparison to BC derived from low-temperature pyrolysis.

Overall, BC has been demonstrated to be an efficient additive to improve all steps of the anaerobic process, from hydrolysis to methanogenesis. Figure 12.1 summarizes the most proven benefits of using BC for AD improvement. Furthermore, the use of BC in AD has been shown to improve the elemental composition of the solid fraction of anaerobic digestate, which can be subsequently used as fertilizer for the improvement of agricultural soils (Fagbohungbe *et al.*, 2017; Masebinu *et al.*, 2019; Shen *et al.*, 2016, 2017). The use of BC for digestate treatment will be discussed in the following section.

12.4 BIOCHAR USE IN ANAEROBIC DIGESTATE TREATMENT

The digestate produced at the end of AD is a rich source of nutrients that can be applied for soil reclamation (Paavola & Rintala, 2008). However, the digestate cannot be applied directly to the soil because of its high pathogenic content and unstable nature. Pathogen annihilation and decomposition of organic matter require at least six to eight months to stabilize the digestate in settling ponds (Costa *et al.*, 2017; Hutchison *et al.*, 2004). Also, such a long stabilization period may lead to an increase in C/N ratio, since denitrification of nitrogenous components causes a decrease in ammonium ion ($NH4^+$), dissolved organic nitrogen (DON), and an increase in nitrates (NO_x) because of nitrification and microbial activity (Paillat *et al.*, 2005; Yoo *et al.*, 1999). Plants absorb nitrogen in the form of inorganic NH_4^+ and NO_x. Moreover, a longer stabilization time often leads to a shortage of storage space and excessive wastewater accumulation further impacting the environment. Dissolved organic nitrogen is necessary to maintain a nitrogen pool for the plants (Nasholm & Persson, 2001; Warren, 2014). Therefore, a reduction in the total nitrogen content of the digestate might act as a huge loss for the utilization of anaerobic digestate for the soil reclamation process.

Organic solid wastes like municipal sludge, kitchen waste, crop residues or industrial waste, accumulate enormous amounts of contaminants like heavy metals, drugs, antibiotics, microplastics (MPs), polycyclic aromatic hydrocarbons (PAHs), and furans such as furfural and 5-hydroxy methyl furfural (5-HMF) that need to be removed from the digestate for direct field applications. Li *et al.*, (2014) demonstrated that supplementation of BC at 75 g/l concentration resulted in 10 g/l removal of furfural from water. Oxidized BC modified with the addition of ruthenium particles stimulates the transformation of furfural into furfuryl alcohol under high temperature and pressure conditions (Bardestani *et al.*, 2020). Although further insights are needed to better understand the role of BC in eliminating furfurals and 5-HMF in AD. (Kim and Karthikeyan (2021) reported that furfural and 5-HMF above or at 1000 and 2000 mg/l, respectively, inhibits biomethane yield by 24.7 and 47.6% in the presence of glucose as a sole substrate. Furan derivatives, furfurals and 5-HMF are noxious contaminants produced during acid-pretreated lignocellulose hydrolysates (Koopman *et al.*, 2010) as well as during high-temperature pre-treatment of food wastes (Kim *et al.*, 2011). Elimination of these inhibitors has been reported by several groups via ether extraction/alkaline precipitation/enzymatic digestion using laccases (Mussatto & Roberto, 2004; Palmqvist & Hahn-Hagerdal, 2000). However, very few studies have been performed using BC for the removal of these derivatives.

BC supplementation enhances AD process stability against the inhibitory compounds together with the removal of refractory contaminants. According to Zhuang *et al.* (2009), heavy metals like Cu, Zn, As, Pb (Inyang *et al.*, 2016) are the major contaminants that prevent AD process stability and treatment. Aggregation of heavy metals on biosolids present in waste may range from 0.5 to 2 TS% (Pathak *et al.*, 2009). Yesil *et al.* (2021) reported the total amount of Cr, Cu, Ni and Zn in sewage sludge to be 1353.5, 706.5, 582.0, and 1645.5 mg/kg, respectively. Similarly, antibiotics like tetracyclines, quinolones and fluoroquinolones are abundant, up to 20 000 μg/kg, in livestock manure and sludge (Gaballah *et al.*, 2021). Global annual consumption of antibiotics is approximately $\geq$0.2 million tonnes, out of which maximum is contributed by veterinary drugs and growth factors (Wise, 2002) and due to the low absorption percentage of these compounds, more than 50% ends up in sewage. Table 12.1

Table 12.1 Effect of antibiotics present in waste at a molecular level.

Antibiotics	Substrate	Methane production	Mechanism of action	References
Roxithromycin (upto 1000 μg/l)	Dewatered sludge	Decreased methane production from 163 ml/g VS to 150 ml/g VS	Inhibits protein synthesis	Ni *et al.* (2020)
Norfloxacin (100 mg/l)	Dewatered sludge	Inoculation ratio above 1/1 inhibits methane production	Inhibits bacterial DNA gyrase	Zhi *et al.* (2019)
Oxytetracycline (100 mg/l)		Enhanced methane production by 1–25%	Inhibits protein synthesis	
Sulfadimethoxine (100 mg/l)		Enhanced methane production by 39–52%	Blocks folic acid pathway	
Chlortetracycline (≤1.28 mg/l)	Dairy manure	Inhibited methane production by 12–33% at >0.64 mg/l	Inhibits protein synthesis	Andriamanohiarisoamanana *et al.* (2020)
Oxytetracycline (≤1.28 mg/l)		Inhibited methane production by 20–32%		
Ceftiofur (≤250 mg/l)	Dairy manure	Inhibited methane production by 20% at >50 mg/l	Inhibits bacterial cell wall synthesis	Flores-Orozco *et al.* (2020)

represents the adverse effects of antibiotics on the molecular functioning of an organism. Furthermore, a comparative account of the use of engineered BC (pristine and composite BC) for antibiotics adsorption is given in Table 12.2.

On the other hand, microplastics (MPs), reported to be present up to 3160 particles per liter in wastewater, more than 90% of them could be retained in sewage sludge during wastewater treatment (Alavian Petroody *et al.*, 2021; He *et al.*, 2021). Of the several MPs present in sludge (28–89% polyester, 4–51% polyethylene, 4–35% terephthalate and 3–30% polyamide), fiber is present in copious amounts causing eco-toxicity to microbial communities and human health (He *et al.*, 2021; Kosuth *et al.*, 2018; Sun *et al.*, 2019). Additionally, incomplete combustion of fossil fuels generates polyaromatic hydrocarbons (PAHs), that get adsorbed and precipitated in the sewage during treatment in WWTPs (Needham & Ghosh, 2019; Sun *et al.*, 2019). Toxic phenolic and furan derivatives produced during pre-treatment of lignocellulosic biomass hinders the downstream processing and thus, are required to be washed out for sustaining the bioprocess. Toxic furfurals and 5-HMFs can be removed by ether extraction/alkaline precipitation/enzymatic metabolism using laccases (Mussatto & Roberto, 2004; Palmqvist & Hahn-Hagerdal, 2000). Thus, such refractory compounds must be removed/treated prior to the usage of anaerobic digestate. These constraints intimidate the use of some pragmatic process to ensure an unobjectionable, stable, and nutrient-rich digestate production. One such method is the augmentation of carbon-based conductive material – the BC. BC, synthesized via processes like pyrolysis and gasification of organic matter (agricultural wastes and municipal sewage sludge) (Rajapaksha *et al.*, 2019), is of great potential to reduce the environmental and climate impacts and establish a circular economy. BC can be used for a wide range of applications such as soil amendment, building material, compost, carbon-based conductive material for enhancing AD, and many more (Cha *et al.*, 2016; Kah *et al.*, 2017; Pan *et al.*, 2019 Woolf *et al.*, 2010). It interacts with various compounds via covalent or electrostatic interactions (Figure 12.2). BC has also been evaluated in agricultural and horticultural applications as well as a component of chemical fertilizers (Glaser *et al.*, 2002; Pandit *et al.*, 2018). It has also been reported to mitigate heavy metals in soil (Qin *et al.*, 2020; Rizwan *et al.*, 2016, 2018). Table 12.3 shows the heavy metals adsorption capacity of BC derived from various sources.

BC, as a waste-derived conductive carbon material, enhances the robustness of the digestion process against inhibitory compounds (antibiotics, MPs, humic acids, furfurals). BC aids in adsorbing these refractory substances, maintaining pH and improving the transfer of electrons between microbial communities via a process referred to as DIET (Gahlot *et al.*, 2020). It has been reported to enhance biogas production by 35% in ammonia-stressed bioreactors (Yan *et al.*, 2021).

Physico-chemical properties of BC such as large surface area, catalytic nature, strong surface charge, etc. (Huggins *et al.*; 2016; El-Naggar *et al.*, 2019; Mulligan *et al.*, 2009; Shaheen *et al.*, 2019), allow rapid electron transfers with the substrate and ultimately enhance biodegradation of organic matter. On a porous matrix, like BC, adsorption can occur via any of the following processes:

Table 12.2 Adsorption of antibiotics on engineered BC.

Antibiotic	Pristine BC				Composite BC			
	Adsorption conditions	**BC dosage**	**Q_{max} (maximum adsorption capacity)**	**Reference**	**Adsorption conditions**	**BC dosage**	**Q_{max} (maximum adsorption capacity)**	**References**
Sulfamethazine	72 h, 25°C	1 g/l	20.56 mg/g	Rajpaksha *et al.* (2015)	48 h, 25°C	1 g/l (BC+GO)	6.5 mg/g	Huang *et al.* (2017)
Sulfamethoxazole	72 h, 25°C	10–15 mg/l	1.9 mg/g	Xie *et al.* (2014)	24 h, 25°C	0.2 g/l (BC+CZF)	212.8 mg/g	Heo *et al.* (2019)
Sulfadiazine	48 h, 25°C	0.08–0.5 g/l	1.27 mg/g	Li *et al.* (2019)				
Sulfapyridine	72 h, 25°C	10–15 mg/l	1.5 mg/g	Xie *et al.* (2014)	24 h, 25°C	0.4 g/l (BC-HC-CNTs)	86% removal	Inyang *et al.* (2015)
Tetracycline	48 h, 25°C	0.08–0.5 g/l	5.53 mg/g	Li *et al.* (2019)	12 h	2 g/l (BC-MMT)	4.2 mg/g	Premarathna *et al.* (2019)
Tetracycline	12 h, 25°C	0.8–4 g/l	7.22 mg/g	Dai *et al.* (2020)	24 h, 30°C	0.8 g/l (IP-BC-Fe)	102 mg/g	Chen *et al.* (2019)
Tetracycline	5 d, 25°C	0.1 g/l	372.31 mg/g	Jang and Kan (2019)	6 h	0.5 g/l	31.50%	Fu *et al.* (2017)
Tetracycline	48 d, 25°C	0.1 g/l	35.9 mg/g	Tan *et al.* (2019)	24 h, 25°C	0.67 g/l (BC-Mg/Al)	125	Tan *et al.* (2016)

Table 12.3 Adsorption capacities of BC from different sources for heavy metals.

BC source	Pyrolysis temperature (°C)	Heavy metal	Adsorption capacity (mg/g)	References
Dairy manure	200	Pb	140.9	Cao *et al.* (2009)
Sewage sludge (anaerobically digested)	600		51.4	Inyang *et al.* (2011)
Oak bark	450	Cd	7.4	Mohan *et al.* (2014)
Flax shive	200		39.1	El-Shafey *et al.* (2002)
Soybean stalk	700	Hg	0.7	Kong *et al.* (2011)
Flax shive	200	Ni	33.4	El-Shafey *et al.* (2002)
Sugar beet tailings	300	Cr	123	Dong *et al.* (2011)
Dried sewage sludge	650		3	Otero *et al.* (2009)
Corn straw	600	Zn	11	Chen *et al.* (2011)
Bone char	500		34.7	Cheung *et al.* (2000)
Flex shive	160/200	Co	17.3/31.2	El-Shafey *et al.* (2002)
Dried sewage sludge	650	Cu	6.7	Otero *et al.* (2009)
Corn straw	600		12.5	Chen *et al.* (2011)

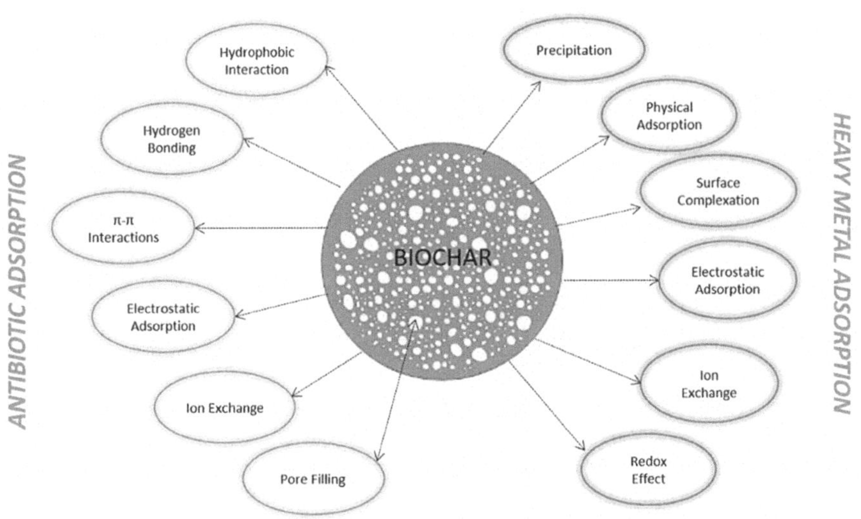

Figure 12.2 Molecular interactions upon adsorption of antibiotics and heavy metals on BC.

Table 12.4 Efficiency of corn straw and hardwood-based BC in removing and immobilization of microplastic beads.

BC type	Elements (w/w %)			Specific surface area (by EGME m^2/g)	Pore volume (ml)	MPs sphere removal efficiency	Inference
	Carbon	Oxygen	Nitrogen				
C300	70	18.3	3.1	808.3 ± 81.9	47.28	~97%	Complex honeycomb structure in corn straw BC synthesized at 500°C and hardwood BC leads to highest removal and immobilization of microplastic spheres.
C400	81.3	11.7	19.1	609 ± 83.5	46.14	~100%	
C500	68.7	19.1	14.1	177.5 ± 16.3	44.49	~100%	
Hardwood	84.7	14.1	0.3	292.8 ± 69.3	34.89	~100%	

Source: Wang *et al.* (2020a).

(1) transfer of adsorbate from the solution to the BC, (2) diffusion of adsorbate through the liquid film surrounding the BC, (3) diffusion of the adsorbate from BC surface to the active sites, and (4) interactions between adsorbate and BC. BC is advantageous over other catalytic processes (such as metal catalysis, photocatalysis, base catalysis) in the way that it is inexpensive, easily available, and eco-friendly. BC has been reported to enhance catalysis of persulfate during the digestate stabilization process (Hussain *et al.*, 2017; Jiang *et al.*, 2018). The liquid fraction of digestate is rich in nutrients that can be used as a fertilizer for soil reformation, especially when the digester is present near farms so that the overall transportation costs are reduced. However, these direct field applications might be impractical with respect to higher transportation costs and environmental hazards when the plant is distantly located (Fuchs & Drosg *et al.*, 2013). Thus, volume reduction for concentration of nutrients is a mandate for utilization of such a valuable digestate as a natural filler of nutrients in depleted soils. AD acts as an advantageous biochemical process explored to synthesize renewable energy source in the form of biogas (biomethane) from organic waste as well as nutrient recycling (Bareha *et al.*, 2018). Germination index (GI) is considered as an indicator of the quality and toxicity of the digestate. Steadiness of anaerobic digestate in stabilization ponds with GI value of more than 50% represents a stable effluent. Chen *et al.*, (2021) studied the synergistic effect of BC and persulfate (BC/PS) on the stabilization of the digestate. The results showed that BC/PS ameliorates the stabilization efficiency by higher microbial inactivation, particulate total nitrogen biodegradation, and reduced nitrate generation. At GI >50%, the stabilization time of the digestate shortened from 30 days to 11 days. Depending upon the protocol and modification method for BC synthesis, the final product can be categorized into modified and composite BC. BC composites can be either BC-metal compounds or BC-clays or BC-carbon compounds (Aljeboree & Alshirifi, 2018; Kyzas *et al.*, 2015; Rashed, 2013). Wang *et al.* (2020a) demonstrated the potential of BC (corn stover-C300, C400, C500 and hardwood BC) for improving removal of microplastic spheres from tertiary treatment system in WWTPs. They concluded that the removal efficiency of the filter to microplastic spheres was ≥95%, higher than sand filters (60–80%) with particle loss of ≤0.5% when the samples were flushed back with distilled water. Table 12.4 shows the efficiency of corn straw and hardwood-based BC in removing and immobilization of microplastic beads (10 μm diameter).

12.5 CONCLUSIONS

The word digestate refers to a set of a heterogeneous organic matter generated from urban and centralized AD plants receiving waste from various sources like industries, municipal, domestic, food, or livestock. Evaluation of digestate treatment processes is necessary for its complete valorization and utilization, keeping in mind the qualitative, economic and environmental aspects of the applied methodologies. One such method used for digestate pretreatment is BC supplementation. The heterogeneity in properties of BC such as vibrant physico-chemical and biological nature, synthesis from diverse feedstocks and treatment conditions, makes BC an appropriate carbon-based material to be used for improving the quality of the digestate and reducing the quantity of contaminants present in it. BC with some strong oxidant empowers the AD process. Advanced oxidation processes could accelerate biodegradation of organic matter and microbial inactivation (Chen *et al.*, 2021). Reduced nitrate generation in the digestate decreases the chances of its leaching into the soil, thereby, decreasing probability of groundwater pollution. Such BC augmented strong oxidation processes could help in solving the problems of wastewater turnover, large fluctuation of wastewater volume and insufficient pond space caused due to livestock expansion. Retention of refractory contaminants such as heavy metals, PAHs, microplastics, drugs, antibiotics, industrial toxicants, etc., in the digestate limits its direct application. Inherent advantages of BC can be further deciphered to render enhancement in AD performance. BC, undoubtedly, stimulates AD performance in case of organic solid waste (OSW) by improving biomethane production and reduction/removal of pollutants (Qi *et al.*, 2021) ascribed to its properties like large surface area for microbial attachments and its conductive behavior. However, the pollutants adsorbed by BC needs to be further processed/removed in order to stabilize the digestate to fulfill biosafety norms with respect to its final application. Elimination of contaminants is obligatory, especially in case of antibiotics (Homem & Santos, 2011), since it may lead to proliferation of bacterial communities with antibiotic resistance genes that might breed to serious ecological concerns. Despite of these promising assets, the benefits of BC remain highly debated. Development of new strategies is needed to combat the effects of ever-increasing OSWs generation. Further efforts are required to synthesize modified and advanced BC that could stimulate AD performance as well as produce digestate free from refractory pollutants, for direct application to soil reformation. BC addition to AD has proved to benefit the overall process. However, BC type and dosage are key elements for an efficient AD process. Furthermore, interactions between feedstock materials at specific AD operation conditions and the BC used should be meticulously studied for each process. BC enhanced both methane production and digestate quality. Thus, BC properties such as the sorption of contaminants and immobilization of microbial cells make it a promising approach for digestate treatment, either during or after AD. It is noteworthy that BC is a cost-effective scenario compared to other additives due to its production from renewable biomass. Therefore, recent studies have been focused on integrating slow pyrolysis and AD processes for a closed-loop efficient system for biomass valorization. Nevertheless, further effort in future research is still needed to reach a successful integrative process from an economic and sustainable perspective. In a nutshell, BC, as an adsorbent, helps in attenuating microbial toxicity, eliminating/mitigating refractory compounds and enhancing biogas quantity. It also helps in mediating faster electron transfer between syntropic microbes through DIET that ultimately stimulates AD process.

REFERENCES

Alavian Petroody S. S., Hashemi S. H., van Gestel C. A. M. (2021). Transport and accumulation of microplastics through wastewater treatment sludge processes. *Chemosphere*, **278**, 130471.

Aljeboree A. M. and Alshirifi A. N. (2018). Adsorption of pharmaceuticals as emerging contaminants from aqueous solutions on to friendly surfaces such as activated carbon: a review. *Journal of Pharmaceutical Sciences and Research*, **10**(9), 2252–2257.

Ambaye T. G., Rene E. R., Nizami A. S., Dupont C., Vaccari M. and van Hullebusch E. D. (2021). Beneficial role of biochar addition on the anaerobic digestion of food waste: a systematic and critical review of the operational parameters and mechanisms. *Journal of Environmental Management*, **290**, 112537, https://doi.org/10.1016/J.JENVMAN.2021.112537

Andriamanohiarisoamanana F. J., Ihara I., Yoshida G. and Umetsu K. (2020). Kinetic study of oxytetracycline and chlortetracycline inhibition in the anaerobic digestion of dairy manure. *Bioresource Technology*, **315**, 123810, https://doi.org/10.1016/j.biortech.2020.123810

Bardestani R., Biriaei R. and Kaliaguine S. (2020). Hydrogenation of furfural to furfuryl alcohol over Ru particles supported on mildly oxidized biochar. *Catalysts*, **10**(8), 934.

Bareha Y., Girault R., Jimenez J. and Trémier A. (2018). Characterization and prediction of organic nitrogen biodegradability during anaerobic digestion: a bioaccessibility approach. *Bioresource Technology*, **263**, 425–436.

Cao X., Ma L., Gao B. and Harris W. (2009). Dairy-manure derived biochar effectively sorbs lead and atrazine. *Environmental Science & Technology*, **43**(9):3285–3291, https://doi.org/10.1021/es803092k

Cha J. S., Park S. H., Jung S. C., Ryu C., Jeon J. K., Shin M. C. and Park Y. K. (2016). Production and utilization of biochar: a review. *Journal of Industrial and Engineering Chemistry*, **40**, 1–15, https://doi.org/10.1016/j.jiec.2016.06.002

Chen X., Chen G., Chen L., Chen Y., Lehmann J., McBride M. B. and Hay A. G. (2011). Adsorption of copper and zinc by biochars produced from pyrolysis of hardwood and corn straw in aqueous solution. *Bioresource Technology*, **102**(19), 8877–8884, https://doi.org/10.1016/j.biortech.2011.06.078

Chen J., Jiang X., Tang X., Sun Y. and Zhou L. (2021). Use of biochar/persulfate for accelerating the stabilization process and improving nitrogen stability of animal waste digestate. *Science of the Total Environment*, **757**, 144158.

Chen M., Tao X., Wang D., Xu Z., Xu X., Hu X., Xu N. and Cao X. (2019). Facilitated transport of cadmium by biochar-Fe_3O_4 nanocomposites in water-saturated natural soils. *Science of the Total Environment*, **684**, 265–275, https://doi.org/10.1016/j.scitotenv.2019.05.326

Cheng Q., Xu C., Huang W., Jiang M., Yan J., Fan G., Zhang J., Chen K., Xiao B. and Song G. (2020). Improving anaerobic digestion of piggery wastewater by alleviating stress of ammonia using biochar derived from rice straw. *Environmental Technology & Innovation*, **19**, 100948, https://doi.org/10.1016/J.ETI.2020.100948

Cheung C., Porter J. and McKay G. (2000). Sorption kinetics for the removal of copper and zinc from effluents using bone char. *Separation and Purification Technology*, **19**(1–2), 55–64, https://doi.org/10.1016/S1383-5866(99)00073-8

Chong C. C., Cheng Y. W., Ishak S., Lam M. K., Lim J. W., Tan I. S., Show P. L. and Lee K. T. (2022). Anaerobic digestate as a low-cost nutrient source for sustainable microalgae cultivation: a way forward through waste valorization approach. *Science of the Total Environment*, **803**, 150070, https://doi.org/10.1016/J.SCITOTENV.2021.150070

Codignole Luz F., Cordiner S., Manni A., Mulone V. and Rocco V. (2018). Biochar characteristics and early applications in anaerobic digestion – a review. *Journal of Environmental Chemical Engineering*, **6**, 2892–2909, https://doi.org/10.1016/J.JECE.2018.04.015

Costa A., Gusmara C., Gardoni D., Zaninelli M., Tambone F., Sala V. and Guarino M. (2017). The effect of anaerobic digestion and storage on indicator microorganisms in swine and dairy manure. *Environmental Science and Pollution Research*, **24**, 24135–24146, https://doi.org/10.1007/s11356-017-0011-5

Cui Y., Mao F., Zhang J., He Y., Tong Y. W. and Peng Y. (2021). Biochar enhanced high-solid mesophilic anaerobic digestion of food waste: cell viability and methanogenic pathways. *Chemosphere*, **272**, 129863, https://doi.org/10.1016/J.CHEMOSPHERE.2021.129863

Dai Y., Li J. and Shan D. (2020). Adsorption of tetracycline in aqueous solution by biochar derived from waste Auricularia auricula dregs. *Chemosphere*, **238**, 124432.

Dong X., Ma L. Q. and Li Y. (2011). Characteristics and mechanisms of hexavalent chromium removal by biochar from sugar beet tailing. *Journal of Hazardous Materials*, **190**(1–3), 909–915, https://doi.org/10.1016/j.jhazmat.2011.04.008

El-Naggar A., Lee M.-H., Hur J., Lee Y. H., Igalavithana A. D., Shaheen S. M., Ryu C., Rinklebe J., Tsang D. C. W. and Ok Y. S. (2019). Biochar-induced metal immobilization and soil biogeochemical process: An integrated mechanistic approach. *Science of the Total Environment*, **698**, 134112.

El-Shafey E. I., Cox M., Pichugin A. A. and Appleton Q. (2002). Application of a carbon sorbent for the removal of cadmium and other heavy metal ions from aqueous solution. *Journal of Chemical Technology & Biotechnology*, **77**, 429–436, https://doi.org/10.1002/jctb.577

Enders A., Hanley K., Whitman T., Joseph S. and Lehmann J. (2012). Characterization of biochars to evaluate recalcitrance and agronomic performance. *Bioresource Technology*, **114**, 644–653, https://doi.org/10.1016/J.BIORTECH.2012.03.022

Fabbri D. and Torri, C. (2016). Linking pyrolysis and anaerobic digestion (Py-AD) for the conversion of lignocellulosic biomass. *Current Opinion in Biotechnology*, **38**, 167–173, https://doi.org/10.1016/J.COPBIO.2016.02.004

Fagbohungbe M. O., Herbert B. M. J., Hurst L., Ibeto C. N., Li H., Usmani S. Q. and Semple K. T. (2017). The challenges of anaerobic digestion and the role of biochar in optimizing anaerobic digestion. *Waste Management*, **61**, 236–249, https://doi.org/10.1016/J.WASMAN.2016.11.028

Fazeli Sangani M., Abrishamkesh S. and Owens G. (2020). Physicochemical characteristics of biochars can be beneficially manipulated using post-pyrolyzed particle size modification. *Bioresource Technology*, **306**, 123157, https://doi.org/10.1016/J.BIORTECH.2020.123157

Flores-Orozco D., Patidar R., Levin D. B., Sparling R., Kumar A. and Cicek N. (2020). Effect of ceftiofur on mesophilic anaerobic digestion of dairy manure and the reduction of the cephalosporin-resistance gene cmy-2. *Bioresource Technology*, **301**, 122729, https://doi.org/10.1016/j.biortech.2019.122729

Fu D., Chen Z., Xia D., Shen L., Wang Y. and Li Q. (2017). A novel solid digestate-derived biochar-Cu NP composite activating H_2O_2 system for simultaneous adsorption and degradation of tetracycline. *Environmental Pollution*, **221**, 301–310.

Fuchs W. and Drosg B. (2013). Assessment of the state of the art of technologies for the processing of digestate residue from anaerobic digesters. *Water Science and Technology*, **67**(9), 1984–1993,

Gaballah M. S., Guo J., Sun H., Aboagye D., Sobhi M., Muhmood A. and Dong R. (2021). A review targeting veterinary antibiotics removal from livestock manure management systems and future outlook. *Bioresource Technology*, **333**, 125069.

Gahlot P., Ahmed B., Tiwari S. B., Aryal N., Khursheed A., Kazmi A. A. and Tyagi V. K. (2020). Conductive material engineered direct interspecies electron transfer (DIET) in anaerobic digestion: mechanism and application. *Environmental Technology & Innovation*, **20**, 101056, https://doi.org/10.1016/J.ETI.2020.101056

Ghysels S., Ronsse F., Dickinson D. and Prins W. (2019). Production and characterization of slow pyrolysis biochar from lignin-rich digested stillage from lignocellulosic ethanol production. *Biomass and Bioenergy*, **122**, 349–360, https://doi.org/10.1016/J.BIOMBIOE.2019.01.040

Giwa A. S., Xu H., Chang F., Wu J., Li Y., Ali N., Ding S. and Wang K. (2019). Effect of biochar on reactor performance and methane generation during the anaerobic digestion of food waste treatment at long-run operations. *Journal of Environmental Chemical Engineering*, **7**, 103067, https://doi.org/10.1016/J.JECE.2019.103067

Glaser B., Lehmann J. and Zech W. (2002). Ameliorating physical and chemical properties of highly weathered soils in the tropics with charcoal-a review. *Biology and Fertility of Soils*, **35**(4), 219– 230.

He Z. W., Yang W. J., Ren Y. X., Jin H. Y., Tang C. C., Liu W. Z., Yang C. X., Zhou A. J. and Wang A. J. (2021). Occurrence, effect, and fate of residual microplastics in anaerobic digestion of waste activated sludge: A state-of-the-art review. *Bioresource Technology*, **331**, 125035.

Heo J., Yoon Y., Lee G., Kim Y., Han J. and Park C. M. (2019). Enhanced adsorption of bisphenol A and sulfamethoxazole by a novel magnetic $CuZnFe_2O_4$–biochar composite. *Bioresource Technology*, **281**, 179–187.

Homem V. and Santos L. (2011). Degradation and removal methods of antibiotics from aqueous matrices – a review. *Journal of Environmental Management*, **92**(10), 2304–2347, https://doi.org/10.1016/j.jenvman.2011.05.023

Huang H., Tang J., Gao K., He R., Zhao H. and Werner D. (2017). Characterization of KOH modified biochars from different pyrolysis temperatures and enhanced adsorption of antibiotics. *RSC Advances*, **7**(24), 14640–14648.

Huggins T. M., Haeger A., Biffinger J. C. and Ren Z. J. (2016). Granular biochar compared with activated carbon for wastewater treatment and resource recovery. *Water Research*, **94**, 225–232.

Hussain I., Li M., Zhang Y., Li Y., Huang S., Du X., Liu G., Hayat W. and Anwar N. (2017). Insights into the mechanism of persulfate activation with nZVI/BC nanocomposite for the degradation of nonylphenol. *Journal of Chemical Engineering*, **311**, 163–172.

Hutchison M. L., Walters L. D., Avery S. M., Synge B. A. and Moore A. (2004). Levels of zoonoticagents in British livestock manures. *Letters in Applied Microbiology*, **39**(2), 207–214, https://doi.org/10.1111/j.1472-765X.2004.01564.x

Inyang M., Gao B., Ding W., Pullammanappallil P., Zimmerman A. R. and Cao X. (2011). Enhanced lead sorption by biochar derived from anaerobically digested sugarcane bagasse. *Separation Science and Technology*, **46**(12), 1950–1956, https://doi.org/10.1080/01496395.2011.584604

Inyang M., Gao B., Zimmerman A., Zhou Y. and Cao X. (2015). Sorption and cosorption of lead and sulfapyridine on carbon nanotube-modified biochars. *Environmental Science and Pollution Research*, **22**(3), 1868–1876.

Inyang M. I., Gao B., Yao Y., Xue Y., Zimmerman A., Mosa A., Pullammanappallil P., Ok Y. S. and Cao X. (2016). A review of biochar as a low-cost adsorbent for aqueous heavy metal removal. *Critical Reviews in Environmental Science and Technology*, **46**(4), 406–433, https://doi.org/10.1080/10643389.2015.1096880

Jang H. M. and Kan E. (2019). A novel hay-derived biochar for removal of tetracyclines in water. *Bioresource Technology*, **274**, 162–172.

Jiang S., Ling L., Chen W., Liu W., Li D. and Jiang H. (2018). High efficient removal of bisphenol A in a peroxymonosulfate/iron functionalized biochar system: mechanistic elucidation and quantification of the contributors. *Journal of Chemical Engineering*, **359**, 572–583.

Johansen A., Carter M. S., Jensen E. S., Hauggard-Nielsen H. and Ambus P. (2013). Effects of digestate from anaerobically digested cattle slurry and plant materials on soil microbial community and emission of CO_2 and N_2O. *Applied Soil Ecology*, **63**, 36–44, https://doi.org/10.1016/J.APSOIL.2012.09.003

Kah M., Sigmund G., Xiao F. and Hofmann T. (2017). Sorption of ionizable and ionic organic compounds to biochar, activated carbon and other carbonaceous materials. *Water Research*, **124**, 673–692.

Kant Bhatia S., Palai A. K., Kumar A., Kant Bhatia R., Kumar Patel A., Kumar Thakur V. and Yang Y. H. (2021). Trends in renewable energy production employing biomass-based biochar. *Bioresource Technology*, **340**, 125644, https://doi.org/10.1016/J.BIORTECH.2021.125644

Khalid Z. B., Siddique M. N. I., Nayeem A., Adyel T. M., Ismail S. B. and Ibrahim M. Z. (2021). Biochar application as sustainable precursors for enhanced anaerobic digestion: a systematic review. *Journal of Environmental Chemical Engineering*, **9**, 105489, https://doi.org/10.1016/J.JECE.2021.105489

Kim J. R. and Karthikeyan K. G. (2021). Effects of severe pretreatment conditions and lignocellulose-derived furan byproducts on anaerobic digestion of dairy manure. *Bioresource Technology*, **340**, 125632.

Kim H. K., Choi Y. W., Lee E. N., Park J. K., Kim S. G., Park D. J., Kim B. S., Lim Y. T. and Yoon S. (2011). 5-Hydroxymethylfurfural from black garlic extract prevents TNF alpha-induced monocytic cell adhesion to HUVECs by suppression of vascular cell adhesion molecule-1 expression, reactive oxygen species generation and NF-kappaB activation. *Phytotherapy Research*, **25**(7), 965–974.

Kong H., He J., Gao Y., Wu H. and Zhu X. (2011). Cosorption of phenanthrene and mercury (II) from aqueous solution by soybean stalk-based biochar. *Journal of Agricultural and Food Chemistry*, **59**(22), 12116–12123.

Koopman F., Wierckx N., de Winde J. H. and Ruijssenaars H. J. (2010). Identification and characterization of the furfural and 5-(hydroxymethyl) furfural degradation pathways of Cupriavidus basilensis HMF14. *Proceedings of the National Academy of Sciences*, **107**(11), 4919–4924.

Kosuth M., Mason S. A. and Wattenberg E. V. (2018). Anthropogenic contamination of tap water, beer, and sea salt. *PLoS One*, **13**(4), e0194970.

Kumar A. N., Dissanayake P. D., Masek O., Priya A., Ki Lin C. S., Ok Y. S. and Kim S. H. (2021a). Recent trends in biochar integration with anaerobic fermentation: win-win strategies in a closed-loop. *Renewable and Sustainable Energy Reviews*, **149**, 111371, https://doi.org/10.1016/J.RSER.2021.111371

Kumar M., Dutta S., You S., Luo G., Zhang S., Show P. L., Sawarkar A. D., Singh L. and Tsang D. C. W. (2021b). A critical review on biochar for enhancing biogas production from anaerobic digestion of food waste and sludge. *Journal of Cleaner Production*, **305**, 127143, https://doi.org/10.1016/J.JCLEPRO.2021.127143

Kyzas G. Z., Fu J., Lazaridis N. K., Bikiaris D. N. and Matis K. A. (2015). New approaches on the removal of pharmaceuticals from wastewaters with adsorbent materials. *Journal of Molecular Liquids*, **209**, 87–93, https://doi.org/10.1016/j.molliq.2015.05.025

Li Y., Shao J., Wang X., Deng Y., Yang H. and Chen H. (2014). Characterization of modified biochars derived from bamboo pyrolysis and their utilization for target component (furfural) adsorption. *Energy & Fuels*, **28**(8), 5119–5127.

Li C., Zhu X., He H., Fang Y., Dong H., Lü J., Li J. and Li Y. (2019). Adsorption of two antibiotics on biochar prepared in air-containing atmosphere: influence of biochar porosity and molecular size of antibiotics. *Journal of Molecular Liquids*, **274**, 353–361.

Lim E. Y., Tian H., Chen Y., Ni K., Zhang J. and Tong Y. W. (2020). Methanogenic pathway and microbial succession during start-up and stabilization of thermophilic food waste anaerobic digestion with biochar. *Bioresource Technology*, **314**, 123751, https://doi.org/10.1016/J.BIORTECH.2020.123751

Lü F., Luo C., Shao L. and He P. (2016). Biochar alleviates combined stress of ammonium and acids by firstly enriching *Methanosaeta* and then *Methanosarcina*. *Water Research*, **90**, 34–43, https://doi.org/10.1016/j.watres.2015.12.029

Lü F., Liu Y., Shao L. and He P. (2019). Powdered biochar doubled microbial growth in anaerobic digestion of oil. *Applied Energy*, **247**, 605–614, https://doi.org/10.1016/J.APENERGY.2019.04.052

Luo C., Lü F., Shao L. and He P. (2015). Application of eco-compatible biochar in anaerobic digestion to relieve acid stress and promote the selective colonization of functional microbes. *Water Research*, **68**, 710–718, https://doi.org/10.1016/J.WATRES.2014.10.052

Ma J., Pan J., Qiu L., Wang Q. and Zhang Z. (2019). Biochar triggering multipath methanogenesis and subdued propionic acid accumulation during semi-continuous anaerobic digestion. *Bioresource Technology*, **293**, 122026, https://doi.org/10.1016/j.biortech.2019.122026

Ma H., Hu Y., Kobayashi T. and Xu K. Q. (2020). The role of rice husk biochar addition in anaerobic digestion for sweet sorghum under high loading condition. *Biotechnology Reports*, **27**, e00515, https://doi.org/10.1016/J.BTRE.2020.E00515

Masebinu S. O., Akinlabi E. T., Muzenda E. and Aboyade A. O. (2019). A review of biochar properties and their roles in mitigating challenges with anaerobic digestion. *Renewable and Sustainable Energy Reviews*, **103**, 291–307, https://doi.org/10.1016/J.RSER.2018.12.048

Mohan D., Sarswat A., Ok Y. S. and Pittman C. U. (2014). Organic and inorganic contaminants removal from water with biochar, a renewable, low cost and sustainable adsorbent – a critical review. *Bioresource Technology*, **160**:191–202, https://doi.org/10.1016/j.biortech.2014.01.120

Mulligan C. N., Davarpanah N., Fukue M. and Inoue T. (2009). Filtration of contaminated suspended solids for the treatment of surface water. *Chemosphere*, **74**, 779–786.

Mussatto S. I. and Roberto I. C. (2004). Alternatives for detoxification of diluted-acid lignocellulosic hydrolyzates for use in fermentative processes: a review. *Bioresource Technology* **93**(1), 1–10.

Nasholm T. and Persson J. (2001). Plant acquisition of organic nitrogen in boreal forests. *Physiologia Plantarum*, **111**(4), 419–426, https://doi.org/10.1034/j.1399-3054.2001.1110401.x

Needham T. P. and Ghosh U. (2019). Four decades since the ban, old urban wastewater treatment plant remains a dominant source of PCBs to the environment. *Environmental Pollution*, **246**, 390–397.

Ni B. J., Zeng S., Wei W., Dai X. and Sun J. (2020). Impact of roxithromycin on waste activated sludge anaerobic digestion: methane production, carbon transformation and antibiotic resistance genes. *Science of the Total Environment*, **703**, 134899, https://doi.org/10.1016/j.scitotenv.2019.134899

Otero M., Rozada F., Morán A., Calvo L. and García A. (2009). Removal of heavy metals from aqueous solution by sewage sludge based sorbents: competitive effects. *Desalination*, **239**(1–3), 46–57, https://doi.org/10.1016/j.desal.2008.03.005

Paavola T. and Rintala J. (2008). Effects of storage on characteristics and hygienic quality of digestates from four co-digestion concepts of manure and biowaste. *Bioresource Technology*, **99**, 7041–7050, https://doi.org/10.1016/j.biortech.2008.01.005

Paillat J.-M., Robin P., Hassouna M. and Leterme P. (2005). Predicting ammonia and carbon dioxide emissions from carbon and nitrogen biodegradability during animal waste composting. *Atmospheric Environment*, **39**(36), 6833–6842, https://doi.org/10.1016/j.atmosenv.2005.07.045

Palmqvist E. and Hahn-Hägerdal B. (2000). Fermentation of lignocellulosic hydrolysates. II: inhibitors and mechanisms of inhibition. *Bioresource Technology*, **74**(1), 25–33.

Pan J., Ma J., Zhai L., Luo T., Mei Z. and Liu H. (2019). Achievements of biochar application for enhanced anaerobic digestion: A review. *Bioresource Technology*, **292**, 122058.

Pandit N. R., Mulder J., Hale S. E., Martinsen V., Schmidt H. P. and Cornelissen, G. (2018). Biochar improves maize growth by alleviation of nutrient stress in a moderately acidic low-input Nepalese soil. *Science of the Total Environment*, **625**, 1380–1389.

Paritosh K. and Vivekanand V. (2019). Biochar enabled syntrophic action: solid state anaerobic digestion of agricultural stubble for enhanced methane production. *Bioresource Technology*, **289**, 121712, https://doi.org/10.1016/J.BIORTECH.2019.121712

Pathak A., Dastidar M. G., and Sreekrishnan T. R. (2009). Bioleaching of heavy metals from sewage sludge: a review. *Journal of Environmental Management*, **90**(8), 2343–2353.

Premarathna K. S. D., Rajapaksha A. U., Adassoriya N., Sarkar B., Sirimuthu N. M., Cooray A., Ok Y. S. and Vithanage M. (2019). Clay-biochar composites for sorptive removal of tetracycline antibiotic in aqueous media. *Journal of Environmental Management*, **238**, 315–322.

Qi Q., Sun C., Cristhian C., Zhang T., Zhang J., Tian H., He Y. and Tong Y. W. (2021). Enhancement of methanogenic performance by gasification biochar on anaerobic digestion. *Bioresource Technology*, **330**, 124993, https://doi.org/10.1016/J.BIORTECH.2021.124993

Qian K., Kumar A., Patil K., Bellmer D., Wang D., Yuan W. and Huhnke R. L. (2013). Effects of biomass feedstocks and gasification conditions on the physiochemical properties of char. Energies, **6**(8), 3972–3986, https://doi.org/10.3390/en6083972

Qian K., Kumar A., Zhang H., Bellmer D. and Huhnke R. (2015). Recent advances in utilization of biochar. *Renewable and Sustainable Energy Reviews*, **42**, 1055–1064, https://doi.org/10.1016/J.RSER.2014.10.074

Qin Y., Yin X., Xu X., Yan X., Bi F. and Wu W. (2020). Specific surface area and electron donating capacity determine biochar's role in methane production during anaerobic digestion. *Bioresource Technology*, **303**, 122919, https://doi.org/10.1016/J.BIORTECH.2020.122919

Qiu L., Deng Y. F., Wang F., Davaritouchaee M. and Yao Y. Q. (2019). A review on biochar-mediated anaerobic digestion with enhanced methane recovery. *Renewable and Sustainable Energy Reviews*, **115**, 109373, https://doi.org/10.1016/J.RSER.2019.109373

Rajapaksha A. U., Ok Y. S., El-Naggar A., Kim H., Song F., Kang S. and Tsang Y. F. (2019). Dissolved organic matter characterization of biochars produced from different feedstock materials. *Journal of Environmental Management*, **233**, 393–399, https://doi.org/10.1016/j.jenvman.2018.12.069

Rajapaksha A. U., Vithanage M., Ahmad M., Seo D. C., Cho J. S., Lee S. E., Lee S. S. and Ok Y. S. (2015). Enhanced sulfamethazine removal by steam-activated invasive plant-derived biochar. *Journal of Hazardous Materials*, **290**, 43–50.

Rasapoor M., Young B., Asadov A., Brar R., Sarmah A. K., Zhuang W. Q. and Baroutian S. (2020). Effects of biochar and activated carbon on biogas generation: a thermogravimetric and chemical analysis approach. *Energy Conversion and Management*, **203**, 112221, https://doi.org/10.1016/J.ENCONMAN.2019.112221

Rashed M. N. (2013). Adsorption technique for the removal of organic pollutants from water and wastewater. *Organic Pollutants-Monitoring, Risk and Treatment*, **7**, 167–194.

Riau V., Burgos L., Camps F., Domingo F., Torrellas M., Antón A. and Bonmatí A. (2021). Closing nutrient loops in a maize rotation. Catch crops to reduce nutrient leaching and increase biogas production by anaerobic co-digestion with dairy manure. *Waste Management*, **126**, 719–727, https://doi.org/10.1016/J.WASMAN.2021.04.006

Rizwan M., Ali S., Qayyum M. F., Ibrahim M., Zia-ur-Rehman M., Abbas T. and Ok Y.S. (2016). Mechanisms of biochar-mediated alleviation of toxicity of trace elements in plants: a critical review. *Environmental Science and Pollution Research*, **23**(3), 2230–2248.

Rizwan M., Ali S., Abbas T., Adrees M., Zia-ur-Rehman M., Ibrahim M., Abbas F., Qayyum M. F. and Nawaz R. (2018). Residual effects of biochar on growth, photosynthesis and cadmium uptake in rice (Oryza sativa L.) under Cd stress with different water conditions. *Journal of Environmental Management*, **206**, 676–683.

Romero-Güiza M. S., Vila J., Mata-Alvarez J., Chimenos J. M. and Astals S. (2016). The role of additives on anaerobic digestion: a review. *Renewable and Sustainable Energy Reviews*, **58**, 1486–1499, https://doi.org/10.1016/J.RSER.2015.12.094

Shaheen S. M., El-Naggar A., Wang J., Hassan N. E. E., Niazi N. K., Wang H., Tsang D. C. W., Ok Y. S., Bolan N. and Rinklebe J. (2019). Biochar as an (Im)mobilizing agent for the potentially toxic elements in contaminated soils. In: Biochar from Biomass and Waste, Ok Y. S., Tsang D. C. W., Bolan N., Novak J. M. (eds.), Elsevier, UK, 255–274. ISBN 9780128117293.

Shen Y., Linville J. L., Ignacio-de Leon P. A. A., Schoene R. P. and Urgun-Demirtas M. (2016). Towards a sustainable paradigm of waste-to-energy process: enhanced anaerobic digestion of sludge with woody biochar. *Journal of Cleaner Production*, **135**, 1054–1064, https://doi.org/10.1016/J.JCLEPRO.2016.06.144

Shen Y., Forrester S., Koval J. and Urgun-Demirtas M. (2017). Yearlong semi-continuous operation of thermophilic two-stage anaerobic digesters amended with biochar for enhanced biomethane production. *Journal of Cleaner Production*, **167**, 863–874, https://doi.org/10.1016/J.JCLEPRO.2017.05.135

Sun C., Liu F., Song Z., Wang J., Li Y., Pan Y., Sheng T. and Li L. (2019). Feasibility of dry anaerobic digestion of beer lees for methane production and biochar enhanced performance at mesophilic and thermophilic temperature. *Bioresource Technology*, **276**, 65–73, https://doi.org/10.1016/j.biortech.2018.12.105

Sunyoto N. M. S., Zhu M., Zhang Z. and Zhang D. (2016). Effect of biochar addition on hydrogen and methane production in two-phase anaerobic digestion of aqueous carbohydrates food waste. *Bioresource Technology*, **219**, 29–36, https://doi.org/10.1016/j.biortech.2016.07.089

Sunyoto N. M. S., Zhu M., Zhang Z. and Zhang D. (2017). Effect of biochar addition and initial pH on hydrogen production from the first phase of two-phase anaerobic digestion of carbohydrates food waste. *Energy Procedia*, **105**, 379–384, https://doi.org/10.1016/j.egypro.2017.03.329

Suthar S., Sharma B., Kumar K., Rajesh Banu J. and Tyagi V. K. (2022). Enhanced biogas production in dilute acid-thermal pretreatment and cattle dung biochar mediated biomethanation of water hyacinth. *Fuel*, **307**, 121897, https://doi.org/10.1016/J.FUEL.2021.121897

Tan X., Liu S., Liu Y., Gu Y., Zeng G., Cai X., Yan Z., Yang C., Hu X. and Chen B. (2016). One-pot synthesis of carbon supported calcined-Mg/Al layered double hydroxides for antibiotic removal by slow pyrolysis of biomass waste. *Scientific Reports*, **6**(1), 1–12.

Tan Z., Zhang X., Wang L., Gao B., Luo J., Fang R., Zou W. and Meng, N. (2019). Sorption of tetracycline on H_2O_2-modified biochar derived from rape stalk. *Environmental Pollutants and Bioavailability*, **31**(1), 198–207.

Tiwari S. B., Dubey M., Ahmed B., Gahlot P., Khan A. A., Rajpal A., Kazmi A. A. and Tyagi V. K. (2021). Carbon-based conductive materials facilitated anaerobic co-digestion of agro waste under thermophilic conditions. *Waste Management*, **124**, 17–25, https://doi.org/10.1016/J.WASMAN.2021.01.032

Tripathi M., Sahu J. N. and Ganesan P. (2016). Effect of process parameters on production of biochar from biomass waste through pyrolysis: a review. *Renewable and Sustainable Energy Reviews*, **55**, 467–481, https://doi.org/10.1016/J.RSER.2015.10.122

Wang G., Li Q., Dzakpasu M., Gao X., Yuwen C. and Wang X. C. (2018a). Impacts of different biochar types on hydrogen production promotion during fermentative co-digestion of food wastes and dewatered sewage sludge. *Waste Management*, **80**, 73–80, https://doi.org/10.1016/J.WASMAN.2018.08.042

Wang G., Li Q., Gao X. and Wang X. C. (2018b). Synergetic promotion of syntrophic methane production from anaerobic digestion of complex organic wastes by biochar: performance and associated mechanisms. *Bioresource Technology*, **250**, 812–820, https://doi.org/10.1016/J.BIORTECH.2017.12.004

Wang G., Li Q., Li Y., Xing Y., Yao G., Liu Y., Chen R. and Wang X. C. (2020a). Redox-active biochar facilitates potential electron transfer between syntrophic partners to enhance anaerobic digestion under high organic loading rate. *Bioresource Technology*, **298**, 122524, https://doi.org/10.1016/J.BIORTECH.2019.122524

Wang Z., Sedighi M. and Lea-Langton A. (2020b). Filtration of microplastic spheres by biochar: removal efficiency and immobilisation mechanisms. *Water Research*, **184**, 116165, https://doi.org/10.1016/j.watres.2020.116165

Wang C., Liu Y., Wang C., Xing B., Zhu S., Huang J., Xu X. and Zhu L. (2021a). Biochar facilitates rapid restoration of methanogenesis by enhancing direct interspecies electron transfer after high organic loading shock. *Bioresource Technology*, **320**, 124360, https://doi.org/10.1016/J.BIORTECH.2020.124360

Wang J., Zhao Z. and Zhang Y. (2021b). Enhancing anaerobic digestion of kitchen wastes with biochar: link between different properties and critical mechanisms of promoting interspecies electron transfer. *Renewable Energy*, **167**, 791–799, https://doi.org/10.1016/J.RENENE.2020.11.153

Warren C. R. (2014). Organic N molecules in the soil solution: what is known, what is unknown and the path forwards. *Plant and Soil*, **375**(1), 1–19.

Weber K. and Quicker P. (2018). Properties of biochar. *Fuel*, **217**, 240–261, https://doi.org/10.1016/J.FUEL.2017.12.054

Wei W., Guo W., Ngo H. H., Mannina G., Wang D., Chen X., Liu Y., Peng L. and Ni B. J. (2020). Enhanced high-quality biomethane production from anaerobic digestion of primary sludge by corn stover biochar. *Bioresource Technology*, **306**, 123159, https://doi.org/10.1016/J.BIORTECH.2020.123159

Wise R. (2002). Antimicrobial resistance: priorities for action. *Journal of Antimicrobial Chemotherapy*, **49**(4), 585–586.

Woolf D., Amonette J. E., Street-Perrott F. A., Lehmann J. and Joseph S. (2010). Sustainable biochar to mitigate global climate change. *Nature Communications*, **1**(1), 1–9.

Xie M., Chen W., Xu Z., Zheng S. and Zhu, D. (2014). Adsorption of sulfonamides to demineralized pine wood biochars prepared under different thermochemical conditions. *Environmental Pollution*, **186**, 187–194.

Xu Q., Liao Y., Cho E. and Ko J. H. (2020). Effects of biochar addition on the anaerobic digestion of carbohydrate-rich, protein-rich, and lipid-rich substrates. *Journal of the Air & Waste Management Association*, **70**, 455–467, https://doi.org/10.1080/10962247.2020.1733133

Xu J., Li C., Zhu N., Shen Y. and Yuan H. (2021). Alleviating the nitrite stress on anaerobic ammonium oxidation by pyrolytic biochar. *Science of the Total Environment*, **774**, 145800, https://doi.org/10.1016/J.SCITOTENV.2021.145800

Yan M., Zhu X., Treu L., Ravenni G., Campanaro S.,Goonesekera E. M., Ferrigno R., Jacobsen C. S., Zervas A., Angelidaki I., and Fotidis I. A. (2021). Comprehensive evaluation of different strategies to recover methanogenic performance in ammonia-stressed reactors. *Bioresource Technology*, **336**, 125329.

Yesil H., Molaey R., Calli B. and Tugtas A. E. (2021). Removal and recovery of heavy metals from sewage sludge via three-stage integrated process. *Chemosphere*, **280**, 130650.

Yoo H., Ahn K. H., Lee H. J., Lee K.-H., Kwak Y. J. and Song K. G. (1999). Nitrogen removal from synthetic wastewater by simultaneous nitrification and denitrification (SND) via nitrite in an intermittently-aerated reactor. *Water Research*, **33**(1), 145–154, https://doi.org/10.1016/S0043-1354(98)00159-6

Zhao L., Cao X., Mašek O. and Zimmerman A. (2013). Heterogeneity of biochar properties as a function of feedstock sources and production temperatures. *Journal of Hazardous Materials*, **256–257**, 1–9, https://doi.org/10.1016/J.JHAZMAT.2013.04.015

Zhao L., Wang Z., Ren H. Y., Chen C., Nan J., Cao G. L., Yang S. S. and Ren N. Q. (2021a). Residue cornstalk derived biochar promotes direct bio-hydrogen production from anaerobic fermentation of cornstalk. *Bioresource Technology*, **320**, 124338, https://doi.org/10.1016/J.BIORTECH.2020.124338

Zhao W., Yang H., He S., Zhao Q. and Wei L. (2021b). A review of biochar in anaerobic digestion to improve biogas production: performances, mechanisms and economic assessments. *Bioresource Technology*, **341**, 125797, https://doi.org/10.1016/J.BIORTECH.2021.125797

Zhao Z., Cao Y., Li S. and Zhang Y. (2021c). Effects of biowaste-derived biochar on the electron transport efficiency during anaerobic acid orange 7 removal. *Bioresource Technology*, **320**, 124295, https://doi.org/10.1016/J.BIORTECH.2020.124295

Zhi S., Li Q., Yang F., Yang Z. and Zhang K. (2019). How methane yield, crucial parameters and microbial communities respond to the stimulating effect of antibiotics during high solid anaerobic digestion. *Bioresource Technology*, **283**, 286–296, https://doi.org/10.1016/j.biortech.2019.03.083

Zhuang P., McBride M. B., Xia H., Li N. and Li Z. (2009). Health risk from heavy metals via consumption of food crops in the vicinity of Dabaoshan mine, South China. *Science of Total Environment*, **407**(5), 1551–1561, https://doi.org/10.1016/j.biortech.2019.03.083

doi: 10.2166/9781789062755_0293

Chapter 13

Hydrothermal post-treatment of anaerobic digestate

Donato Scrinzi, Roberta Ferrentino, Luca Fiori and Gianni Andreottola*
Department of Civil, Environmental and Mechanical Engineering, University of Trento, Trento, Italy
**Corresponding author e-mail: donato.scrinzi@unitn.it*

ABSTRACT

Currently, anaerobic digestion (AD) is the most used process to stabilize the sewage sludge produced in a wastewater treatment plant. Although AD allows achieving a reduction from 30% up to 50% of the sludge volume to be disposed of and to produce biogas and biomethane usable as an energy source, the digested sludge still contains a large amount of compounds that could be valorized. Furthermore, there is a strong pressure to apply innovative solutions able to ensure a high level of sludge reduction. Among all the possible sludge reduction technologies, hydrothermal carbonization (HTC) of municipal sludge has recently been recognized as a promising technology to efficiently reduce waste volume, recover bioenergy and nutrients, such as phosphorus and nitrogen. Thus, this chapter aims to address the main advantages of the combined HTC and AD treatment, considering a plant scheme where HTC is implemented as post-treatment of AD. In this configuration, the main advantages are: (1) sludge reduction; (2) enhancement in biogas and biomethane production thanks to the recirculation to the AD unit of the liquid and solid products of the HTC treatment; and (3) the possibility to recover valuable products from the solid product of the HTC in the form of phosphorous and nitrogen. Moreover, the chapter presents an overview of the possible application schemes related to technical and economical aspects of AD and HTC integrated systems.

Keywords: Anaerobic digestion post-treatment, biogas enhancement, hydrothermal carbonization, nutrients recovery, sludge reduction, sludge valorization

13.1 INTRODUCTION

The management of waste biomass is one of the critical challenges around the world, still not completely solved. Many countries, particularly in Europe, do not allow any reuse of sewage sludge digestate (SSD) in agricultural soils, mainly due to its phytotoxicity assessed in many studies (Coelho *et al.*, 2018; Yoshida *et al.*, 2018), thus making SSD one of the most problematic biowaste to manage. The continuous growth of urban population together with the increasingly tight (European) regulations on the disposal of sewage sludge have recently raised

Editor(s): Vinay Kumar Tyagi, Kaoutar Aboudi and Cigdem Eskicioglu

the need for developing innovative processes and technologies for a proper, but sustainable management. Nowadays, sewage sludge treatment plant managers and the scientific community in the sector are focusing, on the one hand, on the development of innovative technologies for sewage sludge reduction and, on the other, on its environmentally safe disposal and/or valorization. Integrating anaerobic digestion (AD) with other processes to recover high-value by-products from SSD is today a key global issue. Among hydrothermal processes (HTPs) which can help towards this objective, hydrothermal carbonization (HTC) is considered a promising opportunity (Aragón-Briceño *et al.*, 2021a; Gherghel *et al.*, 2019) for an efficient and sustainable sludge minimization and valorization, following the principle of the circular economy.

The aim of this chapter is to highlight the main advantages of the implementation of the HTC process as post-treatment of the AD and to report technical and economic feasibility assessments of possible plant configurations for the application of the combined process at an industrial scale.

13.2 WHAT ARE HTPs?

HTPs are thermochemical treatments applied to feedstocks heated with water, which can be present as both original moisture or external adding. In this manner, differently from other dry thermochemical processes (such as pyrolysis), HTPs does not require a drying pre-treatment. Therefore, wet residual biowastes could be used as feedstocks of the HTPs that were assessed as promising technologies for waste management.

Depending on the reaction temperature, thus on the severity of the process, HTPs can be classified as described in Table 13.1 (Biller & Ross, 2016).

Thermal hydrolysis (TH) is a well-established technology which is usually performed in two steps, a heating step at 140–180°C, up to 10 bar, for 30 up to 60 min, followed by a rapid decompression to ambient pressure, ensuring both sterilization and solid fragmentation.

When the reaction temperature is higher than 180°C, the process shifts to HTC. In this case, a set of reactions occurs in parallel to hydrolysis, dehydration, and decarboxylation, converting the biomass solids into a carbonaceous solid product, named hydrochar (HC), a liquid product, called HTC liquor (HTCL) and also HTC gas, composed mainly of carbon dioxide.

Depending on the severity of the process (temperature in the range 180–250°C, and duration from some minutes to many hours) and on physicochemical properties of the feedstock (typology, pH, organic matter content, nutrients and metals content, potential presence of a catalyst), the final properties of the HTC products vary, opening the road for a huge number of possible applications, ranging from nutrients and high-value organic compounds recovery to advanced materials production. The advantages of coupling AD and HTC for SSD treatment are explained in detail in the following sections.

For temperatures over 250–280°C, the hydrothermal conditions favor the recombination of fragments and new organic compounds in solution into a bio-oil or bio-crude, through condensation, cyclization and polymerization. In subcritical conditions, under 374°C and 22.1 MPa, the process is called

Table 13.1 Types of hydrothermal treatments for biowaste conversion.

Process	Temperature range	Related saturation pressure	Description	Main product
Thermal hydrolysis (TH)	<180°C	<1 MPa	Hydrolysis is the main reaction	Slurry
Hydrothermal carbonization (HTC)	180–250°C	1–4 MPa	Many reactions occur in parallel together with hydrolysis, dehydration, decarboxylation	Hydrochar (solid)
Hydrothermal liquefaction (HTL)	250–370°C	4–22 MPa	Recombination mechanisms of reactive fragments (condensation, cyclization, and polymerization)	Bio-crude (liquid)
Hydrothermal gasification (HTG) or Supercritical water gasification (SCWG)	>370°C	>22 MPa	Steam-reforming, water-gas shift, and methanation	Syngas (gaseous)

Source: Adapted from Biller & Ross (2016) and Aragón-Briceño *et al.* (2021a).

hydrothermal liquefaction, HTL; conversely, in supercritical conditions, it is called hydrothermal gasification (HTG). In this latter case, the production of gas, called syngas, and the solid reduction are enhanced.

However, the high thermal needs and the pressures required at more severe HTPs, as for HTL and HTG, make it more difficult to implement and manage these processes at full scale rather than lab- and bench-scale.

13.3 ADVANTAGES OF HTC APPLIED TO SEWAGE SLUDGE ANAEROBIC DIGESTATE

Many studies around the world are assessing the performances of HTC applied to all kinds of biomasses, especially residual biomasses. This section collects the experimental evidence on the main advantages of using the HTC as post-treatment of SSD. The plant configuration, visible in Figure 13.1, is analyzed and used as a reference for the following discussion as a plant upgrading scenario for a sludge line in a typical wastewater treatment plant (WWTP), where AD is coupled with HTC (here thereof AD-HTC).

In this scenario, the primary and secondary sludge are collected in sludge lines and sent together to an anaerobic digester. The digestate produced can be subjected to a thickening process in order to reach a total solids (TS) content of 10%, a suitable concentration for the HTC process. On the one hand, the dewatered liquid can be circulated back to the water line, on the other, the thickened SSD is sent to an HTC unit. The product of the HTC process, called

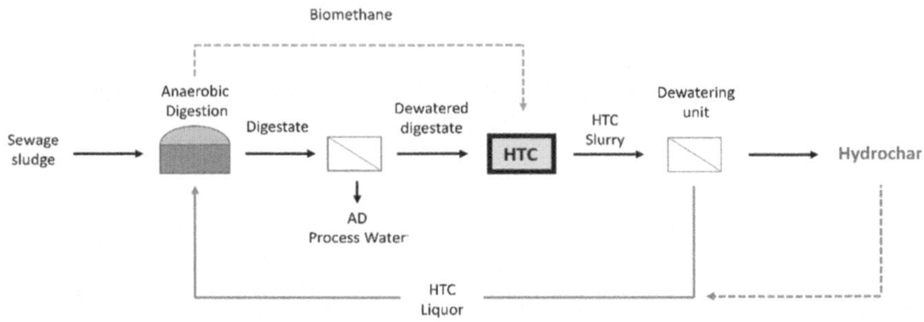

Figure 13.1 Scheme of the integration of the AD with the HTC process: AD–HTC scenario. The green biomethane stream represents a heat input to HTC. [Adapted from Ferrentino *et al.* (2020), Creative Commons Attribution License CC BY 4.0].

slurry, is dewatered to produce a solid, called hydrochar (up to 60%TS) and a liquor, called HTC liquor (HTCL). Finally, both hydrochar and HTCL, in different amounts, can be recirculated back to the AD process to enhance the production of biogas and biomethane that can be converted into thermal and electrical energy usable to support HTC energy requirement.

The main advantages deriving from the integration of AD-HTC are as follows (Nuchdang *et al.*, 2018):

1. Contribute to reduce the digestate volume, in the final form of hydrochar, due to both solids yield reduction and dewaterability improvement. The reduction of digestate volume promotes the reduction of the costs for sludge management and disposal (see §14.2.1) which is the first cost item of a WWTP. Moreover, the hydrochar is sterilized, avoiding any biological contamination issue due to the presence of pathogens.
2. Improve the biogas and biomethane production, where HTCL and eventually part of the hydrochar are recycled back to AD (see §14.2.2).
3. Favor the recovery of valuable products from sludge where nutrients, mainly P and N, can be extracted from hydrochar to produce fertilizers usable in land applications (see §14.2.3).

Given these advantages, the HTC process implemented as post-treatment of AD could strongly contribute to valorize the excess sludge trying to use all the resources available within it by following, as much as possible, the principle of the circular economy.

13.3.1 Effects on digestate volume reduction

One of the main effects of the HTC applied to SSD is related to sludge volume reduction. This effect is of particular interest for waste management since it contributes to solve the economical and environmental problems of sludge disposal. Sludge volume reduction is assessed through the evaluation of the solid yield after the HTC treatment. As reported by several literature studies, as process severity increases, namely increasing temperature and residence time, the solid yield in general decreases resulting in a remarkable sludge volume reduction. Recently, Liu *et al.* (2021), reviewing several literature works, reported that HTC studies showed a mean solid yield of 60.2 (22.6 10% percentile, 84.9 90% percentile), highlighting that there is a considerable variation due to the severity of the reaction. For example, Berge *et al.* (2011), using SSD as feedstock for an HTC process at 250°C for 20 h, obtained a very low solid yield of 47.1%. Moreover, Kim *et al.* (2014), investigating the HTC treatment to convert SSD into clean solid fuels, reported different values of solid yield corresponding to temperatures between 180°C and 280°C. The results revealed that at a lower temperature (i.e., 180°C) the solid yield was equal to 93.9%; on the contrary, increasing the temperature up to 220°C, the solid yield decreased to 88.7%. A further temperature increase to 250°C and 280°C led to a decrease in the solid yield resulting equal to, respectively, 83.4% and 80.4%. Thus, this study confirmed that increasing the process severity, in particular the temperature, has a significant effect on the solid yield. These results agree with those reported by Aragón-Briceño *et al.* (2017), who evaluated the effect of temperature on solid yield for hydrochar derived from HTC of SSD. The results showed that the hydrochar yields obtained were equal to 73.42% at 220°C and 68.79% at 250°C.

In addition to the sludge reduction due to the solid yield, the HTC post-treatment of the SSD enhances the solid reduction by means of an enhancement in sludge dewaterability. As reported in several literature studies, sludge dewaterability could be improved by thermal treatments at a temperature higher than 150°C (Bougrier *et al.*, 2008; Neyens & Baeyens, 2003; Nyktari *et al.*, 2017). Presently, HTC is considered one of the most promising technologies for solving the problem of sludge dewatering. HTC treatment leads to the breaking of the sludge flocs, less hydrophilic functional groups, small surface area, lower viscosity of water, and gas formation, finally resulting in an increase of free water (Wang *et al.*, 2019). Thus, HTC process is able to convert bound water, chemically bound to the particles and stored within them, to free water thus enhancing dewaterability.

Escala *et al.* (2012) performed HTC of digested sludge at 205°C for a residence time equal to 7 h. Dewaterability of the HTC product was evaluated by centrifugation tests (4000 rpm for 10 min) and then mechanical pressing (4 MPa at room temperature). The results revealed that the moisture content of the treated sludge decreased from 80% up to 48%. This result was confirmed by the study of Kim *et al.* (2014), who applied HTC to treat digested sludge at 180, 220 and 250°C for 30 min and evaluated the dewaterability by means of capillary suction time (CST) measurements and centrifugation (4000 rpm for 10 min at room temperature). The CST of the raw sludge was equal to 72 s and decreased to 15.3, 10.9 and 10.1 s, respectively, at 180°C, 220°C and 250°C. Moreover, centrifugation tests showed a reduction of the moisture content from 96% (raw sludge) to 81% (sludge treated at 200°C).

Ahmed *et al.* (2021) highlighted the dependence of dewaterability on HTC treatment duration, at the fixed HTC temperature of 190°C. The dewaterability was evaluated considering CST measurements. The CST decreased from 278 ± 26 s for the raw slurry to 204 ± 33 s after 30 min of HTC, revealing a very poor improvement in sludge dewaterability. On the contrary, when the

HTC treatment duration increased to 1 hour, a strong reduction in the CST value was observed (25 ± 2 s), meaning an increase in dewaterability by a factor of 7. Moreover, when increasing the HTC temperature up to 2 and 3 h, the corresponding CST values were 24 ± 3 s and 16.3 ± 1.4 s. Thus, HTC treatment of 2 and 3 h did not cause a further enhancement of dewaterability, such as to justify the increase in reaction time. Hence, results of the CST measurements recommended treatment duration of 1 h for the mechanical separation of hydrochar from the HTCL while lower treatment times could be used when there is no need for a solid/liquid separation after the HTC treatment, meaning a post utilization of the whole HTC slurry.

13.3.2 Effects on biogas production enhancement

As reported above, one of the main advantages of using HTC as post-treatment of the AD is the enhancement in the biogas and biomethane production due to the feeding of HTC by-products to the AD unit. The amount of biogas produced depends on sludge characteristics and the main HTC treatment conditions. Aragón-Briceño *et al.* (2017), feeding the AD with slurry after HTC at 220 and 250°C, reported an increase in the biomethane production between 17% and 58% compared with the control. Moreover, feeding the AD only with HTCL, an increase between 29% and 58% in the biomethane production was observed compared to the control sample. Thus, the biomethane enhancement of the HTCL was similar to that of the HTC slurry. Ahmed *et al.* (2021), using a mixture (TS 10% ± 0.6) of digestate and dewatered sludge, evaluated the increase in the specific biogas production of the HTCL – by means of biochemical methane potential (BMP) tests – obtained at different HTC treatment times: 30 min, 1, 2 and 3 h. The results reported specific biogas yields equal to 235 ± 9, 157 ± 9, 146 ± 23 and 135 ± 1 mL biogas/g COD_{added} for the HTCLs obtained at 30 min, 1, 2 and 3 h HTC treatment times, respectively, corresponding to a methane content of 63.0%, 66.4%, 65.8% and 63.7% of the produced biogas. Moreover, the TCOD inlet removal was equal to 35%, 37%, 45% and 44% at 30 min, 1, 2 and 3 h HTC treatment times, respectively. Thus, results revealed a strong reduction of the biogas production passing from an HTC treatment time of 30 min up to 1 h. On the contrary, a little decrease was highlighted when the HTC treatment time passed from 1 to 3 h. Hence, the biogas and methane yields are affected by HTC reaction severity: increasing the reaction severity, a decrease in the biogas production could be observed (Gaur *et al.*, 2020). The results of Ahmed *et al.* (2021) were consistent with those of Parmar and Ross (2019) who, performing HTC of digestate sludge at 200°C and 1 h treatment time, obtained a methane production of 137.7 mL CH_4/g COD_{added}.

In addition to the HTCL, a portion of the hydrochar could be fed to the AD unit, in order to further increase the biogas production. Ferrentino *et al.* (2020) investigated the improvement in biomethane production due to the recycling to the AD of HTCL and hydrochar derived from HTC of digestate and dewatered sludge mixture at 190°C for 1 h. The results showed that the specific biomethane production reached 102 ± 3 mL CH_4/g COD_{added} when 5 g of HTCL was fed to AD (inoculum 67 g) in addition to primary (21 g) and secondary sludge (7 g). In this case, the methane production was almost doubled compared to that of AD of primary and secondary sludge accounting for 55 ± 20 mL CH_4/g COD_{added}. Moreover, the specific methane production increased, even more, when both HTCL and hydrochar were added to the AD. Adding 6% of hydrochar, compared to the total feedstock biomasses (primary sludge, secondary sludge and HTCL), a slight increase in biomethane production, equal to 113 ± 11 mL CH_4/g COD_{added}, was observed compared to the one with only HTCL. On the contrary, the biomethane production increased up to 187 ± 18 mL CH_4/g COD_{added} when 45% of hydrochar was added. This means that some biodegradable molecular organic compounds were present in the hydrochar and could contribute to enhancing biogas and biomethane production in the AD process.

13.3.3 Nutrients recovery

Since HTC strongly affects the distribution of nutrients among its products, the integrated AD-HTC process can favor the shift from standard treatment plants to biorefineries, thus enhancing the production of valuable materials usable as fertilizer. An important property of HTPs is the sterilization of the products, which guarantees the pathogens inactivation during nutrient recovery and subsequent agricultural application (Muhmood *et al.*, 2021). The application of HTC from a nitrogen and phosphorus perspective has been recently investigated (Aragón-Briceño *et al.*, 2021a), including the influence in nutrient transformation due to the variation of process water pH (Zheng *et al.*, 2022).

13.3.3.1 Phosphorus recovery

Many recent studies tried to summarize the possible reaction pathways of phosphorus during HTC of SSD. Ovsyannikova *et al.* (2019) performed an experimental analysis on phosphate behavior under HTC (180, 220, and 260°C under 1, 2, and 4 h of reaction time in batch reactor) of SSD at two different dry matter to water ratio (B/W), namely 0.2 and 0.1 w/w. The results showed that most of the organic phosphate was converted into inorganic forms during HTC at all reaction conditions. The P is part of a complex panel of reactions involving the metals contained in the feedstock, thus the P availability changes after the process. NaOH-extractable P decreases (Al- and Fe-bound) in favor to stable insoluble Ca-bound forms, immobilized into the solid hydrochar. However, the content of total P into the solid phase increases at increasing HTC temperatures. Other studies (Volpe *et al.*, 2020; Wang *et al.*, 2020) confirmed the previous statements, finding that almost 90% of total P can be recovered as phosphate salts after HTC and acid extraction of SSD hydrochar. Zheng *et al.* (2020) assessed the effect of adjusting the molar ratio of Ca/P for enhancing the formation of hydroxyapatite (biocompatible with soil and long-term fertilizer) from SSD through HTC. As a matter of fact, the addition of more than 4% in weight of CaO powder was sufficient to convert all the initial inorganic phosphate into Ca-bound phosphates. This fact can be particularly useful in all the contexts in which sludge use for agricultural purposes is allowed.

Gerner *et al.* (2021) studied the effect of the acidification point after HTC (200°C for 4 h); in particular, they tested P recovery through acid stripping (PS, Figure 13.2a) of HTC slurry compared to the acid leaching (PL, Figure 13.2b) of the hydrochar after solid/liquid separation, using sulfuric acid as solvent

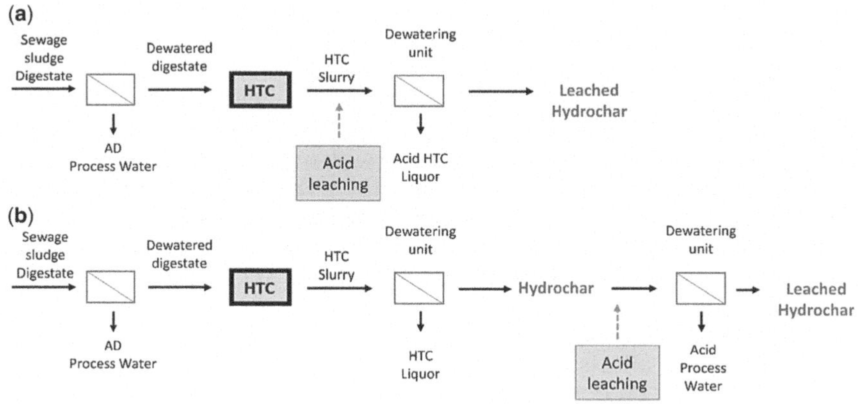

Figure 13.2 Two scenarios for acid leaching before (a) or after (b) HTC slurry dewatering [Adapted from Gerner *et al.* (2021) and Lucian *et al.* (2021), Creative Common Attribution License C BY 4.0].

(pH 2). They found that about 14 kg H_2SO_4 were consumed per each kg of P recovered in both the cases studied. The results of recovered nutrients in the liquid phase for PS and PL were around 84% and 71% for phosphorus, and 53% and 54% for nitrogen, respectively. A similar study performed on P recovery from agro-industrial digestate with an industrial-scale HTC reactor was done by Lucian *et al.* (2021). The acids tested were chloridic, sulfuric, and citric acid applied to wet hydrochar after centrifugation, or before filtration with also additives (perlite or cellulose) to promote cake formation. Data showed that phosphorus was successfully recovered through struvite precipitation with a recovery mass yield between 51 and 98 wt% of the P contained in the HTC filtrates. The best recovery yields were achieved when starting from centrifuged hydrochar treated with 0.5 M citric acid (90.9 wt%) and 0.5 M sulfuric acid (97.9 wt%), leading to, respectively, total P recovery rates of approximately 84.6 and 91.0 wt% in comparison with the initial P content of the agro-industrial feedstock.

13.3.3.2 Nitrogen recovery

Around 80% of ammonia produced is globally used as fertilizer (International Plant Nutrition Institute, 2015). According to Kyriakou *et al.* (2020), ammonia production has a strong impact all over the world, representing about 1%–2% of global energy consumption and 1.44% of global CO_2 emissions. Hence, there is a necessity of recovering ammonia applying innovative processes. He *et al.* (2015) proposed a full analysis of the pathways under near-critical HTPs, together with an integrated HTP-ammonia recovery through air stripping of the liquor. N removal efficiency achieved 76.9% and a concentrated ammonium solution was obtained. Moreover, the addition of CaO as a catalyst for enhanced hydrolysis and cracking led to 86.4% N removal efficiency from solid residues.

The coupled catalyzed HTC-air stripping helped to recover 62.0% of the total N present in SSD.

In a more recent study, Hämäläinen *et al.* (2021) proposed an HTC integrated plant with ammonia recovery and use of hydrochar as fuel or as a soil amendment, testing different reaction temperatures in mild conditions, 210, 230, and 250°C, for 0.5 or 1 h and SSD with two different solids contents, namely 15 and 25% TS. The authors found out that HTC at 210°C for 0.5 h was the mildest condition for hydrochar production, but the temperature of 230°C could be regarded as the optimum temperature, balancing the methane production potential and a minimum ammonium content to enable ammonium recovery. A simulation applied to a centralised biogas plant resulted in an increase in annual biogas production and ammonium recovery by 5% and 25%, respectively. Moreover, the hydrochar annually produced could be used as fuel to produce 83 GJ/year or to direct 350 tons P/year to agriculture.

13.4 TECHNO-ECONOMICAL FEASIBILITY OF AD-HTC IMPLEMENTATION AT INDUSTRIAL SCALE

In this paragraph, three techno-economic case studies on the integration of the HTC process in the sludge line as a post-treatment of the AD process are presented. The first case study refers to a new WWTP (named Trento 3) under construction in the province of Trento (Italy) having a capacity of 300 000 population equivalent (PE) and an influent flow rate of 96 000 m^3/d. An analysis of two scenarios where AD and HTC processes were integrated was proposed by Scrinzi *et al.* (2020). In the first scenario proposed, the HTC process was fed with thickened SSD produced inside the WWTP Trento 3 (Figure 13.3a), while, in the second scheme, dewatered SSDs coming from other local WWTPs located in the surrounding areas were also added as feedstock for the HTC process, thus upgrading the plant as a centralized sludge treatment facility (Figure 13.3b). After HTC process at 190°C for 1 h, the yields of solid, liquid and gaseous were equal to 72.8, 26.0, and 1.2%, respectively, for a centrifuged SSD with 10% TS (Ferrentino *et al.*, 2020). At these operating parameters, 90.4% on a dry basis of the initial total phosphorus could be segregated into the solid product at the end of the HTC process (Volpe *et al.*, 2020). The final TS of hydrochar was considered equal to 60% of dry matter after a dewatering process. The results of the simulation reported a significant effect on sludge reduction equal to −70% and −75%, respectively, for scenarios 1 and 2. Moreover, the HTCL was recirculated to the AD, causing an increase in the digester flow of +22.9% and +59.1%, respectively, implying an increment in methane production of +15.5% and +40%, respectively, for the two scenarios, considering a value of COD in the HTC liquor equal to 42.5 kg COD/m^3 (Merzari *et al.*, 2019) and a methane production yield equal to 0.144 m^3 CH_4/kg COD_{added} (Wirth *et al.*, 2015). An economic assessment was performed, considering the extra costs deriving for HTC operation (31 €/ton HC, from TerraNova® Ultra, TerraNova Energy GmbH) together with the gains from increased methane production (average

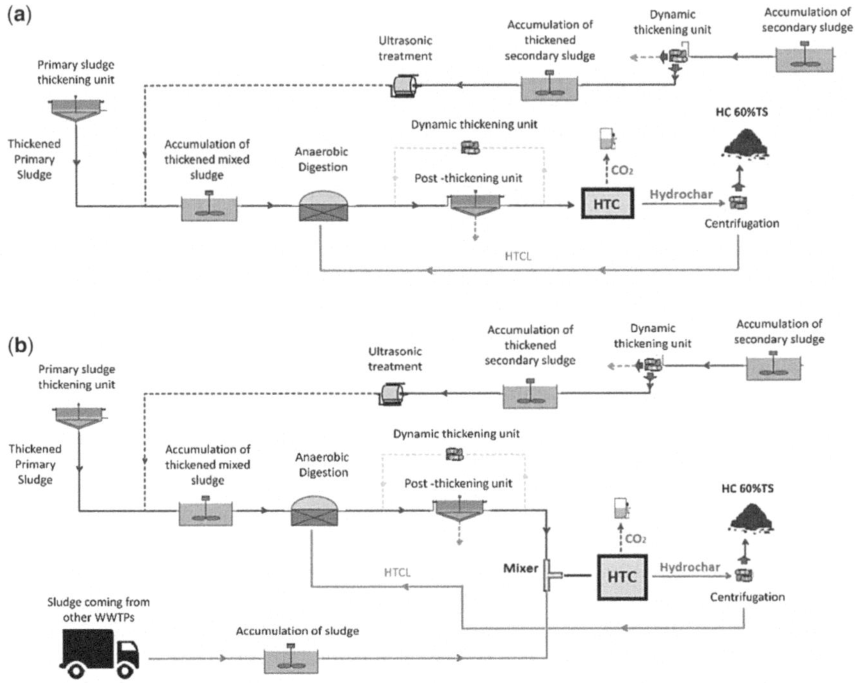

Figure 13.3 Scheme of Trento WWTP: (a) scenario 1 and (b) scenario 2. Adapted from Scrinzi *et al.* (2020).

price of methane was assumed equal to 0.30 €/m^3), the savings from the reduced sludge transportation, and the reduced final hydrochar disposal cost (102.4 €/ton – A: the reference year 2019, 150 €/ton – B: expected sludge disposal cost). The savings resulted of being equal to 31 and 1226 k€/year, respectively, for scenarios 1 and 2 (A), becoming 439 and 2507 k€/year in the case of increased disposal price (B). Furthermore, the possibility of nutrient recovery was assessed, considering only the external sludge aliquot (the only one with up-to-date data). The input parameter was 70% of P recoverable as phosphate salt with acid leaching from hydrochar and subsequent sodium hydroxide addition, and removal of total Kjeldahl nitrogen (TKN) of 37% with ammonium stripping at a pH of 9, and 53% increasing the pH up to 10, from stirred HTCL, with a final ammonium sulfate precipitation. Therefore, the estimate of the amount of phosphorous salt was around 103 ton P/year, while the nitrogen recovered as ammonium sulfate was estimated to be 140 ton N/year with pH 9, reaching up to 200 ton N/year with pH 10. These amounts will be even higher considering also the contribution from the internal sludge produced from the under construction Trento 3 WWTP.

In conclusion, both scenarios allowed for a reduced sludge production with a notable increase in the produced biogas, due to the HTC liquor recirculation to the AD, up to 2.5 M€/year of management savings, and possible nutrients recovery as valuable fertilizers.

An interesting treatment scheme was reported by Gerner *et al.* (2021), who proposed an economic feasibility assessment on the application of HTC at a WWTP with AD, and a following phosphorus acid extraction with two alternative methods, as proposed in Figure 13.2. Hence, these schemes promoted the recycling of phosphorus from sewage sludge that will be mandatory in Switzerland by 2026 and in Germany by 2029. The investment costs and annual fixed costs, given by the HTC-company GRegio Energie AG, Switzerland, included the HTC process, the liquid–solid separation, drying and briquetting of the hydrochar. Annual fixed costs were based on Swiss rates capital costs and calculated on a payback period of 10 years with 2% annual interest. Only the costs (based on lab experiments) of in-house acid consumption for the P-extraction were considered, while industrial-scale plant costs and/or fertilizer production were not included, as well as transport and process water treatment costs were neglected. Input parameters of the evaluation were 11 200 tons/year of SSD at 15% TS, treated at 200°C for 4 h in the continuous industrial-scale HTC plant from GRegio Energie AG (reactor volume 5.6 m^3). The results of the study showed that the specific overall costs for HTC implementation were 608 USD per ton TS. Both methods assumed equal costs of 3.96 USD per kg P recovered due to acid consumption, equivalent to usage of sulfuric acid per dry ton SSD of 406 kg for PS and 349 kg for PL. However, considering that the average disposal cost (including transportation and drying) per dry ton sewage sludge ranged in 2001 between 669 and 882 USD (cement plant or incineration), reaching in 2021 the cost of 1173 USD (incineration at WWTP Werdhölzli, Zurich, Switzerland), the study claims HTC as an economically feasible and competitive process for SSD treatment and nutrient recovery.

Aragón-Briceño *et al.* (2021b) reported a complete AD-HTC-AD (Figure 13.4) system tested under six operational conditions. In particular, three different

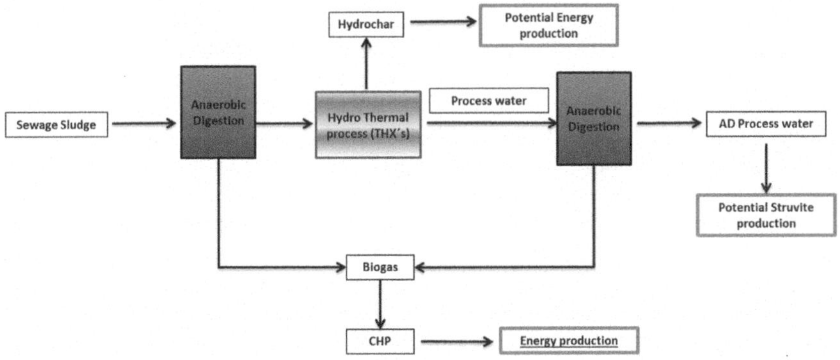

Figure 13.4 Scheme of the AD-HTC-AD scenario for energy and struvite recovery. Reprinted with permission from Aragón-Briceño *et al.* (2021b) [License N. 5198300713161 Nov. 29, 2021].

kinds of SSD (primary, secondary, and a mixture) were considered as feedstock and treated at two different HTC reaction temperatures, 160°C and 250°C for 0.5 h. The simulation found that the overall total solid reduction ranged in the scenarios from 47% to 68%. The energy balance was positive in all the proposed scenarios, with a net energy production of 14–188 kWh per ton SSD treated, or in the range 104–270 kWh per ton SSD treated, when the hydrochar was considered as a fuel source. From the economic point of view, based on the UK rates, the implementation of the process in a WWTP with a treatment capacity of 100 000 PE caused a total profit ranging from 390 to 614 k€/year (corresponding to 26–41 €/ton of SSD treated). The total gain included the electricity production from biomethane and hydrochar combustion and the recovery of struvite as valuable material. Since primary sludge has an organic content higher than secondary sludge, the solutions with primary or mixed sludge are preferable. Furthermore, the reaction temperature of 250°C permitted the highest proceeds due to high recovery of struvite, considering an average price of 475.5 €/ton struvite.

13.5 CONCLUSIONS AND FUTURE PERSPECTIVES

HTPs are perfect candidates to upgrade large-scale plants treating moist waste biomass, such as digestate and sludge since they operate with water without the need for drying and are highly tunable technologies. HTC promotes a mean solid yield of around 60% (in the range of 20–80%), depending on the feedstock properties and the reaction temperature and duration. Together with the improved dewaterability of the slurry, the solids reduction helps in reducing the digested sludge streams to be disposed of. Experimental tests on the biomethane potential of HTC liquor demonstrated that up to 140 mL CH_4/g COD_{added} can be recovered from HTCL recirculation, even up to 190 mL CH_4/g COD_{added} when a fraction of hydrochar is added. Moreover, the integration of HTC in a WWTP enables the possibility of in-house nutrient recovery in the framework of circular economy, in particular ammonium and phosphate. They can be reclaimed separately or with a combined process, such as for struvite precipitation for agricultural use. In both cases, increased HTC reaction temperatures generate an accumulation of ammonium (mainly in the liquid phase) and phosphate (>90% of initial P into the solid phase). The recovery of other higher added-value by-products can be conceived in a biorefinery concept, like polyphenols and other organic compounds. From an economic point of view, many studies assessed the techno-economic feasibility at a large scale of the implementation of AD-HTC in WWTPs, with benefits deriving from reduced residues, higher biomethane production, eventual struvite production, and use of hydrochar as fuel for fossil coal substitution. Since the properties of the HTPs products are highly dependent on the feedstock characteristics and the operating conditions, the design of new treating plants or the refurbishment of existing plants must be properly assessed on a case-by-case basis, taking into account the main purposes, the technically feasible possibilities, and the evolving legislation and market prices.

REFERENCES

Ahmed M., Andreottola G., Elagroudy S., Negm M. S. and Fiori L. (2021). Coupling hydrothermal carbonization and anaerobic digestion for sewage digestate management: influence of hydrothermal treatment time on dewaterability and bio-methane production. *Journal of Environmental Management*, **281**, 111910, https://doi.org/10.1016/j.jenvman.2020.111910

Aragón-Briceño C. I., Pozarlik A. K., Bramer E. A., Niedzwiecki L., Pawlak-Kruczek H. and Brem G. (2021a). Hydrothermal carbonization of wet biomass from nitrogen and phosphorus approach: a review. *Renewable Energy*, **171**, 401–415, Elsevier Ltd, https://doi.org/10.1016/j.renene.2021.02.109

Aragón-Briceño C., Ross A. B. and Camargo-Valero M. A. (2017). Evaluation and comparison of product yields and bio-methane potential in sewage digestate following hydrothermal treatment. *Applied Energy*, **208**, 1357–1369, https://doi.org/10.1016/j.apenergy.2017.09.019

Aragón-Briceño C. I., Ross A. B. and Camargo-Valero M. A. (2021b). Mass and energy integration study of hydrothermal carbonization with anaerobic digestion of sewage sludge. *Renewable Energy*, **167**, 473–483, Elsevier Ltd, https://doi.org/10.1016/j.renene.2020.11.103

Berge N. D., Ro K. S., Mao J., Flora J. R. V., Chappell M. A. and Bae S. (2011). Hydrothermal carbonization of municipal waste streams. *Environmental Science and Technology*, **45**(13), 5696–5703, American Chemical Society, https://doi.org/10.1021/es2004528

Biller P. and Ross A. B. (2016). Production of biofuels via hydrothermal conversion. In: Handbook of Biofuels Production: Processes and Technologies, 2nd edn. Woodhead Publishing, pp. 509–547. ISBN 9780081004555. https://doi.org/10.1016/B978-0-08-100455-5.00017-5

Bougrier C., Delgenès J. P. and Carrère H. (2008). Effects of thermal treatments on five different waste activated sludge samples solubilisation, physical properties and anaerobic digestion. *Chemical Engineering Journal*, **139**(2), 236–244, https://doi.org/10.1016/j.cej.2007.07.099

Coelho J. J., Prieto M. L., Dowling S., Hennessy A., Casey I., Woodcock T. and Kennedy N. (2018). Physical-chemical traits, phytotoxicity and pathogen detection in liquid anaerobic digestates. *Waste Management*, **78**, 8–15, Pergamon, https://doi.org/10.1016/j.wasman.2018.05.017

Escala M., Zumbühl T., Koller C., Junge R. and Krebs R. (2012). Hydrothermal carbonization as an energy-efficient alternative to established drying technologies for sewage sludge: a feasibility study on a laboratory scale. *Energy and Fuels*, **27**(1), 454–460, American Chemical Society, https://doi.org/10.1021/ef3015266

Ferrentino R., Merzari F., Fiori L. and Andreottola G. (2020). Coupling hydrothermal carbonization with anaerobic digestion for sewage sludge treatment: influence of HTC liquor and hydrochar on biomethane production. *Energies*, **13**(23), 6262, MDPI AG, https://doi.org/10.3390/en13236262

Gaur R. Z., Khoury O., Zohar M., Poverenov E., Darzi R., Laor Y. and Posmanik R. (2020). Hydrothermal carbonization of sewage sludge coupled with anaerobic digestion: integrated approach for sludge management and energy recycling. *Energy Conversion and Management*, **224**, 113353, Elsevier Ltd, https://doi.org/10.1016/j.enconman.2020.113353

Gerner G., Meyer L., Wanner R., Keller T. and Krebs R. (2021). Sewage sludge treatment by hydrothermal carbonization: feasibility study for sustainable nutrient recovery and fuel production. *Energies*, **14**(9), 2697, Multidisciplinary Digital Publishing Institute, https://doi.org/10.3390/en14092697

Gherghel A., Teodosiu C. and De Gisi S. (2019). A review on wastewater sludge valorisation and its challenges in the context of circular economy. *Journal of Cleaner Production*, **228**, 244–263, Elsevier Ltd, https://doi.org/10.1016/j.jclepro.2019.04.240

Hämäläinen A., Kokko M., Kinnunen V., Hilli T. and Rintala J. (2021). Hydrothermal carbonisation of mechanically dewatered digested sewage sludge – energy and nutrient recovery in centralised biogas plant. *Water Research*, **201**, 117284, Pergamon, https://doi.org/10.1016/j.watres.2021.117284

He C., Wang K., Yang Y., Amaniampong P. N. and Wang J. Y. (2015). Effective nitrogen removal and recovery from dewatered sewage sludge using a novel integrated system of accelerated hydrothermal deamination and air stripping. *Environmental Science and Technology*, **49**(11), 6872–6880, https://doi.org/10.1021/acs.est.5b00652

International Plant Nutrition Institute. (2015). Ammonia. *Nutrient Source Specifics*, No. 10, Ref # 10070. Available at: http://www.ipni.net/specifics-en.

Kim D., Lee K. and Park K. Y. (2014). Hydrothermal carbonization of anaerobically digested sludge for solid fuel production and energy recovery. *Fuel*, **130**, 120–125, Elsevier Ltd, https://doi.org/10.1016/j.fuel.2014.04.030

Kyriakou V., Garagounis I., Vourros A., Vasileiou E. and Stoukides M. (2020). An electrochemical Haber–Bosch process. *Joule*, **4**(1), 142–158, Cell Press, https://doi.org/10.1016/j.joule.2019.10.006

Liu H., Basar I. A., Nzihou A. and Eskicioglu C. (2021). Hydrochar derived from municipal sludge through hydrothermal processing: a critical review on its formation, characterization, and valorization. *Water Research*, **199**, 117186, Pergamon, https://doi.org/10.1016/j.watres.2021.117186

Lucian M., Merzari F., Gubert M., Messineo A. and Volpe M. (2021). Industrial-scale hydrothermal carbonization of agro-industrial digested sludge: filterability enhancement and phosphorus recovery. *Sustainability*, **13**, 9343, Multidisciplinary Digital Publishing Institute, https://doi.org/10.3390/su13169343

Merzari F., Langone M., Andreottola G. and Fiori L. (2019). Methane production from process water of sewage sludge hydrothermal carbonization. A review. Valorising sludge through hydrothermal carbonization. *Critical Reviews in Environmental Science and Technology*, **49**(11), 947–988, Taylor and Francis Inc., https://doi.org/10.1080/10643389.2018.1561104

Muhmood A., Wang X., Dong R., Xiao H. and Wu S. (2021). Quantitative characterization and effective inactivation of biological hazards in struvite recovered from digested poultry slurry. *Water Research*, **204**, 117659, Pergamon, https://doi.org/10.1016/j.watres.2021.117659

Neyens E. and Baeyens J. (2003). A review of thermal sludge pre-treatment processes to improve dewaterability. *Journal of Hazardous Materials*, **98**, 51–67, https://doi.org/10.1016/S0304-3894(02)00320-5

Nuchdang S., Frigon J. C., Roy C., Pilon G., Phalakornkule C. and Guiot S. R. (2018). Hydrothermal post-treatment of digestate to maximize the methane yield from the anaerobic digestion of microalgae. *Waste Management*, **71**, 683–688, Pergamon, https://doi.org/10.1016/j.wasman.2017.06.021

Nyktari E., Danso-Boateng E., Wheatley A. and Holdich R. (2017). Anaerobic digestion of liquid products following hydrothermal carbonisation of faecal sludge at different reaction conditions. *Desalination and Water Treatment*, **91**, 245–251, Desalination Publications, https://doi.org/10.5004/dwt.2017.20782

Ovsyannikova E., Arauzo P. J., Becker G. and Kruse A. (2019). Experimental and thermodynamic studies of phosphate behavior during the hydrothermal carbonization of sewage sludge. *Science of the Total Environment*, **692**, 147–156, Elsevier B.V., https://doi.org/10.1016/j.scitotenv.2019.07.217

Parmar K. R. and Ross A. B. (2019). Integration of hydrothermal carbonisation with anaerobic digestion; opportunities for valorisation of digestate. *Energies*, **12**(9), 1586, https://doi.org/10.3390/en12091586

Scrinzi D., Ferrentino R., Baù E., Fiori L. and Andreottola G. (2020). Hydrothermal carbonization scenarios for sewage sludge reduction and nutrients recovery. SIDISA 2020 – XI International Symposium on Environmental Engineering.

TerraNova Energy GmbH. https://terranova-energy.com/en/ (accessed 11 November 2021).

Volpe M., Fiori L., Merzari F., Messineo A. and Andreottola G. (2020). Hydrothermal carbonization as an efficient tool for sewage sludge valorization and phosphorous recovery. *Chemical Engineering Transactions*, **80**, 199–204, Italian Association of Chemical Engineering – AIDIC, https://doi.org/10.3303/CET2080034

Wang L., Chang Y. and Li A. (2019). Hydrothermal carbonization for energy-efficient processing of sewage sludge: a review. *Renewable and Sustainable Energy Reviews*, **108**, 423–440, https://doi.org/10.1016/j.rser.2019.04.011

Wang H., Yang Z., Li X. and Liu Y. (2020). Distribution and transformation behaviors of heavy metals and phosphorus during hydrothermal carbonization of sewage sludge. *Environmental Science and Pollution Research*, **27**(14), 17109–17122, Springer, https://doi.org/10.1007/s11356-020-08098-4

Wirth B., Reza T. and Mumme J. (2015). Influence of digestion temperature and organic loading rate on the continuous anaerobic treatment of process liquor from hydrothermal carbonization of sewage sludge. *Bioresource Technology*, **198**, 215–222, Elsevier Ltd, https://doi.org/10.1016/j.biortech.2015.09.022

Yoshida H., ten Hoeve M., Christensen T. H., Bruun S., Jensen L. S. and Scheutz C. (2018). Life cycle assessment of sewage sludge management options including long-term impacts after land application. *Journal of Cleaner Production*, **174**, 538–547, Elsevier, https://doi.org/10.1016/j.jclepro.2017.10.175

Zheng X., Ye Y., Jiang Z., Ying Z., Ji S., Chen W., Wang B. and Dou B. (2020). Enhanced transformation of phosphorus (P) in sewage sludge to hydroxyapatite via hydrothermal carbonization and calcium-based additive. *Science of the Total Environment*, **738**, 139789, https://doi.org/10.1016/j.scitotenv.2020.139786

Zheng X., Shen M., Ying Z., Feng Y., Wang B. and Dou B. (2022). Correlating phosphorus transformation with process water during hydrothermal carbonization of sewage sludge via experimental study and mathematical modelling. *Science of the Total Environment*, **807**, 150750, Elsevier, https://doi.org/10.1016/j.scitotenv.2021.150750

doi: 10.2166/9781789062755_0309

Chapter 14

Anaerobic digestate for polyhydroxyalkanoates (bioplastics precursors) production

M. R. Atelge[1], Hamdi Muratcobanoglu[2], David Krisa[3] and A. E. Atabani[*,4]*

[1]Department of Mechanical Engineering, Faculty of Engineering, Siirt University, 56100 Siirt, Turkey
[2]Department of Environmental Engineering, Faculty of Engineering, Niğde Ömer Halisdemir University, 51240 Niğde, Turkey
[3]UBC Bioreactor Technology Group, School of Engineering, The University of British Columbia, Okanagan Campus, 3333 University Way, Kelowna, BC V1V 1V7, Canada
[4]Alternative Fuels Research Laboratory (AFRL), Energy Division, Department of Mechanical Engineering, Faculty of Engineering, Erciyes University, 38039 Kayseri, Turkey
**Corresponding author email: aeatabani@gmail.com*

ABSTRACT

Petroleum-based plastic has grown at an average annual rate of 8.5% since 1950 when petroleum-based plastic was 1.5 million metric tons. 8.3 billion metric tons of plastic has been produced since 1950. It was estimated that plastic production will be around 600 million metric tons annually by the mid-21st century. By 2050, the estimated cumulative growth of plastic waste created may approach 25 billion metric tons based on the high demand for using plastic. Biopolymers have been proposed as a potential solution to this problem; however, they should fulfill the basic criteria, high quality, and durability while in use, but have high degradability at the end of their lifetime. Replacing petroleum-based plastics with bio-based alternatives provides a more sustainable approach to plastic production because of the reduced associated carbon emissions from petroleum extraction and processing. Poly (hydroxyalkanoates) (PHAs) are entirely biodegradable and biocompatible polyesters in comparison to other biopolymers. Moreover, high-cost carbon sources are one of the main challenges in the aspect of production costs and scaling-up production. The cost of carbon resources is responsible for $\sim$40% of the total production cost of PHAs due to artificial sources. The residual as liquid and solid parts of digestate from the AD process can be used for feedstock to produce PHAs due to abundance and low cost. The digestate is both cost-effective and a good source of carbon, both of which are necessary for the synthesis of PHAs. This chapter focuses on the production of PHAs from liquid and solid parts of the digestate, its recovery, purification, characterization, and applications.

Keywords: PHAs, bioplastic, biorefinery, anaerobic digestate

Editor(s): Vinay Kumar Tyagi, Kaoutar Aboudi and Cigdem Eskicioglu

14.1 INTRODUCTION

Plastics are a daily used product in human beings' life due to their higher resistance ability against chemical and physical impacts; therefore, their usage has increased. Sixty million metric tons of petroleum-based plastic was produced across the world in 1980. Production of plastic was 187 million metric tons in 2000, 265 million in 2010, and 348 million in 2017 as shown in Figure 14.1 (Chalmin, 2019). Since 1950, when it was 1.5 million metric tons, it has grown at an average annual rate of 8.5%. 8.3 billion metric tons of plastic has been produced since 1950. The International Energy Agency estimates that by the mid-21st century, annual production will be around 600 million metric tons (Chalmin, 2019). Specifications typically requested from plastic such as resistance ability of chemical and physical impacts becomes a problem at the end of their usage, and that causes one of the major pollution problems being plastic pollution. By 2050, the estimated cumulative growth of mostly plastic waste created may approach 25 billion metric tons due to the high demand for using plastic. Biopolymers have been presented as a solution to this issue, but they should fulfill the basic criteria of high quality and durability while in use, but have high degradability at the end of their lifetime. The production of petroleum-based plastic and biodegradable plastics is estimated to expand at a rate of 13% per year from 2020 to 2023, when combined with the shift toward sustainability using non-petroleum-based plastics (Meereboer *et al.*, 2020). Between 2010 and 2017, production capacity for bio-based poly(ethylene) (Bio-PE), bio-based poly (ethylene terephthalate) (Bio-PET), poly (lactic acid) (PLA), and poly (hydroxyalkanoate) (PHAs) increased by ~22%, 10 000%, 300%, and 41%, respectively (Meereboer *et al.*, 2020). Because of the reduced associated carbon emissions from petroleum extraction and

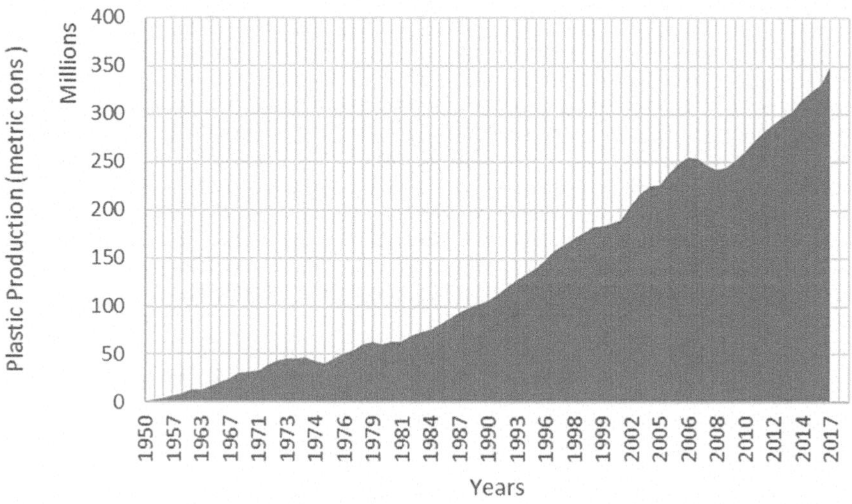

Figure 14.1 Plastic production between 1950 and 2017 (adapted from Chalmin, 2019).

processing, replacing petroleum-based plastics with bio-based alternatives provides a more sustainable approach to plastic production. When compared to other biopolymers, PHAs are entirely biodegradable and biocompatible polyesters, with the maximum degradability (De Donno Novelli *et al.*, 2021). PHAs have a high range of potential industrial applications from packing to medical equipment production and are an eco-friendly approach due to their renewability, biodegradability, and almost carbon neutral properties (Kovalcik *et al.*, 2017). The development of PHAs as synthetic plastic alternatives promotes the achievement of the United Nations' Sustainable Development Goals 6 related to water sanitation, 9 related to innovation, and 12 related to responsible manufacturing (United-Nation, 2020).

PHAs have been produced from a variety of carbon sources via fermentation process by different microorganisms. Biopolymers with varying chemo-physical characteristics were obtained during this process and were dependent on bacterial cultures and carbon resources. Biopolymers have enormous application areas such as medical equipment, packaging, molded goods, paper coatings, non-woven textiles, adhesives, films, and performance enhancers, therefore, they can be replaced with petroleum-based plastic (Andreeßen & Steinbüchel, 2019; Bugnicourt *et al.*, 2014; De Donno Novelli *et al.*, 2021). However, commercialization of PHAs is now hampered by high production costs, as industrial-scale manufacturing of PHAs is mostly completed using pure and designed bacterial cultures, which necessitate absolutely sterile operating conditions (Meereboer *et al.*, 2020). Additionally, extraction and purification processes increase production cost of PHAs (Mannina *et al.*, 2020; Rodriguez-Perez *et al.*, 2018). Moreover, high-cost carbon sources are one of the main challenges when it comes to production costs and scaling-up production. The cost of carbon resources is responsible for ~40% of the total production cost of PHAs due to artificial sources (De Donno Novelli *et al.*, 2021; Kourmentza *et al.*, 2017). Therefore, researchers have been dedicated to find suitable and abundant carbon sources for PHA production in the past decade. Among abundant carbon resources, industrial waste streams have gained attention, especially the biogas sector. That is because biogas production through an anaerobic digestion (AD) process may answer two issues which are waste disposal and to meet increasing energy demand due to population growth.

The growing population has increased the energy demand in the last century. The production of energy to meet this incremental demand pushes human beings to find sustainable and renewable energy sources because traditional energy production from fossil fuels cause irreversible damages to the earth such as global warming and climate change. Renewable energy systems (RES) can include solar, wind, hydro, geothermal, and biomass. Among them, biomass energy has caught the attention of policy makers and researchers due to two major benefits which are energy security and waste management in the last two decades. Biomass energy can be produced through AD methods which are defined where organic waste resources can be converted into methane as an energy carrier and carbon dioxide in an oxygen-free environment by microorganisms.

AD is a way to produce bio-methane and bio-hydrogen which are forms of renewable energy. Developed countries including many European counties are the leaders for biogas production and its utilization according to their established biogas facilities. In Europe, the biogas production facilities increased from 6227 to 18 943 between 2009 and 2019 (EBA, 2020). If this growth rate remains, it is projected that biogas production may reach an equivalent of 467 TWh by 2030 and 1020 TWh by 2050 (EBA, 2020). AD process does not only produce biogas but it also produces digestate. Disposal of digestate should be paid attention to because it increases with the production of biogas. The researchers investigated to find a path to utilize the digestate from a biorefinery. Several researchers reported that digestate can be used as a bio-fertilizer. However, recent reports showed that digestate can be converted into other value-added products such as a substrate for algae cultivation, nitrogen and phosphorus extraction, energy storage and conversion, synthesis of various carbon and doped metal nanoparticles for photocatalytic activity, and production of biopolymers (Selvaraj *et al.*, 2020). Among them, biopolymers are the most attractive product because they can replace synthetic polymers (plastics) that are petroleum-based.

PHAs can be replaced with petroleum-based plastic by lowering cost using efficient bacterial strains, downstream processing, and lower-cost feedstocks such as digestate. Due to the multiple benefits of the AD process, the residual as digestate liquid and solid parts can be used for feedstock to produce PHAs due to abundance and low cost. The digestate is both cost-effective and a good source of carbon, both of which are necessary for the synthesis of PHAs. This chapter focuses on the production of PHAs from the digestate, its recovery, purification, characterization, and applications.

14.2 POLYHYDROXYALKANOATES PRODUCTION

Polyhydroxyalkanoates (PHAs) are hydroxyalkanoic acid polyesters synthesized intracellularly by diverse bacteria as energy and carbon storage materials under stress situations. PHAs are aliphatic polyesters that are well known for their biodegradability and bacteria-based production processes. There are more than 91 distinct polyhydroxyalkanoic acid components, which are using PHAs production, and the number is constantly expanding (Meereboer *et al.*, 2020). An infinite number of PHAs can be made based on the possible combinations of monomer units. PHAs are bacterial polyesters which consist of three hydroxyalkanoic acids (HA acids) as shown in Figure 14.2 (Gahlawat, 2019). They are created and stored as an internal energy storage material by many bacteria when nutrients such as nitrogen, phosphorus, or sulfur are limited and there is an excess carbon source available. Enriching the PHA accumulating microorganisms in mixed microbial community using a feast and famine strategy, this creates stress on microorganisms and make them tend to accumulate PHA at intercellular levels (Battista *et al.*, 2020). PHAs are great reserve materials because they do not have an impact on osmotic pressure.

PHAs are often produced commercially by recombinant *Escherichia coli* (Jung *et al.*, 2019), although they can also be produced by a variety of different bacteria such as *Aeromonas* (Mozejko-Ciesielska *et al.*, 2019), *Azotobacter* (Mok *et al.*, 2021), *Cupriavidus* (Benesova *et al.*, 2017), *Clostridium* (Hassan

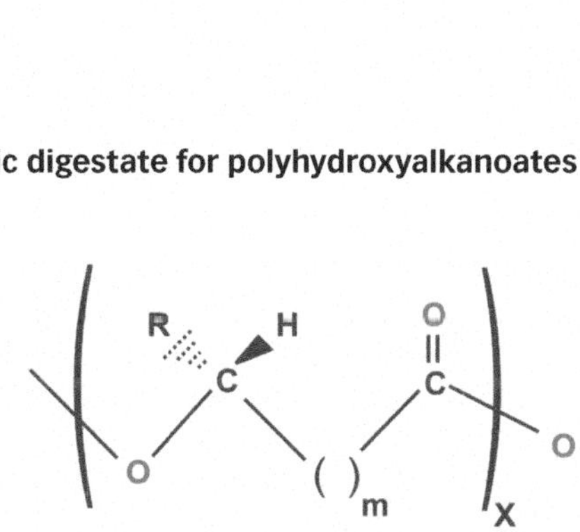

X varies from 100 to 30,000

R = Hydrogen Poly (3-hydroxpropionate) R = Propyl Poly (3-hydroxyhexanoate
R = Ethyl Poly (3-hydroxyvalerate) R = Pentyl Poly (3-hydroxyoctanoate)
R = Methyl Poly (3-hydroxybutyrate)

Figure 14.2 Typical structure of PHAs (Gahlawat, 2019).

et al., 2019), *Methylobacterium* (Wang *et al.*, 2020), and *Pseudomonas* (Wang *et al.*, 2020). *Pseudomonas* is the only bacteria known to make PHAs with a lengthy carbon chain. 3-Hydroxybutyrate synthase of acetoacetyl-CoA polymerizes 3-hydroxybutyrate monomers, but other chemical routes exist such as ring-opening polymerization of β-butyrolactone (Meereboer *et al.*, 2020). As 3-hydroxybutyrate process, 3-hydroxybutyrate-*co*-3-hydroxyvalerate is produced in the presence of propionic acid.

The 3-hydroxybutyrate and 3-hydroxyvalerate units are formed via enzymatic copolymerization of acetoacetyl-CoA and 3-ketovaleryl-CoA, respectively. PHAs, on the other hand, become a very expensive product when compared to commercial petroleum-based polymers with equivalent qualities because added nutrition and energy expenses account for more than 75% of the final cost (Wang & Chen, 2017). Metabolic pathways for the synthesis of PHAs are shown in Figure 14.3.

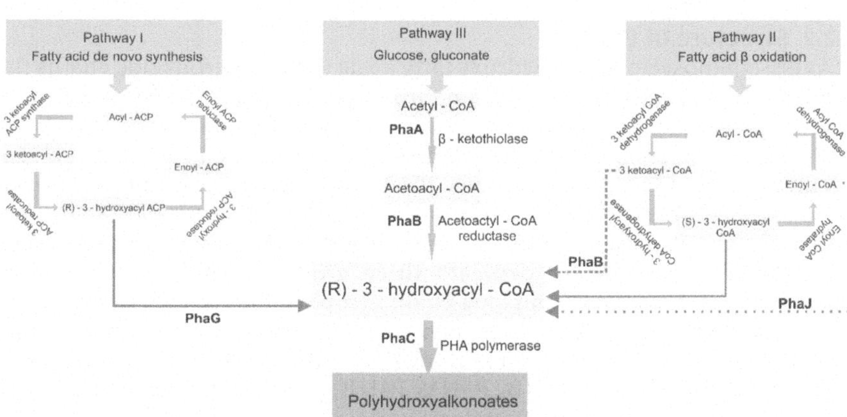

Figure 14.3 Major metabolic pathways for synthesis of PHAs (adapted from De Donno Novelli *et al.*, 2021; Mozejko-Ciesielska *et al.*, 2019; Pakalapati *et al.*, 2018).

14.2.1 Microbial production

Within PHA production systems, waste can be employed as carbon sources and microbial inoculum. PHA production with mixed microbial community (MMC) was studied with waste-activated sludge (WAS) from secondary wastewater treatment in municipal facilities as digestate (Tu *et al.*, 2019; Werker *et al.*, 2018). It was reported that different types of PHAs were accumulated and recovered from WAS as 0.605 g which is 6.04% of sludge by solid weight (Kumar *et al.*, 2018). The laboratory-scale combination of PHA generation and nitrogen removal has also been described for prospective integration in municipal wastewater treatment, as well as in a full-scale industrial wastewater treatment plant. Because carbohydrates are ideally stored as glycogen by MMCs, but volatile fatty acid is stored as PHAs, the organic content of feedstock supplied from waste water plant is generally transformed to volatile fatty acids by various acidogenic fermentation processes (Anterrieu *et al.*, 2014; Basset *et al.*, 2016). Moreover, this makes the digestate or effluents from sectors such as sludge and solid waste treatment, food waste, and biohydrogen generation attractive options for PHA production (Guerra-Blanco *et al.*, 2018; Guventurk *et al.*, 2020; Wijeyekoon *et al.*, 2018).

Many studies have been conducted with PHA production from digestate liquid, many of which are rich in VFA as shown in Table 14.1. PHA contents of dry media (microorganism) vary from 6.6% to 93.3% largely depends on VFA concentration of digestate liquid; used microorganism species (pure culture, mixed culture, etc.); and producing strategy like batch, semi-continuous, or feast–famine approach. The literature indicates that in the term of digestate management, the liquid part of the digestate can be used as a feedstock for PHA production; however, the solid portion still remains as a problem to be managed. To solve this problem, the solid portion can be used as a soil conditioner (Badagliacca *et al.*, 2020), feedstock for gasification, and incineration processes to contribute to the circular economy (Mayer *et al.*, 2021).

14.2.2 Structure of PHAs

PHAs are composed of (R)-3-hydroxy fatty acids with numerous side chains that have been found, although in certain circumstances the hydroxy substitution could be present at the 4th or 5th position in PHA (Kumar *et al.*, 2019). The rectus (R) structure when the alkyl (R) group ranges from C1 to C13 carbon can be found in most PHA configurations. PHAs and their co-polymers are further categorized as three different kinds based on the length of the alkyl side chain, which includes aromatic, halogenated, epoxidized, or other branching monomers (Liong, 2015). They are divided into three groups related to their carbon structure as short, medium, and long-chain hydroxyalkanoic acids. Short-chain hydroxyalkanoic acids (PHA*scl*), which have C3–C5, have an alkyl side chain and can be produced by bacteria such as *Ralstonia eutropha*, and *Alcaligenes latus*. Their products are P3HB, P3HV, P4HB, and co-polymer of P(3HB-*co*-3HV). For the medium chain, *Pseudomonas oleovorans*, *Pseudomonas putida*, *Pseudomonas mendocina*, and *Pseudomonas sensu stricto* create medium-chain hydroxyalkanoic acids (PHA*mcl*), which have between C6 and C14, with an alkyl side chain such as P3HH*x*, P3HO, and co-polymer of 3HH*x*-*co*-3HO.

Table 14.1 Production of PHAs from liquid and solid digestate and their production methods.

Material types	Microorganism species	Methods	Content of PHAs	References
Digestate liquor	*Cupriavidus necator*	Fermentation	90%	Passanha *et al.* (2013)
VFA from paper wastewater anaerobic digestate	*Cupriavidus necator*	Fermentation	57%	Al Battashi *et al.* (2021)
VFA from anaerobic digestate	*Pseudomonas putida*	Fermentation	93.3%	Cerrone *et al.* (2014)
Digestate, chicken manure and waste oil combination	*Cupriavidus necator*	Fermentation	75%	Altun (2019)
VFA from AD of MSW	Mixed culture	Fermentation	20%	Papa *et al.* (2020)
Digestate liquor	Mixed culture	Fermentation	59.5%	Jia *et al.* (2014)
Digestate supernatant + mineral medium	*Synechocystis salina*	Photofermentation	6.6%	Kovalcik *et al.* (2017)
Digestate supernatant + OfMSW + WAS	Mixed culture	Aerobic	48%	Valentino *et al.* (2020)

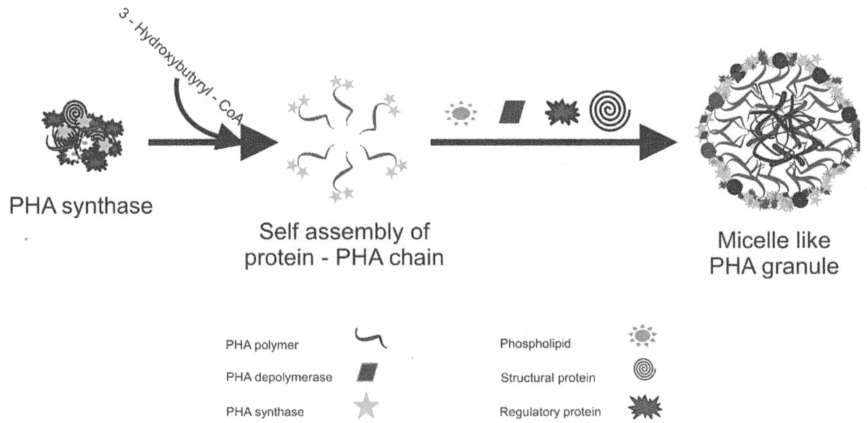

Figure 14.4 Schematic of PHA formation in bacteria (adapted from Pakalapati *et al.*, 2018).

For long chain, *Shewanella oneidensis* and *Aureispira marina* are capable of producing long chain (>C14) hydroxyalkanoic acids (PHA*lcl*) (Kumar *et al.*, 2019; Rai *et al.*, 2011).

Poly(3-hydroxybutyrate-*co*-3-hydroxyhexanoate) is a hybrid polymer combined of short- and medium carbon monomeric units. Chemical modification is caused by differences in the size and chemical composition of the side-chain substituents of PHAs, and these qualities could lead to a variety of uses for PHAs and their copolymers in a variety of disciplines. Because of their chemical structure and material properties, the presence of monomeric units in PHAs may vary. PHA structure is characterized by the kind of carbon substrates to generate the appropriate monomer units, which can be homo-polymeric or hetero-polymeric (Koller, 2020). Furthermore, PHA*scl* has been proven to be hard crystalline solids, but PHA*lcl* has been demonstrated to be elastomeric due to its chemical structure and monomeric present fractions. Moreover, the PHAs and PHBs are more similar to polypropylene and have superior moisture and UV resilience (Koller, 2020; Liong, 2015). PHAs are eco-friendly due to their monomeric nature and chemical composition, as well as their copolymeric structure. PHAs are not only environmentally biodegradable into soil and air but also biocompatible. Moreover, PHAs can be used for piezoelectric materials (Kumar *et al.*, 2019). The schematic of PHA formation in bacteria is represented in Figure 14.4.

14.2.3 Applications of PHAs

PHAs are intracellular carbon and energy storage elements that are obtained from a range of biodegradable materials in the absence of microbes. PHAs are biodegradable, non-toxic, and biocompatible polymers that are ecofriendly and also polymerized to a great degree of crystalline, optically dynamic, isotactic, piezoelectric, and insoluble in aqueous media (Kumar *et al.*, 2019; Meereboer *et al.*, 2020; Meixner *et al.*, 2018; Morgan-Sagastume *et al.*, 2020;

Venkateswar Reddy *et al.*, 2015). PHAs with such unique properties have a wide range of applications in a variety of fields, leading to the replacement of highly aggressive petroleum-based products. Initially, PHAs were widely used in hustling films, containers, and paper coatings, as well as other similar applications, analogous to traditional plastics that include disposable items such as razors, utensils, diapers, cosmetic containers, cleanser jugs, shampoo bottles, and cups, demonstrating that PHAs can act as a potent plastic material (Kumar *et al.*, 2019). In the United States, Elastomeric Metabolix PHAs were made, and U.S. Food and Drug Administration approved can be used with food. PHA is used in the production of foil, film, and diaphragm. *Aeromonas hydrophilais* produces 3HB-3HHx, which is widely used in shock wave sensor, ultrasonic detector, headphone, loudspeaker, and other products (Anjum *et al.*, 2016; Meereboer *et al.*, 2020). Moreover, PHA and its derivative products with paper materials are widely used as raw materials for the manufacture of environmentally friendly and rapidly biodegradable bags, as well as personal hygiene products such as diapers, napkins, sanitary towels, and tampons (Vasmara & Marchetti, 2016). Furthermore, stereoregular PHAs are excellent chiral precursors that are used in the synthesis of optically active chemicals that are used in a variety of applications including as a biodegradable carrier for long-term medications, insecticides, and herbicides (Tu *et al.*, 2019; Vasmara & Marchetti, 2016). PHA can be used as osteosynthetic material in bone development devices such as biodegradable screws, rivets, bone plating systems, bone plates for cartilage and bone fixation, bone dowels, blood vessel replacements, and orthopedic pins in medication areas because of its important piezoelectric property (Williams & Martin, 2002). However, because of its slow biodegradation and great hydrodynamic stability in sterile tissues, PHA is used in a limited number of medicinal, therapeutic, and pharmaceutical applications (Kumar *et al.*, 2019; Williams & Martin, 2002). Furthermore, plant-derived PHA can be employed in large-scale chemical manufacturing by depolymerization followed by direct esterification. It was discovered that the esterification products including PHA*mcl* and PHB*mcl* could be used as biofuels (Zhang *et al.*, 2009). It can be seen based on the above discussion that enormous application areas can utilize PHA and therefore their applications continue to increase.

14.2.4 Enrichment of PHAs

Proteobacteria, Actinobacteria, and *Bacteroidetes* are some of the most common bacterial communities discovered after enrichment. In most cases, the yield of PHAs remain the same during the enrichment of PHA-accumulating bacteria, because this is accomplished by allowing them to consume the PHAs after it has accumulated and been stored (Valentino *et al.*, 2017). Enrichment processes of PHAs could be inhibited due to the expansion of bacteria that feed on residual organic sugars after VFA depletion, usage of carbohydrate substrates, and high population of *Proteobacterium*, which is a recognized predator of PHA producers (De Donno Novelli *et al.*, 2021; Valentino *et al.*, 2017). In all circumstances, the enhanced MMC's quality deteriorates, lowering its subsequent effectiveness for PHA accumulation. Either using fermented effluents or selecting a limiting factor such as the electron acceptor, substrate and nutrients can be solved (De Donno Novelli *et al.*, 2021).

Also, once the VFAs have been diminished, the supernatant could be released to eliminate any leftover organic carbon (Kourmentza *et al.*, 2017). Enrichment process cost is significantly reduced when the waste stream is used as feedstocks and MMC is used as the inoculum. Aerobic dynamic feeding is an extra cost for enrichment that should be eliminated with the combining of anaerobic procedures (Mannina *et al.*, 2020).

14.2.5 Anaerobic/aerobic cycles

Among the enrichment process, anaerobic/aerobic cycles are one of promising methods. With this approach, bacteria are chosen for PHA accumulation by switching between anaerobic and aerobic environments (Kourmentza *et al.*, 2017). Using this approach, PHA accumulation by microorganisms was first seen during increased biological phosphate elimination. Phosphorous and glycogen accumulating microorganisms gather PHAs during the anaerobic phase, driven by electron-acceptor limiting circumstances (Dias *et al.*, 2006). To enrich PHA accumulation with photoheterotrophic bacteria, lighted conditions were used to investigate the full anaerobic enrichment process at the batch reactor. They used MMC as inoculum with the substrate that consisted of simulated effluents from dark fermentation by mixing acetate and butyrate in a 1:1 ratio (Guerra-Blanco *et al.*, 2018). The pH buffering capacity was increased with a high concentration of ammonia till the exponential growth stage, and then N-limitation was applied in the second phase. The result revealed that *Clostridium*, *Pseudomonas*, and *Rhodopseudomonas* microorganisms, which have high capacity of PHA accumulation, were observed to be 29%, 8%, and 5% of the consortia and PHA storage reached 44% of dry cell weight (Guerra-Blanco *et al.*, 2018). If co-polymers of PHB and PHV are existent with PHAs, higher resistance and flexibility can be produced from this biopolymer (Arcos-Hernández *et al.*, 2013). However, the cost of the enrichment process should be reduced by eliminating the aerobic part and using entirely anaerobic conditions (De Donno Novelli *et al.*, 2021).

14.2.6 PHA separation and purification

The procedures for recovering PHAs from non-PHAs' cell mass is described in two pathways in the literature. The first one explains that the breaking of the bacterium cell wall that encases PHAs and then PHAs are released. The second one defines the dissolving of PHAs in a suitable solvent that can allow the movement of PHAs through the cell membrane (De Donno Novelli *et al.*, 2021; Mannina *et al.*, 2020). Isolation and purification of PHAs add considerably to the total cost of PHAs' synthesis. The intracellular storage of PHAs and the limited amount produced by each cell are the key reasons for high recovery costs (Kosseva & Rusbandi, 2018). As a result, numerous criteria should be addressed when choosing a suitable PHA recovery strategy. Koller (2020) described those criteria as the cell membrane fragility, types and quantities of PHAs, high-purity demand, the effect of extraction technique for PHAs, total cost, and effects on the environment. The most well investigated procedures for PHA recovery were classified into three groups as digestion of non-PHA cell, chemical extraction, and mechanical methods (Mannina *et al.*, 2020). Applying heat, mechanical pretreatment, and chemical extraction cause weakened cell walls, therefore, extraction of the polymer becomes possible with a greater recovery rate (De Donno Novelli *et al.*, 2021). Chemical extraction is currently the most popular method in the industrial application process due to the demand of a high purity, even though this method is not cost-effective and ecofriendly. Researchers have recently focused on figuring out downstream processes which are more cost-effective and environmentally friendly (Fernández-Dacosta *et al.*, 2015; Hajnal *et al.*, 2016; López-Abelairas *et al.*, 2015).

14.2.7 Biological downstream process

The novel biological approaches have been recently developed using insects and animals. These insects and animals consume lyophilized cells which belong to *Cupriavidus necator* H16 that contain PHAs (Kosseva & Rusbandi, 2018). Murugan *et al.* (2016) discovered that by ingesting PHAs from cell walls, the mealworm larvae beetle *Tenebrio molitor* may retrieve them. The PHAs were extracted and freeze-dried according to usual protocols and contained 54% of a dry weight cell. The obtained PHAs were nearly pure after washing with distilled water, detergent, and heat. When compared to the chloroform extracted PHAs, the molecular weight and combination of PHAs did not alter (Murugan *et al.*, 2016). Another biological approach was established for recovering PHAs from *Cupriavidus necator* H16 cell walls. *Cupriavidus necator* H16 were freeze-dried and fed to laboratory rats and the test animal feces contained 82–97 of P(3HB) (Kunasundari *et al.*, 2017). Washing the feces with water and/or low concentrations of detergent eliminated the remaining contaminants, resulting in high purity P(3HB). P(3HB) produced by this approach has similar molecular weights and thermal characteristics to P(3HB) recovered from bacterial cells using chloroform (Kunasundari *et al.*, 2017). Solvents and chemicals were not required using this method. Therefore, the methods should now be investigated for industrial applications.

14.2.8 Economic aspect of PHAs

A life-cycle analysis is an environmental management tool that is frequently applied to examine overall environmental impacts. The life-cycle analyses of PHA and PHB generation from industrial leftovers are available in the literature (Fernández-Dacosta *et al.*, 2015; Koller *et al.*, 2017). Another study reported the environmental and economic advantages of a biopolymer production method with a double optimization process (Koller *et al.*, 2013). The results revealed that high mechanical energy was required during the fermentation step, and a low production rate of PHAs led to substantial environmental effects (Koller *et al.*, 2013). For downstream processes, alkali treatment for cell disruption was determined to be the most promising technique after an integrated approach that included both economic and environmental aspects (Fernández-Dacosta *et al.*, 2015). The conclusion drawn was that wastewater-based P3HB has been suggested as a viable alternative to pure culture PHAs from sugar, however, the price of biopolymers is still far away from petroleum-based plastic (Fernández-Dacosta *et al.*, 2015).

For the commercial success of PHA production systems, fermentation conditions should be regulated very well. The cost of PHAs is heavily related to substrate price, PHAs' production yield, and cost of downstream processing. The cost of generating PHAs on a commercial scale is between 4000 and 15 000 USD per megaton depending on the types of biopolymers. On the market, PHAs are now evaluated in the range of 5–6 USD per kg, and its global market value is estimated to be around 94 million USD by 2021 (De Donno Novelli *et al.*, 2021). The value of PHAs obtained from *Alcaligenes eutrophus* is approximately 16 USD per kg, which is 18 times more expensive than the polymer polypropylene obtained from recombinant *Escherichia coli*. As a result, the cost of PHAs is reduced to around 4 USD per kg, which is comparable to other recyclable polymers such as aliphatic polyesters and poly (3-hydroxybutyric acid) P3HB (Kumar *et al.*, 2019).

14.3 CONCLUSION

PHAs are helpful and valuable materials that can be employed in a variety of industrial domains due to their biodegradability, biocompatibility, and nontoxicity. They are thought to be the most potential future alternative to petroleum-based polymers. However, the high manufacturing costs are still a major issue in the commercialization of PHAs because added nutrition and energy expenses account for more than 75% of the final cost. The cost of nutrition is responsible for ~40% of the total production cost of PHAs due to artificial sources. The residual as liquid and solid parts of digestate from the AD process can be used for feedstock to produce PHAs due to abundance and low cost. Many studies conducted with PHA production from digestate liquid are rich in VFA. PHA contents of dry media (microorganism) varying 6.6–93.3% often depend on VFA concentration of digestate liquid; used microorganism species (pure culture, mixed culture, etc.); and producing strategies like batch, semi-continuous, or feast–famine approach. In terms of digestate management, the liquid part of the digestate can be used as a feedstock for PHA production; however, the solid portion still remains a problem to be managed. To solve this problem, the solid portion can be used as a soil conditioner, feedstock for gasification, and incineration processes to contribute to the circular economy. The conclusion can be drawn that the solid portion of the digestate should now be investigated for suitability of PHA production.

REFERENCES

Al Battashi H., Al-Kindi S., Gupta V. K. and Sivakumar N. (2021). Polyhydroxyalkanoate (PHA) production using volatile fatty acids derived from the anaerobic digestion of waste paper. *Journal of Polymers and the Environment*, **29**(1), 250–259, https://doi.org/10.1007/s10924-020-01870-0

Altun M. (2019). Polyhydroxyalkanoate production using waste vegetable oil and filtered digestate liquor of chicken manure. *Preparative Biochemistry & Biotechnology*, **49**(5), 493–500, https://doi.org/10.1080/10826068.2019.1587626

Andreeßen C. and Steinbüchel A. (2019). Recent developments in non-biodegradable biopolymers: precursors, production processes, and future perspectives. *Applied*

Microbiology and Biotechnology, **103**(1), 143–157, https://doi.org/10.1007/s00253-018-9483-6

Anjum A., Zuber M., Zia K. M., Noreen A., Anjum M. N. and Tabasum S. (2016). Microbial production of polyhydroxyalkanoates (PHAs) and its copolymers: a review of recent advancements. *International Journal of Biological Macromolecules*, **89**, 161–174, https://doi.org/10.1016/j.ijbiomac.2016.04.069

Anterrieu S., Quadri L., Geurkink B., Dinkla I., Bengtsson S., Arcos-Hernandez M., Alexandersson T., Morgan-Sagastume F., Karlsson A., Hjort M., Karabegovic L., Magnusson P., Johansson P., Christensson M. and Werker A. (2014). Integration of biopolymer production with process water treatment at a sugar factory. *New Biotechnology*, **31**(4), 308–323, https://doi.org/10.1016/j.nbt.2013.11.008

Arcos-Hernández M. V., Laycock B., Donose B. C., Pratt S., Halley P., Al-Luaibi S., Werker A. and Lant P. A. (2013). Physicochemical and mechanical properties of mixed culture polyhydroxyalkanoate (PHBV). *European Polymer Journal*, **49**(4), 904–913, https://doi.org/10.1016/j.eurpolymj.2012.10.025

Badagliacca G., Petrovičovà B., Pathan S. I., Roccotelli A., Romeo M., Monti M. and Gelsomino A. (2020). Use of solid anaerobic digestate and no-tillage practice for restoring the fertility status of two Mediterranean orchard soils with contrasting properties. *Agriculture, Ecosystems & Environment*, **300**, 107010–107010, https://doi.org/10.1016/j.agee.2020.107010

Basset N., Katsou E., Frison N., Malamis S., Dosta J. and Fatone F. (2016). Integrating the selection of PHA storing biomass and nitrogen removal via nitrite in the main wastewater treatment line. *Bioresource Technology*, **200**, 820–829, https://doi.org/10.1016/j.biortech.2015.10.063

Battista F., Frison N., Pavan P., Cavinato C., Gottardo M., Fatone F., Eusebi A. L., Majone M., Zeppilli M., Valentino F., Fino D., Tommasi T. and Bolzonella D. (2020). Food wastes and sewage sludge as feedstock for an urban biorefinery producing biofuels and added-value bioproducts. *Journal of Chemical Technology & Biotechnology*, **95**(2), 328–338, https://doi.org/10.1002/jctb.6096

Benesova P., Kucera D., Marova I. and Obruca S. (2017). Chicken feather hydrolysate as an inexpensive complex nitrogen source for PHA production by *Cupriavidus necator* on waste frying oils. *Letters in Applied Microbiology*, **65**(2), 182–188, https://doi.org/10.1111/lam.12762

Bugnicourt E., Cinelli P., Lazzeri A. and Alvarez V. (2014). Polyhydroxyalkanoate (PHA): review of synthesis, characteristics, processing and potential applications in packaging. *Express Polymer Letters*, **8**(11), 791–808, https://doi.org/10.3144/expresspolymlett.2014.82

Cerrone F., Choudhari S. K., Davis R., Cysneiros D., O'Flaherty V., Duane G., Casey E., Guzik M. W., Kenny S. T., Babu R. P. and O'Connor K. (2014). Medium chain length polyhydroxyalkanoate (mcl-PHA) production from volatile fatty acids derived from the anaerobic digestion of grass. *Applied Microbiology and Biotechnology*, **98**(2), 611–620, https://doi.org/10.1007/s00253-013-5323-x

Chalmin P. (2019). Field actions science reports the history of plastics: from the capitol to the Tarpeian rock. Field actions science reports. *The Journal of Field Actions*, **2019**(Special Issue 19), 6–11.

De Donno Novelli L., Moreno Sayavedra S. and Rene E. R. (2021). Polyhydroxyalkanoate (PHA) production via resource recovery from industrial waste streams: a review of techniques and perspectives. *Bioresource Technology*, **331**, 124985–124985, https://doi.org/10.1016/j.biortech.2021.124985

Dias J. M. L., Lemos P. C., Serafim L. S., Oliveira C., Eiroa M., Albuquerque M. G. E., Ramos A. M., Oliveira R. and Reis M. A. M. (2006). Recent advances in polyhydroxyalkanoate production by mixed aerobic cultures: from the substrate to the final product. *Macromolecular Bioscience*, **6**(11), 885–906, https://doi.org/10.1002/mabi.200600112

EBA. (2020). European Biogas Association Annual Report 2020.

Fernández-Dacosta C., Posada J. A., Kleerebezem R., Cuellar M. C. and Ramirez A. (2015). Microbial community-based polyhydroxyalkanoates (PHAs) production from wastewater: techno-economic analysis and ex-ante environmental assessment. *Bioresource Technology*, **185**, 368–377, https://doi.org/10.1016/j.biortech.2015.03.025

Gahlawat G. (2019). Polyhydroxyalkanoates biopolymers: production strategies.

Guerra-Blanco P., Cortes O., Poznyak T., Chairez I. and García-Peña E. I. (2018). Polyhydroxyalkanoates (PHA) production by photoheterotrophic microbial consortia: effect of culture conditions over microbial population and biopolymer yield and composition. *European Polymer Journal*, **98**, 94–104, https://doi.org/10.1016/j.eurpolymj.2017.11.007

Guventurk A., Ozturk D., Ozyildiz G., Ayisigi E., Guven D., Zengin G. E., Tas D. O., Olmez-Hanci T., Pala-Ozkok I., Yagci N., Insel G. and Cokgor E. (2020). Determination of the potential of pickle wastewater as feedstock for biopolymer production. *Water Science and Technology*, **81**(1), 21–28, https://doi.org/10.2166/wst.2020.060

Hajnal I., Chen X. and Chen G.-Q. (2016). A novel cell autolysis system for cost-competitive downstream processing. *Applied Microbiology and Biotechnology*, **100**(21), 9103–9110, https://doi.org/10.1007/s00253-016-7669-3

Hassan E. A., Abd-Alla M. H., Zohri A. N. A., Ragaey M. M. and Ali S. M. (2019). Production of butanol and polyhydroxyalkanoate from industrial waste by *Clostridium beijerinckii* ASU10. *International Journal of Energy Research*, **43**(8), 3640–3652, https://doi.org/10.1002/er.4514

Jia Q., Xiong H., Wang H., Shi H., Sheng X., Sun R. and Chen G. (2014). Production of polyhydroxyalkanoates (PHA) by bacterial consortium from excess sludge fermentation liquid at laboratory and pilot scales. *Bioresource Technology*, **171**(1), 159–167, https://doi.org/10.1016/j.biortech.2014.08.059

Jung H. R., Yang S. Y., Moon Y. M., Choi T. R., Song H. S., Bhatia S. K., Gurav R., Kim E. J., Kim B. G. and Yang Y. H. (2019). Construction of efficient platform *Escherichia coli* strains for polyhydroxyalkanoate production by engineering branched pathway. *Polymers*, **11**(3), 1–14.

Koller M. (2020). The Handbook of Polyhydroxyalkanoates, Three Volume Set, 1st edn. Taylor & Francis Group, United States.

Koller M., Sandholzer D., Salerno A., Braunegg G. and Narodoslawsky M. (2013). Biopolymer from industrial residues: life cycle assessment of poly(hydroxyalkanoates) from whey. *Resources, Conservation and Recycling*, **73**, 64–71, https://doi.org/10.1016/j.resconrec.2013.01.017

Koller M., Maršálek L., de Sousa Dias M. M. and Braunegg G. (2017). Producing microbial polyhydroxyalkanoate (PHA) biopolyesters in a sustainable manner. *New Biotechnology*, **37**, 24–38, https://doi.org/10.1016/j.nbt.2016.05.001

Kosseva M. R. and Rusbandi E. (2018). Trends in the biomanufacture of polyhydroxyalkanoates with focus on downstream processing. *International Journal of Biological Macromolecules*, **107**(PartA), 762–778, https://doi.org/10.1016/j.ijbiomac.2017.09.054

Kourmentza C., Plácido J., Venetsaneas N., Burniol-Figols A., Varrone C., Gavala H. N. and Reis M. A. M. (2017). Recent advances and challenges towards sustainable polyhydroxyalkanoate (PHA) production. *Bioengineering*, **4**(4), 55–55, https://doi.org/10.3390/bioengineering4020055

Kovalcik A., Meixner K., Mihalic M., Zeilinger W., Fritz I., Fuchs W., Kucharczyk P., Stelzer F. and Drosg B. (2017). Characterization of polyhydroxyalkanoates produced by *Synechocystis salina* from digestate supernatant. *International Journal of Biological Macromolecules*, **102**, 497–504, https://doi.org/10.1016/j.ijbiomac.2017.04.054

Kumar M., Ghosh P., Khosla K. and Thakur I. S. (2018). Recovery of polyhydroxyalkanoates from municipal secondary wastewater sludge. *Bioresource Technology*, **255**, 111–115, https://doi.org/10.1016/j.biortech.2018.01.031

Kumar G., Ponnusamy V. K., Bhosale R. R., Shobana S., Yoon J. J., Bhatia S. K., Rajesh Banu J. and Kim S. H. (2019). A review on the conversion of volatile fatty acids to polyhydroxyalkanoates using dark fermentative effluents from hydrogen production. *Bioresource Technology*, **287**, 121427–121427, https://doi.org/10.1016/j.biortech.2019.121427

Kunasundari B., Arza C. R., Maurer F. H. J., Murugaiyah V., Kaur G. and Sudesh K. (2017). Biological recovery and properties of poly(3-hydroxybutyrate) from *Cupriavidus necator* H16. *Separation and Purification Technology*, **172**, 1–6, https://doi.org/10.1016/j.seppur.2016.07.043

Liong M.-T. (2015). Beneficial Microorganisms in Agriculture, Aquaculture and Other Areas. Springer International Publishing, Switzerland. Vol. **29**.

López-Abelairas M., García-Torreiro M., Lú-Chau T., Lema J. M. and Steinbüchel A. (2015). Comparison of several methods for the separation of poly(3-hydroxybutyrate) from *Cupriavidus necator* H16 cultures. *Biochemical Engineering Journal*, **93**, 250–259, https://doi.org/10.1016/j.bej.2014.10.018

Mannina G., Presti D., Montiel-Jarillo G., Carrera J. and Suárez-Ojeda M. E. (2020). Recovery of polyhydroxyalkanoates (PHAs) from wastewater: a review. *Bioresource Technology*, **297**, 122478–122478, https://doi.org/10.1016/j.biortech.2019.122478

Mayer F., Bhandari R. and Gäth S. A. (2021). Life cycle assessment on the treatment of organic waste streams by anaerobic digestion, hydrothermal carbonization and incineration. *Waste Management*, **130**, 93–106, https://doi.org/10.1016/j.wasman.2021.05.019

Meereboer K. W., Misra M. and Mohanty A. K. (2020). Review of recent advances in the biodegradability of polyhydroxyalkanoate (PHA) bioplastics and their composites. *Green Chemistry*, **22**(17), 5519–5558, https://doi.org/10.1039/D0GC01647K

Meixner K., Kovalcik A., Sykacek E., Gruber-Brunhumer M., Zeilinger W., Markl K., Haas C., Fritz I., Mundigler N., Stelzer F., Neureiter M., Fuchs W. and Drosg B. (2018). Cyanobacteria biorefinery – production of poly(3-hydroxybutyrate) with *Synechocystis salina* and utilisation of residual biomass. *Journal of Biotechnology*, **265**, 46–53, https://doi.org/10.1016/j.jbiotec.2017.10.020

Mok P.-S., Chuah J.-A., Najimudin N., Liew P.-W.-Y., Jong B.-C. and Sudesh K. (2021). In vivo characterization and application of the PHA synthase from *Azotobacter vinelandii* for the biosynthesis of polyhydroxyalkanoate containing 4-hydroxybutyrate. *Polymers*, **13**(10), 1576–1576, https://doi.org/10.3390/polym13101576

Morgan-Sagastume F., Bengtsson S., De Grazia G., Alexandersson T., Quadri L., Johansson P., Magnusson P., and Werker A. (2020). Mixed-culture polyhydroxyalkanoate (PHA) production integrated into a food-industry effluent biological treatment: a pilot-scale evaluation. *Journal of Environmental Chemical Engineering*, **8**(6), 104469–104469, https://doi.org/10.1016/j.jece.2020.104469

Mozejko-Ciesielska J., Marciniak P. and Szacherska K. (2019). Polyhydroxyalkanoates synthesized by *Aeromonas* species: trends and challenges. *Polymers*, **11**(8), 1–16, https://doi.org/10.3390/polym11081328

Murugan P., Han L., Gan C.-Y., Maurer F. H. J. and Sudesh K. (2016). A new biological recovery approach for PHA using mealworm, *Tenebrio molitor*. *Journal of Biotechnology*, **239**, 98–105, https://doi.org/10.1016/j.jbiotec.2016.10.012

Pakalapati H., Chang C. K., Show P. L., Arumugasamy S. K. and Lan J. C. W. (2018). Development of polyhydroxyalkanoates production from waste feedstocks and applications. *Journal of Bioscience and Bioengineering*, **126**(3), 282–292, https://doi.org/10.1016/j.jbiosc.2018.03.016

Papa G., Pepè Sciarria T., Carrara A., Scaglia B., D'Imporzano G. and Adani F. (2020). Implementing polyhydroxyalkanoates production to anaerobic digestion of organic fraction of municipal solid waste to diversify products and increase total energy recovery. *Bioresource Technology*, **318**, 124270–124270, https://doi.org/10.1016/j.biortech.2020.124270

Passanha P., Esteves S. R., Kedia G., Dinsdale R. M. and Guwy A. J. (2013). Increasing polyhydroxyalkanoate (PHA) yields from *Cupriavidus necator* by using filtered digestate liquors. *Bioresource Technology*, **147**, 345–352, https://doi.org/10.1016/j.biortech.2013.08.050

Rai R., Keshavarz T., Roether J. A., Boccaccini A. R. and Roy I. (2011). Medium chain length polyhydroxyalkanoates, promising new biomedical materials for the future. *Materials Science and Engineering: R: Reports*, **72**(3), 29–47, https://doi.org/10.1016/j.mser.2010.11.002

Rodriguez-Perez S., Serrano A., Pantión A. A. and Alonso-Fariñas B. (2018). Challenges of scaling-up PHA production from waste streams. A review. *Journal of Environmental Management*, **205**, 215–230, https://doi.org/10.1016/j.jenvman.2017.09.083

Selvaraj P. S., Periasamy K., Suganya K., Ramadass K., Muthusamy S., Ramesh P., Bush R., Vincent S. G. T. and Palanisami T. (2020). Novel resources recovery from anaerobic digestates: current trends and future perspectives. *Critical Reviews in Environmental Science and Technology*, **0**(0), 1–85.

Tu W., Zhang D. and Wang H. (2019). Polyhydroxyalkanoates (PHA) production from fermented thermal-hydrolyzed sludge by mixed microbial cultures: the link between phosphorus and PHA yields. *Waste Management*, **96**, 149–157, https://doi.org/10.1016/j.wasman.2019.07.021

United-Nation. (2020). The Sustainable Development Goals Report 2020. The Department of Economic and Social Affairs of United Nations, New York.

Valentino F., Morgan-Sagastume F., Campanari S., Villano M., Werker A. and Majone M. (2017). Carbon recovery from wastewater through bioconversion into biodegradable polymers. *New Biotechnology*, **37**, 9–23, https://doi.org/10.1016/j.nbt.2016.05.007

Valentino F., Lorini L., Gottardo M., Pavan P. and Majone M. (2020). Effect of the temperature in a mixed culture pilot scale aerobic process for food waste and sewage sludge conversion into polyhydroxyalkanoates. *Journal of Biotechnology*, **323**, 54–61, https://doi.org/10.1016/j.jbiotec.2020.07.022

Vasmara C. and Marchetti R. (2016). Biogas production from biodegradable bioplastics. *Environmental Engineering and Management Journal*, **15**(9), 2041–2048, https://doi.org/10.30638/eemj.2016.220

Venkateswar Reddy M., Kotamraju A. and Venkata Mohan S. (2015). Bacterial synthesis of polyhydroxyalkanoates using dark fermentation effluents: comparison between pure and enriched mixed cultures. *Engineering in Life Sciences*, **15**(6), 646–654, https://doi.org/10.1002/elsc.201500016

Wang Y. and Chen G. Q. (2017). Polyhydroxyalkanoates: Sustainability, Production, and Industrialization. Sustainable Polymers From Biomass. Wiley VCH, Weinheim, Germany, 14.

Wang J., Tan H., Li K. and Yin H. (2020). Two-stage fermentation optimization for poly-3-hydroxybutyrate production from methanol by a new *Methylobacterium* isolate from oil fields. *Journal of Applied Microbiology*, **128**(1), 171–181, https://doi.org/10.1111/jam.14463

Werker A., Bengtsson S., Korving L., Hjort M., Anterrieu S., Alexandersson T., Johansson P., Karlsson A., Karabegovic L., Magnusson P., Morgan-Sagastume F., Sijstermans L., Tietema M., Visser C., Wypkema E., van der Kooij Y., Deeke A. and Uijterlinde C. (2018). Consistent production of high quality PHA using activated sludge harvested from full scale municipal wastewater treatment – PHARIO. *Water Science and Technology*, **78**(11), 2256–2269, https://doi.org/10.2166/wst.2018.502

Wijeyekoon S., Carere C. R., West M., Nath S. and Gapes D. (2018). Mixed culture polyhydroxyalkanoate (PHA) synthesis from nutrient rich wet oxidation liquors. *Water Research*, **140**, 1–11, https://doi.org/10.1016/j.watres.2018.04.017

Williams S. F. and Martin D. P. (2002). Applications of PHAs in medicine and pharmacy. *Biopolymers*, **4**, 91–127.

Zhang X., Luo R., Wang Z., Deng Y. and Chen G.-Q. (2009). Application of (R)-3-hydroxyalkanoate methyl esters derived from microbial polyhydroxyalkanoates as novel biofuels. *Biomacromolecules*, **10**(4), 707–711, https://doi.org/10.1021/bm801424e

doi: 10.2166/9781789062755_0327

Chapter 15

Fate of emerging contaminants in anaerobic digestate

Ahmed Tawfik[1,], Vinay Kumar Tyagi[2] and Dominique Patureau[3]*

[1]National Research Centre, Water Pollution Research Department 12622, Dokki, Cairo, Egypt
[2]Environmental Hydrology Division, National Institute of Hydrology, Roorkee, 247667, India
[3]INRAE, University Montpellier, LBE, 102 Avenue des étangs, F-11000 Narbonne, France
**Corresponding author e-mail: prof.tawfik.nrc@gmail.com*

ABSTRACT

Recovering of valuable products from wastewater treatment processes are the backbone of a circular economy. The sludge contains a huge diversity of contaminants such as linear alkylbenzene sulphonates, phthalates, pesticides, nonylphenol ethoxylates (NPEs), estrogens and pharmaceuticals, whose presence is due to anthropogenic activities. Emerging contaminants (ECs) are mainly removed from the bulk liquid during the biological treatment process of various wastewaters and adsorbed onto the sludge, which needs further treatment processes. ECs are abundant in sludge and present at concentrations from μg to mg/kg of dry weight (DW). The safe reuse of sludge depends on their EC content and their fate and impact after sludge disposal. Anaerobic digestion of sludge partially removed ECs from it, but this removal is limited by the contaminants' sorption onto organic matter and the microbial degrading capacity. In order to overcome this limitation, sludge pretreatment may be useful. This chapter gathers information on contaminant occurrence in sludge and digestate underlying their fate during anaerobic digestion. The role of sludge pretreatments to mitigate ECs' fate during anaerobic digestion is depicted. Depending on the compound and the type of process, pretreatment could increase ECs' solubility or reduce their molecular size, along with reducing the hydrophobicity of the sludge and subsequently enhancing EC biodegradation and the biomethanization process.

Keywords: Emerging contaminants, Enzyme activities, Inhibition effect, Pretreatment, Sludge

15.1 INTRODUCTION

Wastewater treatment plants produce huge quantities of sludge containing emerging contaminants (ECs) (Dubey *et al.*, 2021). Indeed, detergents like linear alkylbenzene sulphonates and nonylphenol ethoxylates (NPEs), plasticizers like phthalates and bisphenols, pesticides, hormones, flame retardants, pharmaceuticals and personal care products (PPCPs) have been quantified in sewage sludge (Clarke and Smith, 2011; Mailler *et al.*, 2017) (Figure 15.1 and Figure 15.2). ECs are mainly removed from the bulk liquid during the biological treatment process of various wastewaters by sorption onto the sludge (Castiglioni *et al.*, 2006; Barret *et al.*, 2010). The ECs share low solubility, large molecular structure and a hydrophobic nature, which make them easily sorbed by sludge in the mixed liquor (Ma and Lin, 2011). In many countries, biosolids are used as fertilizers (McClellan and Halden, 2010; (Kelessidis and Stasinakis *et al.*, 2012) because doing so positively contributes to the enhancement of soil properties and addresses EU policy on sustainability and recycling. However, the widespread environmental and recognized (eco)toxic effects of ECs (Arslan *et al.*, 2017) may hinder the disposal of agricultural biosolids due to the potential negative impact on the soil and crops (Sabourin *et al.*, 2012). Indeed, ECs have been quantified in each environmental compartment, that is, water, soil, air and biota (Arslan *et al.*, 2017). ECs, provided alone on soil, affect biodegradation processes in the soil (Carter, Garman *et al.*, 2014), earthworm growth (Carter, Harris *et al.*, 2014), animal grazing (Duarte-Davidson and Jones, 1996) and groundwater quality (Topp *et al.*, 2010; Abbasi *et al.*, 2012). In addition, they accumulate in economic crops (Carter, Harris *et al.*, 2014; Prosser *et al.*, 2014; Wu *et al.*, 2015). However, long-term field experiments on biosolid disposal at standard agronomic rates showed low soil accumulation and very low EC transfer to leachates and plants, and based on the risk quotient, the estimated

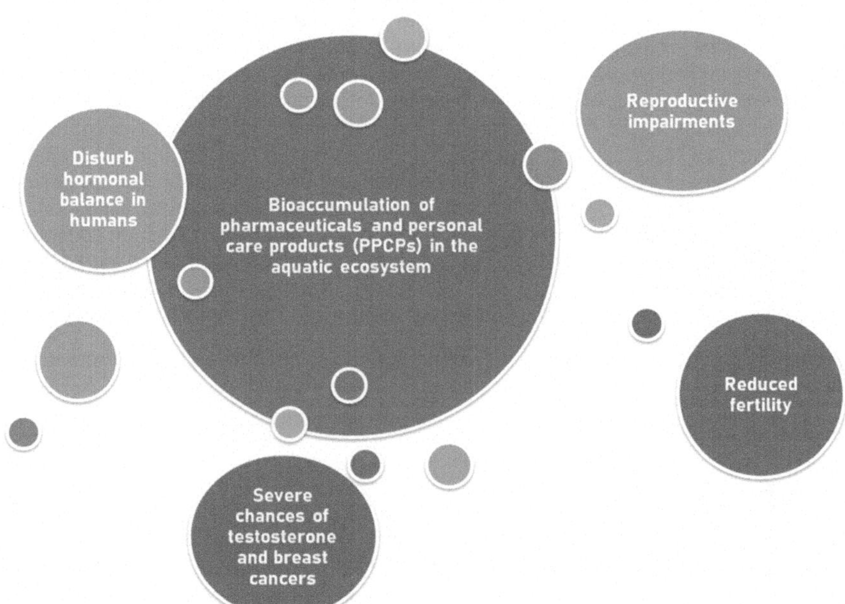

Figure 15.1 The effects of dangerous ECs on human health.

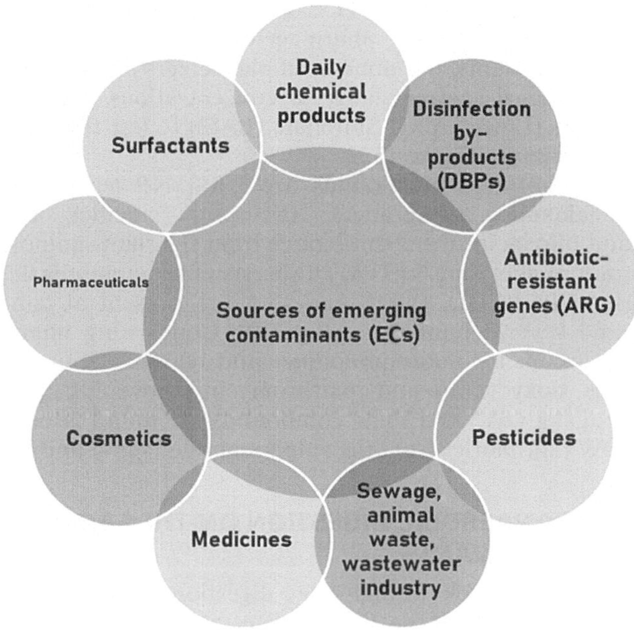

Figure 15.2 Categories of emerging contaminants (ECs).

ecotoxicological risks for different soil organisms were low (Sabourin *et al.*, 2012; Bourdat-Deschamps *et al.*, 2017). The anaerobic digestion (AD) process is mainly employed for bioenergy production from the excess sludge. Pretreatment of the sludge containing ECs prior to the AD process is necessary to (1) increase sludge solubility and biodegradability, (2) break down the ECs and (3) increase their availability to finally enhance their transformation. The main objective of this chapter is to comprehensively discuss the occurrence of ECs in sludge and digestate, and their fate during anaerobic digestion. A focus is on low-cost pretreatment processes that could mitigate the fate of ECs in anaerobic digestion.

15.2 EMERGING CONTAMINANTS IN SLUDGE

The recovery of sludge and/or biosolids from wastewater treatment plants has recently become challenging in the context of a circular economy. These biosolids have broad characteristics in terms of their chemical, biological and physical composition. Their carbon and nutrient content renders them interesting for recovery and reutilization as fertilizers. However, their EC content could be a big barrier for their reutilization. Many papers refer to the quantification of ECs in sludge (Mailler *et al.*, 2017; Sertillanges *et al.*, 2020). As for wastewater, the sludge content reflects anthropogenic activities. Indeed, a huge diversity of organic compounds, including hydrocarbons (Polycyclic aromatic hydrocarbons

(PAHs), polychlorinated biphenyls (PCBs), polychlorinated dibenzo-p-dioxins (PCDDs)), surfactants, solvents, pharmaceuticals, personal care products, pesticides, flame retardants, hormones and plasticizers has been detected and quantified in municipal sewage sludge at concentrations varying from g/kg DW for surfactants (Linear Alkyl Sulfonate (LAS)) to less than μg/kg DW for sulfamethoxypyridazine (Figure 15.3).

The most concentrated compounds are LAS, NP (surfactants), DEHP (plasticizer), galaxolide (fragrance), triclosan (biocide), ciprofloxacin, norfloxacin and ofloxacin (three antibiotics from the fluoroquinolone family) with concentrations above mg/kg DW. The persistent compounds (Figure 15.3a) such as PAH, PCB, PCDD, PBDE and PFA are present at concentrations between 10 and several hundred μg/kg DW. Concerning pharmaceuticals (Figure 15.3b), except for fluoroquinolones and tylosine (above mg/kg DW) and tetracycline, doxycycline and roxithromycin (above 100 μg/kg DW), all are below 100 μg/kg DW. And some compounds are found at concentrations below μg/kg DW (antibiotics from the sulphonamide family and hormones).

15.3 EFFECT OF ANAEROBIC DIGESTION ON THE FATE OF EMERGING CONTAMINANTS

One way to observe the role of anaerobic digestion on EC fate is to look at the concentration levels of these ECs in digestates. Data retrieved from the literature on digested sludge are summarized in Figure 21.4. The number of data is low compared to that for raw sludge, and for some compounds, such as PCA and PCN, no concentration was found or a concentration below the limit of quantification was provided, such as for antibiotics from the sulfonamide and tetracycline families. The range of concentrations for each compound is quite similar to the one observed for raw sludge, meaning that residual concentrations from g/kg DW for LAS to μg/kg DW for hormones and sulfonamides are found in digested sludge (Figure 15.4a). As these ECs are present at low to very low concentrations, it is usually assumed that these compounds could not sustain growth and are removed thanks to cometabolism and the presence of another substrate. Indeed, the microbial and functional diversity of anaerobic ecosystems may contribute to the biotransformation of these compounds. As shown by many authors (Ghattas *et al.*, 2017; Patureau *et al.*, 2021), some compounds are quite well removed during anaerobic digestion. This is the case for naproxen, atenolol, loratidine, miconazole, tramadol, domperidone, azithromycin, trimethoprim, tylosin and some sulfonamides. However, antibacterial agents like triclocarban and triclosan are poorly removed (Wang *et al.*, 2017; Yan, Meng *et al.*, 2019). This is also the case for persistent compounds such as PAH, PCB and DEHP, for which the removal rate is usually less than 30% (Benabdalha El-Hadj *et al.*, 2007b). However, when summarizing data on EC removal, it is also observed that the range of removal for one compound could be very large depending on the studies and the experimental conditions (Patureau *et al.*, 2021). Sometimes there are contradictory results, as have been observed for hormones that are demonstrated to be well degraded under anaerobic digestion by Carballa *et al.* (2007) but not by Muller *et al.* (2010). The

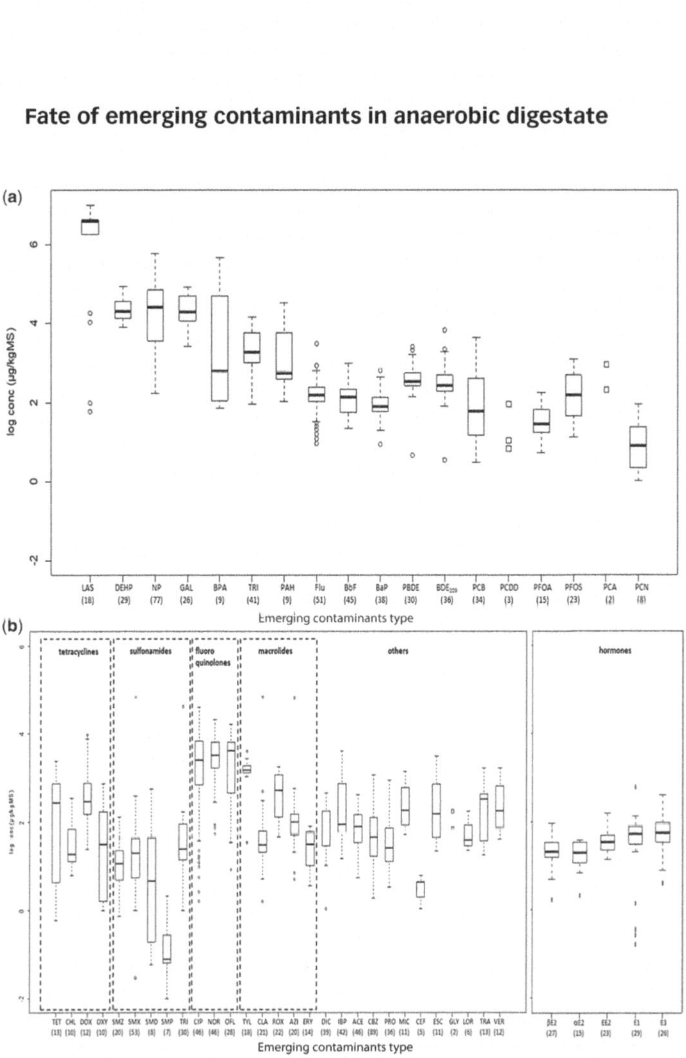

Figure 15.3 Concentration ranges of ECs in raw sludge for (a) persistent compounds and (b) pharmaceuticals and hormones. The number of data points used to create the boxplots are in parentheses. The box corresponds to the 25th and 75th percentiles, the whiskers to the 10th and 90th percentiles and the solid line to the median; outliers are depicted by individual points. Whisker-box plots were drawn if more than three values were found. Legend of molecules: (a) linear alkylbenzene sulfonate (LAS, sum of 4), diethyl hexyl phthalate (DEHP), nonylphenols (NP), galaxolide (GAL), bisphenolA (BPA), triclosan (TRI), polycyclic aromatic hydrocarbons (PAH, sum of 16), fluoranthene (Flu), benzo(b)fluoranthene (BbF), benzo(a) pyrene (BaP), polybromodiphenylether (PBDE), bromodiphenylether 209 (BDE_{209}), polychlorobiphenyls (PCB, sum of 7), polychlorodibenzodioxine (PCDD in ng I-TEQ/kg DM), perfluoro octanoic acid (PFOA), perfluorooctane sulfonate (PFOS), polychloro n-alcane (PCA), polychloronaphtahlene (PCN); (b) tetracycline (TET), chlortetracycline (CHL), doxycycline (DOX), oxytetracycline (OXY), sulfamethazine (SMZ), sulfamethoxazole (SMX), sulfadiazine (SDZ), sulfamethoxypyridazine (SMP), trimethoprim (TRI), ciprofloxacin (CIP), norfloxacin (NOR), ofloxacin (OFL), tylosine (TYL), clarithromycine (CLA), roxythromycine (ROX), azithromycine (AZI), erythromycin (ERY), diclofenac (DIC), ibupropfen (IBP), acetaminophen (ACE), carbamazepine (CBZ), propranolol (PRO), miconazole (MIC), cefoperazone (CEF), escitalopram (ESC), glybencyclamide (GLY), loratidine (LOR), tramadol (TRA), verapamil (VER).

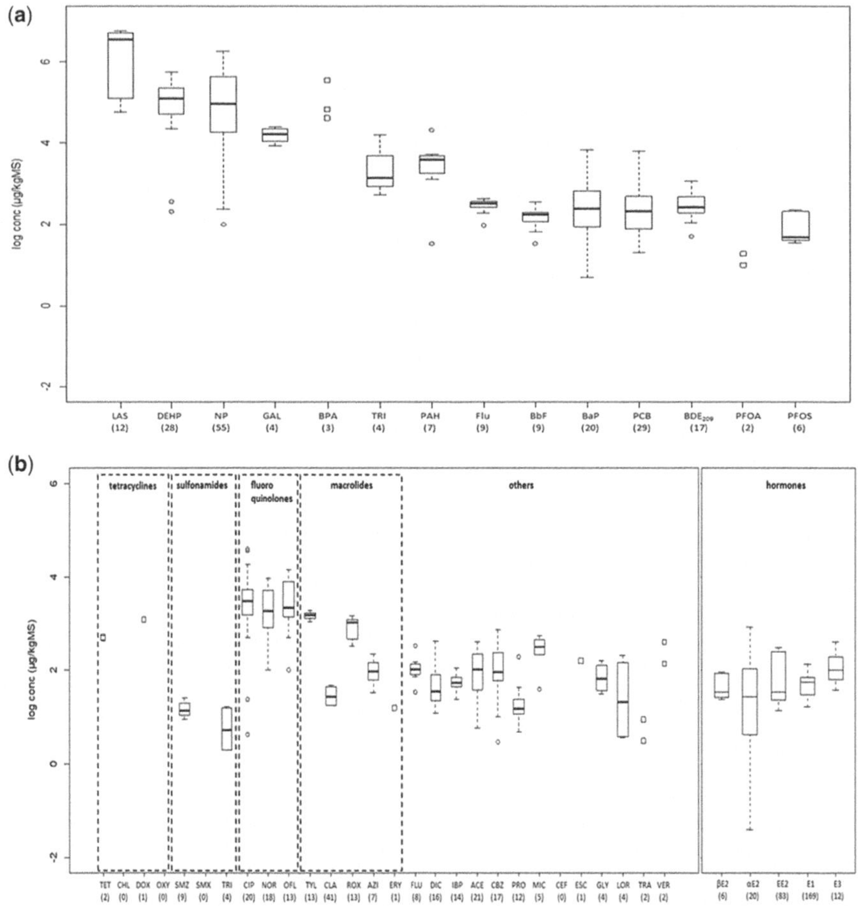

Figure 15.4 Concentration ranges of ECs in *digested* sludge for (a) persistent compounds and (b) pharmaceuticals and hormones. The number of data points used to create the boxplots is in parentheses. The box corresponds to the 25th and 75th percentiles, the whiskers to the 10th and 90th percentiles and the solid line to the median; outliers are depicted by individual points. Whisker-box plots were drawn if more than three values were found. Legend of molecules: (a) linear alkylbenzene sulfonate (LAS, sum of 4), diethyl hexyl phthalate (DEHP), nonylphenols (NP), galaxolide (GAL), bisphenolA (BPA), triclosan (TRI), polycyclic aromatic hydrocarbons (PAH, sum of 16), fluoranthene (Flu), benzo(b)fluoranthene (BbF), benzo(a)pyrene (BaP), bromodiphenylether 209 (BDE_{209}), polychlorobiphenyls (PCB, sum of 7), perfluoro octanoic acid (PFOA), perfluorooctane sulfonate (PFOS); (b) tetracycline (TET), chlortetracycline (CHL), doxycycline (DOX), oxytetracycline (OXY), sulfamethazine (SMZ), sulfamethoxazole (SMX), trimethoprim (TRI), ciprofloxacin (CIP), norfloxacin (NOR), ofloxacin (OFL), tylosine (TYL), clarithromycine (CLA), roxythromycine (ROX), azithromycine (AZI), erythromycin (ERY), fluoxetine (FLU), diclofenac (DIC), ibupropfen (IBP), acetaminophen (ACE), carbamazepine (CBZ), propranolol (PRO), miconazole (MIC), cefoperazone (CEF), escitalopram (ESC), glybencyclamide (GLY), loratidine (LOR), tramadol (TRA), verapamil (VER).

transformation of ECs in sludge during anaerobic digestion is mainly limited by the (bio)availability of the compounds that are strongly sorbed to the organic matter of the sludge thanks to hydrophobic or charged interactions (Figure 15.4b). Breaking down these interactions may facilitate EC transformation. Pretreatment of the sludge that enhances organic matter solubilization such as ozonation, ultrasound and thermal lysis may be a good option to improve EC removal. Sorption of bisphenol A (BPA) during the fermentation process takes place in sludge via strong binding to extracellular polymeric substances (Yan, Zhu *et al.*, 2019), and subsequently biodegradation mechanism removal (Zhao *et al.*, 2008). Triclocarban (TCC) and triclocarban (TCS) were partitioned onto the sludge as reported earlier by Zhou *et al.* (2019). The removal of TCS occurs mainly due to the binding of TCS in sludge by extracellular polymeric substances (EPS) (Zhu *et al.*, 2019).

The presence of endocrine-disrupting chemicals (EDCs) (4-nonylphenol (4-NP) and triclocarban (TCC)) in the sludge improved the activities and diversity of the microbial community and subsequently enhanced the solubilisation and acidogenesis process (Duan *et al.*, 2016; Wang *et al.*, 2017). Nevertheless, acidogenic microbes had the ability to biodegrade 4-NP in the activated sludge (Duan *et al.*, 2016; Duan *et al.*, 2018). TCC and TCS were highly accumulated in the fermented sludge. Wang *et al.* (2017) found that TCC was slightly removed from sludge during the anaerobic digestion of activated sludge. This indicates the risk of TCC and TCS in the fermented sludge still causing severe pollution. TCC and TCS should be removed from fermented sludge prior to disposal. TCC and TCS accumulated in the sludge mainly during the anaerobic digestion process due to sorption to the solid materials (Zhou *et al.*, 2019). The fermentation of the sludge by acidogenesis minimized the levels of retinoids by only 3.3% and EDCs by only 1.4% in the supernatant, which confirms the low contribution of anaerobic digestion to the removal of ECs from sludge (Zhou *et al.*, 2020). Lin *et al.* (2017) reported that acidogenic fermentation of the sludge could effectively reduce the level of ECs in the supernatant so that it is favorable for reutilization for resource recovery. Further research is needed on improving the removal of EDCs from the solid phase of sludge to minimize their health risk and environmental damage. The supernatant (liquid) digestate resulting from anaerobic digestion of sludge containing emerging pollutants had the highest concentration of 147 ug/L for sunscreen octocrylene and 58.6 ug/L for acetaminophen (paracetamol) (Ali *et al.*, 2019). However, lower fractions of octocrylene (>600 ng/g ww), and tris(1-chloro-2-propyl) phosphate (>500 ng/gww) were found in solid digestates. The total contaminants of emerging concern (CECs) were 1411 ng/g in solid digestates and 354 ng/g in liquid digestates. The behavior of the Irbesartan, benzoylecgonine and Venlafaxine in the sludge after mesophillic and thermophillic anaerobic digestion was reported by Boix *et al.* (2016). Apparently, benzoylecgonine and Irbesartan in the sludge were biodegraded by a fermentation process. Venlafaxine concentration was significantly decreased in the aqueous phase and increased in the solid phase. Concentrations of Irbesartan, benzoylecgonine and Venlafaxine in the solid phase were higher than in the aqueous phase, suggesting that they were highly adsorbed onto the bio solids.

15.4 THE FATE OF EMERGING POLLUTANTS DURING THE ANAEROBIC DIGESTION OF SLUDGE

Sludge solubilization could be highly achieved by pretreatment processes. This might rupture the microbial cells, solubilize the coarse suspended solids or increase the soluble chemical oxygen demand (COD), resulting in the possible degradation of refractory compounds and in increased mineralization of organic matter, thereby improving the methanogenesis process in the presence of the emerging pollutants in the sludge (Mohapatra *et al.*, 2011; Mohapatra *et al.*, 2012; Samaras *et al.*, 2014). Pretreatment of the sludge reduces the required hydrolysis time of anaerobic digestion (Carballa *et al.*, 2007). The pretreatment process could increase the solubility of ECs in the sludge and subsequently enhance their biodegradation and also the biomethanization process (Zhang *et al.*, 2021). Sonication and ozonation were effective in solubilization of sludge, resulting in a high removal of ECs from the sludge (Mohapatra *et al.*, 2012).

The electro-oxidation process aims to effectively mineralize the organic matter in the sludge. A direct current is continuously applied in an electrochemical cell. Electrochemical conversion and/or combustion of organics (phenol and hormones) take place due to the generation of free radicals, particularly hydroxyl radicals (OH) (Rivera-Utrilla *et al.*, 2013). Boron-doped diamond electrodes were reported to be the most active anodes for oxidation and mineralization of emerging pollutants (Chen, 2004). Electro-oxidation using boron-doped diamond electrodes oxidizing sludge achieved nonylphenol, 4,4-(propane-2,2-diyl)diphenol and 5-chloro-2-(2,4-dichlorophenoxy)phenol removals of 89%, 73% and 82%, respectively, at a pH value of 3 and current density of 40 mA/cm^2 for 1 hour. The volatile solids of the sludge were mineralized by 23% and COD was removed by 27% (Barrios *et al.*, 2015). At a pH of 3, there was 42% mineralization of emerging pollutants achieved, which decreased to 25% and 14% for pH values of 5 and 7, respectively (Barrios *et al.*, 2015). Moreover, zeta potential decreased from –17 to –10 mV due to mineralization and destroyed and/or destabilized negatively colloidal particles, improving sludge quality. COD removal was maximized at a level of 31% at a current density of 40 mA/cm and significantly dropped at 10 mA/cm. Titanium (Ti) anodes removed pathogens and enhanced dewatering of sludge (Drogui *et al.*, 2012). Emerging pollutants (ECs) are destabilized and destroyed by hydroxyl radicals (•OH) released by electrodes. The •OH radicals react with emerging pollutants (Eq. 15.1).

$$RH + {}^{\circ}OH \rightarrow {}^{\circ}R + H_2O \tag{15.1}$$

The electrochemical cell produces hydroxyl radicals that react with ECs in the sludge. The EC radicals further react with oxygen or hydroxyl radicals to generate oxidative by-products (Eqs. 15.2 and 15.3).

$${}^{\circ}R + O_2 \rightarrow Products \tag{15.2}$$

$${}^{\circ}R + {}^{\circ}OH \rightarrow Products \tag{15.3}$$

The oxidation of 4,4-(propane-2,2-diyl)diphenol (BPA) by ▪OH radicals breaks the adjacent bonds (methyl bridge) into phenol and iso-propylene alcohol (Figure 15.5). The iso-propylene alcohol molecule is further oxidized by ▪OH, producing short chain carboxylic acid molecules. Catechol, hydroquinone and resorcinol compounds might be produced from the hydroxylation of the phenolic ring. Hydroquinone was the main by-product in earlier studies (Gözmen *et al.*, 2003). Subsequently, quinone is formed by a dehydrogenation reaction with ▪OH radicals. Phenolic ring molecules are mainly oxidized and converted into small fragmented by-products, resulting in short chain aliphatic acids. Maleic, oxalic, fumaric, formic and acetic acids are the main aliphatic acids. These by-products could be efficiently utilized by anaerobes for bioenergy productivity. However, extending the reaction time will allow the attack of ▪OH on those acids, resulting in carbon dioxide for complete degradation of BPA (Gözmen *et al.*, 2003).

Carboxylic acids are the main by-products from the degradation of chlorophenols as shown in Figure 15.6. Electrochemical process by conductive-diamond completely removed Triclosan from sludge (Gözmen *et al.*, 2003).

Ultrasonication is an efficient mechanical pretreatment process of sludge to enhance its biodegradability (Pilli *et al.*, 2011). Ultrasonication certainly enhances the digestibility of sludge by disrupting its physical, biological and

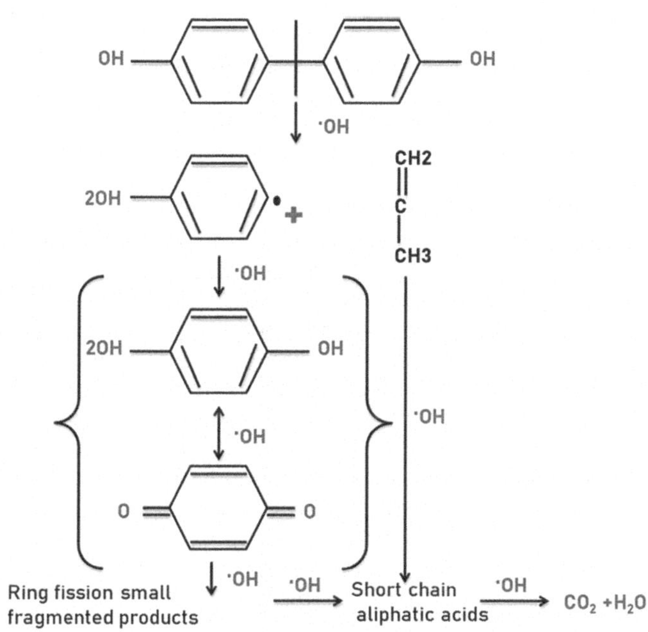

Figure 15.5 The anodic oxidation reaction mechanism of 4,4-(propane-2,2-diyl)diphenol (BPA) with BDD electrode (Martín de Vidales *et al.*, 2013).

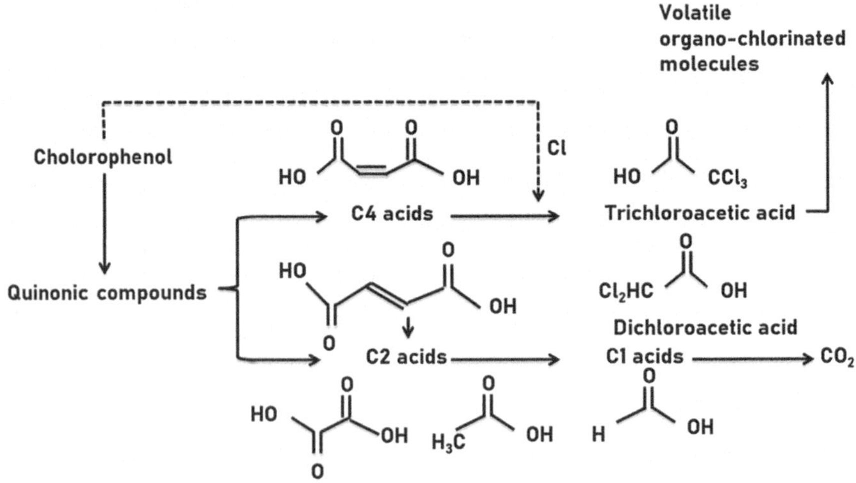

Figure 15.6 The anodic oxidation reaction mechanism of 5-chloro-2-(2,4-dichlorophenoxy) phenol (Triclosan) with BDD electrode (Martín de Vidales *et al.*, 2013).

chemical properties. The degree of sludge disintegration is mainly affected by the sonication parameters and sludge characteristics. Full-scale applications of ultrasonication systems have efficiently demonstrated a 50% increase in biogas productivity. Moreover, the energy balance revealed that the ratio of the net energy gain to electricity consumption by the ultrasound system was 2.5. Shimizu *et al.* (1993) found that sonication of sludge prior to anaerobic digestion (AD) highly increased biogas productivity by 40% at short hydraulic retention time (HRT). Moreover, the digestibility of the sludge was improved by 60%. Ultrasonication of the sludge induced cavitation of the bacterial cells, which lysed the microbial cell walls and released the intracellular enzyme components into the aqueous medium. This resulted in an increase of volatile solids (VS) and subsequently improved bioenergy productivity. The sonication of sludge improved the hydrolysis and methanogenesis processes as reported earlier by Mao and Show (2007). The rate of hydrolysis was increased by 19% to 75% for digesters supplied with sonicated sludge at sonication densities of 0.18 to 0.52 W/mL. This was mainly attributed to the acceleration of hydrolysis of organics into soluble by-products and promotion of the methanogenesis process. Ultrasonic pretreatment of sludge improved polycyclic aromatic hydrocarbon (PAH) diffusion into the aqueous phase, resulting in a reduction of its content during the anaerobic digestion process (Benabdallah El-Hadj *et al.*, 2007a). Naphthalene was removed from the sonicated sludge in the mesophilic anaerobic digester. Nevertheless, pyrene removal remained unaffected by the pretreatment process due to its high hydrophobicity. After the combined pretreatment process (ultrasound+Fenton), 75.52% to 84.40% of polycyclic aromatic hydrocarbons (PAHs) and 16.32% to 31.13% of sludge organic matter (SOM) was removed (Ning *et al.*, 2018).

Thermal hydrolysis is a low-cost process and is mainly used for pretreatment of sludge prior to anaerobic digestion processes (Suárez-Iglesias *et al.*, 2017). This process is used for solubilizing organics (polysaccharides and lipids) and fragments of ECs so that they can be easily biodegraded by anaerobes. Thermal hydrolysis partially removed micropollutants from sewage sludge (Taboada-Santos, *et al.*, 2019); however, anaerobic metabolism and biotransformation did not occur. McNamara *et al.* (2012) found that thermal hydrolysis partially removed nonylphenol or nonylphenol ethoxylates by volatilization. Thermal hydrolysis of sludge at 130°C for 60 min reduced the concentration of some antibiotics by 58% to 90% (Zhang and Li, 2018). However, erythromycin and roxithromycin concentration increased by 38% to 88% due to the lysis of the intracellular and/or disintegration of sludge cells. Biogas productivity of 357 ± 24 L (N) CH_4/kg (VS) was achieved from sludge after thermal pretreatment process (Taboada-Santos *et al.*, 2019). Naproxen was completely removed by the anaerobic digestion process (Alvarino *et al.*, 2014). Ibuprofen, diazepam, roxithromycin, citalopram, trimethoprim, galaxolide and tonalide were partially biotransformed 30% to 80% by anaerobes (Gonzalez-Gil *et al.*, 2016; Yang *et al.*, 2017; Taboada-Santos *et al.*, 2019). These results confirmed that ECs are partially or totally removed by thermal pretreatment followed by anaerobic digestion. The insolubilized ECs are adsorbed onto the solids, creating a protective barrier against their removal. Zhang and Li (2018) found that thermal sludge pretreatment at 130°C for 60 min increased antibiotic removal by 58% to 90% during anaerobic digestion. However, erythromycin and roxithromycin increased by 38% to 88% in the bulk liquid. Naproxen was completely removed after thermal and anaerobic digestion of sludge (Taboada-Santos *et al.*, 2019). Ozonation + AD efficiently removed galaxolide (HHCB) and tonalide (AHTN) from sludge (Carballa *et al.*, 2008). Musk fragrances such as HHCB and AHTN were biotransformed and biodegraded by specific microorganisms (*Erysipelotrichaceae* and *Comamonadaceae* families).

15.5 CONCLUDING REMARKS

The ECs share low solubility, large molecular structure and a hydrophobic or charged nature, which make them able to be easily sorbed on sludge in a mixed liquor. Depending on specific country policies, sludge could be spread on soil to increase its fertility and crop growth. Fifty percent of digested sludge (equivalent to 3.5 million tons of dry sludge per year) is reused as soil fertilizer. However, the occurrence of ECs in digested sludge may cause a barrier for reuse in agriculture due to their potential persistence and the negative effects on the environment and human health. Sludge solubilization could be highly achieved by pretreatment processes. This might rupture the microbial cells, solubilize the coarse suspended solids and/or increase the COD solubility, resulting in a high degradation efficiency of refractory compounds, thereby improving the methanogenesis process in the presence of the emerging pollutants in the sludge. Sonication and ozonation are effective for solubilization of sludge, potentially resulting in the highest removal of ECs. Pretreatment of the sludge reduces the required hydrolysis time, increases the solubility of ECs and subsequently enhances their biodegradation while enhancing the biomethanization process.

ACKNOWLEDGMENTS

The authors are very grateful to the Science, Technology & Innovation Funding Authority (STIFA)–Egypt (project ID: 26271, 41591), the Academy of Scientific Research and Technology (ASRT)–Egypt (code: call no. 2/2019/ASRT-Nexus) and the Imhotep project for partially financially supporting the research. The first author is grateful to the National Research Centre–Egypt for partially supporting the research (project ID: 12030202).

REFERENCES

Abbasi T., Tauseef S. M. and Abbasi S. A. (2012). Anaerobic digestion for global warming control and energy generation: An overview. *Renewable and Sustainable Energy Reviews*, **16**(5), 3228–3242. http://dx.doi.org/10.1016/j.rser.2012.02.046

Ali A. M., Nesse A. S., Eich-greatorex S., Sogn T. A., Aanrud S. G., Bunæs A. A., Lyche J. L. and Kallenborn R. (2019). Organic contaminants of emerging concern in Norwegian digestates from biogas production. *Environmental Science: Processes and Impacts*, **21**(9), 1498–1508.

Alvarino T., Suarez S., Lema J. M. and Omil F. (2014). Understanding the removal mechanisms of PPCPs and the influence of main technological parameters in anaerobic UASB and aerobic CAS reactors. *Journal of Hazardous Materials*, **278**, 506–513. http://dx.doi.org/10.1016/j.jhazmat.2014.06.031

Arslan H., Doğanay E., Karataş E., Ünlü M. A. and Ahmed H. M. A. (2017). Effect of low-level Laser therapy on postoperative pain after root canal retreatment: a preliminary placebo-controlled, triple-blind, Randomized Clinical Trial. *Journal of Endodontics*, **43**(11), 1765–1769.

Barret M., Carrère H., Latrille E., Wisniewski C. and Patureau D. (2010) Micropollutant and sludge characterization for modeling sorption equilibria. *Environmental Science and Technology*, **44**(3), 1100–1106.

Barrios J. A., Becerril E., De León C., Barrera-Díaz C. and Jiménez B. (2015). Electrooxidation treatment for removal of emerging pollutants in wastewater sludge. *Fuel*, **149**, 26–33. http://dx.doi.org/10.1016/j.fuel.2014.10.055

Benabdallah El-Hadj T., Dosta J., Márquez-Serrano R. and Mata-Álvarez J. (2007a). Effect of ultrasound pretreatment in mesophilic and thermophilic anaerobic digestion with emphasis on naphthalene and pyrene removal. *Water Research*, **41**(1), 87–94.

Benabdallah El-Hadj T., Dosta J., Torres R. and Mata-Álvarez J. (2007b). PCB and AOX removal in mesophilic and thermophilic sewage sludge digestion. *Biochemical Engineering Journal*, **36**(3), 281–287.

Boix C., Ibáñez M., Fabregat-Safont D., Morales E., Pastor L., Sancho J. V., Sánchez-Ramírez J. E. and Hernández F. (2016). Behaviour of emerging contaminants in sewage sludge after anaerobic digestion. *Chemosphere*, **163**, 296–304.

Bourdat-Deschamps M., Ferhi S., Bernet N., Feder F., Crouzet O., Patureau D., Montenach D., Moussard G. D., Mercier V., Benoit P. and Houot, S. (2017). Fate and impacts of pharmaceuticals and personal care products after repeated applications of organic waste products in long-term field experiments. *Science of the Total Environment*, **607–608**, 271–280. http://dx.doi.org/10.1016/j.scitotenv.2017.06.240

Carballa M., Fink G., Omil F., Lema J. M. and Ternes T. (2008). Determination of the solid-water distribution coefficient (Kd) for pharmaceuticals, estrogens and musk fragrances in digested sludge. *Water Research*, **42**(1–2), 287–295.

Carballa M., Manterola G., Larrea L., Ternes T., Omil F. and Lema J. M. (2007). Influence of ozone pre-treatment on sludge anaerobic digestion: Removal of pharmaceutical and personal care products. *Chemosphere*, **67**(7), 1444–1452.

Carter L. J., Garman C. D., Ryan J., Dowle A., Bergström E., Thomas-Oates J. and Boxall A. B. A. (2014). Fate and uptake of pharmaceuticals in soil-earthworm systems. *Environmental Science and Technology*, **48**(10), 5955–5963.

Carter L. J., Harris E., Williams M., Ryan J. J., Kookana R. S. and Boxall A. B. A. (2014). Fate and uptake of pharmaceuticals in soil-plant systems. *Journal of Agricultural and Food Chemistry*, **62**(4), 816–825.

Castiglioni S., Bagnati R., Fanelli R., Pomati F., Calamari D. and Zuccato E. (2006). Removal of pharmaceuticals in sewage treatment plants in Italy. *Environmental Science and Technology*, **40**(1), 357–363.

Chen G. (2004). Electrochemical technologies in wastewater treatment. *Separation and Purification Technology*, **38**(1), 11–41.

Clarke B. O. and Smith S. R. (2011) Review of "emerging" organic contaminants in biosolids and assessment of international research priorities for the agricultural use of biosolids. *Environment International*, **37**(1), 226–247. http://dx.doi.org/10.1016/j.envint.2010.06.004.

Drogui P., Bureau M.-A., Mercier G. and Blais J. F. (2012). Effectiveness of electrooxidation process for stabilizing and conditioning of urban and industrial wastewater sludge. *Water Environment Research*, **85**(1), 35–43.

Duan X., Wang X., Xie J., Feng L., Yan Y., Wang F. and Zhou Q. (2018). Acidogenic bacteria assisted biodegradation of nonylphenol in waste activated sludge during anaerobic fermentation for short-chain fatty acids production. *Bioresource Technology*, **268**, 692–699. https://doi.org/10.1016/j.biortech.2018.08.053

Duan X., Wang X., Xie J., Feng L., Yan Y. and Zhou Q. (2016). Effect of nonylphenol on volatile fatty acids accumulation during anaerobic fermentation of waste activated sludge. *Water Research*, **105**, 209–217. http://dx.doi.org/10.1016/j.watres.2016.08.062

Duarte-Davidson R. and Jones K. C. (1996). Screening the environmental fate of organic contaminants in sewage sludge applied to agricultural soils: II. The potential for transfers to plants and grazing animals. *Science of the Total Environment*, **185**(1–3), 59–70.

Dubey M., Mohapatra S., Tyagi V. K., Suthar S. and Kazmi A. A. (2021). Occurrence, fate, and persistence of emerging micropollutants in sewage sludge treatment. *Environmental Pollution*, **273**, 116515. https://doi.org/10.1016/j.envpol.2021.116515

Ghattas A. K., Fischer F., Wick A. and Ternes T. A. (2017). Anaerobic biodegradation of (emerging) organic contaminants in the aquatic environment. *Water Research*, **116**, 268–295. http://dx.doi.org/10.1016/j.watres.2017.02.001

Gonzalez-Gil L., Papa M., Feretti D., Ceretti E., Mazzoleni G., Steimberg N., Pedrazzani R., Bertanza G., Lema J. M. and Carballa M. (2016). Is anaerobic digestion effective for the removal of organic micropollutants and biological activities from sewage sludge? *Water Research*, **102**, 211–220. http://dx.doi.org/10.1016/j.watres.2016.06.025

Gözmen B., Oturan M. A., Oturan N. and Erbatur O. (2003). Indirect electrochemical treatment of bisphenol A in water via electrochemically generated Fenton's reagent. *Environmental Science and Technology*, **37**(16), 3716–3723.

Kelessidis A. and Stasinakis A. S. (2012). Comparative study of the methods used for treatment and final disposal of sewage sludge in European countries. *Waste Management*, **32**(6), 1186–1195. http://dx.doi.org/10.1016/j.wasman.2012.01.012.

Lin L., Li R. hong, Li Y., Xu J. and Li X. yan. (2017). Recovery of organic carbon and phosphorus from wastewater by Fe-enhanced primary sedimentation and sludge fermentation. *Process Biochemistry*, **54**, 135–139. http://dx.doi.org/10.1016/j.procbio.2016.12.016

Ma Y. S. and Lin J. G. (2011). Sono-alkalization pretreatment of sewage sludge containing phthalate acid esters. *Journal of Environmental Science and Health: Part A. Toxic/Hazardous Substances and Environmental Engineering*, **46**(9), 980–988.

Mailler R., Gasperi J., Patureau D., Vulliet E., Delgenes N., Danel A., Deshayes S., Eudes V., Guerin S., Moilleron R., Chebbo G. and Rocher V. (2017). Fate of emerging and priority micropollutants during the sewage sludge treatment: case study of Paris conurbation. Part 1: Contamination of the different types of sewage sludge. *Waste Management*, **59**, 379–393. http://dx.doi.org/10.1016/j.wasman.2016.11.010.

Mao T. and Show K.-Y. (2007). Influence of ultrasonication on anaerobic bioconversion of sludge. *Water Environment Research*, **79**(4), 436–441.

Martín de Vidales M. J., Sáez C., Cañizares P. and Rodrigo M. A. (2013). Removal of triclosan by conductive-diamond electrolysis and sonoelectrolysis. *Journal of Chemical Technology and Biotechnology*, **88**(5), 823–828.

McClellan K. and Halden R. U. (2010). Pharmaceuticals and personal care products in archived U.S. biosolids from the 2001 EPA national sewage sludge survey. *Water Research*, **44**(2), 658–668. http://dx.doi.org/10.1016/j.watres.2009.12.032.

McNamara P. J., Wilson C. A., Wogen M. T., Murthy S. N., Novak J. T. and Novak P. J. (2012). The effect of thermal hydrolysis pretreatment on the anaerobic degradation of nonylphenol and short-chain nonylphenol ethoxylates in digested biosolids. *Water Research*, **46**(9), 2937–2946. http://dx.doi.org/10.1016/j.watres.2012.03.015

Mohapatra D. P., Brar S. K., Tyagi R. D., Picard P. and Surampalli R. Y. (2012). Ferro-sonication and partial ozonation pre-treatment and biotransformation of wastewater sludge for degradation of bisphenol A: Rheology studies. *Chemical Engineering Science*, **81**, 20–27. http://dx.doi.org/10.1016/j.ces.2012.06.054

Mohapatra D. P., Brar S. K., Tyagi R. D. and Surampalli R. Y. (2011). Concomitant degradation of bisphenol A during ultrasonication and Fenton oxidation and production of biofertilizer from wastewater sludge. *Ultrasonics Sonochemistry*, **18**(5), 1018–1027. http://dx.doi.org/10.1016/j.ultsonch.2011.03.013

Morales E., Pastor L., Sancho J. V. and Boix C. (2016). Behaviour of emerging contaminants in sewage sludge after anaerobic digestion. **163**, 296–304.

Muller M., Combalbert S., Delgenès N., Bergheaud V., Rocher V., Benoît P., Delgenès J. P., Patureau D. and Hernandez-Raquet G. (2010). Occurrence of estrogens in sewage sludge and their fate during plant-scale anaerobic digestion. *Chemosphere*, **81**(1), 65–71. http://dx.doi.org/10.1016/j.chemosphere.2010.06.062

Ning X., Qi H., Li R., Jin Y., McNutt M. A. and Yin, Y. (2018). Synthesis and antitumor activity of novel 2, 3-didithiocarbamate substituted naphthoquinones as inhibitors of pyruvate kinase M2 isoform. *Journal of Enzyme Inhibition and Medicinal Chemistry*, **33**(1), 126–129. https://doi.org/10.1080/14756366.2017.1404591.

Patureau D., Mailler R., Delgenes N., Danel A., Vulliet E., Deshayes S., Moilleron R., Rocher V. and Gasperi J. (2021). Fate of emerging and priority micropollutants during the sewage sludge treatment–Part 2: Mass balances of organic contaminants on sludge treatments are challenging. *Waste Management*, **125**, 122–131. https://doi.org/10.1016/j.wasman.2021.02.034

Pilli S., Bhunia P., Yan S., LeBlanc R. J., Tyagi R. D. and Surampalli R. Y. (2011). Ultrasonic pretreatment of sludge: A review. *Ultrasonics Sonochemistry*, **18**(1), 1–18. http://dx.doi.org/10.1016/j.ultsonch.2010.02.014

Prosser R. S., Trapp S. and Sibley P. K. (2014). Modeling uptake of selected pharmaceuticals and personal care products into food crops from biosolids-amended soil. *Environmental Science and Technology*, **48**(19), 11397–11404.

Rivera-Utrilla J., Sánchez-Polo M., Ferro-García M. Á., Prados-Joya G. and Ocampo-Pérez R. (2013). Pharmaceuticals as emerging contaminants and their removal from water: A review. *Chemosphere*, **93**(7), 1268–1287.

Sabourin L., Duenk P., Bonte-Gelok S., Payne M., Lapen D. R. and Topp E. (2012). Uptake of pharmaceuticals, hormones and parabens into vegetables grown in soil fertilized with municipal biosolids. *Science of the Total Environment*, **431**, 233–236. http://dx.doi.org/10.1016/j.scitotenv.2012.05.017.

Samaras V. G., Stasinakis A. S., Thomaidis N. S., Mamais D. and Lekkas T. D. (2014). Fate of selected emerging micropollutants during mesophilic, thermophilic and temperature co-phased anaerobic digestion of sewage sludge. *Bioresource Technology*, **162**, 365–372. http://dx.doi.org/10.1016/j.biortech.2014.03.154

Sertillanges N., Haudin C. S., Bourdat-Deschamps M., Bernet N. Serre V., Danel A., Houot S and Patureau D. (2020). Process type is the key driver of the fate of organic micropollutants during industrial scale treatment of organic wastes. *Science of the Total Environment*, **734**, 139108. https://doi.org/10.1016/j.scitotenv.2020.139108

Shimizu T., Kudo K. and Nasu Y. (1993). Anaerobic waste-activated sludge digestion: A bioconversion mechanism and kinetic model. *Biotechnology and Bioengineering*, **41**(11), 1082–1091.

Suárez-Iglesias O., Urrea J. L., Oulego P., Collado S. and Díaz M. (2017). Valuable compounds from sewage sludge by thermal hydrolysis and wet oxidation: A review. *Science of the Total Environment*, **584–585**, 921–934. http://dx.doi.org/10.1016/j.scitotenv.2017.01.140

Taboada-Santos A., Braz G. H. R., Fernandez-Gonzalez N., Carballa M. and Lema J. M. (2019). Thermal hydrolysis of sewage sludge partially removes organic micropollutants but does not enhance their anaerobic biotransformation. *Science of the Total Environment*, **690**, 534–542. https://doi.org/10.1016/j.scitotenv.2019.06.492

Topp E., Metcalfe C. D., Boxall A. B. and Lapen D. R. (2010). Transport of PPCPs and veterinary medicines from agricultural fields following application of biosolids or manure. *ACS Symposium Series*, **1048**, 227–240.

Wang Y., Wang D., Liu Y., Wang Q., Chen F., Yang Q., Li X., Zeng G. and Li H. (2017). Triclocarban enhances short-chain fatty acids production from anaerobic fermentation of waste activated sludge. *Water Research*, **127**, 150–161. https://doi.org/10.1016/j.watres.2017.09.062

Wu X., Dodgen L. K., Conkle J. L. and Gan J. (2015). Plant uptake of pharmaceutical and personal care products from recycled water and biosolids: A review. *Science of the Total Environment*, **536**, 655–666. http://dx.doi.org/10.1016/j.scitotenv.2015.07.129

Yan Z. run, Meng H. shan, Yang X. yuan, Zhu Y. ying, Li X. yan, Xu J. and Sheng G. ping. (2019). Insights into the interactions between triclosan (TCS) and extracellular polymeric substance (EPS) of activated sludge. *Journal of Environmental Management*, **232**, 219–225. https://doi.org/10.1016/j.jenvman.2018.11.059

Yan Z. run, Zhu Y. ying, Meng H. shan, Wang S. yuan, Gan L. hong, Li X. yan, Xu J. and Zhang W. (2019). Insights into thermodynamic mechanisms driving bisphenol A (BPA) binding to extracellular polymeric substances (EPS) of activated sludge.

Science of the Total Environment, **677**, 502–510. https://doi.org/10.1016/j.scitotenv.2019.04.413

Yang S., McDonald J., Hai F. I., Price W. E., Khan S. J. and Nghiem L. D. (2017). The fate of trace organic contaminants in sewage sludge during recuperative thickening anaerobic digestion. *Bioresource Technology*, **240**, 197–206. http://dx.doi.org/10.1016/j.biortech.2017.02.020

Zhang X. and Li R. (2018). Variation of Antibiotics in Sludge Pretreatment and Anaerobic Digestion Processes: Degradation and Solid-Liquid Distribution. https://doi.org/10.1016/j.biortech.2018.01.100

Zhang Y., Wu D., Su Y. and Xie B. (2021). Occurrence, influence and removal strategies of mycotoxins, antibiotics and microplastics in anaerobic digestion treating food waste and co-digestive biosolids: A critical review. *Bioresource Technology*, **330**(January), 124987. https://doi.org/10.1016/j.biortech.2021.124987

Zhao J., Li Y., Zhang C., Zeng Q. and Zhou Q. (2008). Sorption and degradation of bisphenol A by aerobic activated sludge. *Journal of Hazardous Materials*, **155**(1–2), 305–311.

Zhou G., Lin L., Li X., Mei K. and Leung Y. (2020). Removal of emerging contaminants from wastewater during chemically enhanced primary sedimentation and acidogenic sludge fermentation. *Water Research*, **175**, 115646. https://doi.org/10.1016/j.watres.2020.115646

Zhou G. J., Li X. Y. and Leung K. M. Y. (2019). Retinoids and oestrogenic endocrine disrupting chemicals in saline sewage treatment plants: Removal efficiencies and ecological risks to marine organisms. *Environment International*, **127**(March), 103–113. https://doi.org/10.1016/j.envint.2019.03.030

Zhu Q., Jia J., Wang Y., Zhang K., Zhang H., Liao C. and Jiang G. (2019). Spatial distribution of parabens, triclocarban, triclosan, bisphenols, and tetrabromobisphenol A and its alternatives in municipal sewage sludges in China. *Science of the Total Environment*, **679**, 61–69. https://doi.org/10.1016/j.scitotenv.2019.05.059

doi: 10.2166/9781789062755_0343

Chapter 16

Single-cell protein production using anaerobic digestate

Rongfen Chen[1,2] and Yan Zhou[1,3,*]

[1]Advanced Environmental Biotechnology Centre, Nanyang Environment & Water Research Institute, Nanyang Technological University, 1 Cleantech Loop, Singapore 637141, Singapore
[2]Interdisciplinary Graduate School, Nanyang Technological University, 50 Nanyang Avenue, 639798, Singapore
[3]School of Civil and Environmental Engineering, Nanyang Technological University, 50 Nanyang Avenue, Singapore 639798, Singapore
*Corresponding author e-mail: zhouyan@ntu.edu.sg

ABSTRACT

This chapter discusses the production of single-cell microbial protein using anaerobic digestate. Quick overview of conventional treatment methods of anaerobic digestate was presented, followed by introduction of several protein-rich unicellular microbes (namely, purple non-sulphur bacteria, aerobic heterotrophic bacteria, and microalgae). Benefits and challenges of single-cell production using anaerobic digestate were later deliberated to assess the reliability and practicability of a potential integration of single-cell protein production and anaerobic digestion.

Keywords: Anaerobic digestate management, anaerobic digestion, microbial protein, single-cell protein.

16.1 INTRODUCTION

Increasing population and anthropogenic activities are adding tremendous amount of waste into the natural system (Eurostat, 2018; Gustavsson *et al.*, 2011; Tyagi *et al.*, 2018). The organic-rich waste, if left untreated, could lead to environmental pollution in open dump sites (Hoornweg & Bhada-Tata, 2012; Huang *et al.*, 2016). However, the 'misplaced resource' could be utilized given the right treatment methods. Anaerobic digestion (AD) breaks down complex organic molecules in the absence of oxygen (Sawatdeenarunat *et al.*, 2015). End-products of AD are in three different phases, namely, biogas, soluble short-chain acids, and residual solids. Energy and resource recovery are achieved through AD of the organic-rich waste (Monlau *et al.*, 2015; Morero *et al.*, 2015). Energy-rich biogas from AD is a renewable energy source for heat and/or electricity production. On the other hand, nutrient-rich digestate (both liquid

Editor(s): Vinay Kumar Tyagi, Kaoutar Aboudi and Cigdem Eskicioglu

and solid) could be transformed into fertilizers for landscaping and agriculture (Tyagi *et al.*, 2018; Verbeeck *et al.*, 2021). It was estimated that utilization of both biogas (for energy production) and digestate (as fertilizer) contributed to an overall reduction of CO_2 emission by 230–340 kg CO_2/t of biowaste (Cecchi *et al.*, 2011). Nevertheless, there are complications in pushing up the energy and material recovery efficiency. 15–60% volume composition of CO_2 in the biogas (Baciocchi *et al.*, 2013) lowered the heating value of the biogas (Bharathiraja *et al.*, 2018). Owing to inconsistent composition, supply shortage and unstable physical condition, realized value from AD digestate-derived fertilizer fell short below 20% of its intrinsic value (Verbeeck *et al.*, 2021). Furthermore, mineralization of organic waste through AD and subsequent application of digestate-derived fertilizer in agriculture for food production are not the most efficient ways to reuse the nutrient contents (Ritala *et al.*, 2017). Agriculture is the largest freshwater user (Mekonnen & Hoekstra, 2014) and a major contributor for global anthropogenic greenhouse gas emissions (Vermeulen *et al.*, 2012). Furthermore, global climate changes are also bringing more frequent extreme weather events worldwide, compounding the climate-dependent agricultural activities. Alternatives to digestate-to-fertilizer-to-food system are essential for the next-generation AD digestate treatments. Single-cell protein (SCP) production is an emerging technology for high value-added nutrient recovery from biodegradable biowaste though the term 'SCP' was first introduced in 1968 (Sharif *et al.*, 2021). SCP is a bulk of protein-rich dried cells generated from algae, yeast, bacteria, and fungi (Ritala *et al.*, 2017). Proteins extracted from the microbial biomass could be edible directly or indirectly (Zha *et al.*, 2021), serving as a potentially viable alternative to plant-based protein (Hadi & Brightwell, 2021). Contrary to plant-based protein production, SCP is not limited by climate, soil characteristics, and availability of arable land (Hülsen *et al.*, 2018). SCP production by bacteria was characterized by rapid growth rate and high protein content (Gao *et al.*, 2007). Furthermore, nutrient losses due to leaching, runoff and volatilization (Galloway *et al.*, 2003) in crop farming could be avoided. In terms of policy and compliance, SCP has been approved by the European Union for livestock use (European_Commission, 2017), paving the way for wider commercial application as animal feed (Hülsen *et al.*, 2018). Synthesis of SCP uses simpler organic molecules and inorganic N and P as building blocks (Sharif *et al.*, 2021). To lower the production cost, upcycling of organic carbon from biowaste for SCP production (Matassa *et al.*, 2020) 'kills two birds with one stone'. The use of organic carbon source from waste material, such as industrial and agricultural, is common in SCP production (Gervasi *et al.*, 2018). For example, fruit waste (Nasseri *et al.*, 2011), potato starch processing waste (Liu *et al.*, 2014), urban biowaste (Khoshnevisan *et al.*, 2019), and lignocellulosic waste (Mohammadi *et al.*, 2016) have been used as substrates for SCP production. The properties and compositions of biowaste could vary from type to type or from time to time within the same type. Fluctuating substrate conditions could impede the multiplication rate of biomass and the quality of the biomass produced. Hence, AD can be integrated with SCP to produce a stable stream of simpler organic molecules for downstream protein synthesis in an inexpensive manner.

16.2 CONVENTIONAL TREATMENTS OF ANAEROBIC DIGESTATE

In aerobic digestion, inorganic forms of nitrogen and phosphorus were released alongside with breakdown and conversion of organic carbon compounds into biogases. Due to the lack of autotrophic metabolism, inorganic nitrogen and phosphorus are not completely removed but are mostly retained in the digestate (Noike *et al.*, 2004). Regardless of the feeding material, AD digestate is characterized by high nitrogenous content (Xia & Murphy, 2016). Therefore, liquid AD digestate generally requires post-treatments to produce ready-to-discharge high-quality effluent (Chernicharo, 2006). Normally, there are two different pathways for the post-treatment: residual nutrients are either removed or redirected for further nutrient recovery (Figure 16.1).

To remove the residual carbon, nitrogen and phosphorus, liquid AD digestate could be pumped back into mainstream wastewater treatment system. For example, NH_4^+ in the digestate is removed via the two-stage nitrification–denitrification process. Full nitrification occurred under aerobic condition, leading to intensive energy demand and high operational cost (Akizuki *et al.*, 2021). An attempt to reduce aeration cost by integrating biofilter with natural convection of air was tested (Hunter & Deshusses, 2020). Nevertheless, oxygen limitation could not be avoided after some period of operation. Alternatively, excessive aeration demand from conventional activated sludge processes could be alleviated through the energy-neutral anaerobic ammonium oxidation (anammox) process. Complete autotrophic nitrogen removal which combines partial nitritation with anammox could reduce 60% aeration needs and production of biosolids (Magrí *et al.*, 2013). However, slow growth rate of anammox bacteria translates to long start-up period. Furthermore, sole removal of residual nutrient in AD digestate weakens the recoverability of useful resources from the digestate.

Removal of residual nutrient in AD digestate focuses on one dimension of waste management that is to protect the receiving waters and soils from

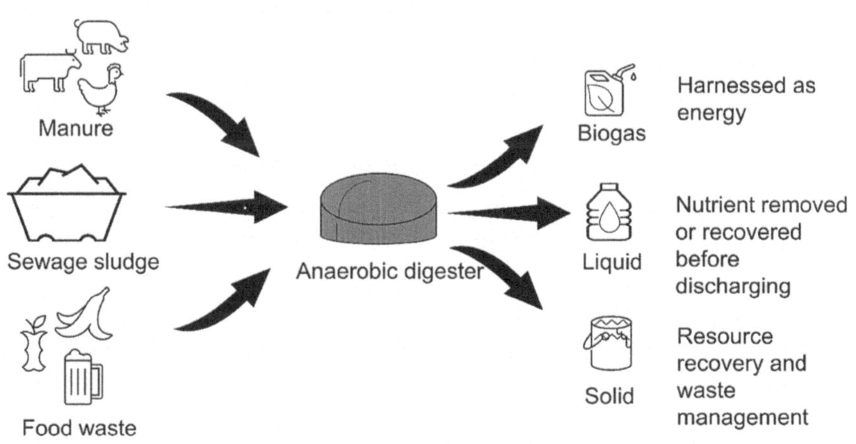

Figure 16.1 Schematic of AD and the fate of its products traditionally.

harmful effects upon discharging back to the environment. However, there has been a paradigm shift in waste management from removal-focused to recovery-centric. This is in line with the concept of circular economy for which the value of materials and resources is kept in the economy for as long as possible (Stahel, 2016). Hence, separating residual nutrient from the liquid phase and redirecting them for further resource recovery are gaining popularity. Phase separation of ammonia from AD digestate could be achieved through various methods: gas stripping, adsorption, membrane process, and struvite precipitation. Gas stripping involves a conversion of ammonium ions to free ammonia and a diffusion of free ammonia from liquid to the stripping gas (Baldi *et al.*, 2018). Ammonia recovery from stripping is primarily dependent on temperature, pH, and gas flowrate (Palakodeti *et al.*, 2021). Fouling of stripping column and unintentional ammonia release could compromise the process (De la Rubia *et al.*, 2010). Adsorption employs materials such as zeolite, activated carbon and biochar to concentrate residual nutrient from the liquid phase (Kizito *et al.*, 2016). One unique advantage for adsorption using biochar is that the adsorbent material can be generated from pyrolytic product of the solid digestate (Lehmann & Joseph, 2015). Membrane contactor process separates the feeding and receiving streams with a partially permeable membrane while allowing cross-diffusion of ammonia through the pores (Hasanoğlu *et al.*, 2010). Dosing of chemical on the receiving side, such as dilute sulphuric acid, to maintain inter-phase diffusion is needed. Struvite is a crystalline mineral with a chemical formula of $MgNH_4PO_4{\cdot}6H_2O$. Struvite precipitation is a promising approach to recycle N and P from AD digestate as it is a slow-releasing multi-nutrient fertilizer (Talboys *et al.*, 2016). Given the right molar ratio, pH, reaction time, and temperature, as much as 94% of P and half as much N could be recovered in the form of struvite (Rahman *et al.*, 2014).

There is a clear divergence between the removal-focused and recovery-centric routes. The removal-focused route emphasizes the biological metabolism of microorganisms while the recovery-centric route relies heavily on physical and chemical means. It seems that resource recovery could only be done by concentrating them straight from the liquid phase. Residual nutrients in the AD digestate, after recovery, are mostly in the inorganic form and ready to be applied as fertilizers.

16.3 SCP PRODUCTION BY UNICELLULAR MICROBES

Several types of unicellular microorganisms have been identified for SCP production. Figure 16.2 depicts three main microbes for SCP production and their interactions. Depending on their energy source, they could be divided into photosynthetic (e.g. microalgae or purple non-sulphur bacteria (PNSB)) or non-photosynthetic (e.g. heterotrophic bacteria). Alternatively, they could also be classified as autotrophs (microalgae) and heterotrophs (PNSB and heterotrophic bacteria) based on the different carbon molecules as substrates. The above-mentioned microorganisms tap on a wide range of carbon substrates for their production of microbial protein, allowing for a close amalgamation with AD process.

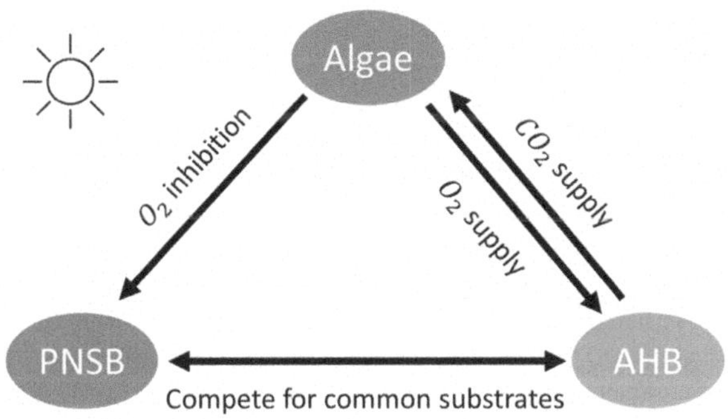

Figure 16.2 Interaction among three main unicellular microbes, namely, microalgae, PNSB, and AHB.

Carbon dioxide fixation ability of microalgae is particularly relevant in the context of the warming globe due to the rising atmospheric CO_2 level. Plus, the main compositions of biogas from AD are CH_4 (60–70% v/v) and CO_2 (30–40% v/v) (Matassa *et al.*, 2020). Wang *et al.* (2020) studied the use of microalgae-methanotroph coculture to transform biogas from AD into more value-added products while removing nutrients from the digestate. In a coculture of 3 microalgae:1 methanotroph fed with 70% CH_4 and 30% CO_2, total nitrogen and phosphorus recovery were achieved alongside with improved nutrient recovery from the liquid digestate. In another similar batch setup using 250 mL serum bottles, Zha *et al.* (2021) reported that methanotrophic *Methylomans* and *Methylophilus* spp. could be cultivated efficiently together on gas and liquid products from sewage sludge AD.

Other than utilizing gaseous substrates, both aerobic heterotrophic bacteria (AHB) and PNSB could survive on a wide range of short-chain organic compounds in the liquid phase. AHB, sometimes is also known as aerobic activated sludge produced from wastewater treatment processes. In fact, most of the studies on SCP production have been conducted on organic containing wastewater, with a clear purpose to supplement the water treatment goals (Alloul *et al.*, 2018; Nasseri *et al.*, 2011; Ravindra, 2000). For example, growth of AHB using aquaculture waste resulted in 90% removal of inorganic nitrogen and 80% ortho-phosphate, with the aid of organic carbon supplement (Schneider *et al.*, 2006). AHB is rich in protein (38–60 g protein per 100 g total suspended solid) and other beneficial vitamins (e.g. vitamins B1, B2, B6 and B12) (Vriens *et al.*, 1989). PNSB is metabolically versatile, depending on the availability of light and oxygen. Under anaerobic condition and in the presence of light of near-infrared range, PNSB could grow using volatile fatty acids (VFAs) (Lu *et al.*, 2019) and simple sugar molecules (e.g. glucose, sucrose and lactose) (Keskin & Hallenbeck, 2012). Biomass yield of PNSB on simple substrates could go as high as 1.0 g-COD/g-$COD_{removed}$ (Alloul *et al.*, 2019) with appealingly high protein content (41–60 g protein per 100 g total suspended solid) and quality vitamins such as B1, B2, B3, B5, B6, B9, E and biotin (Sasaki *et al.*, 1998).

Despite some recent studies which employed PNSB to treat agri-industrial (Hülsen *et al.*, 2018) and domestic wastewater (Hülsen *et al.*, 2016), majority of the PNSB reports in existing literature used synthetic or axenic media and pure culture, such as *Rhodobacter sphaeroides* and *Rhodopseudomonas palustris*.

Microalgae, AHB and PNSB demand different conditions for their microbial protein synthesis (Figure 16.2). Microalgae are primarily photoautotrophs notwithstanding the ability of some species to grow heterotrophically (Zheng *et al.*, 2012). Autotrophic nature of algae was evident from the improved nitrogen and phosphorus removal in swine slurry after CO_2 dosage (Taiganides, 1992). Without CO_2 addition, the source of inorganic carbon for algae was derived from AHB who tapped on the O_2 produced by the algae (Borowitzka, 1998), bringing about a potential collaboration between both microbes. In the meantime, the need for illumination is common for both microalgae and PNSB. In a comparison treating five agri-industrial wastewaters, PNSB saw moderate nutrient removal but higher biomass yield while microalgae-containing consortium saw better removal but lower biomass yield (Hülsen *et al.*, 2018). AHB and PNSB, on the other hand, were treated as two separate types of SCP because of their diverging oxic conditions. SCP protein production by PNSB was impeded in the presence of oxygen supply (dissolved oxygen $\leq$0.5 mg-O_2/L) owing to the metabolic shift from photoheterotrophic to aerobic heterotrophic (Capson-Tojo *et al.*, 2021). There is a trade-off between rate and microbial protein qualities. AHB grows rapidly under aerobic conditions while PNSB is slow but with better conversion efficiency. Higher cost associated with anaerobic PNSB photoreactor also needed to be addressed (Van Haandel & Van Der Lubbe, 2007). A blend of AHB and PNSB could achieve a balance between microbial protein production rate and quality in sequential photo- and chemotrophic reactors (Alloul *et al.*, 2021b).

16.4 INTEGRATING SCP PRODUCTION WITH ANAEROBIC DIGESTATE MANAGEMENT

Although the production of SCP using methane could be dated back to the 1960s (Bewersdorff & Dostálek, 1971) and protein content in methanotrophs could go as high as 70% by weight (Øverland *et al.*, 2011), this chapter will focus on the integration of SCP production with the management of the AD digestate instead. More specifically, this chapter explores the use of AD as a mean of pre-treatment for organic-rich wastewater or biosolids for subsequent high-quality SCP production via unicellular microorganisms.

Organic carbon containing wastewater, such as red meat processing wastewater (Hülsen *et al.*, 2020) and food and beverage wastewater (Muys *et al.*, 2020), have been used for SCP production after physical separation of unwanted biosolids. The obtained biomass yield of PNSB using a wide range of agri-industrial and domestic wastewater has been comparable to those cultured using synthetic media (Capson-Tojo *et al.*, 2020). Different types of waste products were characterized differently: monosaccharide and disaccharide-rich waste (e.g. banana waste and spoiled date palm fruits), starch-rich waste (e.g. wheat and rice brans), structure polysaccharide-rich waste (e.g. apple pomace and citrus pulp), and protein and lipid-rich waste (e.g. soybean meal) (Finco *et al.*, 2017).

However, raw wastewater consists of varying mixture whose composition changes constantly with time, operating conditions, and many other factors. A fluctuating quality of raw wastewater deeply compromises the selectivity of desirable biomass and predictability of reactor performance. From an economic standpoint, consistent quality of SCP produced from waste material is the key for large-scale applications (Verstraete *et al.*, 2016). Meanwhile, complex organic molecules, in AD, could be broken down to a cluster of short-chain VFAs (such as butyrate, propionate, and acetate) whose toxicity to microbes were well defined (Sun & O'Riordan, 2013). Production of VFAs via hydrolysis and acidogenesis levels off the variability of the organic composition in the liquid phase, ensuring consistency of the organic compound complexity in the influent for SCP production. In addition, anaerobic pre-treatment removes harmful pathogens (Wagner *et al.*, 2008) which might be found in the raw wastewater. Capson-Tojo *et al.* (2020) further added that, in the case of PNSB, a substrate that is more reduced than the biomass (e.g. butyric acid) would serve a dual role as both carbon source and electron donor for CO_2 fixation. In an attempt to map out the effects of acetate, propionate, butyrate, and a mix of VFAs on the growth and biomass yield of PNSB, Alloul *et al.* (2019) observed that VFAs mixture, which could be produced from AD of longer chain organic molecules, accelerated the growth of PNSB by 1.3–2.5 times. Therefore, a synergetic integration between AD pre-treatment and subsequent SCP production could yield more beneficial outcomes.

To realize the fullest potential of such an integration, there is a need to enhance the VFA production in the AD pre-treatment process. The conversion of complex organic molecules to methane gas in AD involves two phases (acid-forming phase and methane-forming phase) or four stages: hydrolysis, acidogenesis, acetogenesis, and methanogenesis. Interestingly, the acid-forming and methane-forming microbes show significant differences in their physiology, nutrient needs, growth kinetics, and response to the environment (Adekunle & Okolie, 2015). For example, most acid-forming microbes favour mesophilic condition (30–40°C) while thermophilic (50–60°C) results in higher biogas production (Gou *et al.*, 2014; Liu *et al.*, 2017). When food waste was used as substrate, higher organic loading rate favoured hydrolysis/acidogenesis over acetogenesis/methanogenesis (Nagao *et al.*, 2012). Food waste with higher nitrogen content produces excessive ammonia in the digestate and eventually leads to inhibitory effects on methanogens (Shi *et al.*, 2017). Operating pH too plays a role in the production of VFAs in AD process as methanogens typically grow at pH of 6.7–7.5. Through stepwise pH increment in thermophilic sludge fermentation, it was reported that maximum acidification occurred at pH 8.9 with a maximum VFAs production of 423.22 ± 25.49 mg COD/g VSS (Chen *et al.*, 2017). Furthermore, methane-forming microbes have slower growth kinetics and thus could be suppressed with shorter hydraulic retention time (Adekunle & Okolie, 2015).

Another area of integration between AD and SCP production lies in the management of municipal solid waste (MSW). Typically, there was around 50%

organic fraction in MSW (referred to as OFMSW) (Tyagi *et al.*, 2018), which could be harnessed as thermal or electrical energy via AD process with an energy conversion of 47.01 J/g COD/d (Marti *et al.*, 2008). The low recovery efficiency from biogas attracts new ideas in the direct transformation of OFMSW into more valuable commodities. Recently, a three-step 'capture–ferment–upgrade' concept for the management of OFMSW has been in the centre of discussion recently (Alloul *et al.*, 2018). The three-step approach distinguishes itself from other conventional treatment methods with its focus on the role of VFAs in place of biogas. Different from the liquid-fed AD processes, hydrolysis is the rate-determining step in biosolid AD reactions. Sludge pre-treatment using mechanical, thermal, and biological methods might be needed. Mechanical pre-treatment increases the specific surface area of the biosolid and therefore ensuring better contact between substrate and anaerobic microbes (Carrère *et al.*, 2010). Thermal pre-treatment promotes the hydrolysis process due to disintegration of the cytomembrane of the substrate (Prorot *et al.*, 2011). Biological pre-treatment involves the inoculation of microorganisms and enzymes to promote hydrolysis of substrate. The amount of VFAs obtained could increase with the aid of aerobic pre-fermentation of food waste (Sahu *et al.*, 2017). In addition, co-digestion of food waste and waste activated sludge saw higher accumulation of VFAs at pH 4.3–5.3 (Vidal-Antich *et al.*, 2021).

Wu *et al.* (2019) solubilized a mixture of primary and secondary sludge from wastewater treatment plants into a fermented liquor consisted of 47.6% VFAs. Subsequently, the fermented liquor was fed into a microbial fuel cell (MFC) for in-situ power generation and nutrient recovery. Polyhydroxyalkanoates (PHA) and polyphosphates were found to be stored inside the cell while the presence of PNSB taxon *Rhodocyclaceae* was confirmed by high-throughput sequencing. Thereafter, PNSB taxon *Rhodopseudomonas palustris* was selected and enriched with NIR illumination from the effluent of the MFC to understand its pathways and mechanisms for SCP production. In the presence of light, organic carbon, ammonium, and phosphorus were transformed to protein, poly-β-hydroxybutyrate (PHB) and polyphosphates in the cell. 0.89 g COD biomass/g COD substrate was obtained with 65% of the biomass being protein (Wu *et al.*, 2021) that consisted 15 identified amino acids. Similarly, Allegue *et al.* (2020) designed a combined approach, couple thermal hydrolysis with AD and photo-fermentation to treat OFMSW. PNSB in the photo-fermentation phase consumed up to 80% of the soluble organic compounds received from the liquid AD digestate.

The emphasis of the integration between AD processes and SCP production lies in the role of VFAs. VFAs are intermediate products in AD which has long been seen as undesirable (Yuan & Zhu, 2016). On the contrary, VFAs are perfect substrates for stable and fast growth of unicellular SCP producing microorganisms. These opposing views on VFA means that there will be a paradigm shift in the focus of digested products when it means integration with SCP productions. Retaining a large portion of organic compound in the form of VFA along with other nutrients in the liquid phase will no longer be a problem but a welcoming outcome instead. Over decades and centuries, scientists have invested considerable efforts in exploring options to boost the biogas production yield from ADs. The biogas-boosting approaches may soon become obsolete against the backdrops of a rising need to upcycle waste in the form of more valuable SCPs. Foreseeably, there will be research needs to look at ways to improve VFA production and to produce desirable mix of VFAs for different SCP-producing microorganisms.

16.5 CHALLENGES FACED BY SCP PRODUCTION FROM AD DIGESTATE

SCP production from waste represents a progressive resource upcycling approach to deal with both the increasing waste production and the rising demand for food. Based on some estimation, cost of substrates used for SCP production accounts for 45–75% of the total cost (Sharif *et al.*, 2021). By tapping on the biodegradable waste products as substrates for growth, production cost of SCP will be reduced. As a potential feed for aquaculture, livestock and poultry, SCP is rich in a diverse mix of essential amino acids such as lysine and methionine which lack in most animal or plant-based proteins (Gao *et al.*, 2012). With the aid of AD pre-treatment, SCP production could operate round the clock with comparable feeding conditions. Nevertheless, there remain several challenges and bottlenecks in SCP production, with or without AD pre-treatment, in wide-scale applications.

Safety aspects of SCP as food or as animal feed remain largely unknown, particularly related to allergens, toxins, and heavy metals. For instance, β-chain of the C-phycocyanin protein in cyanobacteria has been identified as a main allergen (Lang-Yona *et al.*, 2018). In microalgae, toxins could be accumulated if the environment is contaminated with toxin-secreting cyanobacteria. Contamination by heavy metal, either through passive surface sorption or active metabolic uptake (Markou *et al.*, 2018), becomes a major concern when waste products of no clear origin is used as substrate for growth. For the fast-growing bacterial species, high nucleic acid content of 15–16% (Ravindra, 2000) could also compromise its nutritional quality as purines degrade into uric acids (Hadi & Brightwell, 2021). It is necessary to reduce the nucleic acid content in bacterial biomass through physical and chemical treatment.

Despite higher protein contents in bacterial biomass, there are drawbacks in the design of reactor to improve the efficiency in light illumination for anaerobic photobioreactor and in aeration for aerobic bioreactor. Growth of PNSB is light-dependent. Adequate and effective supply of NIR radiation determines the nutrient removal and biomass yield. The cost of artificial lighting raised the production cost of PNSB biomass and making it less attractive for large-scale application (Capson-Tojo *et al.*, 2020). Tapping on daytime sunlight is gaining attention as a potential solution looking forward though it brings in uncertainty factors such as weather. Furthermore, NIR radiation could be absorbed by water easily. Reactor configuration needs to be carefully considered to achieve optimal light delivery throughout the reactor. Oxygen is indispensable for the growth of AHB. Traditionally, supply of oxygen to the aerobic microbe is done through oxygenating the water with air bubbles. Standard oxygen transfer efficiency for bubbling aeration is only 20–30% (Li, 2018).

Extraction and post-treatment of the produced biomass are also important for large-scale applications. Biomass harvested from the bioreactor typically consists of sizeable amount of water content. Concentration of PNSB in a raceway reactor was about 0.81 ± 0.04 g TSS/L (Alloul *et al.*, 2021a), while concentration of AHB could go beyond 10 g TSS/L (Alloul *et al.*, 2021b). Aerobic heterotrophs, with a better settleability, could be preliminarily concentrated through gravity. However, membrane process might be needed to concentrate the less settleable PNSB. The fact that the concentrated biomass is for direct or indirect edible purposes also limits the choice of dewatering options. In sludge dewatering, addition of coagulants and flocculants is the most prevalent conditioning method when cost is of concern (Wei *et al.*, 2018). Dosing of coagulants and flocculants does not seem appropriate unless the coagulants and flocculants used are safe and edible.

16.6 CONCLUSION

This chapter provides an overview of the conventional treatment methods for anaerobic digestate and outlined the potential integration of SCP production and anaerobic digestate management. SCP production from waste material is a promising alternative to tackle the problems of rising waste production and increasing demand for protein. Successful application of SCP production from waste material allows some carbons to bypass the slow natural carbon cycle and upcycles quickly into more valuable protein contents. AD could serve as a pre-treatment step to convert various forms of waste organic compounds into a mix of volatile fatty acids and other necessary nutrients for the growth of protein-producing bacteria. Coupling AD with SCP production improves the predictability of influent into subsequent SCP production.

REFERENCES

Adekunle K. F. and Okolie J. A. (2015). A review of biochemical process of anaerobic digestion. *Advances in Bioscience and Biotechnology*, **6**(03), 205–212, https://doi.org/10.4236/abb.2015.63020

Akizuki S., Cuevas-Rodríguez G. and Toda T. (2021). Nitrification of anaerobic digestate using a consortium of microalgae and nitrifiers in an open photobioreactor with moving bed carriers. *Chemosphere*, **263**, 127948, https://doi.org/10.1016/j.chemosphere.2020.127948

Allegue L. D., Puyol D. and Melero J. A. (2020). Novel approach for the treatment of the organic fraction of municipal solid waste: coupling thermal hydrolysis with anaerobic digestion and photo-fermentation. *Science of the Total Environment*, **714**, 136845, https://doi.org/10.1016/j.scitotenv.2020.136845

Alloul A., Cerruti M., Adamczyk D., Weissbrodt D. G. and Vlaeminck S. E. (2021a). Operational strategies to selectively produce purple bacteria for microbial protein in raceway reactors. *Environmental Science & Technology*, **55**(12), 8278–8286.

Alloul A., Muys M., Hertoghs N., Kerckhof F.-M. and Vlaeminck S. E. (2021b). Cocultivating aerobic heterotrophs and purple bacteria for microbial protein in sequential photo-and chemotrophic reactors. *Bioresource Technology*, **319**, 124192, https://doi.org/10.1016/j.biortech.2020.124192

Alloul A., Ganigué R., Spiller M., Meerburg F., Cagnetta C., Rabaey K. and Vlaeminck S. E. (2018). Capture–ferment–upgrade: a three-step approach for the valorization of sewage organics as commodities. *Environmental Science & Technology*, **52**(12), 6729–6742, https://doi.org/10.1021/acs.est.7b05712

Alloul A., Wuyts S., Lebeer S. and Vlaeminck S. E. (2019). Volatile fatty acids impacting phototrophic growth kinetics of purple bacteria: paving the way for protein production on fermented wastewater. *Water Research*, **152**, 138–147, https://doi.org/10.1016/j.watres.2018.12.025

Baciocchi R., Carnevale E., Costa G., Gavasci R., Lombardi L., Olivieri T., Zanchi L. and Zingaretti D. (2013). Performance of a biogas upgrading process based on alkali absorption with regeneration using air pollution control residues. *Waste Management*, **33**(12), 2694–2705, https://doi.org/10.1016/j.wasman.2013.08.022

Baldi M., Collivignarelli M. C., Abbà A. and Benigna I. (2018). The valorization of ammonia in manure digestate by means of alternative stripping reactors. *Sustainability*, **10**(9), 3073, https://doi.org/10.3390/su10093073

Bewersdorff M. and Dostálek M. (1971). The use of methane for production of bacterial protein. *Biotechnology and Bioengineering*, **13**(1), 49–62, https://doi.org/10.1002/bit.260130104

Bharathiraja B., Sudharsana T., Jayamuthunagai J., Praveenkumar R., Chozhavendhan S. and Iyyappan J. (2018). Biogas production – a review on composition, fuel properties, feed stock and principles of anaerobic digestion. *Renewable and Sustainable Energy Reviews*, **90**, 570–582, https://doi.org/10.1016/j.rser.2018.03.093

Borowitzka M. A. (1998). Limits to growth. In: Wastewater Treatment with Algae. Springer, Berlin, Heidelberg, pp. 203–226.

Capson-Tojo G., Batstone D. J., Grassino M., Vlaeminck S. E., Puyol D., Verstraete W., Kleerebezem R., Oehmen A., Ghimire A. and Pikaar I. (2020). Purple phototrophic bacteria for resource recovery: challenges and opportunities. *Biotechnology Advances*, **43**, 107567, https://doi.org/10.1016/j.biotechadv.2020.107567

Capson-Tojo G., Lin S., Batstone D. J. and Hülsen T. (2021). Purple phototrophic bacteria are outcompeted by aerobic heterotrophs in the presence of oxygen. *Water Research*, **194**, 116941, https://doi.org/10.1016/j.watres.2021.116941

Carrère H., Dumas C., Battimelli A., Batstone D. J., Delgenes J. P., Steyer J.-P. and Ferrer I. (2010). Pretreatment methods to improve sludge anaerobic degradability: a review. *Journal of Hazardous Materials*, **183**(1–3), 1–15, https://doi.org/10.1016/j.jhazmat.2010.06.129

Cecchi F., Bolzonella D., Pavan P., Macé S. and Mata-Alvarez J. (2011). Anaerobic digestion of the organic fraction of municipal solid waste for methane production: research and industrial application.

Chen Y., Jiang X., Xiao K., Shen N., Zeng R. J. and Zhou Y. (2017). Enhanced volatile fatty acids (VFAs) production in a thermophilic fermenter with stepwise pH increase – investigation on dissolved organic matter transformation and microbial community shift. *Water Research*, **112**, 261–268, https://doi.org/10.1016/j.watres.2017.01.067

Chernicharo C. D. (2006). Post-treatment options for the anaerobic treatment of domestic wastewater. *Reviews in Environmental Science and Bio/Technology*, **5**(1), 73–92, https://doi.org/10.1007/s11157-005-5683-5

De la Rubia MÁ, Walker M., Heaven S., Banks C. J. and Borja R. (2010). Preliminary trials of in situ ammonia stripping from source segregated domestic food waste digestate using biogas: effect of temperature and flow rate. *Bioresource Technology*, **101**(24), 9486–9492, https://doi.org/10.1016/j.biortech.2010.07.096

European_Commission (2017). Commission Regulation (EU) 2017/1017 of 15 June 2017 amending Regulation (EU) No 68/2013 on the Catalogue of feed materials. In: *Official Journal of the European Union*, Eurostat, the Statistical Office of the European Union, pp. 48–119.

Eurostat E. (2018). Sewage Sludge Production and Disposal. Luxembourg.

Finco A. M. d. O., Mamani L. D. G., Carvalho J. C. d., de Melo Pereira G. V., Thomaz-Soccol V. and Soccol C. R. (2017). Technological trends and market perspectives for production of microbial oils rich in omega-3. *Critical Reviews in Biotechnology*, **37**(5), 656–671, https://doi.org/10.1080/07388551.2016.1213221

Galloway J. N., Aber J. D., Erisman J. W., Seitzinger S. P., Howarth R. W., Cowling E. B. and Cosby B. J. (2003). The nitrogen cascade. *Bioscience*, **53**(4), 341–356, https://doi.org/10.1641/0006-3568(2003)053[0341:TNC]2.0.CO;2

Gao L., Chi Z., Sheng J., Ni X. and Wang L. (2007). Single-cell protein production from Jerusalem artichoke extract by a recently isolated marine yeast *Cryptococcus aureus* G7a and its nutritive analysis. *Applied Microbiology and Biotechnology*, **77**(4), 825–832, https://doi.org/10.1007/s00253-007-1210-7

Gao Y., Li D. and Liu Y. (2012). Production of single cell protein from soy molasses using *Candida tropicalis*. *Annals of Microbiology*, **62**(3), 1165–1172, https://doi.org/10.1007/s13213-011-0356-9

Gervasi T., Pellizzeri V., Calabrese G., Di Bella G., Cicero N. and Dugo G. (2018). Production of single cell protein (SCP) from food and agricultural waste by using *Saccharomyces cerevisiae*. *Natural Product Research*, **32**(6), 648–653, https://doi.org/10.1080/14786419.2017.1332617

Gou C., Yang Z., Huang J., Wang H., Xu H. and Wang L. (2014). Effects of temperature and organic loading rate on the performance and microbial community of anaerobic co-digestion of waste activated sludge and food waste. *Chemosphere*, **105**, 146–151, https://doi.org/10.1016/j.chemosphere.2014.01.018

Gustavsson J., Cederberg C., Sonesson U., Van Otterdijk R. and Meybeck A. (2011). Global Food Losses and Food Waste. FAO, Rome.

Hadi J. and Brightwell G. (2021). Safety of alternative proteins: technological, environmental and regulatory aspects of cultured meat, plant-based meat, insect protein and single-cell protein. *Foods*, **10**(6), 1226, https://doi.org/10.3390/foods10061226

Hasanoğlu A., Romero J., Pérez B. and Plaza A. (2010). Ammonia removal from wastewater streams through membrane contactors: experimental and theoretical analysis of operation parameters and configuration. *Chemical Engineering Journal*, **160**(2), 530–537, https://doi.org/10.1016/j.cej.2010.03.064

Hoornweg D. and Bhada-Tata P. (2012). What a waste: a global review of solid waste management.

Huang W., Wang Z., Guo Q., Wang H., Zhou Y. and Ng W. J. (2016). Pilot-scale landfill with leachate recirculation for enhanced stabilization. *Biochemical Engineering Journal*, **105**, 437–445, https://doi.org/10.1016/j.bej.2015.10.013

Hülsen T., Barry E. M., Lu Y., Puyol D., Keller J. and Batstone D. J. (2016). Domestic wastewater treatment with purple phototrophic bacteria using a novel continuous photo anaerobic membrane bioreactor. *Water Research*, **100**, 486–495, https://doi.org/10.1016/j.watres.2016.04.061

Hülsen T., Hsieh K., Lu Y., Tait S. and Batstone D. J. (2018). Simultaneous treatment and single cell protein production from agri-industrial wastewaters using purple phototrophic bacteria or microalgae – a comparison. *Bioresource Technology*, **254**, 214–223, https://doi.org/10.1016/j.biortech.2018.01.032

Hülsen T., Sander E. M., Jensen P. D. and Batstone D. J. (2020). Application of purple phototrophic bacteria in a biofilm photobioreactor for single cell protein production: biofilm vs suspended growth. *Water Research*, **181**, 115909, https://doi.org/10.1016/j.watres.2020.115909

Hunter B. and Deshusses M. A. (2020). Resources recovery from high-strength human waste anaerobic digestate using simple nitrification and denitrification filters. *Science of the Total Environment*, **712**, 135509, https://doi.org/10.1016/j.scitotenv.2019.135509

Keskin T. and Hallenbeck P. C. (2012). Hydrogen production from sugar industry wastes using single-stage photofermentation. *Bioresource Technology*, **112**, 131–136, https://doi.org/10.1016/j.biortech.2012.02.077

Khoshnevisan B., Tsapekos P., Zhang Y., Valverde-Pérez B. and Angelidaki I. (2019). Urban biowaste valorization by coupling anaerobic digestion and single cell protein production. *Bioresource Technology*, **290**, 121743, https://doi.org/10.1016/j.biortech.2019.121743

Kizito S., Wu S., Wandera S. M., Guo L. and Dong R. (2016). Evaluation of ammonium adsorption in biochar-fixed beds for treatment of anaerobically digested swine slurry: experimental optimization and modeling. *Science of the Total Environment*, **563–564**, 1095–1104, https://doi.org/10.1016/j.scitotenv.2016.05.149

Lang-Yona N., Kunert A. T., Vogel L., Kampf C. J., Bellinghausen I., Saloga J., Schink A., Ziegler K., Lucas K. and Schuppan D. (2018). Fresh water, marine and terrestrial cyanobacteria display distinct allergen characteristics. *Science of the Total Environment*, **612**, 767–774, https://doi.org/10.1016/j.scitotenv.2017.08.069

Lehmann J. and Joseph S. (2015). Biochar for Environmental Management: Science, Technology and Implementation. Routledge, London.

Li Q. (2018). Pilot-scale plant application of membrane aerated biofilm reactor (MABR) technology in wastewater treatment.

Liu B., Li Y., Song J., Zhang L., Dong J. and Yang Q. (2014). Production of single-cell protein with two-step fermentation for treatment of potato starch processing waste. *Cellulose*, **21**(5), 3637–3645, https://doi.org/10.1007/s10570-014-0400-6

Liu C., Wang W., Anwar N., Ma Z., Liu G. and Zhang R. (2017). Effect of organic loading rate on anaerobic digestion of food waste under mesophilic and thermophilic conditions. *Energy & Fuels*, **31**(3), 2976–2984, https://doi.org/10.1021/acs.energyfuels.7b00018

Lu H., Zhang G., Zheng Z., Meng F., Du T. and He S. (2019). Bio-conversion of photosynthetic bacteria from non-toxic wastewater to realize wastewater treatment and bioresource recovery: a review. *Bioresource Technology*, **278**, 383–399, https://doi.org/10.1016/j.biortech.2019.01.070

Magrí A., Béline F. and Dabert P. (2013). Feasibility and interest of the anammox process as treatment alternative for anaerobic digester supernatants in manure processing – an overview. *Journal of Environmental Management*, **131**, 170–184, https://doi.org/10.1016/j.jenvman.2013.09.021

Markou G., Wang L., Ye J. and Unc A. (2018). Using agro-industrial wastes for the cultivation of microalgae and duckweeds: contamination risks and biomass safety concerns. *Biotechnology Advances*, **36**(4), 1238–1254, https://doi.org/10.1016/j.biotechadv.2018.04.003

Marti N., Bouzas A., Seco A. and Ferrer J. (2008). Struvite precipitation assessment in anaerobic digestion processes. *Chemical Engineering Journal*, **141**(1–3), 67–74, https://doi.org/10.1016/j.cej.2007.10.023

Matassa S., Papirio S., Pikaar I., Hülsen T., Leijenhorst E., Esposito G., Pirozzi F. and Verstraete W. (2020). Upcycling of biowaste carbon and nutrients in line with consumer confidence: the "full gas" route to single cell protein. *Green Chemistry*, **22**(15), 4912–4929, https://doi.org/10.1039/D0GC01382J

Mekonnen M. M. and Hoekstra A. Y. (2014). Water footprint benchmarks for crop production: a first global assessment. *Ecological Indicators*, **46**, 214–223, https://doi.org/10.1016/j.ecolind.2014.06.013

Mohammadi M., Samadi S. and Najafpour Darzi G. (2016). Production of single cell protein from sugarcane bagasse by *Saccharomyces cerevisiae* in tray bioreactor. *International Journal of Engineering*, **29**(8), 1029–1036.

Monlau F., Sambusiti C., Ficara E., Aboulkas A., Barakat A. and Carrère H. (2015). New opportunities for agricultural digestate valorization: current situation and perspectives. *Energy & Environmental Science*, **8**(9), 2600–2621, https://doi.org/10.1039/C5EE01633A

Morero B., Groppelli E. and Campanella E. A. (2015). Life cycle assessment of biomethane use in Argentina. *Bioresource Technology*, **182**, 208–216, https://doi.org/10.1016/j.biortech.2015.01.077

Muys M., Papini G., Spiller M., Sakarika M., Schwaiger B., Lesueur C., Vermeir P. and Vlaeminck S. E. (2020). Dried aerobic heterotrophic bacteria from treatment of food and beverage effluents: screening of correlations between operation parameters and microbial protein quality. *Bioresource Technology*, **307**, 123242, https://doi.org/10.1016/j.biortech.2020.123242

Nagao N., Tajima N., Kawai M., Niwa C., Kurosawa N., Matsuyama T., Yusoff F. M. and Toda T. (2012). Maximum organic loading rate for the single-stage wet anaerobic digestion of food waste. *Bioresource Technology*, **118**, 210–218, https://doi.org/10.1016/j.biortech.2012.05.045

Nasseri A., Rasoul-Amini S., Morowvat M. and Ghasemi Y. (2011). Single cell protein: production and process. *American Journal of Food Technology*, **6**(2), 103–116, https://doi.org/10.3923/ajft.2011.103.116

Noike T., Goo I., Matsumoto H. and Miyahara T. (2004). Development of a new type of anaerobic digestion process equipped with the function of nitrogen removal. *Water Science and Technology*, **49**(5–6), 173–179, https://doi.org/10.2166/wst.2004.0751

Øverland M., Borge G., Vogt G., Schøyen H. and Skrede A. (2011). Oxidative stability and sensory quality of meat from broiler chickens fed a bacterial meal produced on natural gas. *Poultry Science*, **90**(1), 201–210, https://doi.org/10.3382/ps.2010-00784

Palakodeti A., Azman S., Rossi B., Dewil R. and Appels L. (2021). A critical review of ammonia recovery from anaerobic digestate of organic wastes via stripping. *Renewable and Sustainable Energy Reviews*, **143**, 110903, https://doi.org/10.1016/j.rser.2021.110903

Prorot A., Julien L., Christophe D. and Patrick L. (2011). Sludge disintegration during heat treatment at low temperature: a better understanding of involved mechanisms with a multiparametric approach. *Biochemical Engineering Journal*, **54**(3), 178–184, https://doi.org/10.1016/j.bej.2011.02.016

Rahman M. M., Salleh M. A. M., Rashid U., Ahsan A., Hossain M. M. and Ra C. S. (2014). Production of slow release crystal fertilizer from wastewaters through struvite crystallization – a review. *Arabian Journal of Chemistry*, **7**(1), 139–155, https://doi.org/10.1016/j.arabjc.2013.10.007

Ravindra P. (2000). Value-added food: single cell protein. *Biotechnology Advances*, **18**(6), 459–479, https://doi.org/10.1016/S0734-9750(00)00045-8

Ritala A., Häkkinen S. T., Toivari M. and Wiebe M. G. (2017). Single cell protein – state-of-the-art, industrial landscape and patents 2001–2016. *Frontiers in Microbiology*, **8**, 2009, https://doi.org/10.3389/fmicb.2017.02009

Sahu N., Deshmukh S., Chandrashekhar B., Sharma G., Kapley A. and Pandey R. (2017). Optimization of hydrolysis conditions for minimizing ammonia accumulation in two-stage biogas production process using kitchen waste for sustainable process development. *Journal of Environmental Chemical Engineering*, **5**(3), 2378–2387, https://doi.org/10.1016/j.jece.2017.04.045

Sasaki K., Tanaka T. and Nagai S. (1998) Use of photosynthetic bacteria for the production of SCP and chemicals from organic wastes. In: Bioconversion of Waste Materials to Industrial Products. Springer, Boston, MA, pp. 247–292.

Sawatdeenarunat C., Surendra K., Takara D., Oechsner H. and Khanal S. K. (2015). Anaerobic digestion of lignocellulosic biomass: challenges and opportunities. *Bioresource Technology*, **178**, 178–186, https://doi.org/10.1016/j.biortech.2014.09.103

Schneider O., Sereti V., Machiels M. A., Eding E. H. and Verreth J. A. (2006). The potential of producing heterotrophic bacteria biomass on aquaculture waste. *Water Research*, **40**(14), 2684–2694, https://doi.org/10.1016/j.watres.2006.05.008

Sharif M., Zafar M. H., Aqib A. I., Saeed M., Farag M. R. and Alagawany M. (2021). Single cell protein: sources, mechanism of production, nutritional value and its uses in aquaculture nutrition. *Aquaculture*, **531**, 735885, https://doi.org/10.1016/j.aquaculture.2020.735885

Shi X., Lin J., Zuo J., Li P., Li X. and Guo X. (2017). Effects of free ammonia on volatile fatty acid accumulation and process performance in the anaerobic digestion of two typical bio-wastes. *Journal of Environmental Sciences*, **55**, 49–57, https://doi.org/10.1016/j.jes.2016.07.006

Stahel W. R. (2016). The circular economy. *Nature*, **531**(7595), 435–438, https://doi.org/10.1038/531435a

Sun Y. and O'Riordan M. X. (2013). Regulation of bacterial pathogenesis by intestinal short-chain fatty acids. *Advances in Applied Microbiology*, **85**, 93–118, https://doi.org/10.1016/B978-0-12-407672-3.00003-4

Taiganides E. P. (1992). Pig Waste Management and Recycling: the Singapore Experience. IDRC, Ottawa, ON, CA.

Talboys P. J., Heppell J., Roose T., Healey J. R., Jones D. L. and Withers P. J. (2016). Struvite: a slow-release fertiliser for sustainable phosphorus management? *Plant and Soil*, **401**(1–2), 109–123, https://doi.org/10.1007/s11104-015-2747-3

Tyagi V. K., Fdez-Güelfo L., Zhou Y., Álvarez-Gallego C., Garcia L. R. and Ng W. J. (2018). Anaerobic co-digestion of organic fraction of municipal solid waste (OFMSW): progress and challenges. *Renewable and Sustainable Energy Reviews*, **93**, 380–399, https://doi.org/10.1016/j.rser.2018.05.051

Van Haandel A. and Van Der Lubbe J. (2007). Handbook Biological Waste Water Treatment – Design and Optimisation of Activated Sludge Systems. Webshop Wastewater Handbook, Leidschendam.

Verbeeck K., De Vrieze J., Pikaar I., Verstraete W. and Rabaey K. (2021). Assessing the potential for up-cycling recovered resources from anaerobic digestion through microbial protein production. *Microbial Biotechnology*, **14**(3), 897–910, https://doi.org/10.1111/1751-7915.13600

Vermeulen S. J., Campbell B. M. and Ingram J. S. (2012). Climate change and food systems. *Annual Review of Environment and Resources*, **37**, 195–222, https://doi.org/10.1146/annurev-environ-020411-130608

Verstraete W., Clauwaert P. and Vlaeminck S. E. (2016). Used water and nutrients: recovery perspectives in a 'panta rhei'context. *Bioresource Technology*, **215**, 199–208, https://doi.org/10.1016/j.biortech.2016.04.094

Vidal-Antich C., Perez-Esteban N., Astals S., Peces M., Mata-Alvarez J. and Dosta J. (2021). Assessing the potential of waste activated sludge and food waste co-fermentation for carboxylic acids production. *Science of the Total Environment*, **757**, 143763, https://doi.org/10.1016/j.scitotenv.2020.143763

Vriens L., Nihoul R. and Verachtert H. (1989). Activated sludges as animal feed: a review. *Biological Wastes*, **27**(3), 161–207, https://doi.org/10.1016/0269-7483(89)90001-3

Wagner A. O., Gstraunthaler G. and Illmer P. (2008). Survival of bacterial pathogens during the thermophilic anaerobic digestion of biowaste: laboratory experiments and in situ validation. *Anaerobe*, **14**(3), 181–183, https://doi.org/10.1016/j.anaerobe.2008.03.004

Wang J., Roberts N., Hilliard M. and He Q. P. (2020). A microalgae-methanotroph coculture is a promising platform for fuels and chemical production from wastewater. *Frontiers in Energy Research*, **230**, 563352.

Wei H., Gao B., Ren J., Li A. and Yang H. (2018). Coagulation/flocculation in dewatering of sludge: a review. *Water Research*, **143**, 608–631, https://doi.org/10.1016/j.watres.2018.07.029

Wu D., Sun F., Chua F. J. D., Lu D., Stuckey D. C. and Zhou Y. (2019). In-situ power generation and nutrients recovery from waste activated sludge – long-term performance and system optimization. *Chemical Engineering Journal*, **361**, 1207–1214, https://doi.org/10.1016/j.cej.2019.01.001

Wu D., Zhang L., Le C., Wang L. and Zhou Y. (2021). Pathways and mechanisms of single-cell protein production: carbon and nutrient transformation. *ACS ES&T Water*, **1**(5), 1313–1320, https://doi.org/10.1021/acsestwater.1c00084

Xia A. and Murphy J. D. (2016). Microalgal cultivation in treating liquid digestate from biogas systems. *Trends in Biotechnology*, **34**(4), 264–275, https://doi.org/10.1016/j.tibtech.2015.12.010

Yuan H. and Zhu N. (2016). Progress in inhibition mechanisms and process control of intermediates and by-products in sewage sludge anaerobic digestion. *Renewable and Sustainable Energy Reviews*, **58**, 429–438, https://doi.org/10.1016/j.rser.2015.12.261

Zha X., Tsapekos P., Zhu X., Khoshnevisan B., Lu X. and Angelidaki I. (2021). Bioconversion of wastewater to single cell protein by methanotrophic bacteria. *Bioresource Technology*, **320**(Pt A), 124351, https://doi.org/10.1016/j.biortech.2020.124351

Zheng Y., Chi Z., Lucker B. and Chen S. (2012). Two-stage heterotrophic and phototrophic culture strategy for algal biomass and lipid production. *Bioresource Technology*, **103**(1), 484–488, https://doi.org/10.1016/j.biortech.2011.09.122

IWA PUBLISHING

doi: 10.2166/9781789062755_0359

Chapter 17

Anaerobic digestate: pollutants, ecotoxicology, and legislation

Karel Diéguez-Santana[1,2], Cristhian Chicaiza-Ortiz[3,4,5], Jingxin Zhang[3] and Washington Logroño[6,7,8]*

[1]Facultad de Ciencias de la Vida, Universidad Regional Amazónica IKIAM, Tena, Napo 150150, Ecuador
[2]Department of Organic and Inorganic Chemistry, University of Basque Country UPV/EHU, 48940 Leioa, Spain
[3]Biomass to Resources Group, Universidad Regional Amazónica IKIAM, Tena, Napo 150150, Ecuador
[4]China-UK Low Carbon College, Shanghai Jiao Tong University, Shanghai 201306, China
[5]Programa de posgrado en Cambio Climático, Sustentabilidad y Desarrollo, Área de ambiente y sustentabilidad, Universidad Andina Simón Bolívar, Quito, Ecuador
[6]Department of Environmental Science and Bioeconomy, Sustainable Environment Research Division, ECARI–Ecuadorian Centre for Advanced Research and Innovation, Ecuador
[7]Department of Environmental Microbiology, Helmholtz Centre for Environmental Research–UFZ, 04318 Leipzig, Germany
[8]Grupo de Investigación y Desarrollo para el Ambiente y Cambio Climático (GIDAC-CEAA), Facultad de Ciencias, Escuela Superior Politécnica de Chimborazo, 060155 Riobamba, Ecuador
**Corresponding author e-mail: cristhianchicaiza@hotmail.com*

ABSTRACT

Anaerobic digestion (AD) offers enormous potential towards a sustainable bioeconomy. AD is suitable for waste management of various sources and generates value in multiple sectors such as electricity, heating, or agriculture through biogas or fertilizer production. However, attention must be paid to cheap fertilizers such as digestates because these digested fractions contain heavy metals, (nano) microplastics, hormones, and other chemical compounds that, when accessible, can be detrimental to humans, animals, plants and the environment. Digestates originating from manure substrates should be monitored to control the spread of antimicrobial resistance genes to the environment. Digestate toxicity can be a problem for aquatic and terrestrial organisms as determined by bioassays to detect the adverse consequences of digestate applied to the soil. Anaerobic digestate must meet specified quality requirements prior to utilization in the soil or crops to prevent dangers to human health and the environment. The digestate is available in three forms: whole, liquid, and solid, and can be applied to the soil as a final disposal location if it complies with applicable regulatory standards and is classified as a 'product'.

Keywords: Anaerobic digestion, digestate, ecotoxicity, emerging contaminants, legislation

Editor(s): Vinay Kumar Tyagi, Kaoutar Aboudi and Cigdem Eskicioglu

17.1 INTRODUCTION

Anaerobic digestion (AD) can significantly contribute to paving the way towards a sustainable economy. This technology produces biogas from organic waste, which includes manure, crop residues, urban solid waste, sewage sludge, and so on (Tsapekos *et al.*, 2021). The digestate is the main by-product of the AD process, which is regarded as an organic fertilizer (Gross *et al.*, 2021). With appropriate management, AD can reduce greenhouse gas (GHG) emissions and water, soil, and air pollution (Jiang *et al.*, 2012). Moreover, within the context of the sustainable development goals (SDG), AD makes a significant contribution to guaranteeing cheap and sustainable energy (SDG 7), while also contributing to additional objectives (SDG 6, 9, 13, 15) (Tsapekos *et al.*, 2021). Several AD plants have been installed worldwide, resulting in a local surplus of its main by-product, digestate (Prapaspongsa *et al.*, 2010). For instance, the European Union (EU) administration has adopted many AD facilities towards a more energy-efficient system and low-carbon emissions by utilizing biogas instead of fossil fuels (Huopana *et al.*, 2013). However, the development of other energy carriers such as hydrogen or methanol is rapidly growing. Countries like India consider AD as a promising technological approach to manage biowaste within its municipalities (Gross *et al.*, 2021). However, digestate planning and management are still inadequate in most cases. Another aspect is the subsidies to support this technology, for instance, in China, the current subsidies should also be based on output rather than building costs (Zheng *et al.*, 2020). The digestate is a rich source of nutrients such as nitrogen, phosphorus, and potassium (Möller & Müller, 2012). It can provide economic revenues that encourage its usage as a soil conditioner and fertilizer (Prapaspongsa *et al.*, 2010). However, because these digested fractions contain pathogens and heavy metals that can be harmful to humans and other organisms when bioavailable (Järup, 2003; Jomova & Valko, 2011), environmental agencies and governments must establish safety standards and monitor digestate compositions in order to value waste as a resource for end users (Peng & Pivato, 2017).

Researchers have recently focused their attention on the various pollutants found in digestates. For example, different metals, pesticides, mycotoxins (Jiang *et al.*, 2018), or antimicrobial agents found in environmental matrices derived from manure digestate that enhance the abundance of antimicrobial-resistant bacteria (AMRB) and antimicrobial resistance genes (AMRGs) (Congilosi & Aga, 2021). Additionally, in the digestate derived from the AD process of urban solid waste or biosolids, fractions of (nano) microplastics have been identified to threaten ecosystems. They may represent a risk to human, animal, and plant health (Mohammad Mirsoleimani Azizi *et al.*, 2021).

17.2 DIGESTATE: POLLUTANTS

The origin of the residues used in the AD process is mainly organic. These substrates come from animal and crop production, the food sector, urban–industrial waste, and sewage sludge (Baştabak & Koçar, 2020; Sen *et al.*, 2016). The composition and origin of waste affect the macro and micronutrient content and the presence of certain contaminants such as heavy metals, traces of drugs, and other contaminants that might affect the quality of the digestate (Baştabak & Koçar, 2020).

17.2.1 Heavy metals

Various studies examined the feasibility of digestate recovery for agricultural purposes and found that high heavy metals content in the digestate limited its utilization (Chaher *et al.*, 2021). Moreover, heavy metals have a negative effect on aquatic ecosystems since they may enter food chains and impact higher life forms via biomagnification (Mohammadi *et al.*, 2019). In contrast, when heavy metals concentrations are significantly low, some alternatives are effective, such as hydrothermal carbonization and diluted digestate cake as a start-up seed and strengthening methane generation (Reza *et al.*, 2016). Therefore, heavy metal stability or removal is critical for recycling; the immobilization process is the most common approach due to its simplicity and economic effectiveness (Mohammadi *et al.*, 2019). A comprehensive list of heavy metals is revised, including Hg, Cd, Cr, Ni, among others. Table 17.1 summarizes the major physical–chemical features of digestates characterized in the literature. They are divided into mesophilic and thermophilic AD processes. The main substrates used are substrates derived from animal wastes such as animal by-products (ABP), animal sewage (AS), cattle slurry (CS), cattle manure (CM), chicken manure (ChM), manure (M), pig manure (PM), and pig slurry (PS). Different biomass or crop wastes such as herbaceous biomass (HB), beet leaves (BL), cereal bran (CB), corn (Co), corn silage (CoS), garden waste (GW), groats (Gr), grass silage (GS), maize silage (MS), olive oil cake (OOC), *Triticale* silages (TS), vegetable waste (VW) as well as another crops waste (OCW). The mixture of waste includes agro-industrial residues (AIR), food waste (FW), kitchen waste (KW), sewage sludge (SS), and municipal solid waste (MSW). The parameters C, N, P, S, Na, Mg, K, Ca, Fe, and Al, are expressed in percentage (%). At the same time, heavy metals (Zn, Cu, Pb, Cd, Ni, Cr, and As) are expressed in mg kg^{-1}.

Most studies reported primary and secondary nutrients useful for fertilization and application in crops. However, some studies quantified the presence of potentially toxic heavy metals such as Zn, Cu, Cd, Ni, Pb, Cr, Hg (Opatokun *et al.*, 2017; Pampillón-González *et al.*, 2017). These values depend on the substrate used and the conditions for the AD process.

17.2.2 Emerging contaminants

Contaminants in AD digestates are not only limited to heavy metals and undigested organic loads. They also include residual contaminants in a broad spectrum of biological waste, including emerging contaminants, pathogens discharged through solid or aqueous waste streams. Emerging contaminants such as pharmaceutical and personal care product residues (Chen *et al.*, 2014); endocrine disruptors, antimicrobial-resistant pathogens (Gondim-Porto *et al.*, 2016); microplastics (Mahon *et al.*, 2017), and other persistent organic compounds can be identified in digestates obtained from animal manure, MSW or SS (Longhurst *et al.*, 2019).

Table 17.1 Physical–chemical properties of different digestates.

Digestate origin[a]	Composition (dry basis)																		Reference
	Moi [b]	C	N	P	S	Na	Mg	K	Ca	Fe	Al	Zn	Cu	Pb	Cd	Ni	Cr	As	
	%											mg kg^{-1}							
Mesophilic																			
AIR+HB	–	38.5	1.21	1.82	0.29	0.13	0.85	2.83	3.92	0.19	–	–	–	–	–	–	–	–	Calamai *et al.* (2020)
AS+CM+ MS+TS+CB	–	43	1.3	0.53	0.14	0.18	0.36	1.2	1.08	0.06	0.04	35	11	–	–	1	1	–	Monlau *et al.* (2016)
BL+OCW	94		1.3	0.4		1	0.4	4.2				5	1	–	–	–	–	–	Gunnarsson *et al.* (2011)
CM+AIR	91.9	37.5	2.7	0.89		0.829	0.78	3.45	4.48			30.86	11.98	–	–	–	–	–	Alburquerque *et al.* (2012)
Co	–	40.7	2.4	–	–	–	–	–	–	–	–	–	–	–	–	–	–	–	Dieguez-Alonso *et al.* (2018)
CS+CM	–	52.1	2.33	2.26	0.65	0.58	1.27	1.32	4.1	1.28	0.19	–	–	–	–	–	–	–	Garlapalli *et al.* (2016)
CS+GS+ CM	–	50.6	4.22		0.52	–	–	–	–	–	–	–	–	–	–	–	–	–	Rodriguez Correa *et al.* (2017)
CS	–	38	2.05		0.62	0.385	0.6206	1.25	0.98	–	–	–	–	–	–	–	–	–	Zhang *et al.* (2018)
CM	89.7	56.2	0.65	1.81				2.1				168.59	31.787	–	–	–	–	0.58	Jin and Chang (2011)
CM			0.96	1.14		0.039	0.015	0.158	0.088		0.02	–	–	–	–	–	–	–	Jin and Chang (2011)
FW	90.2	42.1	5.81	1.97	–	–	0.27	0.62	3.17	0.93				10		20	40	0.98	Opatokun *et al.* (2017)
G+OOC+ TS+ChM	–	42.5	1.4	–	0.14	–	–	–	–	–	–	–	–	–	–	–	–	–	Monlau *et al.* (2016)
AIR+HB	76.2	46.7	1.2	–	0.5	–	–	–	–	–	–	–	–	–	–	–	–	–	Miliotti *et al.* (2020)

MS+CM+GS	93.5	50.5	3.6	1.35	0.5	–	0.94	5.23	2.01	0.3	0.52	–	–	–	–	–	–	–	Zhao *et al.* (2018)
PM	89.1	43.11	0.75	1.03	–	–	–	3.2	–	–	–	189.54	149.91	–	–	–	–	2.385	Jin and Chang (2011)
PS+ABP	97.25	42	2.48	0.14				1.42	0.154	–	–	24.72	2.91	–	–	–	–	–	Alburquerque *et al.* (2012)
PS	91.2	43.6	0.69				24266		3.73	0.59		3873	1796	7.2	3.8	16.9	10.1	3.1	Pampillón-GonzáLez *et al.* (2017)
SS	83.5	32.7	5.1	4.58	1	0.16	0.32	0.28	3.02	1.58	–	–	–	–	–	–	–	–	Marin-Batista *et al.* (2020)
Thermophilic																			
CM+FW	24.5	37.8	1.09	0.59	–	–	–	0.32	–	0.88	–	–	–	–	–	–	–	–	Rodriguez Alberto *et al.* (2019)
CM+MS+GS	–	40.3	2.1	0.94	0.3	0.08	0.74	3.67	2.53	0.23	0.25	–	–	–	–	–	–	–	Cao *et al.* (2019a, 2019b)
C+CS+M+VW	–	39.8	1.23	0.79	0.29	0.08	0.47	–	2	0.10	–	–	–	–	–	–	–	–	Calamai *et al.* (2020)
KW+GW	–	34.3	1.9	0.39	0.2	0.66	0.54	1.8	4.29	0.66	0.92	–	–	–	–	–	–	–	Cao *et al.* (2019a, 2019b)
CM	–	42.6	2	0.71	0.4	0.16	0.54	2.84	1.71	0.11	0.1	–	–	–	–	–	–	–	Cao *et al.* (2019a, 2019b)
MS	–	44.1	3.2	1.2	0.3	0.3	0.7	1.6	1.3	2.3	–	–	–	–	–	–	–	–	Parmar and Ross (2019)
MS	–	46.6	1.1	–	0.08	–	–	–	–	–	–	–	–	–	–	–	–	–	Cao *et al.* (2020)
MSW	–	24.1	1.5	0.7	0.2	0.9	1.4	1.6	10.4	3.2	–	–	–	–	–	–	–	–	Parmar and Ross (2019)
VW+GW	–	29.5	2	2.6	0.3	0.5	0.8	0.7	4.3	3	–	–	–	–	–	–	–	–	Parmar and Ross (2019)

[a]**Digestate origin.** Animal by-products (ABP), animal sewage (AS), cattle slurry (CS), cattle manure (CM), chicken manure (ChM), manure (M), pig manure (PM), and pig slurry (PS), herbaceous biomass (HB), beet leaves (BL), cereal bran (CB), corn (Co), corn silage (CoS), garden waste (GW), groats (Gr), grass silage (GS), maize silage (MS), olive oil cake (OOC), *Triticale* silages (TS), vegetable waste (VW), other crops waste (OCW), agro-industrial residues (AIR), food waste (FW), kitchen waste (KW), sewage sludge (SS), and municipal solid waste (MSW).
[b]**Moi:** Moisture, expressed in percentage.

17.2.2.1 Antibiotics

Antibiotics can be introduced directly into the environment when they are excreted by grazing animals or with the application of manure or SS as organic fertilizer in agricultural fields. In the AD processes, different classes of antibiotics have been identified, such as tetracyclines, sulphonamides, fluoroquinolones, β-lactams, and macrolides (Congilosi & Aga, 2021). Many of them are managed as drugs in animal husbandry. These drugs have high excretion rates (see Table 17.2) and different mobility and persistence. These compounds can enter the environment on a regular scale with organic fertilization and contribute to the spread and development of antimicrobial resistance (AMR) and associated mobile genetic elements in soils, which are serious threats to human and animal health (Lehmann & Bloem, 2021). The potential of AD to degrade and eliminate antibiotics depends mainly on the concentration and class of antibiotic, the operating conditions of the bioreactor (mainly temperature), the type of raw material, and the source of inoculum (Massé *et al.*, 2014). Biodegradation and biosorption are two dominant mechanisms of antibiotic removal in AD (Zhou *et al.*, 2021a, 2021b). For example, tylosin, a veterinary antibiotic, was degraded through a mesophilic AD process (Hosseini Taleghani *et al.*, 2020). On the other hand, thermophilic AD has reduced the absolute abundance of antibiotic resistance genes (ARGs) (*sulII* (95%), *intI1* (95%), *tnpA* (77%)) (Wang *et al.*, 2021). A new method to improve the removal of antibiotics in an AD process refers to the addition of conductive or nanomaterials to the AD reactor (Zhou *et al.*, 2021a, 2021b). These additives show a higher removal efficiency of antibiotics than conventional AD, which is moderately effective (Zhou *et al.*, 2021a, 2021b). However, the agricultural use of digestates still represents a risk of contamination to the environment. Agricultural plots disturb the soil's microbial ecology for which an advanced AD is required or combined with

Table 17.2 Average annual veterinary antimicrobial excretion rates in cattle, poultry, and pigs.

Antimicrobial class	Manure matrix	Concentrations in manure (ppm)	Reference
Macrolides	Swine	1.6[a]	Angenent *et al.* (2008)
	Poultry	0.0090–14	Ho *et al.* (2014)
	Cattle	0.012–0.029	Wallace and Aga (2016)
Tetracyclines	Swine	0.048–354	Congilosi and Aga (2021)
	Poultry	0.14–18	Zhao *et al.* (2010)
	Cattle	0.0065–0.27	Congilosi and Aga (2021)
Sulphonamides	Swine	0.015–20 0.10–12[a]	Congilosi and Aga (2021)
	Poultry	0.010–91	Zhao *et al.* (2010)
	Cattle	0.0019–0.014	Wallace and Aga (2016)

Adapted from Congilosi and Aga (2021).
[a]Wet weight.

Table 17.3 Hormone excretion rates by different animals based on wet or dry weight manure.

Hormones	kg excreted per year[a]	Average excretion rate (%)[a]	Manure matrix	Concentration in manure (ppb)	Reference
Estrogens	48 530	90	Swine	70.1–518	Xu *et al.* (2018)
			Cattle	6.20–1416	Zheng *et al.* (2008)
Androgens	4350	40	Poultry	44.2–150	Andaluri *et al.* (2012)
			Cattle	1.30–35.9	Congilosi and Aga (2021)
				17.0–203[b]	
Progestagens	278 900	90	Cattle	Up to 196	Zheng *et al.* (2008)
			Poultry	<10.0–391	Ho *et al.* (2014)

[a]Estimated US excretion of hormones by feedlot animals in 2000 and excretion rates (Biswas *et al.* 2013).
[b]Wet weight.
Adapted from Congilosi and Aga (2021).

composting methods that are more effective for eliminating antibiotics and avoiding the entry of substrates with quantities of these drugs.

17.2.2.2 Endocrine disruptors

Hormones are administered to cattle to encourage growth and regulate the reproductive system. According to Noguera-Oviedo and Aga (2016), the levels of estrogens detected in anaerobically digested manure were found to be substantially higher than the levels known to cause endocrine alterations in aquatic and wildlife organisms (can be affected by EDC even in low concentrations, that is, subpart per billion) (Jobling *et al.*, 2006). According to Zheng *et al.* (2013), aerobic exposure improves the biodegradation of estrogens. It would be beneficial to keep the liquid manure in an aeration pond before application to the soil to decrease the introduction of estrogens into the environment; however, CAPEX and OPEX would need to be considered. Table 17.3 summarized the hormone excretion rates by different animals based on wet or dry weight manure.

17.2.2.3 Microplastics

Microplastics are particles less than 5 mm in diameter (Weithmann *et al.*, 2018), while nanoplastics are particles smaller than 0.1 μm (Meixner *et al.*, 2020). Both have become global issues, posing hazards to biota and public health. Concerns have been raised about the presence of microplastics particles in digestate when used as organic fertilizer and composting (Peng *et al.*, 2021; Weithmann *et al.*, 2018). Recently, a study showed that plastic particles remained in organic fertilizers after fermentation and composting (Keller *et al.*, 2020). Therefore, the management strategies in which the digestate is handled may affect microplastics' accumulation and size distribution (Meixner *et al.*, 2020). A common case is that the microplastics in food waste packaging materials prevent food waste from co-digesting (Peng *et al.*, 2021). Another issue is that land-based systems acquire from 4 to 23 times more microplastic than aquatic systems (Meixner *et al.*, 2020). Biodegradation of microplastics can be carried out by microorganisms such as algae, fungi, and bacteria and is an important process to deal with the plastisphere. For instance, *Brevundimonas* and *Sphingobacterium* were reported to degrade polybutylene adipate-*co*-terephthalate (PBAT) and polylactic acid (PLA) under thermophilic AD conditions (Peng *et al.*, 2021). Moreover, some physical strategies are the destruction of microplastics via UV irradiation, weathering, or tillage (Keller *et al.*, 2020). Studies on the fate of microplastics in the environment should consider spatial processes and temporal scales to understand the mechanisms better behind them.

Most of the (nano) microplastics investigated to date have been found to impair AD performance (Mohammad Mirsoleimani Azizi *et al.*, 2021). Their presence leads to several inhibition mechanisms, including (1) release of toxic additives/chemicals, (2) affecting the activities of key enzymes and functional genes (Campanale *et al.*, 2020), (3) production of reactive oxygen species, (4) damage/penetration of microbial cells, (5) alteration of protein structures in granular sludge (Prata *et al.*, 2021). The main generation sources are sludge from sewage treatment plants and urban solid waste without segregating at the source. Although the study developed by Iyare *et al.* (2020) shows that ~70% of microplastics could be removed during preliminary and primary treatment processes in wastewater treatment plants, study limitations and transmission of (nano) microplastics when applied to soils and crops continues to be a major environmental challenge. It was previously noted that (nano) microplastics could adsorb various environmental pollutants, including antibiotics and heavy metals (Mao *et al.*, 2020; Rolsky *et al.*, 2020), or they can also serve as carriers of AMRGs (Dong *et al.*, 2021).

17.2.3 Microbiological safety

Digestates contain a high microbial load and therefore require treatment prior to the application to soil. Total coliforms associated with faecal coliforms are more prevalent in the tubular digester containing guinea-pig manure. Consequently, the high loads of these microorganisms must be reduced to ensure that the digestate can be disposed of in the environment following the permissible limits of the local legislation. Table 17.4 illustrates the treatment of three common feedstocks: pig slurry, guinea-pig manure, and cattle manure, using diverse reactor configurations.

17.2.4 Antimicrobial resistance in anaerobic digestate

Animal manure application to agricultural land is a widespread practice worldwide. The effects of these practices may be reflected at distant time scales

Table 17.4 Different types of digestate and the involved pathogens.

Substrates	Type of digestate	Pathogens × total coliforms	Faecal coliforms	Helminth eggs:	Study
Guinea-pig manure	Tubular biodigester	1.70×10^8 MPN mL^{-1}	1.70×10^7 MPN mL^{-1}	N/A	Garfí *et al.* (2011)
Cattle manure	Mesophilic tubular digester	–	1.06×10^6 CFU g^{-1} TS	24 HH 4 g^{-1} TS	Castro *et al.* (2017)
Pig slurry	Mesophilic–thermophilic digester (lagoon type)	240×10^4 CFU g^{-1} TS,	3.6×10^4 CFU g^{-1} TS	N/A	Pampillón-GonzáLez *et al.* (2017)

and depend on the load and farm size. Antimicrobials used in farming practices can directly influence the abundance of antimicrobial-resistant bacteria (ARB) and AMRGs in manures, and most importantly, the risk of persistence and spread of this threat in soils is high when applied as fertilizer in agricultural land (Marti *et al.*, 2013). Animal manure harbours AMRGs (Sun *et al.*, 2016), and despite the management through AD and composting, it is still an important challenge to solve prior utilization of digestate as a fertilizer additive in agriculture. Therefore, it is of utmost importance to monitor the prevalence and fate of ARGs in anaerobic digestates. Among the strategies to remove ARGs, pretreatment methods prior to the AD process of manures should be considered a feasible alternative to enhance the biogas yield degradation of pollutants (Congilosi & Aga, 2021). Microwave or activated carbon pretreatment has removed up to 95% of ARGs (Congilosi & Aga, 2021). As the efficiency of composting in reducing ARGs is limited, the combination with AD could lead to better removal efficiencies of ARGs (Congilosi & Aga, 2021).

17.3 DIGESTATE: ECOTOXICOLOGY

17.3.1 Aquatic/terrestrial toxicity assays

Digestate toxicity occurs mainly in aquatic and terrestrial species. Table 17.5 shows some relevant toxicity analysis results, tests, indicators, measurements, and results. Within aquatic species, they have been tested on microalgae (*Chlorella sorokiniana*, *Chlamydomonas reinhardtii*, *Raphidocelis subcapitata*), Gram-negative marine bacterium (*Vibrio fischeri*), crutaceans (*Daphnia magna* and *Artemia* sp.), or aquatic freshwater plant (*Lemna minor* L.). It should be highlighted that digestates can have a harmful effect on water bodies because while some species of microalgae are capable of using organic acids such as acetate or butyrate as a source of carbon and/or energy. Markou *et al.* (2018) demonstrate that certain digestate concentrations can inhibit microalgae species.

Table 17.5 Results of various toxicity tests from digestates with aquatic and terrestrial indicators.

Dig.Typ.[a]	Specie	Class	Assays	M. typ[b]	Exp. cond.[c]	Measurement results	Reference
Aquatic species							
AIR+M	*Chlorella sorokiniana*	Microalgae	Petri dishes	Inh	pH=10, T=25°C	Butyrate showed inhibitory effects at concentrations higher than 0.1 gC L^{-1} and 0.25 gC L^{-1}	Markou *et al.* (2018)
AIR+M	*Chlamydomonas reinhardtii*	Microalgae	Petri dishes	Inh	pH=10, T=25°C	Acetate caused inhibition at concentrations above 0.4 g L^{-1}	Markou *et al.* (2018)
OFWW.	*Vibrio fischeri*	Bioluminescent, Gram-negative marine bacterium	Ecotoxicity of the eluate	Inh	% NaCl solution	High significant toxic response due to low stability over 4 days.	Alvarenga *et al.* (2016)
OFWW	*Daphnia magna*	Small planktonic crustacean	Ecotoxicity of the eluate	Inh	t=24 h	Toxic response significant medium–high due to its low stability over 4 days	Alvarenga *et al.* (2016)
CM	*Artemia* sp.	Small crustacean	Complete effluent toxicity	Inh	T=20°C t=48 h.	Concentrations of 100% of digestate inhibited of the eluate solid–liquid mobility 1:50 (v/v)	Pivato *et al.* (2016)
PS	*Raphidocelis subcapitata*	Green algae	Petri dishes	Inh	T=4°C, pH=7.0	Sensitive organism with an EC_{50} of 0.77% and UT of 129.87.	Tigini *et al.* (2016)
PS	*Lemna minor* L.	Aquatic freshwater plant	Petri dishes	Inh	T=4°C, pH=7.0	Developed fronds and biomass, with EC_{50} in the range of 1.02–1.77% and TU in 58.82–98.03.	Tigini *et al.* (2016)

Terrestrial species							
AW + AIR.	*Lepidium sativum*	Plant	Petri dishes	GI	pH > 7.5.	Significant differences between the dry mass of the seedlings between the digestates and the concentrations 20%, 10%, 1% and 0.1%.	Alburquerque *et al*. (2012)
AW + AIR.	*Lactuca sativa*	Plant	Petri dishes	GI	pH > 7.5	Significant differences between the dry mass of the seedlings between the digestates and the concentrations 20%, 10%, 1% and 0.1%.	Alburquerque *et al*. (2012)
CM	*Eisenia foetida*	Annelids	Complete effluent toxicity	Inh	$T = 25°C$, $t = 15$ min	Hormesis trend concerning relative growth and reproduction	Pivato *et al.* (2016)
WW + WiW.	*Lepidium sativum*	Plant	Petri dishes	GI, Bios	$T = 25°C$, $t = 72$ h	GI > 60%, Bios = 20%	Da Ros *et al.* (2018)
WW + WiW.	*Sinapsis alba*	Plant	Petri dishes	GI, Bios	$T = 25°C$, $t = 72$ h	GI > 60%, Bios = 30%	Da Ros *et al.* (2018)
WW + WiW	*Sorghum saccharatum*	Plant	Petri dishes	GI, Bios.	$T = 25°C$, $t = 72$ h	IG = > 60% Bios = 19% at low doses (6.25%)	Da Ros *et al.* (2018)

[a]Dig.Typ: digestate type, agricultural waste (AW), agro-industrial residues (AIR), cattle manure (CM), manure (M), organic fraction of wastewater (OFWW), pig slurry (PS), wastewater (WW), wine waste (WiW).
[b]M. typ.: measurement type, inhibition (Inh), germination index (GI), biostimulation (Bios).
[c]Exp. Cond: experiment conditions.

Similarly, the tests with the *Daphina magna* indicators demonstrated inhibition due to the toxicity of unstabilized organic waste and chemically stabilized with lime but still contained a high load of degraded organic materials in its matrix (Alvarenga *et al.*, 2016). The terrestrial toxicity experiments mainly involved plant species, and the most used test is the germination index/biostimulation. According to various authors, these tests efficiently, simply, quickly, and economically determine the level of toxicity (Alburquerque *et al.*, 2012; Da Ros *et al.*, 2018). In general, these bioassays using organic waste eluates are critical for determining the adverse consequences of organic waste applied to the soil.

According to Boluda *et al.* (2011), the plant bioassay using cress (*Lepidium sativum*) or lettuce (*Lactuca sativa*) is the most popular plant used in phytotoxicity studies due to its high sensitivity. One of the most commonly used methods is the plate or disc bioassay, which is considered an efficient, simple, and economic ecotoxicological assay to assess toxicity risks derived from soil pollution.

The Petri dish-based assay is critical for identifying stable digestates for agricultural applications. The phytotoxic effects during early growth (germination) are identified as being primarily due to salinity; therefore, Na and Cl concentrations, as well as heavy metals (especially Cu and Zn), must be considered to avoid metal accumulation in the soil and salinization following the application of the digestates. On the other hand, soil invertebrates are good bio-indicators because of their continuous exposure to soil contaminants by skin contact, direct ingestion of soil particles and soil water, and food chain transfers. In this sense, earthworm bioassay using earthworm (*E. foetida*) allows the measurement of responses as life-cycle parameters such as survival, growth, and reproduction (Calisi *et al.*, 2011; Pivato *et al.*, 2016). They are regarded as a potential general biomarker that may be directly linked to organism health compared to other biological responses to pollutants. In addition, it has a high sensitivity to pollutant exposure, suggesting its possible applications as a sensitive, simple, and quick general biomarker for monitoring and assessment applications (Calisi *et al.*, 2011).

While the AD system has numerous benefits for agriculture farmers, it has demonstrated the constraints that practitioners must consider while using the digestate. Hence, Figure 17.1 illustrates the process of manure use on a farm, its collection, treatment, and application to crops. The flow distribution of the major pollutants is represented through all of the processes above, leading to environmental impacts on aquatic and terrestrial bodies. Consequently, these effects might reach the flora and fauna that inhabit these ecosystems but are not limited to these physical boundaries.

17.4 ENVIRONMENTAL IMPACTS LINKED TO DIGESTATE

Life-cycle assessment (LCA) is especially suitable for studying multiple aspects, particularly those related to the environment, such as carbon footprint or climate change impact (Morales-Polo *et al.*, 2018). The methodology is based on the ISO 14040 and ISO 14044 standards; the functional unit can

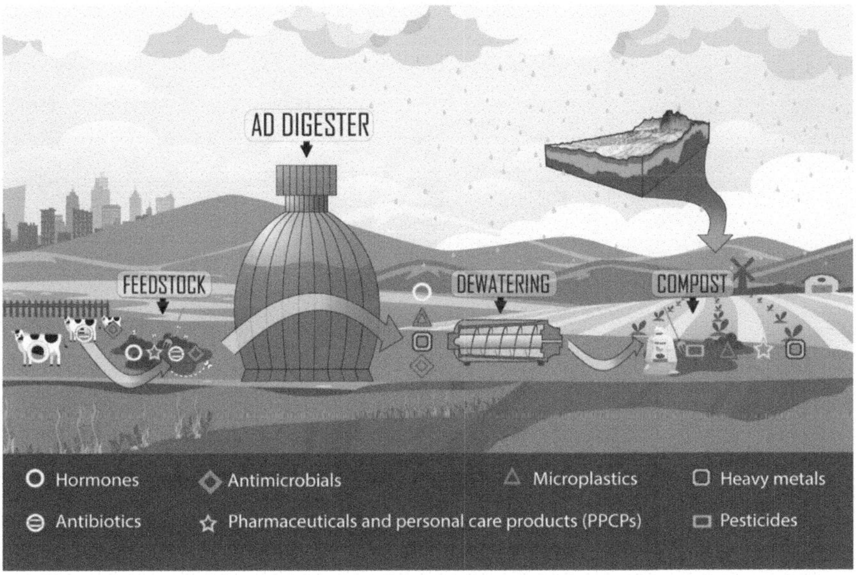

Figure 17.1 Agricultural farm with an AD digester used for the manure treatment and later used as biofertilizer. The fate of some pollutants flowing during the whole process is shown with different symbols.

be 1 kWh of generated electricity or 1 m^3 of biogas, and the software used is specialized, including OpenLCA, Simapro, and the Ecoinvent database. The most commonly evaluated categories are climate change, acidification, eutrophication, and photochemical oxidation (Zhou *et al.*, 2021a, 2021b). Its application includes waste management, AD processes, agriculture crops, and so on (Wei *et al.*, 2018). For instance, it could be implemented to compare biodigesters, measuring CO_2, SO_2, P, N, energy demand, and biodiversity loss, among other parameters (Cherubini *et al.*, 2015). Several feedstock options are represented in Figure 17.2, each of which can be introduced into the system independently or tentatively as a mixture of two or more inputs, resulting in the co-digestion of residues from animals and agriculture crops/leftovers. Additional by-products with a high market interest are identified; for example, the production of biogas with multiple applications is identified, and the production of digestate with agricultural applications and some elements in the construction field are shown. In general terms, a study of LCA is represented at least within these boundaries, though its scope can be expanded to include collection, transportation, and construction, among other stages.

The significance of post-digestion emissions and their link to AD performance has been emphasized as critical elements in reducing net GHG emissions (ca. 75%) and maximizing digestate fertilizer potential (ca. 15%); moreover, the gas-tight digestate storage with residual biogas collection is suggested (Pardo

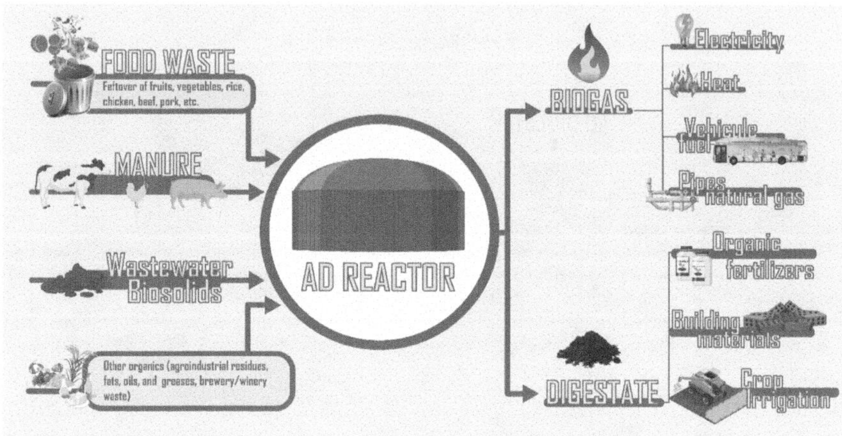

Figure 17.2 Potential inputs and outputs that an AD reactor can manage and different scales.

et al., 2017). It has been demonstrated that upgrading conventional full-scale anaerobic digesters by implementing ultrasonication, adding biochar and inorganic materials as catalysers can reduce GHGs by up to 50% (Mainardis *et al.*, 2021). Another consideration is establishing an AD system according to the local requirements; therefore, small-scale biogas plants are preferred over large AD facilities in areas with a relatively small population (Zhou *et al.*, 2021a, 2021b).

Another strategy for improving digestate management is through slurry tanks, which are a cost-effective method of mitigating the effects of terrestrial acidification and marine eutrophication; a low feed conversion rate reduces environmental impacts associated with the swine supply chain (Cherubini *et al.*, 2015). Despite numerous benefits from the AD process, some environmental impacts include soil depletion due to landfill saturation or increased ground demand for agricultural raw resources, resulting in soil pollution (Morales-Polo *et al.*, 2018).

In developing countries, appropriate management of digestates is feasible; for example, in rural areas in Egypt, fixed-dome digesters are used throughout the operation phase and less so during the construction phase (Ioannou-Ttofa *et al.*, 2021). Additionally, some experiences with cow–buffalo dung and potato waste have been developed in Pakistan. The LCA evaluated 100% cow–buffalo manure (CBM), 100% potato waste (PW), and a mixture of 75% CBM and 25% PW (CBM–PW mixture). It was determined that each 2000 kg of substrate slurry contained a climate change potential of 70, 71 and 149 kg for CBM, PW and CBM–PW mixtures, respectively (Rasheed *et al.*, 2019). A larger study examined low-cost digesters on small-scale farms in Colombia, wherein manure was used as substrate to produce biogas and digestate to replace liquefied petroleum gas

(LPG) and synthetic fertilizer, respectively. It entailed a reduction of 80% in potential environmental impacts associated with manure management, fuel and fertilizer use on farms (Garfí *et al.*, 2019). These low-cost digesters may represent a great tool to improve agricultural waste management and reduce GHG emissions.

Before developing a system architecture that balances sustainability and individual interests, all waste management and recovery parties must agree on a trade-off between the researched metrics. However, no system architecture is economically viable unless an economic analysis is conducted (Cobo *et al.*, 2018). Finally, the interaction between academia, industry and government is critical in the journey towards improved livelihood, economic income generation, and environmental impact reduction. Thus, measuring the sustainability indicators is mandatory, and LCA might continue to represent a suitable tool to better evaluate the deployment of technologies.

17.5 DIGESTATE LEGISLATION

The AD technology produces biogas and digestate; however, the latter needs to be regulated to safeguard the environment and global food chain. Table 17.6 summarizes the main digestate legislations. In the case of the United States, these are found in biosolids and the limits associated with the contaminants are regulated according to USEPA (2018). In European countries, each country has specific regulations and parameters for digestate, compost and manure. These regulations include permissible limits for discharges into soils and water bodies. The report by Saveyn and Eder (2013) groups the main guidelines by country (e.g., the SPCR 120 standard in Sweden (Sverige, 2013), and BSI-PAS-110 in the UK (British Standards Institution, 2014)).

In nations such as China, the digestate has been utilized as a feed supplement for pig, chicken, fish, and shrimp production; however, this option is constrained by national regulations and public approval (Logan & Visvanathan, 2019). Despite these advantages, certain laws restrict the market, and the Ministry of Agriculture opposes standardizing food waste treatment for safety reasons (Freese & Han, 2019).

According to Peng and Pivato (2019), certain quality criteria allow the use of the digestate as a 'product'. These requirements are the input materials, processes and treatment techniques. The digestate is available in three forms: whole, liquid, and solid, and each fraction can be applied to the soil as a final disposal destination once they comply with the relevant regulatory standards and can be classified as a 'product' (Nkoa, 2014; Teglia *et al.*, 2011). Digestate quality standards include hygienic standards, impurities, degree of fermentation, odour, organic matter content, heavy metal content, and declaration parameters. The standards for heavy metals vary slightly between different countries. Among the biological parameters, the digestate used as fertilizer must ensure that the quality of the product has limited pathogens, viruses, and weed seeds (Al Seadi & Lukehurst, 2012).

Table 17.6 Main regulatory criteria for contaminants present in digestates.

Parameters	UM	USA (USEPA, 2018)	BSI PAS, UK (British Standards Institution, 2014)	EU standard[a] (European Commission, 1986; European Commission 1991; European Commission 2000)	Italy[b]	SPCR–120, Sweden (Sverige, 2013)	RAL GZ 245, Germany (Siebert, 2008)	NCh 3375 Chile (INN, 2015)
Physical–Chemical								
Odour							Free from annoying odours.	It must not present unpleasant odours
NO_3^{2-}	mg L^{-1}/ kg ha^{-1} y^{-1}			50 (170 kg N ha^{-1} y^{-1})				
TP	mg L^{-1}			10 (9–98 kg P ha^{-1} y^{-1})				
As	mg kg^{-1}	75		–				55
Cd	mg kg^{-1}	85	1.5	20–40	1.5	1	1.0–3.0	15
Cu	mg kg^{-1}	4300	200	1000–1750	230	600	100–600	667
Cr	mg kg^{-1}	3000	100	–	0.5	100	100–120	167
Hg	mg kg^{-1}	57	1	16–25	1.5	1	1–2	3
Mo	mg kg^{-1}	75		–				
Ni	mg kg^{-1}	420	50	300–400	100	50	50–100	133
Pb	mg kg^{-1}	840	200	750–1200	140	100	100–180	367
Zn	mg kg^{-1}	7500	400	2500–4000	200	800	400–800	1333

Se	mg kg^{-1}	100		–			
Fermentation degree	mg L^{-1}					<4000 mg L^{-1}, expressed as acid equivalent	<4000 mg L^{-1}, expressed as acid equivalent
Volatile fatty acids	g COD/ g VS		0.43				
Organic matter	%					>30%	>30%
Contamination with foreign matter	%		<0.5			<0.5	<0.5
Microbiology							
E. coli			CFUx g^{-1} of fresh matter	≤1.0 × 10^3 CFU g^{-1}	≤1.0 × 10^3 CFU g^{-1}		<1 000 MPN per g of digestate on a dry basis
Salmonella sp.			Absent in 25 g of fresh matter	Absent in 25 g of fresh matter	Absent in 25 g of fresh matter	Absent in 25 g of fresh matter	<3 MPN per g of digestate on a dry basis
Helminth eggs					-		<1 in 4 g of digestate on a dry basis
Ecotoxicity							
Weed germination			Germinating weed plants: 0/1			Less than 2 germinable weeds and germinated plant parts per L	Less than 2 weed propagules per L of digestate.

[a]EU standard: Nitrates: Council Directive 91/676/EEC, Heavy metal: Council Directive 86/278/EEC, Phosphorus. EC. Directive 2000/60/EC, and Microbiology (*E. coli* and *Salmonella* sp) Proposed end-of-waste criteria and Product Quality Requirements for compost and digestate in Saveyn and Eder (2013).
[b]Adapted from Saveyn and Eder (2013). Based on Italian Law on fertilizers (L 748/84; and: 03/98 and 217/06) for BWC/GC/SSC).

17.6 PERSPECTIVES AND CONCLUSIONS

AD offers enormous potential towards a sustainable bioeconomy. AD is suitable for waste management of various sources and generates value in multiple sectors such as electricity, heating, or agriculture through biogas or fertilizer production. However, attention must be paid to cheap fertilizer such as digestates because these digested fractions contain heavy metals, (nano) microplastics, hormones, and other chemical compounds that, when accessible, can be detrimental to humans, animals, plants and the environment. Digestates originating from manure substrates should be monitored to control the spread of antimicrobial resistance genes to the environment. Digestate toxicity can be a problem for aquatic and terrestrial organisms as determined by bioassays to detect the adverse consequences of digestate applied to the soil.

Anaerobic digestate must meet specified quality requirements prior to utilization in the soil or crops to prevent dangers to human health and the environment. The digestate is available in three forms: whole, liquid, and solid, and can be applied to the soil as a final disposal location if it complies with applicable regulatory standards and is classified as a 'product'.

Therefore, governments must establish safety criteria and monitor digestate compositions in the waste valorization chain. There is already legislation regarding digestate management in high-income countries; however, in low-middle income countries, there is none, despite growing interest in using AD as technology for waste management. Furthermore, this chapter aimed to strengthen the importance of appropriately managing digestates in low-middle income countries. Digestates have several potential applications in agriculture, which have been transcendental in farming communities. Further studies should focus on the fate of pollutants and environmental assessment, but it should also include social life-cycle assessment (S-LCA), life-cycle costing (LCC), mass flow analysis (MFA), and exergy flow analysis (EFA).

ACKNOWLEDGEMENT

The authors thank Marco Pérez Celi for his support in the diagramming of the figures in this chapter.

REFERENCES

Alburquerque J. A., de la Fuente C., Ferrer-Costa A., Carrasco L., Cegarra J., Abad M. and Bernal M. P. (2012). Assessment of the fertiliser potential of digestates from farm and agroindustrial residues. *Biomass and Bioenergy*, **40**, 181–189, https://doi.org/10.1016/j.biombioe.2012.02.018

Al Seadi T. and Lukehurst C. (2012). Quality Management of Digestate From Biogas Plants Used as Fertilizer. IEA Bioenergy, United Kingdom.

Alvarenga P., Mourinha C., Farto M., Palma P., Sengo J., Morais M.-C. and Cunha-Queda C. (2016). Ecotoxicological assessment of the potential impact on soil porewater, surface and groundwater from the use of organic wastes as soil amendments. *Ecotoxicology and Environmental Safety*, **126**, 102–110, https://doi.org/10.1016/j.ecoenv.2015.12.019

Andaluri G., Suri R. P. S. and Kumar K. (2012). Occurrence of estrogen hormones in biosolids, animal manure and mushroom compost. *Environmental Monitoring and Assessment*, **184**(2), 1197–1205, https://doi.org/10.1007/s10661-011-2032-8

Angenent L. T., Mau M., George U., Zahn J. A. and Raskin L. (2008). Effect of the presence of the antimicrobial tylosin in swine waste on anaerobic treatment. *Water Research*, **42**(10), 2377–2384, https://doi.org/10.1016/j.watres.2008.01.005

Baştabak B. and Koçar G. (2020). A review of the biogas digestate in agricultural framework. *Journal of Material Cycles and Waste Management*, **22**(5), 1318–1327, https://doi.org/10.1007/s10163-020-01056-9

Biswas S., Shapiro C. A., Kranz W. L., Mader T. L., Shelton D. P., Snow D. D., Bartelt-Hunt S. L., Tarkalson D. D., van Donk S. J., Zhang T. C. and Ensley S. (2013). Current knowledge on the environmental fate, potential impact, and management of growth-promoting steroids used in the US beef cattle industry. *Journal of Soil and Water Conservation*, **68**(4), 325–336, https://doi.org/10.2489/jswc.68.4.325

Boluda R., Roca-Pérez L. and Marimón L. (2011). Soil plate bioassay: An effective method to determine ecotoxicological risks. *Chemosphere*, **84**(1), 1–8, https://doi.org/10.1016/j.chemosphere.2011.02.013

British Standards Institution (2014). PAS BSI.: 110: Specification for whole digestate, separated liquor and separated fibre derived from the anaerobic digestion of source-segregated biodegradable materials. In: Institution, B. S., ed. London, England.

Calamai A., Chiaramonti D., Casini D., Masoni A. and Palchetti E. (2020). Short-term effects of organic amendments on soil properties and maize (*Zea mays* L.) growth. *Agriculture (Switzerland)*, **10**(5), 1–15.

Calisi A., Lionetto M. G. and Schettino T. (2011). Biomarker response in the earthworm *Lumbricus terrestris* exposed to chemical pollutants. *Science of the Total Environment*, **409**(20), 4456–4464, https://doi.org/10.1016/j.scitotenv.2011.06.058

Campanale C., Massarelli C., Savino I., Locaputo V. and Uricchio V. F. (2020). A detailed review study on potential effects of microplastics and additives of concern on human health. *International Journal of Environmental Research and Public Health*, **17**(4), 1212, https://doi.org/10.3390/ijerph17041212

Cao L., Wang J., Xiang S., Huang Z., Ruan R. and Liu Y. (2019a). Nutrient removal from digested swine wastewater by combining ammonia stripping with struvite precipitation. *Environmental Science and Pollution Research*, **26**(7), 6725–6734, https://doi.org/10.1007/s11356-019-04153-x

Cao Z., Jung D., Olszewski M. P., Arauzo P. J. and Kruse A. (2019b). Hydrothermal carbonization of biogas digestate: effect of digestate origin and process conditions. *Waste Management*, **100**, 138–150, https://doi.org/10.1016/j.wasman.2019.09.009

Cao Z., Hülsemann B., Wüst D., Illi L., Oechsner H. and Kruse A. (2020). Valorisation of maise silage digestate from two-stage anaerobic digestion by hydrothermal carbonisation. *Energy Conversion and Management*, **222**, 1–12.

Castro L., Escalante H., Jaimes-Estévez J., Díaz L. J., Vecino K., Rojas G. and Mantilla L. (2017). Low cost digester monitoring under realistic conditions: rural use of biogas and digestate quality. *Bioresource Technology*, **239**, 311–317, https://doi.org/10.1016/j.biortech.2017.05.035

Chaher N. E. H., Nassour A., Hamdi M. and Nelles M. (2021). Digestate post-treatment and upcycling: unconventional moisturizing agent for food waste In-vessel composting. *Waste and Biomass Valorization*, **13**(3), 1–15.

Chen F., Ying G.-G., Ma Y.-B., Chen Z.-F., Lai H.-J. and Peng F.-J. (2014). Field dissipation and risk assessment of typical personal care products TCC, TCS, AHTN and HHCB in biosolid-amended soils. *Science of The Total Environment*, **470–471**, 1078–1086, https://doi.org/10.1016/j.scitotenv.2013.10.080

Cherubini E., Zanghelini G. M., Alvarenga R. A. F., Franco D. and Soares S. R. (2015). Life cycle assessment of swine production in Brazil: a comparison of four manure management systems. *Journal of Cleaner Production*, ***87***, 68–77, https://doi.org/10.1016/j.jclepro.2014.10.035

Cobo S., Dominguez-Ramos A. and Irabien A. (2018). Trade-offs between nutrient circularity and environmental impacts in the management of organic waste. *Environmental Science and Technology*, **52**(19), 10923–10933, https://doi.org/10.1021/acs.est.8b01590

Congilosi J. L. and Aga D. S. (2021). Review on the fate of antimicrobials, antimicrobial resistance genes, and other micropollutants in manure during enhanced anaerobic digestion and composting. *Journal of Hazardous Materials*, **405**, 123634, https://doi.org/10.1016/j.jhazmat.2020.123634

Da Ros C., Libralato G., Ghirardini A. V., Radaelli M. and Cavinato C. (2018). Assessing the potential phytotoxicity of digestate from winery wastes. *Ecotoxicology and Environmental Safety*, **150**, 26–33, https://doi.org/10.1016/j.ecoenv.2017.12.029

Dieguez-Alonso A., Funke A., Anca-Couce A., Rombola A. G., Ojeda G., Bachmann J. and Behrendt F. (2018). Towards biochar and hydrochar engineering-influence of process conditions on surface physical and chemical properties, thermal stability, nutrient availability, toxicity and wettability. *Energies*, **11**(3), 496, https://doi.org/10.3390/en11030496

Dong H., Chen Y., Wang J., Zhang Y., Zhang P., Li X., Zou J. and Zhou A. (2021). Interactions of microplastics and antibiotic resistance genes and their effects on the aquaculture environments. *Journal of Hazardous Materials*, **403**, 123961, https://doi.org/10.1016/j.jhazmat.2020.123961

European Commission (1986). Council Directive 86/278/EEC of 12 June 1986 on the protection of the environment, and in particular of the soil, when sewage sludge is used in agriculture. Official Journal L 181, 04/07/1986 P. 0006–0012.

European Commission (1991). Council Directive 91/676/EEC of 12 December 1991 concerning the protection of waters against pollution caused by nitrates from agricultural sources. Official Journal L 375, 31/12/1991 P. 0001–0008.

European Commission (2000). Directive 2000/60/EC of the European Parliament and of the Council of 23 October 2000 establishing a framework for Community action in the field of water policy. Official Journal L 327, 22/12/2000 P. 0001–0073.

Freese L. and Han S. (2019). From Farm to Table to Energy: Co-digesting China's Urban Food Waste in Wastewater Treatment Plants. Eastern Research Group.

Garfí M., Castro L., Montero N., Escalante H. and Ferrer I. (2019). Evaluating environmental benefits of low-cost biogas digesters in small-scale farms in Colombia: A life cycle assessment. *Bioresource Technology*, ***274***, 541–548, https://doi.org/10.1016/j.biortech.2018.12.007

Garfí M., Gelman P., Comas J., Carrasco W. and Ferrer I. (2011). Agricultural reuse of the digestate from low-cost tubular digesters in rural Andean communities. *Waste Management*, **31**(12), 2584–2589, https://doi.org/10.1016/j.wasman.2011.08.007

Garlapalli R. K., Wirth B. and Reza M. T. (2016). Pyrolysis of hydrochar from digestate: effect of hydrothermal carbonization and pyrolysis temperatures on pyrochar formation. *Bioresource Technology*, **220**, 168–174, https://doi.org/10.1016/j.biortech.2016.08.071

Gondim-Porto C., Platero L., Nadal I. and Navarro-García F. (2016). Fate of classical faecal bacterial markers and ampicillin-resistant bacteria in agricultural soils under Mediterranean climate after urban sludge amendment. *Science of the Total Environment*, **565**, 200–210, https://doi.org/10.1016/j.scitotenv.2016.04.160

Gross T., Breitenmoser L., Kumar S., Ehrensperger A., Wintgens T. and Hugi C. (2021). Anaerobic digestion of biowaste in Indian municipalities: effects on energy, fertilizers, water and the local environment. *Resources, Conservation and Recycling*, **170**, 105569, https://doi.org/10.1016/j.resconrec.2021.105569

Gunnarsson A., Lindén B. and Gertsson U. (2011). Biodigestion of plant material can improve nitrogen use efficiency in a red beet crop sequence. *HortScience*, **46**(5), 765–775, https://doi.org/10.21273/HORTSCI.46.5.765

Ho Y. B., Zakaria M. P., Latif P. A. and Saari N. (2014). Occurrence of veterinary antibiotics and progesterone in broiler manure and agricultural soil in Malaysia. *Science of the Total Environment*, **488–489**, 261–267, https://doi.org/10.1016/j.scitotenv.2014.04.109

Hosseini Taleghani A., Lim T.-T., Lin C.-H., Ericsson A. C. and Vo P. H. (2020). Degradation of veterinary antibiotics in swine manure via anaerobic digestion. *Bioengineering*, **7**(4), 123, https://doi.org/10.3390/bioengineering7040123

Huopana T., Song H., Kolehmainen M. and Niska H. (2013). A regional model for sustainable biogas electricity production: A case study from a Finnish province. *Applied Energy*, **102**, 676–686, https://doi.org/10.1016/j.apenergy.2012.08.018

INN (2015). Instituto Nacional de Normalización. Norma Chilena 3375. Calidad de digestato. In: Normalización, I. N. d., ed. Santiago, Chile, p. 13.

Ioannou-Ttofa L., Foteinis S., Seifelnasr Moustafa A., Abdelsalam E., Samer M. and Fatta-Kassinos D. (2021). Life cycle assessment of household biogas production in Egypt: influence of digester volume, biogas leakages, and digestate valorization as biofertilizer. *Journal of Cleaner Production*, ***286***, 125468, https://doi.org/10.1016/j.jclepro.2020.125468

Iyare P. U., Ouki S. K. and Bond T. (2020). Microplastics removal in wastewater treatment plants: a critical review. *Environmental Science: Water Research & Technology*, **6**(10), 2664–2675, https://doi.org/10.1039/D0EW00397B

Järup L. (2003). Hazards of heavy metal contamination. *British Medical Bulletin*, **68**, 167–182, https://doi.org/10.1093/bmb/ldg032

Jiang B., Lin Y. and Mbog J. C. (2018). Biochar derived from swine manure digestate and applied on the removals of heavy metals and antibiotics. *Bioresource Technology*, **270**, 603–611, https://doi.org/10.1016/j.biortech.2018.08.022

Jiang Y., Heaven S. and Banks C. J. (2012). Strategies for stable anaerobic digestion of vegetable waste. *Renewable Energy*, **44**, 206–214, https://doi.org/10.1016/j.renene.2012.01.012

Jin H. and Chang Z. (2011). Distribution of heavy metal contents and chemical fractions in anaerobically digested manure slurry. *Applied Biochemistry Biotechnology Reports*, **164**(3), 268–282, https://doi.org/10.1007/s12010-010-9133-7

Jobling S., Williams R., Johnson A., Taylor A., Gross-Sorokin M., Nolan M., Tyler C. R., Aerle R. V., Santos E. and Brighty G. (2006). Predicted exposures to steroid estrogens in U.K. rivers correlate with widespread sexual disruption in wild fish populations. *Environmental Health Perspectives*, **114**(Suppl. 1), 32–39, https://doi.org/10.1289/ehp.8050

Jomova K. and Valko M. (2011). Advances in metal-induced oxidative stress and human disease. *Toxicology*, **283**(2–3), 65–87, https://doi.org/10.1016/j.tox.2011.03.001

Keller A. S., Jimenez-Martinez J. and Mitrano D. M. (2020). Transport of nano- and microplastic through unsaturated porous Media from sewage sludge application. *Environmental Science & Technology*, **54**(2), 911–920, https://doi.org/10.1021/acs.est.9b06483

Lehmann L. and Bloem E. (2021). Antibiotic residues in substrates and output materials from biogas plants – implications for agriculture. *Chemosphere*, **278**, 130425, https://doi.org/10.1016/j.chemosphere.2021.130425

Logan M. and Visvanathan C. (2019). Management strategies for anaerobic digestate of organic fraction of municipal solid waste: current status and future prospects.

Waste Management & Research: The Journal for A Sustainable Circular Economy, **37**(1_suppl.), 27–39, https://doi.org/10.1177/0734242X18816793

Longhurst P. J., Tompkins D., Pollard S. J. T., Hough R. L., Chambers B., Gale P., Tyrrel S., Villa R., Taylor M., Wu S., Sakrabani R., Litterick A., Snary E., Leinster P. and Sweet N. (2019). Risk assessments for quality-assured, source-segregated composts and anaerobic digestates for a circular bioeconomy in the UK. *Environment International*, **127**, 253–266, https://doi.org/10.1016/j.envint.2019.03.044

Mahon A. M., O'Connell B., Healy M. G., O'Connor I., Officer R., Nash R. and Morrison L. (2017). Microplastics in sewage sludge: effects of treatment. *Environmental Science & Technology*, **51**(2), 810–818, https://doi.org/10.1021/acs.est.6b04048

Mainardis M., Buttazzoni M., Gievers F., Vance C., Magnolo F., Murphy F. and Goi D. (2021). Life cycle assessment of sewage sludge pretreatment for biogas production: from laboratory tests to full-scale applicability. *Journal of Cleaner Production*, **322**, 129056, https://doi.org/10.1016/j.jclepro.2021.129056

Mao R., Lang M., Yu X., Wu R., Yang X. and Guo X. (2020). Aging mechanism of microplastics with UV irradiation and its effects on the adsorption of heavy metals. *Journal of Hazardous Materials*, **393**, 122515, https://doi.org/10.1016/j.jhazmat.2020.122515

Marin-Batista J. D., Mohedano A. F., Rodríguez J. J. and de la Rubia M. A. (2020). Energy and phosphorous recovery through hydrothermal carbonization of digested sewage sludge. *Waste Management*, **105**, 566–574, https://doi.org/10.1016/j.wasman.2020.03.004

Markou G., Wang L., Ye J. and Unc A. (2018). Using agro-industrial wastes for the cultivation of microalgae and duckweeds: contamination risks and biomass safety concerns. *Biotechnology Advances*, **36**(4), 1238–1254, https://doi.org/10.1016/j.biotechadv.2018.04.003

Marti R., Scott A., Tien Y.-C., Murray R., Sabourin L., Zhang Y. and Topp E. (2013). Impact of manure fertilization on the abundance of antibiotic-resistant bacteria and frequency of detection of antibiotic resistance genes in soil and on vegetables at harvest. *Applied and Environmental Microbiology*, **79**, 5701–5709. https://doi.org/10.1128/AEM.01682-13

Massé D. I., Saady N. M. C. and Gilbert Y. (2014). Potential of biological processes to eliminate antibiotics in livestock manure: an overview. *Animals: an Open Access Journal From MDPI*, **4**(2), 146–163.

Meixner K., Kubiczek M. and Fritz I. (2020). Microplastic in soil–current status in Europe with special focus on method tests with Austrian samples. *AIMS Environmental Science*, **7**(2), 174–191, https://doi.org/10.3934/environsci.2020011

Miliotti E., Casini D., Rosi L., Lotti G., Rizzo A. M. and Chiaramonti D. (2020). Lab-scale pyrolysis and hydrothermal carbonization of biomass digestate: characterization of solid products and compliance with biochar standards. *Biomass and Bioenergy*, **139**, 105593 https://doi.org/10.1016/j.biombioe.2020.105593

Mohammad Mirsoleimani Azizi S., Hai F. I., Lu W., Al-Mamun A. and Ranjan Dhar B. (2021). A review of mechanisms underlying the impacts of (nano)microplastics on anaerobic digestion. *Bioresource Technology*, **329**, 124894, https://doi.org/10.1016/j.biortech.2021.124894

Mohammadi A., Sandberg M., Venkatesh G., Eskandari S., Dalgaard T., Joseph S. and Granström K. (2019). Environmental performance of end-of-life handling alternatives for paper-and-pulp-mill sludge: using digestate as a source of energy or for biochar production. *Energy*, **182**, 594–605, https://doi.org/10.1016/j.energy.2019.06.065

Möller K. and Müller T. (2012). Effects of anaerobic digestion on digestate nutrient availability and crop growth: A review. *Engineering in Life Sciences*, **12**(3), 242–257, https://doi.org/10.1002/elsc.201100085

Monlau F., Francavilla M., Sambusiti C., Antoniou N., Solhy A., Libutti A., Zabaniotou A., Barakat A. and Monteleone M. (2016). Toward a functional integration of anaerobic digestion and pyrolysis for a sustainable resource management. Comparison between solid-digestate and its derived pyrochar as soil amendment. *Applied Energy*, **169**, 652–662, https://doi.org/10.1016/j.apenergy.2016.02.084

Nkoa R. (2014). Agricultural benefits and environmental risks of soil fertilization with anaerobic digestates: A review. *Agronomy for Sustainable Development*, **34**(2), 473–492, https://doi.org/10.1007/s13593-013-0196-z

Noguera-Oviedo K. and Aga D. S. (2016). Chemical and biological assessment of endocrine disrupting chemicals in a full scale dairy manure anaerobic digester with thermal pretreatment. *Science of The Total Environment*, **550**, 827–834, https://doi.org/10.1016/j.scitotenv.2016.01.084

Opatokun S. A., Yousef L. F. and Strezov V. (2017). Agronomic assessment of pyrolyzed food waste digestate for sandy soil management. *Journal of Environmental Management*, **187**, 24–30, https://doi.org/10.1016/j.jenvman.2016.11.030

Pampillón-GonzáLez L., Luna-Guido M., RuÍZ-Valdiviezo V.-M., Franco-Hernández O., Fernández-Luqueño F., Paredes-López O., Hernández G. and Dendooven L. (2017). Greenhouse gas emissions and growth of wheat cultivated in soil amended with digestate from biogas production. *Pedosphere*, **27**(2), 318–327, https://doi.org/10.1016/S1002-0160(17)60319-9

Pardo G., Moral R. and del Prado A. (2017). SIMSWASTE-AD – A modelling framework for the environmental assessment of agricultural waste management strategies: anaerobic digestion. *Science of the Total Environment*, **574**, 806–817. https://doi.org/10.1016/j.scitotenv.2016.09.096

Parmar K. R. and Ross A. B. (2019). Integration of hydrothermal carbonization with anaerobic digestion; opportunities for valorization of digestate. *Energies*, **12**(9), pp. 1586, https://doi.org/10.3390/en12091586

Peng W. and Pivato A. (2019). Sustainable management of digestate from the organic fraction of municipal solid waste and food waste under the concepts of back to earth alternatives and circular economy. *Waste and Biomass Valorisation*, **10**(2), 465–481, https://doi.org/10.1007/s12649-017-0071-2

Peng W., Wang Z., Shu Y., Lu F., Zhang H., Shao L. and He P. (2021). Fate of a biobased polymer via high-solid anaerobic co-digestion with food waste and following aerobic treatment: insights on changes of polymer physicochemical properties and the role of microbial and fungal communities. *Bioresource Technology*, **343**, 126079, https://doi.org/10.1016/j.biortech.2021.126079

Pivato A., Raga R., Lavagnolo M. C., Vanin S., Barausse A., Palmeri L. and Cossu R. (2016). Assessment of compost dosage in farmland through ecotoxicological tests. *Journal of Material Cycles and Waste Management*, **18**(2), 303–317, https://doi.org/10.1007/s10163-014-0333-z

Prapaspongsa T., Christensen P., Schmidt J. H. and Thrane M. (2010). LCA of comprehensive pig manure management incorporating integrated technology systems. *Journal of Cleaner Production*, **18**(14), 1413–1422, https://doi.org/10.1016/j.jclepro.2010.05.015

Prata J., Costa J. and Lopes I. (2021). A One Health perspective of the impacts of microplastics on animal, human and environmental health. *Science of The Total Environment*, **777**, pp. 146094 https://doi.org/10.1016/j.scitotenv.2021.146094

Rasheed R., Yasar A., Wang Y., Tabinda A. B., Ahmad S. R., Tahir F. and Su Y. (2019). Environmental impact and economic sustainability analysis of a novel anaerobic digestion waste-to-energy pilot plant in Pakistan. *Environmental Science and Pollution Research*, **26**(25), 26404–26417, https://doi.org/10.1007/s11356-019-05902-8

Reza M. T., Coronella C., Holtman K. M., Franqui-Villanueva D. and Poulson S. R. (2016). Hydrothermal carbonization of autoclaved municipal solid waste pulp and anaerobically treated pulp digestate. *ACS Sustainable Chemistry & Engineering*, **4**(7), 3649–3658, https://doi.org/10.1021/acssuschemeng.6b00160

Rodriguez Alberto D., Repa K., Hegde S., Miller C. W. and Trabold T. A. (2019). Novel production of magnetite particles via thermochemical processing of digestate from manure and food waste. *IEEE Magnetics Letters*, **7**, 1–15.

Rodriguez Correa C., Bernardo M., Ribeiro R. P. P. L., Esteves I. A. A. C. and Kruse A. (2017). Evaluation of hydrothermal carbonization as a preliminary step for the production of functional materials from biogas digestate. *Journal of Analytical and Applied Pyrolysis*, **124**, 461–474, https://doi.org/10.1016/j.jaap.2017.02.014

Rolsky C., Kelkar V., Driver E. and Halden R. U. (2020). Municipal sewage sludge as a source of microplastics in the environment. *Current Opinion in Environmental Science & Health*, **14**, 16–22, https://doi.org/10.1016/j.coesh.2019.12.001

Saveyn H. and Eder P. (2013). End-of-waste criteria for biodegradable waste subjected to biological treatment (compost and digestate): Technical proposals. EUR 26425. Luxembourg (Luxembourg): Publications Office of the European Union. JRC87124. https://publications.jrc.ec.europa.eu/repository/handle/JRC87124 (Accessed 12 December 2021)

Sen B., Aravind J., Kanmani P. and Lay C.-H. (2016). State of the art and future concept of food waste fermentation to bioenergy. *Renewable and Sustainable Energy Reviews*, **53**, 547–557, https://doi.org/10.1016/j.rser.2015.08.065

Siebert S. (2008). Quality requirements and quality assurance of digestion residuals in Germany. ECN/ORBIT Worshop The future for Anaerobic Digestion of Organic Waste in Europe. Nüremberg, Germany.

Sun W., Qian X., Gu J., Wang X.-J. and Duan M.-L. (2016). Mechanism and effect of temperature on variations in antibiotic resistance genes during anaerobic digestion of dairy manure. *Scientific Reports*, **6**, 30237. https://doi.org/10.1038/srep30237

Sverige A. (2013). Årsrapport Certifierad Återvinning, SPCR 120–produktionsåren 2012, 2011, 2010. Avfall Sverige. Malmo.

Teglia C., Tremier A. and Martel J.-L. (2011). Characterization of solid digestates: part 1, review of existing indicators to assess solid digestates agricultural use. *Waste and Biomass Valorisation*, **2**(1), 43–58, https://doi.org/10.1007/s12649-010-9051-5

Tigini V., Franchino M., Bona F. and Varese G. C. (2016). Is digestate safe? A study on its ecotoxicity and environmental risk on a pig manure. *Science of the Total Environment*, **551–552**, 127–132, https://doi.org/10.1016/j.scitotenv.2016.02.004

Tsapekos P., Khoshnevisan B., Alvarado-Morales M., Zhu X., Pan J., Tian H. and Angelidaki I. (2021). Upcycling the anaerobic digestion streams in a bioeconomy approach: a review. *Renewable and Sustainable Energy Reviews*, **151**, 111635, https://doi.org/10.1016/j.rser.2021.111635

USEPA (2018). United States Environmental Protection Agency. Part 503 – standards for the use or disposal of sewage sludge. Regulatory Determinations for Pollutants in Biosolids. In: Agency, U. S. E. P., ed.

Wallace J. S. and Aga D. S. (2016). Enhancing extraction and detection of veterinary antibiotics in solid and liquid fractions of manure. *Journal of Environmental Quality*, **45**(2), 471–479, https://doi.org/10.2134/jeq2015.05.0246

Wang Y., Pandey P. K., Kuppu S., Pereira R., Aly S. and Zhang R. (2021). Degradation of antibiotic resistance genes and mobile gene elements in dairy manure anerobic digestion. *PLoS ONE*, **16**(8), e0254836.

Wei J., Hao X., van Loosdrecht M. C. M. and Li J. (2018). Feasibility analysis of anaerobic digestion of excess sludge enhanced by iron: A review. *Renewable and Sustainable Energy Reviews*, ***89***, 16–26, https://doi.org/10.1016/j.rser.2018.02.042

Weithmann N., Möller Julia N., Löder Martin G. J., Piehl S., Laforsch C. and Freitag R. (2018). Organic fertilizer as a vehicle for the entry of microplastic into the environment. *Science Advances*, **4**(4), eaap8060, https://doi.org/10.1126/sciadv.aap8060

Xu P., Zhou X., Xu D., Xiang Y., Ling W. and Chen M. (2018). Contamination and risk assessment of estrogens in livestock manure: A case study in Jiangsu province, China. *International Journal of Environmental Research and Public Health*, **15**(1), 125, https://doi.org/10.3390/ijerph15010125

Zhang D., Wang F., Shen X., Yi W., Li Z., Li Y. and Tian C. (2018). Comparison study on fuel properties of hydrochars produced from corn stalk and corn stalk digestate. *Energy*, **165**, 527–536, https://doi.org/10.1016/j.energy.2018.09.174

Zhao L., Dong Y. H. and Wang H. (2010). Residues of veterinary antibiotics in manures from feedlot livestock in eight provinces of China. *Science of the Total Environment*, **408**(5), 1069–1075, https://doi.org/10.1016/j.scitotenv.2009.11.014

Zhao X., Becker G. C., Faweya N., Rodriguez Correa C., Yang S., Xie X. and Kruse A. (2018). Fertilizer and activated carbon production by hydrothermal carbonization of digestate. *Biomass Conversion and Biorefinery*, **8**(2), 423–436, https://doi.org/10.1007/s13399-017-0291-5

Zheng W., Yates S. R. and Bradford S. A. (2008). Analysis of steroid hormones in a typical dairy waste disposal system. *Environmental Science & Technology*, **42**(2), 530–535, https://doi.org/10.1021/es071896b

Zheng W., Zou Y., Li X. and Machesky M. L. (2013). Fate of estrogen conjugate 17α-estradiol-3-sulfate in dairy wastewater: comparison of aerobic and anaerobic degradation and metabolite formation. *Journal of Hazardous Materials*, **258–259**, 109–115, https://doi.org/10.1016/j.jhazmat.2013.04.038

Zheng L., Chen J., Zhao M., Cheng S., Wang L.-P., Mang H.-P. and Li Z. (2020). What could China give to and take from other countries in terms of the development of the biogas industry? *Sustainability*, **12**(4), 1490, https://doi.org/10.3390/su12041490

Zhou H., Yang Q., Gul E., Shi M., Li J., Yang M., Yang H., Chen B., Zhao H., Yan Y., Erdoğan G., Bartocci P. and Fantozzi F. (2021a). Decarbonizing university campuses through the production of biogas from food waste: An LCA analysis. *Renewable Energy*, ***176***, 565–578, https://doi.org/10.1016/j.renene.2021.05.007

Zhou Q., Li X., Wu S., Zhong Y. and Yang C. (2021b). Enhanced strategies for antibiotic removal from swine wastewater in anaerobic digestion. *Trends in Biotechnology*, **39**(1), 8–11, https://doi.org/10.1016/j.tibtech.2020.07.002

doi: 10.2166/9781789062755_0385

Chapter 18

Life-cycle assessment of various anaerobic digestate management technologies

*Sridhar Pilli, Bella Kunnoth, Sagarika Mothe and Polisetty Venkateswara Rao**

Water and Environment Division, Department of Civil Engineering, National Institute of Technology, Warangal, Telangana, India
**Corresponding author e-mail: pvenku@nitw.ac.in*

ABSTRACT

Particularly in India, rich in livestock sector and other established agro-based industries produce tons of organic wastes annually. This has eventually led to development of many small- and large-scale anaerobic digesters treating different kinds of organic wastes. Majority of these organic wastes include wastes from food industries, dairy industries, fruits and vegetable wastes, and agricultural by-products, and so on. The anaerobic digestate released from the digesters have undergone various digestate treatment techniques such as drying, solid–liquid separation, composting, pelletizing, thermal concentration, and so on based on the physical and chemical texture of the digestate. This paper presents a brief idea of various digestate treatment methods employed. The storage and transport of the digestate create lot of nuisances like emission of ammonia and methane along with other economic constraints. Hence, the chosen treatment method should ensure safe disposal of the treated waste with less environmental impact. Life-cycle assessment (LCA) is a management tool for assessing the potential environmental impacts during implementation of a particular project or study. This paper intends to present information on potential environmental impacts from various LCA studies conducted on different anaerobic digestate treatment methods employed. A comprehensive LCA study is necessary to evaluate the net methane and ammonia emissions, net energy gain of a particular method chosen.

Keywords: Anaerobic digestate, life-cycle assessment (LCA), organic wastes, environmental impact, digestate use.

18.1 INTRODUCTION

The expansion of numerous small-, medium- and large-scale biogas plants has led to the production of large amount of digestate demanding appropriate treatment before disposal. Majority of the biogas plants are based on wastes

Editor(s): Vinay Kumar Tyagi, Kaoutar Aboudi and Cigdem Eskicioglu

from food industries, bio wastes and agricultural by-products. Particularly in countries like India, rich in livestock husbandry. The digestate produced is mainly used as a soil amendment because of its rich nutrient content (Khoshnevisan *et al.*, 2018). Digestate can be applied over the land for soil conditioning only after storing it for 3–6 months. This storage for a long time would be problematic. Also, substituting chemical fertilizer with anaerobic digestate creates significant cost loss for farmers due to its lower dry matter (DM) content and large volume (Al Seadi *et al.*, 2013). Moreover, the storage and transport of the digestate led to ammonia and methane emissions, which ultimately contribute to global warming. The application of the digestate on land might result in eutrophication and soil acidification because of its high nutrient contents. Major constituents of different types of the digestate are given in Table 18.1. Besides, over application of the digestate on land at improper times may lead to air and water pollution. Modern digestate management technologies are introduced aiming at its safe disposal as well as reducing environmental impacts during its land application.

Composting (Rizzo *et al.*, 2022), solid–liquid separation (SLS) (Vázquez-Rowe *et al.*, 2015), alkaline and acidic stabilization (Vaneeckhaute *et al.*, 2017), algal nutrient stripping (Xu *et al.*, 2015), membrane filtration (Gienau *et al.*, 2018), drying (Vázquez-Rowe *et al.*, 2015), and ammonia stripping (Yellezuome *et al.*, 2022) are some of the digestate management technologies implemented. All these methods are quite promising in terms of its successful execution and operation. The impacts of these methods on environment are not well studied. Fewer studies are conducted on assessing the potential environmental impacts associated with the implementation of these technologies. Some others have limited system boundaries excluding upstream and anaerobic operations from them. New technology despite having positive influence in some aspects and negative in other aspects make it difficult to optimize the basic parameters. Also, the pros and cons need to be discussed in order to introduce hotspots and the new technology is compared with a baseline option for understanding gaps in earlier types. The environment profile of modern digestate management technologies can be analysed using the life-cycle assessment (LCA). It is an environmental management tool, which evaluates the potential environmental impacts during the production, transport and storage of the lifetime of a product (ISO, 2006). This will help the farming and agricultural stakeholders in choosing the right option. Comparison of different treatment methods will be a support for policy makers in order to select appropriate methods for achieving desirable environmental target, as well as identifying potential environmental hazards followed.

18.2 LCA AND ITS BASIC ELEMENTS

According to the Environmental Protection Agency, LCA is defined as a tool to 'evaluate the environmental effects associated with any given industrial activity from the initial gathering of raw materials from the earth until the point at which all residuals are returned to the earth' or a 'cradle to grave' concept. The

Table 18.1 Physical–chemical characteristics of digestate produced from biogas plants based on different types of feedstock.

Sl. no.	Feed stock	pH	TS (g/L)	TOC (g/kg TS)	TN (g/kg TS)	NH_4-N (g/kg TS)	TP (g/kg TS)	K (g/kg TS)	References
1	Food waste	6.77–6.83	58	38 ± 3	6.9 ± 1.2	–	7.2 ± 1.5	–	Parra-Orobio *et al.* (2021)
2	Cattle manure	5.6	38	465	50	26	13	47	Alburquerque *et al.* (2012)
3	Source separated household waste	7.0	24	412	95	–	12	45	Brod *et al.* (2015)
4	Food waste	8.5	28	374	165	–	12	45	Coelho *et al.* (2018)
5	Cattle manure	7.8	92	355	43	21	7	47	Moller *et al.* (2008)

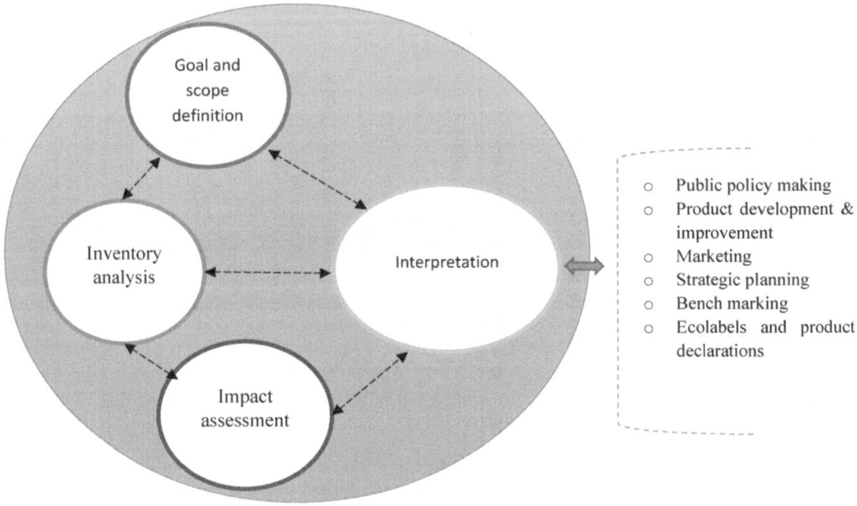

Figure 18.1 Basic elements and outcomes of LCA.

four basic steps of LCA are; goal defining and scoping, inventory analysis, and impact assessment and interpretation (Figure 18.1).

(a) Goal defining and scoping: This is the first stage of LCA where goals and objectives of the work are defined by describing the process or product clearly. The environmental contests in which LCA is to be done are also discussed.
(b) Inventory analysis: This stage comprises identification and quantification of water, energy and other environmental releases like air emissions, solid wastes and waste water.
(c) Impact assessment: Finally, the potential impacts of energy, water and material use are identified.
(d) Interpretation: In this step, the results of inventory analysis and impact assessment are analysed to clearly understand the selected process, service or product without uncertainty.

LCA is typically used for either design purpose or for improvement of products and processes. This involves many phases like; this may involve identifying environmental hotspots in the production processes by considering downstream or upstream routes such as product additives, various energy sources end-of-life management options or by-products consumed in the product's use period. LCA's aiming comparative studies are usually performed to assess one or more differing systems that offer the same type of service. The obtained LCA results can also be used to validate and prove continuous development or for providing information for eco-labelling or environmental rating schemes. LCA can be used in policy making for the development of product supervision and surveillance, and circular economy strategies.

18.3 LCA OF ANAEROBIC DIGESTATE

Various treatment technologies are identified for treating the digestate derived from anaerobic digestion process. The ultimate purpose of these methods is to dispose the digestate safely or to make use of it effectively. The input part of the LCA system commence at the discharge point of the digestate from anaerobic digestion process, followed by the transport between treatment steps and finally disposing fertilizer to field as the output side. The system boundaries were intended to choose in a way comprising all the operations necessary for operating the system. The methodology and LCA framework guidelines by Reference Life Cycle Data System (ILCD) Handbook (Wolf *et al.*, 2010). Detailed information regarding all the assumptions and processes to be carried out during LCA implementation are taken from ISO 14 040 guidelines (ISO, 2006).

The life pathways considered in the LCA of various methods start from the collection or preparation of raw materials to the final application of the treated material to the agricultural land. With the discharge of the digestate from the storage tank of digestion, it is entering the input of the cycle system. Various anaerobic digestate treatment methods are discussed below along with details on the technical system boundary, total process and chain of supply from initial stage to final disposal stage of the digestate. The scope of the study has to be stated first, which in this case is to evaluate the possible environmental impacts of anaerobic digestate treatment methods.

18.3.1 Various digestate treatment options

A brief description of various digestate treatment methods are given below.

18.3.1.1 Conventional digestate treatment

In conventional method, the digestate from anaerobic digestion reactor is kept for a period and then applied on agricultural lands during crop growing season. This storage period varies depending on different cultivation types. The digestate stored is still active and thus continue producing methane. At the same time, ammonia is also released from the digestate. The amount of emissions from the digestate depends on the type of storage system. In a study by Duan *et al.* (2020), a storage time of 3–4 months are chosen for storing digestate produced from a biogas plant treating pig manure. The type of storage system also affects the amount of ammonia emitted. The ammonia emission factor for covered tanks, sealed tanks and open tanks were found to be 5%, 2% and 10%, respectively. Nitrogen emissions from the digestate can be neglected due to its low quantity (Rodhe *et al.*, 2015).

The primary step of conducting LCA is framing goals and setting the scenarios. The direct application of the digestate over agricultural land can be considered as the baseline scenario (BAS) in every digestate treatment option. The BAS is usually accompanied along with other specific treatment options like composting, drying and so on, which are discussed in coming topics (Figure 18.2).

18.3.1.2 Solid–liquid separation

The SLS process aims in separating the digestate into a nitrogen-rich liquid portion and phosphorous-rich dry solid part (Al Seadi *et al.*, 2013). Technologies

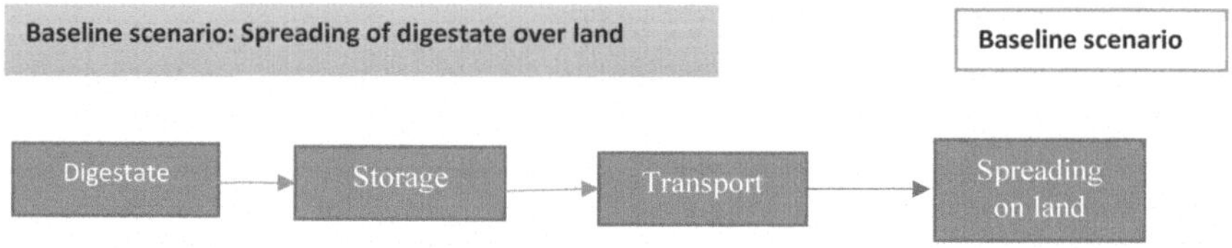

Figure 18.2 Baseline scenario.

such as decanter centrifuge, sieve belt and screw press are some of the widely tested SLS methods. Among these, decanter centrifuge is the most used one in SLS operation of manure digestate. During manure separation process, a lower value of TS can be noticed and a crust-like formation is found on top of the digestate, which releases ammonia emissions into air (Rotz *et al.*, 2007).

18.3.1.3 Composting

Composting consists of three stages; (1) initial pre-treatment of the digestate, (2) composting and (3) making of the soil product. The primary step of pre-treatment involves flocculation and separation of the digestate into solid and liquid takes place. SLS operation is generally applied for managing animal manures as a physical treatment process which helps to remove and separate fibres and coarse solids from it (Jacobson *et al.*, 1999). Separation process can be fastened by adding flocculation substrates to the digestate (Luckert, 2004). The capacity range of the decanter or the SLS device chosen should be noted. The solid matter can be reused depending upon its DM content. The liquid portion usually ends up in the farmland as fertilizer. Usually, sealed windrows are used for composting with intermittent mixing for proper aeration. The material after composting is kept in open storage boxes. It was presumed that around 80% of initial water content and 50% of carbon was lost after composting. Finally, around 57% of DM is left in the heap of the compost. The total ammonia and carbon emissions as CO_2 and CH_4 need to be calculated. Nitrogen emissions were presumed to be; ammonia-20% of NH_4, 8.12% of total carbon as CH_4 and 1.4% of total nitrogen as N_2O.

18.3.1.4 Belt dryer

In belt drying method, the raw digestate is mixed with dried solid digestate to achieve a moisture content of 20%. This process is otherwise called as homogenization, which is used for recharging the belts of dryer. In homogenization, hot air of temperature 85°C is blown over the surface of the substrate. The DM content of the digestate will reach almost 80% after drying. Before transporting to pelletizing unit, it is stored (Rehl & Müller, 2011). Pellet consistency can be increased by adding starch and lime flour.

18.3.1.5 Drum dryer

In drum drying, the digestate is applied continuously over a rotating drum. The drum gets heated as it rotates. Hence, the digestate on the drum also gets dried. A heat of 3 MJ and electricity of 0.3 MJ are required for removing 1 kg of water like this. This heat is supplied from the heat produced by combustion of natural gas in thermal power stations. The substrate is handled as it is done in belt drying.

18.3.1.6 Solar drying

In solar drying method, a greenhouse-dryer powered by solar energy and covered with polycarbonate transparent sheets are employed for drying substrate. The dryer is well loaded with substrate by a pump and then is discharged by a wheel loader. The substrate is mixed automatically in dryer and aerated by fans. The evaporation rate generally lies between 0.6 and 3.50 tons of water per m^2 drying area. According to Burger *et al.* (2006), mixing and aeration of the digestate to a water content of 35% require 200 kWh of electricity. The drying was carried out using solar power alone.

18.3.1.7 Thermal concentration

After SLS operation, the solid and liquid portions of the digestate get separated. The solid fraction is kept onsite and allowed to dry further. Then, it is applied over agricultural land without any treatment. The liquid fraction undergoes vaporization, which is otherwise termed as thermal concentration (TC) where other ingredients and water will separate. This is carried out by heating up the rotary discs using hot steam of around 10 bar pressure and subsequently vaporizing the water from the liquid fraction. When boiling starts, hot sulphuric acid is added to expulse CO_2 and to raise pH 4.4–4.8 and to convert NH_3 to NH_4. The gaseous loss through nitrogen emissions is completely avoided using pH enhancement (Clemens & Wulf, 2005). However due to CO_2 emissions, foaming is produced which can be removed by adding anti-foaming agents. After TC, two products are formed; a condensed steam and a concentrate with high nutrient content. Depending upon the nutrient content of condensate, it can be used for diluting fermentation substrates or can be transported to a region of nutrient deficit.

18.3.1.8 Physico-chemical treatment

In physico-chemical treatment option, the solid part separated from SLS is transported out of the region. The liquid portion is subsequently treated by ultra-filtration by means of membranes and pressure ranging from 2 to 10 bars (Masse *et al.*, 2007; Thörneby *et al.*, 1999). This process separates the liquid part into a permeate and retentate with water containing low and high molecular substances, respectively. Reverse osmosis is used for treating permeate further, where a semi-permeable membrane which can retain dissolved substances using a static pressure of 40–100 bar is employed. These treated retentates and permeates are mixed and stored together before they are transported to the application area. The permeate can also be used as irrigation water (Thörneby *et al.*, 1999).

18.3.1.9 Nutrient recovery

Due to high contents of stable organic carbon and other nutrients, the digestate can be directly applied on agricultural land as a renewable organic fertilizer.

Also, when the organic loads at specific area is found high, it can be further treated in concentrated forms and then translocated to agricultural land in sustainable prices (Rehl & Müller, 2011; Vaneeckhaute *et al.*, 2017). The digestate is an organic substrate rich in phosphorous, potassium and nitrogen. It is normally used for field application where high demand of potassium, nitrogen and phosphorous persists. It can also beapplied for natural production of humus content in soil (Wäger-Baumann, 2007). The phosphorous application on the field is limited to 20 kgP_2O_5 per hectare per area and nitrogen application limited to 170 kgN per hectare per area (Kuhn, 2017). Various options to treat digestate include membrane treatment, drying, stripping and so on. In membrane filtration, the liquid portion of the digestate left after SLS operation is treated by reverse osmosis and ultra-filtration. The concentrate obtained after reverse osmosis is similar to the vinasses obtained from distillation unit which is rich in both micro and macro nutrients (Ledda *et al.*, 2013). In stripping, the digestate undergoes more than one pre-treatment in SLS and then the liquid fraction is sent to a packed-bed tower where ammonia is stripped and transferred physically to aqueous state. This gas steam is then sent to another packed column where absorption of ammonia takes place in an acidic medium producing ammonium sulphate (Bonmatí & Flotats, 2003; Landau, 2015).

18.3.1.10 Ammonia stripping

Ammonia stripping is one of the technologies to reduce nitrogen content in the digestate. Ammonia stripping does not need any valuable chemicals whereas struvite precipitation requires magnesium and phosphate to get a pH of 9.5. Struvite precipitation reduces a maximum of 88% ammonia and maintains a magnesium:ammonium:phosphate:molar ratio of 1.25:1:1 (Miles & Ellis, 1998). Ammonia stripping method helps to retrieve a standardized nitrogen fertilizer product. The liquid fraction that is generated can be utilized to refine other digestate portions in order to achieve standardized nitrogen content.

18.3.1.11 Membrane filtration

The membrane technology can be applied to the liquid portion of the digestate only after significant solids removal. Along with membrane fouling, this is a critical step in membrane purification operations. In the initial SLS process, decanter centrifuges are commonly utilized, and precipitating chemicals are frequently used to improve solids removal. Ultrafiltration is the next phase, followed by reverse osmosis to remove chemical oxygen demand (COD) and ammonia. In most cases, ammonia discharge levels can be achieved by applying three stages of reverse osmosis. The fact that only a small portion of the digestate is clean water is a disadvantage of such membrane purification methods; around half of the digestate is collected as by-products. The third phase of reverse osmosis process can be substituted by ion-exchange process, which collects the ultrafiltration retentate, the reverse osmosis concentrate and the solid fraction. The ultrafiltration retentate is frequently recycled into the biogas plant and/or the SLS phase to decrease the quantity of waste. Membrane purification is both costly and energy intensive.

18.3.1.12 Aerobic treatment

The liquid component of the digestate is processed in an aerobic wastewater treatment facility with other carbon-rich effluents under ideal conditions. In the absence of carbon-rich effluents, an external carbon source is needed to supply since the bacteria require more carbon to complete the denitrification process due to the high ammonia level. Apart from the nitrogen issue, COD and the colour or turbidity of the treated effluent make meeting discharge standards challenging. Furthermore, aerobic wastewater treatment entails a large initial investment as well as ongoing operating expenditures.

18.3.1.13 Anammox process

The Anammox technique (anaerobic ammonium oxidation) can be used to remove ammonia. This is a microbiological process that occurs in a variety of natural environments, as well as the trademarked name of the removal of ammonium method that followed. The Anammox process is carried out by bacteria, the most well-known species being *Planctomyces* and *Pirellula*, and comprises of two phases that occurs in the same reactor: (1) ammonia-oxidizing bacteria accomplish partial nitrification; (2) Anammox bacteria convert the resultant ammonium and nitrite nitrogen dioxide gas and around 15% nitrate (not pictured), according to

$$4NH_4^+ + 3O_2 \rightarrow 2NH_4^+ + 2NO_2^{--} + 4H^+ + 2H_2O$$

$$4NH_4^+ + NO_2^{--} \rightarrow N_2 + 2H_2O$$

18.3.1.14 Evaporation

If enough waste heat is available at the biogas plant, digestate evaporation might be a viable option. Before starting, a large portion of the solids should be removed. This is commonly done by means of technical-scale applications (such as energy crop digestion) by combining vibration sieves with screw press separators to remove fibres and hence prevent heat exchanger blockage. However, a significant quantity of cleaning will be required. A nutrient-rich concentrate is created via evaporation, and the condensate is collected. The goals are to obtain high nitrogen concentrations in the concentrate and low nitrogen concentrations in the condensate. To prevent ammonia evaporation, large volumes of sulphuric acid are used, resulting in exorbitant expenses. Aside from that, the condensate generated does not match the requirements for direct discharge. If a direct discharge is necessary, a reverse osmosis stage can be used to purify the condensate. Finally, because the digestate evaporation necessitates the use of significant chemicals as well as the availability of waste heat, direct discharge levels of the condensate are rarely achieved.

18.3.1.15 Combined processes

SLS processes can be combined with two or refining phases of variable degrees of technological maturity for conversion of the digestate into concentrated fertilizer products. Many countries have agricultural biogas plants with digestate processing facilities in operation, and additional plants are contemplating doing so (Al Seadi *et al.*, 2013). In Germany, several industrial-scale factories use integrated digestate processing ideas. The researchers looked at five full-scale systems for recovering nutrients from anaerobic digestate generated in farm-scale plants. Drying with acidic recovery, stripping with acidic recovery, and membrane separation were all monitored methods. The results revealed high nutrient recovery, with average yields consistently above 50% for both nitrogen and phosphorus (Bolzonella *et al.*, 2018). The entire conditioning process, which includes decantation, ultrafiltration, and reverse osmosis, is an appropriate technological chain for producing well-defined concentrated fertilizer products from the digestate. It creates a solid organic nitrogen/phosphorus fertilizer and a liquid nitrogen/potassium fertilizer in addition to clean water (Gienau *et al.*, 2018).

18.4 DIGESTATE TYPES AND UTILIZATION

Although anaerobic digestion produces biogas which can be utilized as a substitute for natural gas, and produces digestate rich in nutrients which can be used as a fertilizer for agricultural lands. The quality of the digestate depends upon the feedstock/stocks (animal manures like cow manure, chicken manure, pig manure, agricultural biomass like rice straw, wheat straw, corn stover, organic waste like food waste, sewage sludge, organic fraction of municipal solid waste, etc) in anaerobic digestion and post-treatment technologies like composting, drying, physicochemical treatment, ammonia stripping, membrane filtration, aerobic treatment, anammox process, evaporation and combinations of mentioned processes. In digestate, solid, liquid fractions (from mechanical separation) and raw digestate produces fertilizer containing nutrients which contributes to resource preservation. The produced fertilizer type from various processes has been given in Table 18.2. However, the quality of the digestate can be further enhanced through the digestate processes, although processing of the digestate may not be economical as it includes additional maintenance costs. The most sustainable way of utilizing the digestate is by recycling the raw digestate as bio fertilizer. The digestate with sufficient quality assurance will help in achieving several enduring benefits by outweighing the probable risks, when used as fertilizer along with other agricultural practices.

18.5 CASE STUDIES ON LCA OF ANAEROBIC DIGESTATE

The scientific and technical information available on the treatment methods mentioned above is highly variable. Among the processes discussed so far, SLS operation has been broadly employed due to its efficiency of separation and energy consumption, particularly for manure digestate and wastes from other feed stock types (Béline *et al.*, 2008; Turner & Burton, 2003). For nations like with well-established livestock sector often face problems related to excess nutrient release like nitrogen. Many authors have studied about the environmental impacts of these digestate management technologies to assess their efficiency and

Table 18.2 Various digestate processes from different feedstock to anaerobic digestion with their produced form of fertilizers.

Feedstock	Treatment system	Digestate fraction treated	Products	References
Pig liquid manure, sewage sludge and food industry waste	Conventional	Raw digestate	Liquid fertilizer products	Angouria-Tsorochidou *et al.* (2022)
Pig liquid manure, sewage sludge and food industry waste	Centrifugation	Liquid and solid digestates	Solid and liquid fertilizer products	
Pig liquid manure, sewage sludge and food industry waste	Centrifugation, drying of solid digestate, membrane filtration and reverse osmosis unit	Liquid and solid digestates	Solid fertilizer products, UF retentate, RO concentrate	
Pig effluents, chicken manure, energy crops	Dryer	Solid and liquid fractions	Ammonium sulphate and dried organic digestate	Bolzonella *et al.* (2018)
Cow manure, slaughterhouse residues, food waste, energy crops	Dryer	Solid and liquid fractions	Ammonium sulphate and dried organic digestate	
Pig manure, energy crops	Membrane separation	Liquid fraction	Reverse osmosis concentrate, ammonium sulphate, water	
Cow manure, energy crops	Membrane separation	Liquid fraction	Reverse osmosis concentrate, ammonium sulphate, water	
Cow manure, pig manure effluents, energy crops	Stripping	Liquid fraction	Ammonium sulphate	
Corn silage, liquid manure, crop silage, remnants	Separator, decanter centrifuge, ultrafiltration, reverse osmosis	Liquid and solid digestates	Ultra-filtration retentate, reverse osmosis concentrate	Gienau *et al.* (2018)

(Continued)

Table 18.2 Various digestate processes from different feedstock to anaerobic digestion with their produced form of fertilizers (*Continued*).

Feedstock	Treatment system	Digestate fraction treated	Products	References
Digestate from energy crops, organic waste, manure	Conventional	Raw digestate	Fertilizer	Vázquez-Rowe *et al.* (2015)
	Drying and pelletizing	Liquid and solid digestates	Fertilizer pellets	
	Composting	Liquid and solid digestates	Compost fertilizer, ammonium sulphate	
	Biological treatment, reverse osmosis and drying	Liquid and solid digestates	Reverse osmosis concentrate, permeate	
	Ammonia stripping and drying	Liquid and solid digestates	Ammonium sulphate	
Pig manure	Conventional	Liquid and solid digestates	Soil conditioner	Duan *et al.* (2020)
Pig manure	Integrated flocculation–biological contact oxidation technology	Liquid digestate	High turbid liquid	
Pig manure	Integrated flocculation–biological contact oxidation technology with dilution	Liquid digestate	Liquid for microalgae cultivation	
Pig manure	Powder biofertilizers	Liquid digestate	Struvite precipitation	
Household waste	Composting	Liquid and solid digestates	Stabilized digestate, reduction in nitrogenous emissions	Zeng *et al.* (2016)

(*Continued*)

Table 18.2 Various digestate processes from different feedstock to anaerobic digestion with their produced form of fertilizers (*Continued*).

Feedstock	Treatment system	Digestate fraction treated	Products	References
Wastewater sludge	Conventional	Liquid and solid digestates	Fertilizer	Farhad Sakhaee (2020)
Organic household waste	Biopulp	Liquid and solid digestates	Fertilizer with removed rejected fraction	Khoshnevisan *et al.* (2018)
Organic household waste	Screw press	Liquid and solid digestates	Fertilizer with removed rejected fraction	
Organic household waste	Disc screen	Liquid and solid digestates	Fertilizer with removed rejected fraction	
Dairy products and agro-industrial wastes	Composting	Liquid and solid digestates	Liquid fertilizer and solid compost	Kopsahelis *et al.* (2019)
Thickened sludge	High-pressure dewatering	Liquid and solid digestates	Land use	Li and Feng (2018)
Thickened sludge	High-pressure dewatering, thermal drying, pyrolysis	Solid digestate	Biochar	
Food waste	Composting	Solid digestate and sludge from wastewater treatment	Fertilizer	Fei *et al.* (2021)
Food waste	Wastewater treatment	Liquid digestate	Water discharge	

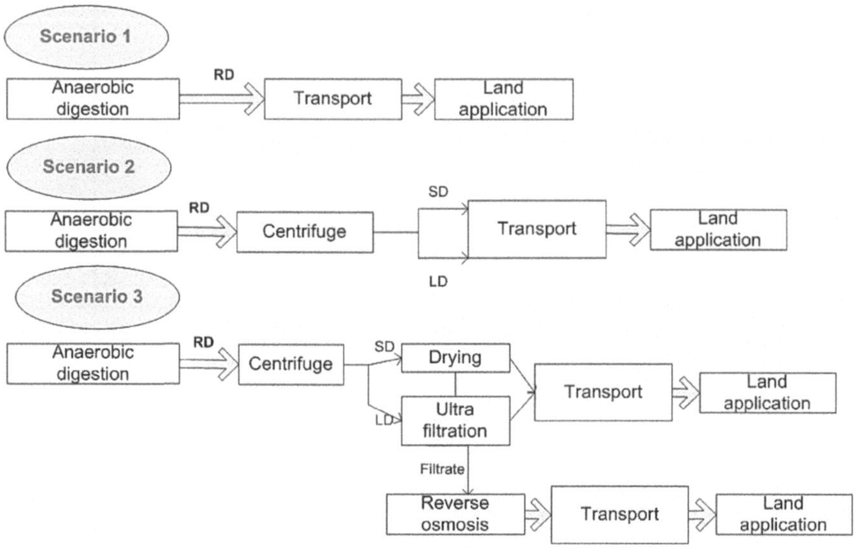

Figure 18.3 Scenario description (RD: raw digestate, LD: liquid digestate, SD: solid digestate).

sustainability in nutrient recycling and other connected environmental impacts. A few of those studies are briefly described below.

Angouria-Tsorochidou *et al.* (2022) looked at three alternative situations; the BAS in which the fresh digestate is distributed uniformly over the agricultural field. The digestate was collected from a mesophilic digester plant treating pig manure mixed with food waste and sewage sludge. The scenarios are depicted in Figure 18.3. Scenario 2 can be described as, first the raw digestate is centrifuged and then undergoes SLS (solid and liquid digestate), which are then dispersed over agricultural land. Scenario 3 involves modelling of a more complex post-treatment system, in which the phase separation of raw digestate is carried out using centrifugation, followed by solid digestate drying and processing of liquid digestate using membrane filtering and a reverse osmosis unit. Among the situations investigated, Scenario 3 allowed for the production of three distinct types of fertilizing products. These products along with some other high-value products may be able to meet a wide range of agronomic demands while also providing enough concentration flexibility to replace synthetic fertilizers. The LCA, however, revealed that scenario 2 had the lowest environmental impact on freshwater eutrophication, global warming, and shortage of mineral resources. While scenario 1 has the lowest impact in terms of terrestrial eco-toxicity, human non-carcinogenic toxicity and fossil resource scarcity, whereas scenario 3 possesses the highest efficiency in terms of soil acidification and eutrophication of marine eco-system.

Another study by Vázquez-Rowe *et al.* (2015) investigated the environmental impacts of digestate treatment technologies, comparing; (1) drying and pelletizing (2), composting, (3) reverse osmosis and drying, biological treatment, and (4) ammonia stripping and drying. The significant environmental impacts of the solid fraction treatment scenarios selected were assessed using the ReCiPe method, used to evaluate the indicators of life-cycle impact (Goedkoop *et al.*, 2009). As in majority of the studies, the BAS was chosen as direct spreading of the digestate on field. Additional scenarios were created based on DM content of the digestate. They are (a) BAS-1, in which a mixture of dried digestate and raw digestate having DM 55.6% is compared with scenario drying and pelletizing, (b) BAS-2, in which the digestate of DM 25.5% is compared with composting scenario, and (c) BAS-3, in which the digestate of DM 8.91% is compared with scenario reverse osmosis and drying, biological treatment, and ammonia stripping and drying. A short-term storage period of 5–7 days was chosen for all treatment methods. The results indicate that BAS-1 and BAS-2 showed environmental impact of 61% and 51%, respectively, higher than that of BAS-3. The reason for this higher impact is linked to higher emissions in BAS-1 and BAS-3, especially particulate matter formation. The scenarios studied imply that the methods employed for pre-treating the digestate will help in achieving sufficient environmental gains compared to direct land application.

Duan *et al.* (2020) studied pig manure digestate using four digestate management technologies: that is spreading the digestate on farmlands to replace chemical fertilizers, (1) integrated flocculation–biological contact oxidation for liquid digestate after mechanical separation of liquid and solid digestate to eliminate the biodegradable organic matter and reduce turbidity, and integrated flocculation–biological contact oxidation for liquid digestate after mechanical separation of liquid and after that, the solid portion is composted well and used as a soil conditioner, (2) combined flocculation–biological contact oxidation with dilution, in which the solid portion of the digestate was undergoing composting and the resulting compost material is used as a soil conditioner, while the liquid fraction was further diluted to achieve an appropriate ammonia level and turbidity suited for culturing microalgae. (3) The third method is the recycling of the liquid portion of digestate into powder-form bio-fertilizers; the extracted powder bio-fertilizer is a concentrated biofertilizer that reduces logistical issues with transportation and application on farmlands. Pig dung was co-digested with microalgae generated by the aforesaid technologies, resulting in a 1.61 MJ/ primary energy savings per ton of pig manure. The scenarios chosen are illustrated in Figure 18.4. When compared to the BAS of direct application of the digestate on agricultural fields, DL scenario was found energetically feasible. DL reduced adverse impacts on the human health harm class by 32% and 116%, respectively, as compared to IFBCO and PBF scenarios. Farhad Sakhaee (2020) performed a life-cycle evaluation on wastewater sludge with a storage capacity of 4 68 000 gallons and a feed rate of 1500–3000 gallons per day for the digester. The university's agricultural area is subjected to the removal of digested sludge from the secondary digester twice a year. The sludge is injected below into the soil and used as a fertilizer. It is recommended that the sludge be cured before it can be used to make a blended soil conditioner. Khoshnevisan *et al.* (2018) tested several ways for converting organic household trash to energy and nutrients. 'Biopulp technology' was the first pre-treatment approach investigated in this investigation. Separator, pulper, reject washing unit,

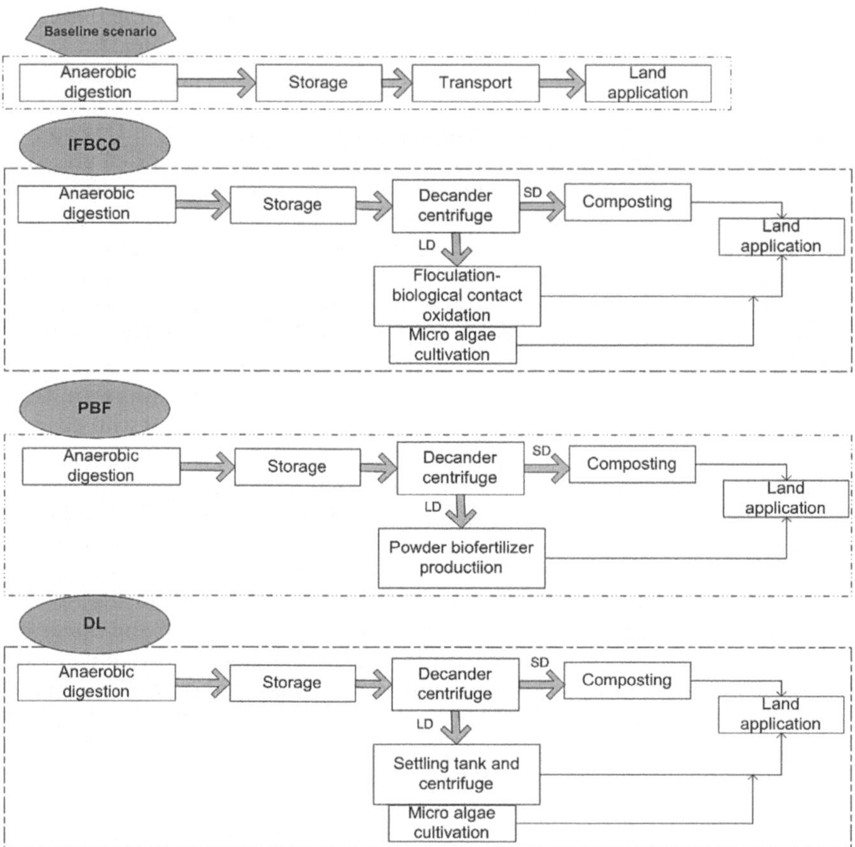

Figure 18.4 Scenario description for study by Duan *et al.* (2020) (LD: liquid digestate, SD: solid digestate).

dewatering unit, and storage tank are some of the components of this technique. The 'Screw press' was the second technique considered in this investigation. Organic household garbage is squeezed through small slots in a metal chamber using this method. The waste streams that exit the chamber form the anaerobic digestion substrate, while the remaining components are channelled through the chamber to be rejected. 'Disc screen' was the most recent technology to be examined for processing organic household trash. The basis of this technique lies in distinguishing light and bigger things from denser and smaller fractions. The separation process occurs on a vertically placed bed with rotating discs arranged in the form of rows at a definite distance apart. Biopulp technology used more power than the other two pre-treatment technologies studied, namely, screw press, disc screen, and bio pulp, but improved fractionation into refuse and substrate aided to obtain a more favourable environmental profile.

In other words, the obtained results demonstrated that proper separation of bio-pulp into refuse and substrate has enhanced bio-pulp technology, and the increased energy requirement for bio-pulping was further offset by the larger quantity of recovered energy and nutrients.

In a gate-to-gate system, Kopsahelis *et al.* (2019) used the LCA on anaerobic co-digestion of dairy products with agro-industrial wastes. The environmental performance of a two-stage (acidogenesis followed by methanogenesis) anaerobic co-digestion process was compared to a single-stage anaerobic co-digestion method in this study. Digestate was divided into liquid and solid fractions, with the liquid portion being utilized as fertilizer straight away and the solid fraction being composted aerobically before being packaged for sale or used to fertilize fields. The author has suggested that the environmental evaluation should be expanded to include a cradle-to-grave examination in future. Li and Feng (2018) has worked on LCA on anaerobic digestion of thickened sludge with high-pressure dewatering produces dewatered sludge directly applied for land use, and pyrolysis along with high-pressure dewatering produces biochar for direct application on land. The first pathway was suggested for sludge having low organic content because of the low raw energy input. Pyrolysis step was recommended for sludge having high organic content because of the low-energy input and the high conversion rate of sludge to energy. Fei *et al.* (2021) has experimented on hydrothermal pre-treatment on food waste on anaerobic digestion with SLS prior to digestion, and compared the solid–liquid combined anaerobic digestion and liquid fraction anaerobic digestion. In the first case, solid fraction fed to composting and liquid fraction led to sewage treatment whereas in second case prior to digestion solid fraction has fed to composting along with solid fraction of the digestate and liquid fraction has sent to sewage treatment. The environmental impact analysis results showed that the contribution of solid–liquid anaerobic digestion to global warming potential (GWP), eutrophication potential (EP), acidification potential (AP), and photochemical ozone creation potential (POCP) was smaller than that of liquid fraction anaerobic digestion. On overall, in anaerobic digestion of food waste, the adoption of solid–liquid anaerobic digestion had a smaller environmental impact (EI) than that of liquid fraction anaerobic digestion and was more useful to the development of environmental sustainability.

18.6 CONCLUSIONS

Biogas production across various countries has experienced rapid growth due to the wide range of benefits received from anaerobic digestion of organic substrates (manure, food waste, sewage sludge, agricultural wastes etc.). Increased number of biogas plants treating these wastes ultimately result in generation of another by-product; digestate. Studies on digestate management options identified several economically feasible technologies for treating digestate rather than applying it on agricultural fields. The conversion technologies used helps in reducing nutrient content in the digestate which ensures safe application of the digestate over farmlands. LCA is an efficient environmental management tool which has the potential to assess the environmental impacts associated with the implementation of various digestate management options discussed in the article. Even though a significant rise in effects related to mineral, energy use and global warming, most analysed studies showed substantial environmental benefits when linked to direct spreading of the digestate, because of the substantial reductions in air emissions, namely carbon dioxide, ammonia and nitrogen oxides. Further LCA studies will help in attaining information regarding the relevance of further developing the digestate treatment options.

REFERENCES

Al Seadi T., Drosg B., Fuchs W., Rutz D. and Janssen R. (2013). Biogas digestate quality and utilization. In: The Biogas Handbook: Science, Production and Applications, https://doi.org/10.1533/9780857097415.2.267

Alburquerque J. A., de la Fuente C. and Bernal M. P. (2012). Chemical properties of anaerobic digestates affecting C and N dynamics in amended soils. *Agriculture, Ecosystems and Environment*, **160**, 15–22, https://doi.org/10.1016/j.agee.2011.03.007

Angouria-Tsorochidou E., Seghetta M., Trémier A. and Thomsen M. (2022). Life cycle assessment of digestate post-treatment and utilization. *Science of the Total Environment*, **815**, 152764, https://doi.org/10.1016/j.scitotenv.2021.152764

Béline F., Daumer M. L., Loyon L., Pourcher A. M., Dabert P., Guiziou F. and Peu P. (2008). The efficiency of biological aerobic treatment of piggery wastewater to control nitrogen, phosphorus, pathogen and gas emissions. *Water Science and Technology*, **57**(12), 1909–1914, https://doi.org/10.2166/wst.2008.316

Bonmatí A. and Flotats X. (2003). Air stripping of ammonia from pig slurry: characterisation and feasibility as a pre- or post-treatment to mesophilic anaerobic digestion. *Waste Management*, **23**(3), 261–272, https://doi.org/10.1016/S0956-053X(02)00144-7

Brod E., Øgaard A. F., Haraldsen T. K. and Krogstad T. (2015). Waste products as alternative phosphorus fertilisers part II: predicting P fertilisation effects by chemical extraction. *Nutrient Cycling in Agroecosystems*, **103**(2), 187–199, https://doi.org/10.1007/s10705-015-9731-4

Burger S., Rohr C. and Vetter H. (2006). Energetische Optimierung der Klärschlammaufbereitung. Im Auftrag EnergieSchweiz Für Infrastrukturanlagen/Bundesamt Für Energie.

Clemens J. and Wulf S. (2005). Reduktion der Ammoniakausgasung aus Kofermentations substraten und Gülle während der Lagerung und Ausbringung durch interne Versaurung mit in NRW anfallenden organischen Kohlenstofffraktionen.

Coelho J. J., Prieto M. L., Dowling S., Hennessy A., Casey I., Woodcock T. and Kennedy N. (2018). Physical-chemical traits, phytotoxicity and pathogen detection in liquid anaerobic digestates. *Waste Management*, **78**, 8–15, https://doi.org/10.1016/j.wasman.2018.05.017

Duan N., Khoshnevisan B., Lin C., Liu Z. and Liu H. (2020). Life cycle assessment of anaerobic digestion of pig manure coupled with different digestate treatment technologies. *Environment International*, **137**, 105522, https://doi.org/10.1016/j.envint.2020.105522

Fei X., Jia W., Chen T. and Ling Y. (2021). Life-cycle assessment of two food waste disposal processes based on anaerobic digestion in China. *Journal of Cleaner Production*, **293**, 126113. https://doi.org/10.1016/j.jclepro.2021.126113

Gienau T., Brüß U., Kraume M. and Rosenberger S. (2018). Nutrient recovery from biogas digestate by optimised membrane treatment. *Waste and Biomass Valorization*, **9**(12), 2337–2347, https://doi.org/10.1007/s12649-018-0231-z

Goedkoop M., Heijungs R., Huijbregts M., Schryver A. De, Struijs J. and Van Zelm R. Van (2009). ReCiPe 2008:A Life Cycle Impact Assessment Method Which Comprises Harmonised Category Indicators at the Midpoint and the Endpoint. Updated RIVM report. RIVM, Bilthoven, Netherlands, **1**, 1–126.

ISO. (2006). Environmental Management: Life Cycle Assessment; Principles and Framework. ISO. https://www.iso.org/standard/37456.html (accessed December 12, 2021)

Jacobson L. D., Moon R., Bicudo J. and Janni K. (1999). Generic environmental impact statement on animal agriculture. A summary of the literature related to air quality and odor (H). *Prepared for the Environmental Quality Board*. Volume 1.

Khoshnevisan B., Tsapekos P., Alvarado-Morales M., Rafiee S., Tabatabaei M. and Angelidaki I. (2018). Life cycle assessment of different strategies for energy and nutrient recovery from source sorted organic fraction of household waste. *Journal of Cleaner Production*, **180**(2018), 360–374, https://doi.org/10.1016/j.jclepro.2018.01.198

Kopsahelis A., Kourmentza C., Zafirib C. and Kornarosa M. (2019). Life cycle assessment (LCA) of end-of-life dairy products (EoL-DPs) valorization via anaerobic co-digestion with agro-industrial wastes for biogas production. *Journal of Chemical Technology & Biotechnology*, **94**, 3687–3697, https://doi.org/10.1002/jctb.6174

Kuhn T. (2017). The revision of the German fertiliser ordinance in 2017. *Food and Resource Economics, Discussion Paper 2017*, **2**, 1–25, http://ageconsearch.umn.edu/record/262054

Landau P. (2015). Annual report 2015. *Bulletin of Medieval Canon Law*, **32**(1), xi–xii, https://doi.org/10.1353/bmc.2015.0008 (accessed November 10, 2021)

Ledda C., Schievano A., Salati S. and Adani F. (2013). Nitrogen and water recovery from animal slurries by a new integrated ultrafiltration, reverse osmosis and cold stripping process: a case study. *Water Research*, **47**(16), 6157–6166, https://doi.org/10.1016/j.watres.2013.07.037

Li H. and Feng K. (2018). Life cycle assessment of the environmental impacts and energy efficiency of an integration of sludge anaerobic digestion and pyrolysis. *Journal of Cleaner Production*, **195**, 476–485. https://doi.org/10.1016/j.jclepro.2018.05.259

Luckert K. (2004). Manual of Mechanical Solid-Liquid Separation. Vulkan-Verlag GmbH.

Masse L., Massé D. I. and Pellerin Y. (2007). The use of membranes for the treatment of manure: a critical literature review. *Biosystems Engineering*, **98**(4), 371–380, https://doi.org/10.1016/j.biosystemseng.2007.09.003

Miles A. and Ellis T. G. (1998). Struvite precipitation potential for nutrient recovery from anaerobically treated wastes. Proceedings of the 998 National Conference on Environmental Engineering, Chicago, Illinois, USA. Available from: http://www.public.iastate.edu/~tge/struvite.pdf [Accessed 12 October 2021].

Moller K., Stinner W., Deuker A. and Leithold G. (2008). Effects of different manuring systems with and without biogas digestion on nitrogen cycle and crop yield in mixed organic dairy farming systems effects of different manuring systems with and without biogas digestion on nitrogen cycle and crop yield in mix. *Nutrient Cycling in Agroecosystems*, **82**(3), 209–232, https://doi.org/10.1007/s10705-008-9196-9

Parra-Orobio B. A., Rotavisky-Sinisterra M. P., Pérez-Vidal A., Marmolejo-Rebellón L. F. and Torres-Lozada P. (2021). Physicochemical, microbiological characterization and phytotoxicity of digestates produced on single-stage and two-stage anaerobic digestion of food waste. *Sustainable Environment Research*, **31**(1), 1–10, https://doi.org/10.1186/s42834-021-00085-9

Rehl T. and Müller J. (2011). Life cycle assessment of biogas digestate processing technologies. *Resources, Conservation and Recycling*, **56**(1), 92–104, https://doi.org/10.1016/j.resconrec.2011.08.007

Rizzo P. F., Young B. J., Pin Viso N., Carbajal J., Martínez L. E., Riera N. I., Bres P. A., Beily M. E., Barbaro L., Farber M., Zubillaga M. S. and Crespo D. C. (2022). Integral approach for the evaluation of poultry manure, compost, and digestate: amendment characterization, mineralization, and effects on soil and intensive crops. *Waste Management*, **139**, 124–135, https://doi.org/10.1016/j.wasman.2021.12.017

Rodhe L. K. K., Ascue J., Willén A., Persson B. V. and Nordberg Å. (2015). Greenhouse gas emissions from storage and field application of anaerobically digested and non-digested cattle slurry. *Agriculture, Ecosystems and Environment*, **199**, 358–368, https://doi.org/10.1016/j.agee.2014.10.004

Rotz C. A., Corson M. S., Chianese D. S., Montes F., Hafner S. D., Bonifacio H. and Coiner C. U. (2007). Integrated farm system model, reference manual, version 2.1. In USDA-Agricultural Research Service.

Sakhaee F. (2020). Anaerobic digestion life cycle assessment (LCA). *Technium: Romanian Journal of Applied Sciences and Technology*, **2**(5), 92–104. https://doi.org/10.47577/technium.v2i5.1066.

Thörneby L., Persson K. and Trägårdh G. (1999). Treatment of liquid effluents from dairy cattle and pigs using reverse osmosis. *Journal of Agricultural and Engineering Research*, **73**(2), 159–170, https://doi.org/10.1006/jaer.1998.0405

Turner C. and Burton C. (2003). Manure Management: Treatment Strategies for Sustainable Agriculture, 2nd ed. Silsoe Research Institute, Silsoe.

Vaneeckhaute C., Lebuf V., Michels E., Belia E., Vanrolleghem P. A., Tack F. M. G. and Meers E. (2017). Nutrient recovery from digestate: systematic technology review and product classification. *Waste and Biomass Valorization*, **8**(1), 21–40, https://doi.org/10.1007/s12649-016-9642-x

Vázquez-Rowe I., Golkowska K., Lebuf V., Vaneeckhaute C., Michels E., Meers E., Benetto E. and Koster D. (2015). Environmental assessment of digestate treatment technologies using LCA methodology. *Waste Management*, **43**, 442–459, https://doi.org/10.1016/j.wasman.2015.05.007

Wäger-Baumann F. (2007). Physical and biological methods for the treatment of the liquid fraction of anaerobic digester.

Wolf M., Chomkhamsri K., Brandao M., Pant R., Ardente F., Pennington D., Manfredi S., De Camillis C. and Goralczyk M. (2010). International reference life cycle data system (ILCD) handbook: review schemes for life cycle assessment. In: Towards Life Cycle Sustainability Management, Finkbeiner M. (eds.), Springer, Dordrecht. https://doi.org/10.1007/978-94-007-1899-9_11

Xu J., Zhao Y., Zhao G. and Zhang H. (2015). Nutrient removal and biogas upgrading by integrating freshwater algae cultivation with piggery anaerobic digestate liquid treatment. *Applied Microbiology and Biotechnology*, **99**(15), 6493–6501, https://doi.org/10.1007/s00253-015-6537-x

Yellezuome D., Zhu X., Wang Z. and Liu R. (2022). Mitigation of ammonia inhibition in anaerobic digestion of nitrogen-rich substrates for biogas production by ammonia stripping: a review. *Renewable and Sustainable Energy Reviews*, **157**, 112043, https://doi.org/10.1016/j.rser.2021.112043

Zeng Y., De Guardia A. and Dabert P. (2016). Improving composting as a post-treatment of anaerobic digestate. *Bioresource Technology*, **201**(November), 293–303. https://doi.org/10.1016/j.biortech.2015.11.013

doi: 10.2166/9781789062755_0405

Chapter 19

Management of anaerobic digestate in a biorefinery advocating circular economy

Larissa C. Ampese[1], William Gustavo Sganzerla[1], Mauro Berni[2], Gilberto Martins[3] and Tânia Forster Carneiro[,1]*

[1]School of Food Engineering, University of Campinas (UNICAMP), Monteiro Lobato St., n.80, Campinas, SP, Brazil
[2]Interdisciplinary Center of Energy Planning (NIPE), University of Campinas (UNICAMP), Campinas, São Paulo, Brazil
[3]Center for Engineering, Modeling and Applied Social Sciences, Federal University of ABC – UFABC, Av. dos Estados, 5001 Santo André, São Paulo, Brazil
**Corresponding author e-mail: taniafc@unicamp.br*

ABSTRACT

The development and application of eco-friendly technologies for waste management and bio-products recovery have been progressing fast over the last couple of decades to alleviate the impacts of the environmental pollution. Anaerobic digestion (AD) promotes the degradation of organic matter into a methane-rich biogas, as a main product. The remaining partially degraded solid and liquid fraction of AD is the digestate, which can be valorized through different technological routes. Within a circular economy framework, it is crucial to be close to the life cycle by the management of the wastes from AD and generate new value-added products wherever possible. Therefore, the successful implementation of AD technology should comprise the management and adequate disposal of the digestate. Based on these premises, this chapter aims to discuss the routes for harnessing AD digestates for biofuels, biochemicals, fertilizers, biochar, and biomaterials production. An industrial biorefinery was proposed for the management of anaerobic digestate, and the associated challenges and opportunities have also been highlighted.

Keywords: Digestate, biofuel, fertilizer, biochar, biomaterials, biorefinery

19.1 INTRODUCTION

The incorrect disposal of solid and liquid wastes causes environmental pollution and the disbursement of additional financial costs (Xu *et al.*, 2018). In this context, eco-friendly technologies for waste management and bio-products recovery have been emerging on the research and development front. The valorization of residues and energy recovery is a crucial element for a modern management of integrated processing systems, which is incidentally the case of a biorefinery

Editor(s): Vinay Kumar Tyagi, Kaoutar Aboudi and Cigdem Eskicioglu

applying the circular economy concept. Figure 19.1 presents the circular economy concept applied to anaerobic digestion (AD), which is a competitive process mediating the degradation of organic matter into methane-rich biogas, promoting energy recovery (Cardoso *et al.*, 2017; Li *et al.*, 2019). Biogas is the principal product generated from AD. AD has been gaining attention because of its versatility in treating solid and liquid wastes (Li *et al.*, 2019).

Moreover, AD embodies adequate technological maturity and is easily adaptable for different substrates. Nevertheless, the formation of a partly stabilized liquid fraction occurs after AD, and this by-product is widely known as digestate (Monlau *et al.*, 2015). The anaerobic digestate can be considered a product with a potential use since it can be valorized through different technological routes (Monfet *et al.*, 2018). Usually, the digestate must be treated before its use. It is also possible to recover nutrients from digestate through mechanical, physicochemical, or biological processes (Monfet *et al.*, 2018). However, identifying sustainable management strategies to handle the digestate represents a key issue (Cesaro, 2021).

Notwithstanding the inherent limitations of waste management from a general perspective, the concept of the circular economy is based on trying to close the life cycle of a process by utilizing the resulting by-products and wastes (Pomponi & Moncaster, 2017). Waste valorization can help solve environmental problems, like soil degradation, erosion, climate change, also diminish the number of residues disposed of in landfills (Ferreira *et al.*, 2021). Within the circular economy framework, the successful implementation of the AD technology should comprise the management and disposal of digestate, thus bringing benefits for cities, industries and promoting job and income creation (Czekała *et al.*, 2020). One of the most common uses of digestate is as

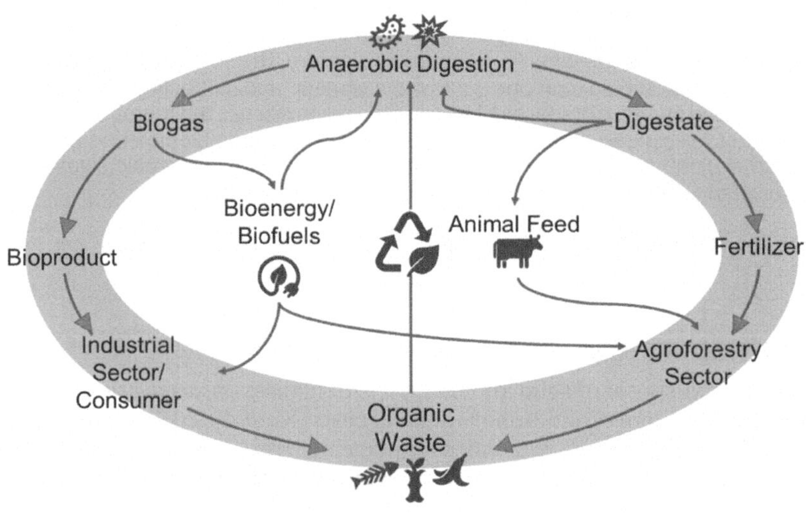

Figure 19.1 Circular economy concept applied to AD.

a soil amendment without prior treatment, or post-treatments like composting or combustion (Serrano *et al.*, 2020). However, several environmental implications of this digestate handling route have been described in the literature. For instance, a digestate with high amounts of organic matter of undigested feedstock can generate greenhouse gas (GHG) emissions (up to 125 g CO_2-eq/kg stored digestate) when disposed of on soil (Logan & Visvanathan, 2019). There are chemical risks of applying the digestate without a previous treatment to remove toxic and pollutant substances, such as ammonia, heavy metals, and pathogens (Coelho *et al.*, 2018). From an agronomic perspective, negative effects on seed germination and plant growth have been described using unstabilized digestate (Serrano *et al.*, 2020). Otherwise, regulations for the desired end use of anaerobic digestate should be introduced to clarify the standards for the applications, protecting human and environmental health (Logan & Visvanathan, 2019). The conversion of industrial organic wastes into bioenergy and digestate, and the further use of the digestate, are all fundamental to ensure nutrients recycling while avoiding GHG emissions and promoting a circular economy.

Based on the above-mentioned points, this chapter brings the relevant concepts regarding the generation of anaerobic digestate, then discusses the main parameters used to characterize this bioresource, and presents different technological routes for its proper management, as well as the main bioproducts obtained following each treatment pathway. Finally, an industrial biorefinery is proposed, where the digestate is used as feedstock for promoting a circular economy. The central objective of this chapter is to supply enough content to show how digestate can be valorized and be transformed into new value-added product, such as biofuel, biochemical, fertilizer, biochar, and biomaterial, all the same discussing the benefits of recovering this potential bioresource. The economic and environmental aspects of anaerobic digestate management in a biorefinery for industrial implementation will also be parsed, and the associated challenges and opportunities in this field highlighted.

19.2 CHARACTERIZATION OF ANAEROBIC DIGESTATE

The digestate is the effluent obtained from AD processes of organic substrates intended for waste stabilization and the concomitant biogas production. This material is found in AD reactors after substrate biodigestion, and still contains nutrients. Depending on its physical–chemical composition, anaerobic digestate can be economically valorized by transforming it into bioproducts (Lima, 2021). After the AD processes, the digestate may still contain high levels of micronutrients and macronutrients, frequently indicated for application in agricultural usage. The digestate quality is related to the absence of phytotoxicity, and evaluating its composition and treatment is also necessary for assessing its final use (Simon & Castilhos Junior, 2020).

Among the parameters which indicate the digestate quality and its viability for bioproducts retrieval, the main one is the operation of the AD reactor. The volumetric production rate of the digestate and its physical–chemical characteristics depend on the substrate (Buller *et al.*, 2021). The AD reactor's operation can be focused on high biogas yields and digestate stabilization based on the hydraulic retention time (HRT) (Kunz *et al.*, 2019). Longer HRTs favor complete substrate degradation, resulting in the digestate with lower organic content. The longer HRTs allow adequate sanitary treatment of the digestate while it brings desirable characteristics to the biodigested substrate for a more efficient final application. Lower HRTs favor the increase in biogas production rates, although the substrate biodigestion is not complete since it stays in the reactor for a short period (Kunz *et al.*, 2019).

Consequently, the digestate can present high organic content which is not completely degraded. Therefore, increased biogas production capacities involve increased digestate production, which must be adequately managed (Lima, 2021). A typical German biogas plant of 500 kW capacity, generating 4000 MWh annually, produces 7600 m^3 of digestate during this period (12 months), and the digestate volume and characteristics depend on the substrates used (Dahlin *et al.*, 2017). According to Dahlin *et al.*, (2017) the digestate's dry and organic matter contents can vary from 1.5 to 45.7% and from 38.6 to 75.4%, respectively, depending on the substrate and reactor operation. The treatment, storage, management and final disposal of this digestate represent a challenge for making biogas an attractive and sustainable energy source, thus promoting an optimized biochemical platform integrating biogas and digestate (Dutta *et al.*, 2021).

Peng *et al.* (2020) showed that systems integrated to AD, based on the circular economy concept involving a sustainable solution for the digestate have been frequently studied worldwide. In these studies, the key parameters considered are life-cycle assessments, new options coming with industrial eco-industrial parks for AD plant's installation, forming a symbiotic relationship that optimizes the use of organic wastes, increases energy efficiency, and reduces GHG emissions (Peng *et al.*, 2020). In addition, industrial ecology aims to restructure the industrial ecosystem in ways for managing and designing linear to closed-loop industrial production and consumption system, providing harmonized relations between ecological and human system, and promoting sustainability (Awan, 2020).

It is important to highlight that for optimizing the uses of the digestate a phase separation in solid and liquid fractions is required since they embody different physical–chemical properties. This separation is relatively simple and can be carried out in filter press units. For biofertilizer production and soil application, the concentration of the nutrient (dry matter) in the digestate can vary from 3.1 to 14.0% for nitrogen, 0.2–3.5% for phosphorus, and 1.9–4.3% for potassium (Dahlin *et al.*, 2017). The macronutrient composition of the solid fraction may present an adequate dosage for agricultural areas, supplying the demand for nutrients of the crops that will be fertilized and avoiding the excessive application of nutrients in the soil and the environmental impacts resulting from this excess (Nkoa, 2014). The dose of organic fertilizer to be applied to the soil, according to Kunz *et al.* (2019), should consider the specific recommendations for the different classes of soil fertility, demand for crops, and their expected yield, content, and agronomic efficiency index of the fertilizer to be used.

It is noteworthy that there are critical environmental limits regarding the excessive application of fertilizers, regardless of mineral or organic origin, since it can otherwise cause significant environmental impacts, mainly due to increased loss of nutrients from the soil and its transfer to the environment (Nkoa, 2014). Furthermore, it should be noted that digestate-derived soil fertilizer can often be poor in biodegradable organic matter. Still, it turns out to be nitrogen and phosphorus-rich effluent since most of the biodegradable organic matter has already been consumed in the process of AD (Logan & Visvanathan, 2019). When agricultural use is limited or inadequate, this nutrient-rich fertilizer has great potential to impact the environment negatively. High concentrations of ammoniacal nitrogen can strongly influence dissolved oxygen dynamics in the medium since to oxidize 1 mg of NH_3, 4.6 mg of O_2 are needed. Furthermore, when in aquatic environments, nitrogen and phosphorus can cause eutrophication (Kunz *et al.*, 2019).

19.3 TECHNOLOGICAL ROUTES FOR ANAEROBIC DIGESTATE MANAGEMENT

Several technological routes can be applied for the management of digestate and production of value-added products. Figure 19.2 shows several processes in a biorefinery plant and the energy, intermediate products, and high-value products generated thereof. From Figure 19.2, it is possible to notice a division where the liquid and solid fractions of digestate follow different paths. The main routes for fertilizer include the treatment of liquid digestate for nitrogen removal followed by distillation; or nutrients recovery followed by evaporation and crystallization, while solid digestate is converted into fertilizer through composting. Another process is to use the liquid digestate for hydroponics offering nutrients to fruits and vegetables.

The route for biofuels includes liquid digestate treatment in microbial fuel cells to energy and the process of algae production and application of their lipids content to biodiesel production. Both the solid and liquid digestates can be recirculated to the AD reactor for methane production enhancement. Also, the solid digestate can be processed for saccharification. The sugars obtained are fermented to ethanol, while the lignin fraction goes to a thermochemical process and generates electricity, and lignin can be purified and used as a biopolymer. Organic acids are obtained by the acid fermentation of solid digestate's sugars. In this chapter, special attention is given to the digestate for the production of biofuels, biochemicals, fertilizer, biochar, and biomaterials, which are individually discussed in the sections below.

19.3.1 Biofuels

Digestate valorization through physicochemical processes stands among the most studied techniques, especially considering the thermochemical production of biofuels or chemicals for the chemical industry or biorefinery (Monfet *et al.*, 2018). Diluted liquid fractions of biomass (algae, macrophytes) are promising for biofuel and the chemical industry (Vaneeckhaute *et al.*, 2017). Some of the possible routes for digestate recovery as energy are through recirculation for

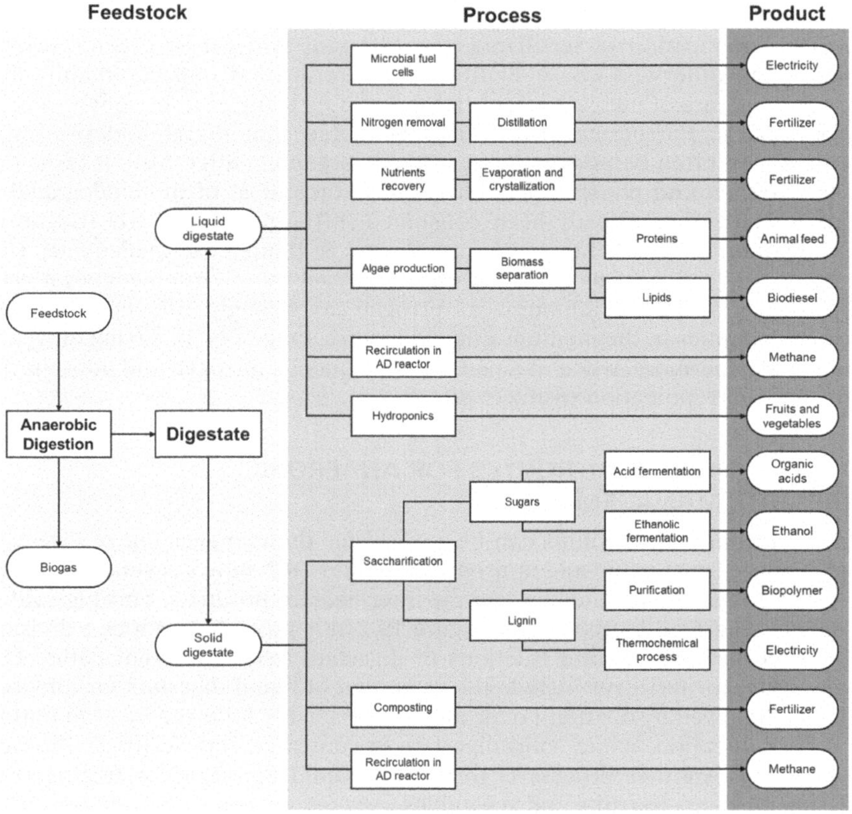

Figure 19.2 Management of anaerobic digestate though different process routes. Drawn based on data in Bakan *et al.* (2021), Kumar *et al.* (2021a), Rekleitis *et al.* (2020) and Sheets *et al.* (2015).

methane production, bioethanol production and thermal conversion processes such as combustion, hydrothermal carbonization (HTC), and pyrolysis (Monlau *et al.*, 2015). The first procedure consists in reintroducing the digestate in the biogas plant in order to obtain more biogas from feedstocks, since it is capable of producing CH_4 while avoiding GHG emissions (Monlau *et al.*, 2015; Sawatdeenarunat *et al.*, 2015). It is worth highlighting that digestate composition depends on the substrate, and digesters operated at higher OLRs contain more biodegradable organic compounds, resulting in higher CH_4 potential (Jiménez-Castro *et al.*, 2020; Sawatdeenarunat *et al.*, 2015).

The alternative usage of digestate to replace freshwater and nutrients for bioethanol (C_2H_5OH) production is promising (Logan & Visvanathan, 2019). The biologic fermentation route promotes the valorization of both the liquid and solid fractions of the digestate (Alrefai *et al.*, 2017). An important advantage the digestate presents for bioethanol production is its enriched cellulose contents, which are easily accessible for microorganisms (Monlau *et al.*, 2015). Still, digested fibers contain recalcitrant structures which impair enzymatic/microbial accessibility for further conversion (Sambusiti *et al.*, 2016). The main products of HTC and pyrolysis are hydrochar with chemical characteristics like those of fossil coals, process water and gas products (HTH), and syngas, bio-oil, and pyrochar during pyrolysis within an oxygen-free atmosphere (Monlau *et al.*, 2015; Wang & Lee, 2021). Usually, the combustion route is applied after drying, and this process is used to transform organic wastes into thermal energy (Fernandez-Lopez *et al.*, 2015). Still, the pre-treatment step to remove moisture from the digestate implies more energy and management costs, giving space for new studies for evaluating strategies which can enable stabilizing the digestate in a self-sustaining process (Serrano *et al.*, 2020). Considering this, Serrano *et al.* (2020) evaluated the self-sustaining smouldering combustion approach based on flameless oxidation as a treatment process for the stabilization of anaerobic digestate, and the study indicated the method as a low energy-intensive technology option for stabilizing digestates, even at high moisture content.

19.3.2 Biochemicals

Another application to digestate management involves using different microorganisms that are capable of utilizing the constituent nutrients to produce bio-based chemicals and products like biosurfactants, biopesticides, and enzymes, which in turn can be channeled towards diverse commercial applications (Kaur *et al.*, 2020). Biosurfactants can be naturally produced by plants and microorganisms, showing higher biodegradability and less toxicity compared to chemical surfactants, thus making them a good substitute for petrochemical-derived surfactants (Guilayn *et al.*, 2020).

The recent study about the production of sophorolipids (a glycolipid) from a biowaste digestate suggests two obstacles for this route, viz., the low pH value needed (around 3.5) and the initial sugar concentration required by the yeast used (Cerda *et al.*, 2019; Guilayn *et al.*, 2020). Regarding biopesticides, they are an eco-friendly alternative to conventional pesticides, since *Bacillus thuringiensis* (or simply 'Bt') is frequently used in biopesticides for insect control (Guilayn *et al.*, 2020), and the digestate is suitable as a nutrient for Bt (Cerda *et al.*, 2019; Guilayn *et al.*, 2020). As mentioned, digestate is a source of enzymes, especially hydrolytic enzymes (Cerda *et al.*, 2019). Lignocellulolytic enzymes (such as endo- and exo-glucanase, xylanase, b-glucosidase, and laccase) produced by fungi have been studied (Musatti *et al.*, 2017). Studies on solid-state fermentation were successful in producing biosurfactants, biopesticides, and enzymes (Cerda *et al.*, 2019), while submerged fermentation demonstrated the possibility of valorizing solid digestate for producing lignocellulolytic enzymes, this route being thus one which can help promote the biodegradation dynamics of lignocellulosic materials (Musatti *et al.*, 2017).

19.3.3 Fertilizer

Depending on the type of digestate, it can contain high contents of nitrogen, phosphorous, and stabilized carbon, and then it is frequently used as a soil amendment and/or fertilizer (Monlau *et al.*, 2015). Consequently, the nutrient recovery from digestate represents a way for the AD plants to meet regulatory drivers and market demands, creating a revenue source for the industry (Vaneeckhaute *et al.*, 2017). The fertilizers obtained from some types of AD digestate present rich organic content and are favorable to plant growth by regulating the soil structure and increasing production efficiency (Baştabak & Koçar, 2020), which is the case of the digestate obtained from food waste AD, which were comparable to commercial inorganic fertilizer (Cheong *et al.*, 2020). Digestates may present nutrient uptake performances close to those of industrial mineral fertilizers or even have better characteristics (Guilayn *et al.*, 2020).

Studies have shown the possibility of using digestate as a bioproduct to help control diseases and pests (Feng *et al.*, 2011). AD digestate can also be applied to hydroponic systems, representing a sustainable and alternate growing media or nutrient solution for baby leaf lettuce (Ronga *et al.*, 2019). Still, it is important to consider the digestate storage on the AD plant until its use for land application, which may cause emissions of CH_4, CO_2, NH_3, and N_2O (Kaur *et al.*, 2020). Another difficulty in digestate usage as fertilizer are the specifications, standards, and regulations conceived for conventional fertilizers and soil amendments. For example, in some situations, the digestate may be considered too wet and nutrient-poor on a wet basis (Guilayn *et al.*, 2020). The application of digestate as a soil conditioner supports the circular economy-sustainable development paradigm of industries producing biogas (Lu & Xu, 2021). The literature brings information both favorable and unfavorable use of digestate as fertilizer, but the characteristics of each material are decisive to either support or avoid the choice of this technological pathway.

19.3.4 Biochar

Biochars are produced via diverse thermochemical processes, such as pyrolysis, gasification, and hydrothermal carbonization, but pyrolysis is the most common route (Kumar *et al.*, 2021a). Biochar is similar to fertilizer due to its highly aromatic structure and a high charge density, carboxylic and phenolic groups, and high specific surface area and porosity (Marcelino, 2020). These characteristics favor carbon sequestration in the soil, increase the low cation-exchange capacity of acidic soils, and sustain the use of digestate-derived biochar in reduction of GHG emissions (Marcelino, 2020). Biochars (including pyrochars/hydrochars) also offer excellent properties as biosorbent (Monlau *et al.*, 2015). Figure 19.3 presents some characteristics that make biochars of great interest for different applications (Marcelino, 2020). They are characterized by high specific surface area, an abundance of surface oxygen functional groups, and their preparation feedstocks are abundant, and the final product presents low cost and good porosity (Monlau *et al.*, 2015; Wang *et al.*, 2019). As a consequence, biochars have been applied in the adsorptive removal of a whole range of diverse micropollutants from wastewater, sediments, soil and others aqueous strata (Wang *et al.*, 2019). After activation and modification, pyrochars/hydrochars can be used for toxic materials removal, as well as heavy metals, organic matters, and antibiotics (Wang & Lee, 2021), by reason of them efficient in bio-adsorbing these contaminants following different but

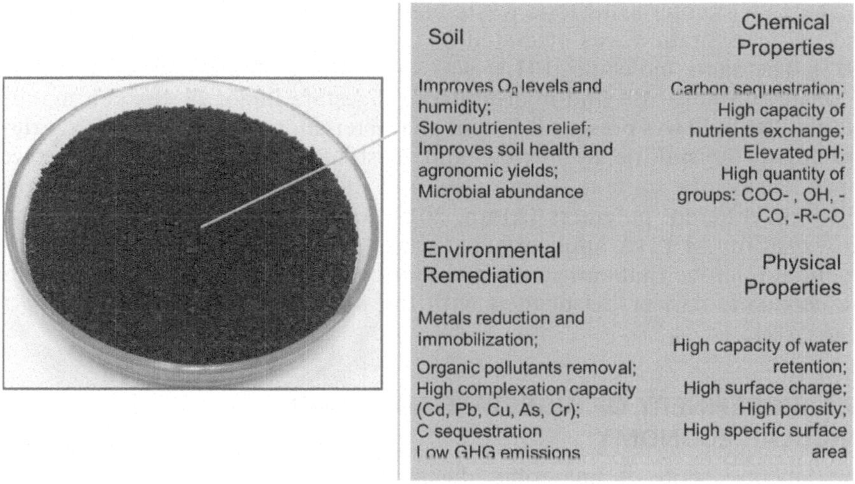

Figure 19.3 Properties and characteristics of biochars making it a potent route for digestate reuse. Data obtained from Marcelino (2020) after translation.

favorable adsorption dynamics and uptake mechanisms (Monlau *et al.*, 2015). According to Wang *et al.*, the characteristics of biochar are influenced mainly by the pyrolysis conditions and type of feedstock (Wang *et al.*, 2019). Some feedstocks already studied include wood, crops, animal manure, and solid wastes. However, biochar obtained from animal manure and solid wastes usually present a relatively small surface area (Kumar *et al.*, 2021b). Biochars attract heightened scientific and public attention as a sustainable strategy and the valorization of digestate into biochar is most apparently suitable for implementation with a biorefinery with diverse feedstocks (Wang *et al.*, 2019).

19.3.5 Biomaterials

Currently, 80% of the polymers worldwide are obtained from fossil fuels (Simonic & Zemljic, 2021). Also, the need for plastics keeps on increasing (Simonic & Zemljic, 2021). Consequentially, bio-based and biodegradable plastics are of great interest since their use will allow reductions in the dependency on fossil fuels and, thereafter reduce the associated environmental impacts (Andreasi Bassi *et al.*, 2021). Anaerobic digestate bears importance in this area of polymer synthesis as well, and this is because it can be used for producing bioplastics (Kaur *et al.*, 2020). One of the most studied bioplastics is polyhydroxyalkanoate (PHA) (Kaur *et al.*, 2020). Some advantages of PHAs are that they are non-toxic, biocompatible, and biodegradable polyesters produced as intracellular storage materials by a variety of Gram-positive and -negative bacteria (Altun, 2019; Wang *et al.*, 2018). PHAs can be used to obtain bio-thermoplastics like poly(3-hydroxybutyrate-*co*-3-hydroxyhexanoate (poly(3HB-*co*-3HHx)), which are biodegradable during AD processes and in aerobic seawater environments (Kaur *et al.*, 2020; Wang *et al.*, 2018). Studies show that PHAs were produced in a batch fermenter using food waste-based digestate from wheat-fed digesters, and the fermentation was mediated by *Cupriavidus necator* (Passanha *et al.*, 2013). The same molecule (PHA) was successfully obtained from a mixture of waste sunflower oil and micro-filtered digestate liquor of chicken manure (Altun, 2019). PHAs present different characteristics ranging from being rigid and highly crystalline to flexible and elastomeric, these being themselves dependent on the carbon source used and the length of the monomers that will form different polymers (Altun, 2019; Mannina *et al.*, 2020). Even with the versatility of PHA applications, further research is required to isolate a potential microbe that can synthesize more varieties of polymers and surpass the economic barriers associated with the production of PHAs (Mohapatra *et al.*, 2021).

19.4 BIOREFINERY OF DIGESTATE ADVOCATING CIRCULAR ECONOMY

The bioconversion of anaerobic digestate into value-added products and bioenergy can be proposed in a biorefinery concept. This concept, which is inherent to circular economy, can be described as an industrial arrangement which is restorative and regenerative by intention and design (Kaur *et al.*, 2018). In addition, the circular economy underpins and aligns itself with the green chemistry and green engineering approaches by actually focusing on renewable processes and products to promote human health and environmental protection (Sheldon, 2016). Adopting AD for bioenergy recovery from the food industry's solid by-products can be a possible approach to transition from the standard linear economy to a circular one (Buller *et al.*, 2021). The circular economy plays an essential role in materials reinsertion in a productive chain, and it is a supportive way to achieve economic growth. The main benefits of a circular economy are also intimately related to the Principles of Green Chemistry (Keijer *et al.*, 2019). For instance, the sixth and seventh principles of green chemistry are associated with improving energy efficiency and the use of renewable feedstocks, respectively (Keijer *et al.*, 2019). Bioenergy recovery from by-products acts to mitigate the environmental side effects caused by the incorrect disposal of such by-products, also avoiding water eutrophication (due to organic matter decomposition and possible leaching in inadequate waste disposal systems) and/or climate change (due to fossil fuel burning).

Additionally, one interesting approach is to place AD within the context of the anaerobic biorefinery to allow an integrated approach which will maximize synergies between technologies and processes (Pérez-Camacho & Curry, 2018). For instance, from the anaerobic biorefinery concept, several bio-products can be generated, such as electricity, heat, biomethane, and fertilizer (Figure 19.4).

Figure 19.5 shows a schematic representation of an anaerobic biorefinery which integrates itself within the circular economy arena. From this schema, it can be observed that the organic wastes generated from the industry and municipalities can be directed towards AD processing systems. The biogas produced can be converted into electricity and heat, which can be used by the industry directly, and thus achieving the replacement of use of fossil fuels.

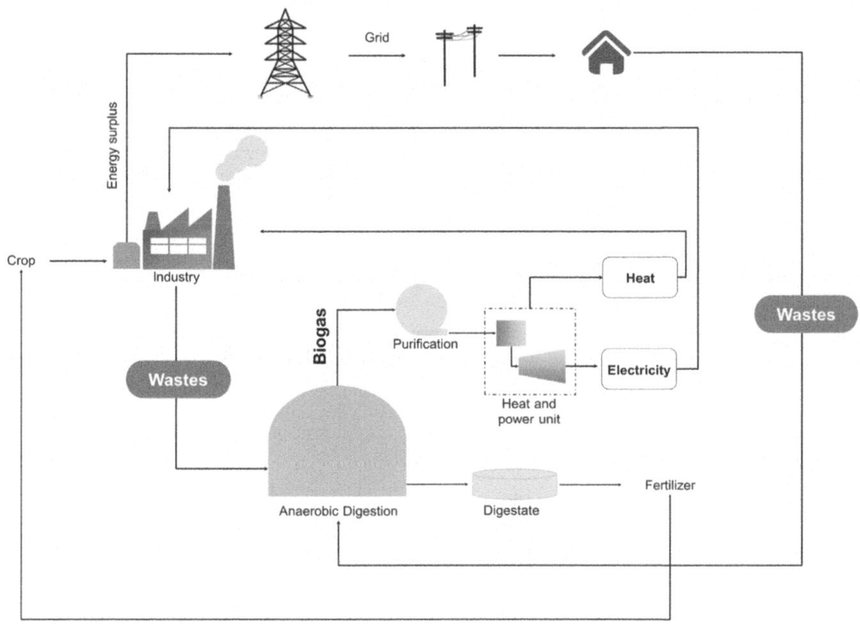

Figure 19.4 Schematic diagram of an anaerobic biorefinery advocating circular economy.

The energy surplus can be diverted to the grid and used by the industry and other consumers. The digestate can be used as fertilizer for crops which are used by the industry (Sganzerla *et al.*, 2021a). The digestate will act as an amendment to the soil and will offset the use of fossil fertilizer partly, but still closing the lifecycle of a given process/product that uses AD and generates digestate (Hagman & Feiz, 2021). Therefore, a biorefinery for solid organic waste management could complete the development of support technologies and infrastructure to advance the circular economy. The food industry could adopt integrating AD reactors as a pretreatment step to the existing liquid treatment systems for methane production.

Figure 19.6 presents a proposal to integrate dry and liquid AD for bioenergy recovery using AD digestate. Such an integrated system presents benefits like reversing production costs, increasing efficiency and competitiveness while reducing environmental side effects (Uludag-Demirer & Demirer, 2021). Depending on the operational condition of solids batch AD reactors, bioenergy could be directly recovered from the biogas produced, and the substrate-reactor after AD (with low solids contents) could be fed to liquid digesters for new digestion. In addition, the biogas produced could be applied to heat and electricity generation for facilities self-consumption, while any eventual energy surplus could be sold to the grid (Sganzerla *et al.*, 2021b). Otherwise, biogas can undergo a purification process to eliminate sulphur compounds and carbon dioxide for the production of biomethane (Ferreira *et al.*, 2021). The

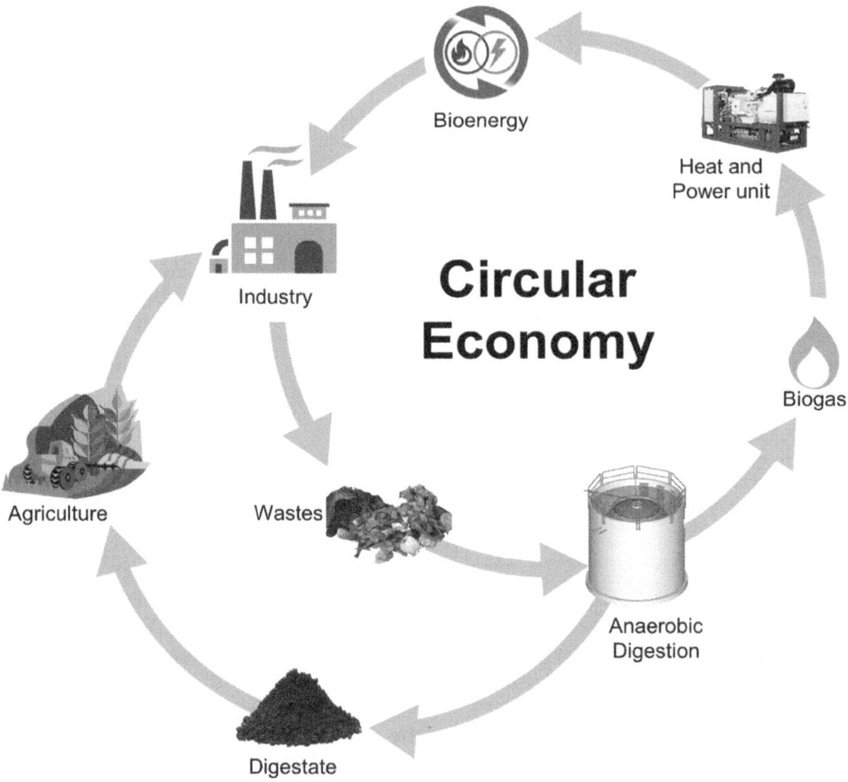

Figure 19.5 A circular economy concept with the implementation of AD.

biomethane can be injected directly into existing natural gas networks to be used in gas ovens or in light and heavy-duty vehicles, which are additional routes to mitigate GHG emissions (Adnan *et al.*, 2019). Hence, taking advantage of the potential energy of waste at a local level, a biogas economy could diversify and contribute to a country's energy matrix renewability without the need to deploy new transmission lines and pipes (Mudhoo *et al.*, 2018; Nunes *et al.*, 2020).

19.5 CHALLENGES AND OPPORTUNITIES

AD carried out with high solids content or carried out in a dry regime, provides an excellent option for recovery of biodegradable solid waste and production of biofuels (hydrogen and methane), fertilizers, and other bio-products, avoiding their disposal in landfills or incineration and promoting the circular economy (Zhang *et al.*, 2021). For example, food industries and municipalities applied conventional AD of the wet type, implementing feeds of low solids content (<15%) (Ferreira *et al.*, 2019). The methane-rich biogas obtained in AD consists of renewable energy that can be transformed into thermal or electrical energies,

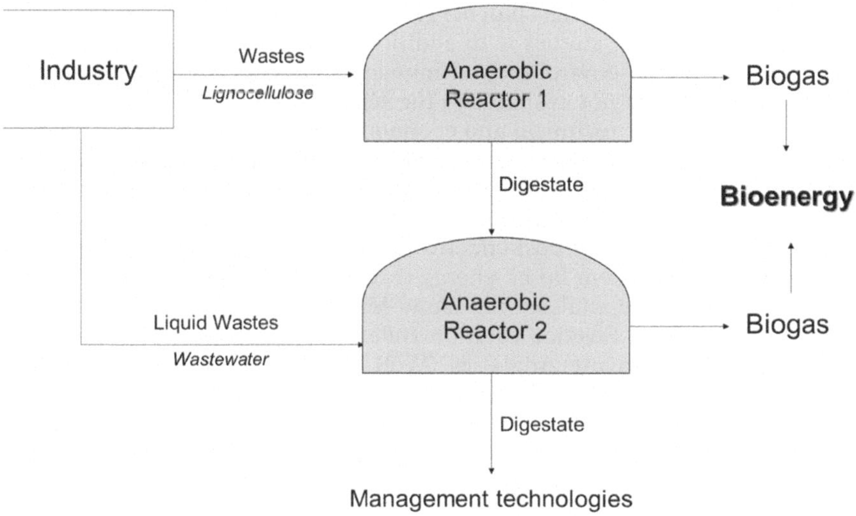

Figure 19.6 Anaerobic reactors integration for bioenergy recovery from the digestate.

and this strategy is especially interesting to valorizate organic wastes from food industries and developing countries.

Future perspectives related to science and technology show a trend in the development of research in AD, presenting technological advances and allowing the development and application of this technology in different areas. Publications on AD in 2020 are 1.8 times higher than the number of publications in 2015 (Ampese *et al.*, 2021). This increase in publications represents a new global trend in the production of knowledge regarding the use of renewable fuels. Several research areas may be related to AD, indicating its multidisciplinary and interdisciplinary character. Many studies in the literature on co-digestion, ammonia inhibition, pig manure, activated waste sludge, volatile fatty acids, biochar, food, and cooking residues and pretreatment are observed, as well as operation with single-stage mesophilic and thermophilic AD or two stages, emerging contaminants, nanomaterials and microplastics (Sganzerla *et al.*, 2021b). Research highlighting factors affecting biogas production, biogas accelerators, the use of inorganic additives, and types of reactors, besides commonly used raw materials are still considered important and poorly developed in countries with high GHG emissions (Cesaro, 2021).

Further discussions on general aspects of AD, advocating the implementation of reactors on an industrial scale for treating lignocellulosic biomass and food residues and water treatment plants, are still scarce topics in the literature. Recent studies show that it is possible to add value to waste from the food industry by implementing a biorefinery system capable of establishing possible industrial routes to produce bio-based products and bioenergy (Jiménez-Castro *et al.*, 2020). Conception of an integrated biorefinery system for the food industry can be based on the unification of the waste management and recovery process of innovative products such as bioenergy and value-added products (Sganzerla *et al.*, 2021a). However, studies with additional details in the economic aspect and dealing with the environmental impacts caused by these processes are still necessary and are not available in the scientific literature. In an overview, detailed studies on the technical and economic dimensions of different systems based on AD on a pilot and industrial scale must take place (Cerda *et al.*, 2019). Finally, by combining the knowledge about characterization and pretreatment of the raw material with modelling and process optimization, one can identify the application of AD as a cost-effective, ecologically correct, and innovative waste management system for bioenergy recovery (Buller *et al.*, 2021).

A brief bibliometric analysis can show scientific production data from the core collection of the Science Citation Index Expanded (SCI-E) – ISI – Web of Science © by Clarivate Analytics (WoS – https://webofknowledge.com/) related to the digestate and your posts. Figure 19.7 shows a network of authors' keywords based on the Web of Science core collection, using '(TS=('digestate' and 'anaerobic digestion')) NOT KP=('digestate' and 'anaerobic digestion')' as a search query and filter publications for articles and reviews. It is observed in Figure 19.7 that the word digestate is widely cited by the articles, with 732 occurrences, showing the research field receives great attention, together with methane and biogas production. This result indicates several alternatives are researched for the valorization of digestate, such as energy recovery from residual organic matter, production of second-generation biogas, and conversion into bioproducts of higher added value. At the same time, for the future studies are expected not only to improve biogas yields and methane content, but also

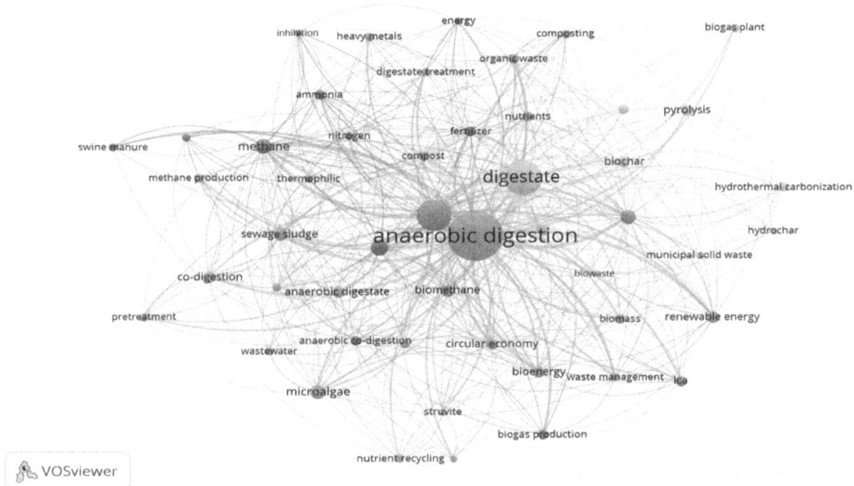

Figure 19.7 Authors' key-words network based on the core collection of Web of Science, using '(TS=('digestate' and 'anaerobic digestion')) NOT KP=('digestate' and 'anaerobic digestion')' as search query and filtering publications for articles and reviews.

to obtain high-quality digestate, capable of recovering, favouring even more the circular economy approach.

The medium- and long-term perspectives for the digestate show that digesting is an environmentally sustainable alternative to reuse and recycle organic waste as a renewable resource. The application as fertilizer for the soil promotes the closed-circuit: symbiosis and circular economy – the use of nutrients and carbon. Digestate applications are also successfully observed in the production of biofuels, biochemists, biochar, and biomaterials. New requirements can be achieved after incentive and new public policies for a bioeconomy, regulatory market milestones, specific marketing standards, and incentives for sustainable waste management.

19.6 CONCLUSION

The digestate comprehends all the effluent from AD, obtained during the process of biogas production, and depending on its physical–chemical composition, it can be economically valorized by transforming it into bioproducts. Some parameters are used to indicate the digestate quality and its viability for bioproducts retrieval, the main one is the operation of the AD reactor. Sustainable solutions for the increasing amounts of the digestate involving the conversion of the digestate into new value-added products is of great interest. For that, several technological routes can be applied in the management of the digestate. The bioconversion of anaerobic digestate into value-added products and bioenergy can be proposed in a biorefinery concept, advocating circular economy. Moreover, bioenergy recovery from by-products acts in environmental side-effects mitigation, favouring even more the circular economy approach and sustainability.

ACKNOWLEDGEMENTS

This study was supported by the Brazilian Science and Research Foundation (CNPq) (productivity grant 302473/2019-0); Coordination for the Improvement of Higher Education Personnel (CAPES, Brazil) (Finance code 001); and São Paulo Research Foundation (FAPESP, Brazil) (grant numbers 2018/14938-4 and 2019/26925-7). The contents of this chapter have also been crosschecked for similarity in the Turnitin software.

REFERENCES

Adnan A. I., Ong M. Y., Nomanbhay S., Chew K. W. and Show P. L. (2019). Technologies for biogas upgrading to biomethane: a review. *Bioengineering*, **6**, 92, https://doi.org/10.3390/BIOENGINEERING6040092

Alrefai R., Benyounis K. Y. and Stokes J. (2017). Integration approach of anaerobic digestion and fermentation process towards producing biogas and bioethanol with zero waste: technical. *Journal of Fundamentals of Renewable Energy and Applications*, **7**(6), 1–6, https://doi.org/10.4172/2090-4541.1000243

Altun M. (2019). Polyhydroxyalkanoate production using waste vegetable oil and filtered digestate liquor of chicken manure. *Preparative Biochemistry & Biotechnology*, **49**, 493–500, https://doi.org/10.1080/10826068.2019.1587626

Ampese L. C., Buller L. S., Monroy Y. M., Garcia M. P., Ramos-Rodriguez A. R. and Forster-Carneiro T. (2021). Macaúba's world scenario: a bibliometric analysis. *Biomass Conversion and Biorefinery*, 1–10, https://doi.org/10.1007/s13399-021-01376-2

Andreasi Bassi S., Boldrin A., Frenna G. and Astrup T. F. (2021). An environmental and economic assessment of bioplastic from urban biowaste. The example of polyhydroxyalkanoate. *Bioresource Technology*, **327**, 1–11, https://doi.org/10.1016/j.biortech.2021.124813

Awan U. (2020). Industrial ecology in support of sustainable development goals, https://doi.org/10.1007/978-3-319-95726-5_18

Bakan B., Bernet N., Bouchez T., Boutrou R., Choubert J.-M., Dabert P., Duquennoi C., Ferraro V., García-Bernet D., Gillot S., Mery J., Rémond C., Steyer J.-P., Trably E. and Tremier A. (2021). Circular economy applied to organic residues and wastewater: research challenges. *Waste and Biomass Valorization*, **13**(2), 1267–1276, https://doi.org/10.1007/s12649-021-01549-0

Baştabak B. and Koçar G. (2020). A review of the biogas digestate in agricultural framework. *Journal of Material Cycles and Waste Management*, **22**(5), 1318–1327, https://doi.org/10.1007/s10163-020-01056-9

Buller L. S., Romero C. W. da S., Lamparelli R. A. C., Ferreira S. F., Bortoleto A. P., Mussatto S. I. and Forster-Carneiro T. (2021). A spatially explicit assessment of sugarcane vinasse as a sustainable by-product. *Science of the Total Environment*, **765**, 1–13, https://doi.org/10.1016/j.scitotenv.2020.142717

Cardoso F., Bomtempo J. V. and Borschiver S. (2017). elaboração de roadmap tecnológico para a produção de biogás a partir de vinhaça (preparation of a technological roadmap for the production of biogas from vinasse). *Prospecções Tecnológicas de Assuntos Específicos*, **10**(3), 328.

Cerda A., Mejias L., Rodríguez P., Rodríguez A., Artola A., Font X., Gea T. and Sánchez A. (2019). Valorisation of digestate from biowaste through solid-state fermentation to obtain value added bioproducts: a first approach. *Bioresource Technology*, **271**, 409–416, https://doi.org/10.1016/j.biortech.2018.09.131

Cesaro A. (2021). The valorization of the anaerobic digestate from the organic fractions of municipal solid waste: challenges and perspectives. *Journal of Environmental Management*, **280**, 1–13, https://doi.org/10.1016/j.jenvman.2020.111742

Cheong J. C., Lee J. T. E., Lim J. W., Song S., Tan J. K. N., Chiam Z. Y., Yap K. Y., Lim E. Y., Zhang J., Tan H. T. W. and Tong Y. W. (2020). Closing the food waste loop: food waste anaerobic digestate as fertilizer for the cultivation of the leafy vegetable, xiao bai cai (*Brassica rapa*). *Science of the Total Environment*, **715**, 136789, https://doi.org/10.1016/J.SCITOTENV.2020.136789

Coelho J. J., Prieto M. L., Dowling S., Hennessy A., Casey I., Woodcock T. and Kennedy N. (2018). Physical-chemical traits, phytotoxicity and pathogen detection in liquid anaerobic digestates. *Waste Management*, **78**, 8–15, https://doi.org/10.1016/j.wasman.2018.05.017

Czekała W., Lewicki A., Pochwatka P., Czekała A., Wojcieszak D., Jóźwiakowski K. and Waliszewska H. 2020. Digestate management in Polish farms as an element of the nutrient cycle. *Journal of Cleaner Production*, **242**, 1–9, https://doi.org/10.1016/j.jclepro.2019.118454

Dahlin J., Nelles M. and Herbes C. (2017). Biogas digestate management: evaluating the attitudes and perceptions of German gardeners towards digestate-based soil amendments. *Resources, Conservation and Recycling*, **118**, 27–38, https://doi.org/10.1016/j.resconrec.2016.11.020

Dutta S., He M., Xiong X. and Tsang D. C. W. (2021). Sustainable management and recycling of food waste anaerobic digestate: a review. *Bioresource Technology*, **341**, 125915, https://doi.org/10.1016/j.biortech.2021.125915

Feng H., Qu G. F., Ning P., Xiong X. F., Jia L. J., Shi Y. K. and Zhang J. (2011). The resource utilization of anaerobic fermentation residue. *Procedia Environmental Sciences*, **11**, 1092–1099, https://doi.org/10.1016/j.proenv.2011.12.165

Fernandez-Lopez M., Puig-Gamero M., Lopez-Gonzalez D., Avalos Ramirez A., Valverde J. and Sanchez-Silva L. (2015). Life cycle assessment of swine and dairy manure: pyrolysis and combustion processes. *Bioresource Technology*, **182**, 184–192, https://doi.org/10.1016/j.biortech.2015.01.140

Ferreira S. F., Buller L. S., Berni M. D., Bajay S. V. and Forster-Carneiro T. (2019). An integrated approach to explore UASB reactors for energy recycling in pulp and paper industry: a case study in Brazil. *Biofuel Research Journal*, **6**, 1039–1045, https://doi.org/10.18331/BRJ2019.6.3.4

Ferreira S. F., Buller L. S., Maciel-Silva F. W., Sganzerla W. G., Berni M. D. and Forster-Carneiro T. (2021). Waste management and bioenergy recovery from açaí processing in the Brazilian Amazonian region: a perspective for a circular economy. *Biofuels, Bioproducts and Biorefining*, **15**, 37–46, https://doi.org/10.1002/bbb.2147

Guilayn F., Rouez M., Crest M., Patureau D. and Jimenez J. (2020). Valorization of digestates from urban or centralized biogas plants: a critical review. *Reviews in Environmental Science and Bio/Technology*, **19**(2), 419–462, https://doi.org/10.1007/S11157-020-09531-3

Hagman L. and Feiz R. (2021). Advancing the circular economy through organic by-product valorisation: a multi-criteria assessment of a wheat-based biorefinery. *Waste and Biomass Valorization*, **12**(11), 6205–6217, https://doi.org/10.1007/s12649-021-01440-y

Jiménez-Castro M. P., Buller L. S., Zoffreo A., Timko M. T. and Forster-Carneiro T. (2020). Two-stage anaerobic digestion of orange peel without pre-treatment: experimental evaluation and application to São Paulo state. *Journal of Environmental Chemical Engineering*, **8**, 1–10, https://doi.org/10.1016/j.jece.2020.104035

Kaur G., Uisan K., Ong K. L. and Ki Lin C. S. (2018). Recent trends in green and sustainable chemistry & waste valorisation: rethinking plastics in a circular economy. *Current Opinion in Green and Sustainable Chemistry*, **9**, 30–39, https://doi.org/10.1016/j.cogsc.2017.11.003

Kaur G., Wong J. W. C., Kumar R., Patria R. D., Bhardwaj A., Uisan K. and Johnravindar D. (2020). Value addition of anaerobic digestate from biowaste: thinking beyond agriculture. *Current Sustainable/Renewable Energy Reports*, **7**(2), 48–55, https://doi.org/10.1007/s40518-020-00148-2

Keijer T., Bakker V. and Slootweg J. C. (2019). Circular chemistry to enable a circular economy. *Nature Chemistry*, **11**(3), 190–195, https://doi.org/10.1038/s41557-019-0226-9

Kumar A. N., Dissanayake P. D., Masek O., Priya A., Ki Lin C. S., Ok Y. S. and Kim S. H. (2021a). Recent trends in biochar integration with anaerobic fermentation: win-win strategies in a closed-loop. *Renewable and Sustainable Energy Reviews*, **149**, 1–12, https://doi.org/10.1016/j.rser.2021.111371

Kumar V., Jaiswal K. K., Tomar M. S., Rajput V., Upadhyay S., Nanda M., Vlaskin M. S., Kumar S. and Kurbatova A. (2021b). Production of high value-added biomolecules by microalgae cultivation in wastewater from anaerobic digestates of food waste: a review. *Biomass Conversion and Biorefinery*, **2021**, 1–18, https://doi.org/10.1007/S13399-021-01906-Y

Kunz A., Steinmetz R. L. R. and Amaral A. C. d. (2019). Fundamentos da digestão anaeróbia, purificação do biogás, uso e tratamento do digestato, https://doi.org/10.21452/978-85-93823-01-5.2019.01

Li Y., Chen Y. and Wu J. (2019). Enhancement of methane production in anaerobic digestion process: a review. *Applied Energy*, **240**, 120–137, https://doi.org/10.1016/j.apenergy.2019.01.243

Lima H. Q. de (2021). Digestato – Aplicações como biofertilizante [WWW Document]. https://energiaebiogas.com.br/digestato-aplicacoes-como-biofertilizante (accessed 10.6.21)

Logan M. and Visvanathan C. (2019). Management strategies for anaerobic digestate of organic fraction of municipal solid waste: current status and future prospects. *Waste Management and Research*, **37**, 27–39, https://doi.org/10.1177/0734242X18816793

Lu J. and Xu S. (2021). Post-treatment of food waste digestate towards land application: a review. *Journal of Cleaner Production*, **303**, 1–12, https://doi.org/10.1016/j.jclepro.2021.127033

Mannina G., Presti D., Montiel-Jarillo G., Carrera J. and Suárez-Ojeda M. E. (2020). Recovery of polyhydroxyalkanoates (PHAs) from wastewater: a review. *Bioresource Technology*, **297**, 1–12, https://doi.org/10.1016/j.biortech.2019.122478

Marcelino I. P., Loss A. and Nogueira Andrade M. A. (2020). Potencialidades do uso do biochar para melhoria dos atributos edáficos (Potentials of the use of biochar to improve edaphic attributes), 2nd edn. In: Redução Do Risco de Desastres E A Resiliência No Meio Rural E Urbano, Panzeri C. G., de Freitas C. M., Mungói C. A., Olivato D, Pereira D, Macedo E. S., Bezerra E. B, Lopes E. S. S., Contel F. B, Barbosa H. A., Zezere J. L., Saad J. C. C., Straubhaar J. D., Ferreira do Vale J. M, Palhares J. M., Sanches Ross J. L., Martin Neto L., Lourenço L., Jatobá L., Ferreira M. D., Silveira M. L., Massambani O., Trajber R., Moreira, R., Luiz Dias Portella S., Fonseca Silveira Massruhá S. M., Lúcia Monteiro V. and Lopes da Silva W. T. (eds.), Centro Paula Souza, São Paulo, pp. 584–601.

Mohapatra S., Vishwakarma K., Joshi N. C., Maity S., Kumar R., Ramchander M., Pattnaik S. and Samantaray D. P. (2021). A review on PHAs: the future biopolymer. *Environmental and Agricultural Microbiology*, **1**, 83–100, https://doi.org/10.1002/9781119525899.ch5

Monfet E., Aubry G. and Ramirez A. A. (2018). Nutrient removal and recovery from digestate: a review of the technology. *Biofuels*, **9**, 247–262, https://doi.org/10.1080/17597269.2017.1336348

Monlau F., Sambusiti C., Ficara E., Aboulkas A., Barakat A. and Carrère H. (2015). New opportunities for agricultural digestate valorization: current situation and perspectives. *Energy and Environmental Science*, **8**, 2600–2621, https://doi.org/10.1039/c5ee01633a

Mudhoo A., Torres-Mayanga P. C., Forster-Carneiro T., Sivagurunathan P., Kumar G., Komilis D. and Sánchez A. (2018). A review of research trends in the enhancement of biomass-to-hydrogen conversion. *Waste Management*, **79**, 580–594, https://doi.org/10.1016/j.wasman.2018.08.028

Musatti A., Ficara E., Mapelli C., Sambusiti C. and Rollini M. (2017). Use of solid digestate for lignocellulolytic enzymes production through submerged fungal fermentation. *Journal of Environmental Management*, **199**, 1–6, https://doi.org/10.1016/j.jenvman.2017.05.022

Nkoa R. (2014). Agricultural benefits and environmental risks of soil fertilization with anaerobic digestates: a review. *Agronomy for Sustainable Development*, **34**(2), 473–492, https://doi.org/10.1007/s13593-013-0196-z

Nunes L. J. R., Causer T. P. and Ciolkosz D. (2020). Biomass for energy: a review on supply chain management models. *Renewable and Sustainable Energy Reviews*, **120**, 1–8, https://doi.org/10.1016/j.rser.2019.109658

Passanha P., Esteves S. R., Kedia G., Dinsdale R. M. and Guwy A. J. (2013). Increasing polyhydroxyalkanoate (PHA) yields from *Cupriavidus necator* by using filtered digestate liquors. *Bioresource Technology*, **147**, 345–352, https://doi.org/10.1016/j.biortech.2013.08.050

Peng W., Lü F., Hao L., Zhang H., Shao L. and He P. (2020). Digestate management for high-solid anaerobic digestion of organic wastes: a review. *Bioresource Technology*, **297**, 1–9, https://doi.org/10.1016/j.biortech.2019.122485

Pérez-Camacho M. N. and Curry R. (2018). Regional assessment of bioeconomy options using the anaerobic biorefinery concept. *Proceedings of Institution of Civil Engineers: Waste and Resource Management*, **171**, 104–113, https://doi.org/10.1680/jwarm.17.00015

Pomponi F. and Moncaster A. (2017). Circular economy for the built environment: a research framework. *Journal of Cleaner Production*, **143**, 710–718, https://doi.org/10.1016/j.jclepro.2016.12.055

Rekleitis G., Haralambous K. J., Loizidou M. and Aravossis K. (2020). Utilization of agricultural and livestock waste in anaerobic digestion (A.D): applying the biorefinery concept in a circular economy. *Energies*, **13**, 1–14, https://doi.org/10.3390/en13174428

Ronga D., Setti L., Salvarani C., de Leo R., Bedin E., Pulvirenti A., Milc J., Pecchioni N. and Francia E. (2019). Effects of solid and liquid digestate for hydroponic baby leaf lettuce (*Lactuca sativa* L.) cultivation. *Scientia Horticulturae*, **244**, 172–181, https://doi.org/10.1016/j.scienta.2018.09.037

Sambusiti C., Monlau F. and Barakat A. (2016). Bioethanol fermentation as alternative valorization route of agricultural digestate according to a biorefinery approach. *Bioresource Technology*, **212**, 289–295, https://doi.org/10.1016/j.biortech.2016.04.056

Sawatdeenarunat C., Surendra K. C., Takara D., Oechsner H. and Khanal S. K. (2015). Anaerobic digestion of lignocellulosic biomass: challenges and opportunities. *Bioresource Technology*, **178**, 178–186, https://doi.org/10.1016/j.biortech.2014.09.103

Serrano A., Wyn H., Dupont L., Villa-Gomez D. K. and Yermán L. (2020). Self-sustaining treatment as a novel alternative for the stabilization of anaerobic digestate. *Journal of Environmental Management*, **264**, 1–7, https://doi.org/10.1016/j.jenvman.2020.110544

Sganzerla W. G., Ampese L. C., Mussatto S. I. and Forster-Carneiro T. (2021a). A bibliometric analysis on potential uses of Brewer's spent grains in a biorefinery for the circular economy transition of the beer industry. *Biofuels, Bioproducts and Biorefining*, **15**(6), 1965–1988, https://doi.org/10.1002/BBB.2290

Sganzerla W. G., Ampese L. C., Parisoto T. A. C. and Forster-Carneiro T. (2021b). Process intensification for the recovery of methane-rich biogas from dry anaerobic digestion of açaí seeds. *Biomass Conversion and Biorefinery*, 1–14, https://doi.org/10.1007/s13399-021-01698-1

Sheets J. P., Yang L., Ge X., Wang Z. and Li Y. (2015). Beyond land application: emerging technologies for the treatment and reuse of anaerobically digested agricultural and food waste. *Waste Management*, **44**, 94–115, https://doi.org/10.1016/j.wasman.2015.07.037

Sheldon R. A. (2016). Green chemistry and resource efficiency: towards a green economy. *Green Chemistry*, **18**(11), 3180–3183, https://doi.org/10.1039/c6gc90040b

Simon F. W. and Castilhos Junior A. B. de. (2020). Avaliação de digestato proveniente de resíduos alimentares por meio de ensaio de germinação. *Revista Gestão & Sustentabilidade Ambiental*, **9**, 329–344, https://doi.org/10.19177/rgsa.v9e02020329-344

Simonic M. and Zemljic F. (2021). Production of bioplastic material from algal biomass. *Chemical Industry and Chemical Engineering Quarterly*, **27**, 79–84, https://doi.org/10.2298/ciceq191024026s

Uludag-Demirer S. and Demirer G. N. (2021). Post-anaerobic treatability and residual biogas potential of digestate. *Biomass Conversion and Biorefinery*, **12**, 1695–1702, https://doi.org/10.1007/s13399-021-01290-7

Vaneeckhaute C., Lebuf V., Michels E., Belia E., Vanrolleghem P. A., Tack F. M. G. and Meers E. (2017). Nutrient recovery from digestate: systematic technology review and product classification. *Waste and Biomass Valorization*, **8**(1), 21–40, https://doi.org/10.1007/s12649-016-9642-x

Wang W. and Lee D. J. (2021). Valorization of anaerobic digestion digestate: a prospect review. *Bioresource Technology*, **323**, 1–15, https://doi.org/10.1016/j.biortech.2020.124626

Wang S., Lydon K. A., White E. M., Grubbs J. B., Lipp E. K., Locklin J. and Jambeck J. R. (2018). Biodegradation of poly(3-hydroxybutyrate-*co*-3-hydroxyhexanoate) plastic under anaerobic sludge and aerobic seawater conditions: gas evolution and microbial diversity. *Environmental Science and Technology*, **52**, 5700–5709, https://doi.org/10.1021/acs.est.7b06688

Wang R. Z., Huang D. L., Liu Y. G., Zhang C., Lai C., Wang X., Zeng G. M., Gong X. M., Duan A., Zhang Q. and Xu P. (2019). Recent advances in biochar-based catalysts: properties, applications and mechanisms for pollution remediation. *Chemical Engineering Journal*, **371**, 380–403, https://doi.org/10.1016/j.cej.2019.04.071

Xu F., Li Y., Ge X., Yang L. and Li Y. (2018). Anaerobic digestion of food waste – challenges and opportunities. *Bioresource Technology*, **247**, 1047–1058, https://doi.org/10.1016/j.biortech.2017.09.020

Zhang Y., Yu Q. and Li J. (2021). Bioenergy research under climate change: a bibliometric analysis from a country perspective. *Environmental Science and Pollution Research*, **28**, 26427–26440, https://doi.org/10.1007/s11356-021-12448-1

Index

C

D

E

F

G

H

I

L

M

R

S

T

V

W

IWA Publishing's authorised EU representative for General Product Safety Regulations is Diane D'Arras, 15 rue Duret, 75116 Paris, France, e-mail: safety@iwap.co.uk.

Printed and bound by CPI Group (UK) Ltd, Croydon, CR0 4YY
16/06/2026
02140885-0005